Methods in Enzymology

Volume 291
CAGED COMPOUNDS

METHODS IN ENZYMOLOGY

EDITORS-IN-CHIEF

John N. Abelson   Melvin I. Simon

DIVISION OF BIOLOGY
CALIFORNIA INSTITUTE OF TECHNOLOGY
PASADENA, CALIFORNIA

FOUNDING EDITORS

Sidney P. Colowick and Nathan O. Kaplan

Methods in Enzymology

Volume 291

Caged Compounds

EDITED BY

Gerard Marriott

MAX PLANCK INSTITUTE FOR BIOCHEMISTRY
MARTINSRIED, GERMANY

ACADEMIC PRESS
San Diego London Boston New York Sydney Tokyo Toronto

This book is printed on acid-free paper.

Copyright © 1998 by ACADEMIC PRESS

All Rights Reserved.
No part of this publication may be reproduced or transmitted in any form or by any means, electronic or mechanical, including photocopy, recording, or any information storage and retrieval system, without permission in writing from the Publisher.
The appearance of the code at the bottom of the first page of a chapter in this book indicates the Publisher's consent that copies of the chapter may be made for personal or internal use, or for the personal or internal use of specific clients. This consent is given on the condition, however, that the copier pay the stated per copy fee through the Copyright Clearance Center, Inc. (222 Rosewood Drive, Danvers, Massachusetts 01923) for copying beyond that permitted by Sections 107 or 108 of the U.S. Copyright Law. This consent does not extend to other kinds of copying, such as copying for general distribution, for advertising or promotional purposes, for creating new collective works, or for resale. Copy fees for pre-1997 chapters are as shown on the chapter title pages. If no fee code appears on the chapter title page, the copy fee is the same as for current chapters.
0076-6879/98 $25.00

Academic Press
15 East 26th Street, 15th Floor, New York, New York 10010, USA
http://www.academicpress.com

Academic Press Limited
24-28 Oval Road, London NW1 7DX, UK
http://www.hbuk.co.uk/ap/

International Standard Book Number: 0-12-182192-7

PRINTED IN THE UNITED STATES OF AMERICA
98 99 00 01 02 MM 9 8 7 6 5 4 3 2 1

Table of Contents

CONTRIBUTORS TO VOLUME 291. ix

PREFACE. xiii

VOLUMES IN SERIES xv

FREDRIC STEWART FAY, H. MAURICE GOODMAN, AND DAVID WARSHAW xxxiii

1. New Photoprotecting Groups: Desyl and *p*-Hydroxyphenacyl Phosphate and Carboxylate Esters — RICHARD S. GIVENS, JÖRG F. W. WEBER, ANDREAS H. JUNG, AND CHAN-HO PARK — 1

2. Synthesis, Photochemistry, and Biological Characterization of Photolabile Protecting Groups for Carboxylic Acids and Neurotransmitters — KYLE R. GEE, BARRY K. CARPENTER, AND GEORGE P. HESS — 30

3. New Caged Groups: 7-Substituted Coumarinylmethyl Phosphate Esters — TOSHIAKI FURUTA AND MICHIKO IWAMURA — 50

4. Caged Fluorescent Probes — T. J. MITCHISON, K. E. SAWIN, J. A. THERIOT, K. GEE, AND A. MALLAVARAPU — 63

5. Biologically Active Peptides Caged on Tyrosine — R. SREEKUMAR, MITSUO IKEBE, FREDRIC S. FAY, AND JEFFERY W. WALKER — 78

6. Light-Directed Activation of Protein Activity from Caged Protein Conjugates — GERARD MARRIOTT, JOHANNES OTTL, MANFRED HEIDECKER, AND DANIELA GABRIEL — 95

7. Caged Peptides and Proteins by Targeted Chemical Modification — HAGAN BAYLEY, CHUNG-YU CHANG, W. TODD MILLER, BRETT NIBLACK, AND PENG PAN — 117

8. Photocleavable Affinity Tags for Isolation and Detection of Biomolecules — JERZY OLEJNIK, EDYTA KRZYMAŃSKA-OLEJNIK, AND KENNETH J. ROTHSCHILD — 135

9. Synthesis and Applications of Heterobifunctional Photocleavable Cross-Linking Reagents — GERARD MARRIOTT AND JOHANNES OTTL — 155

10. Use of Lasers for One- and Two-Photon Photolysis of Caged Compounds — JAMES A. MCCRAY — 175

11. Flash Lamp-Based Irradiation of Caged Compounds — GERT RAPP — 202

12. Fourier Transform Infrared Photolysis Studies of Caged Compounds — VALENTIN CEPUS, CAROLA ULBRICH, CHRISTOPH ALLIN, AGNES TROULLIER, AND KLAUS GERWERT — 223

13. Use of Caged Compounds in Studies of Bioelectronic Imaging and Pattern Recognition — C. W. WHARTON AND R. S. CHITTOCK — 245

14. Use of Caged Nucleotides to Characterize Unstable Intermediates by X-Ray Crystallography — AXEL SCHEIDIG, CHRISTOPH BURMESTER, AND ROGER S. GOODY — 251

15. Photoregulation of Cholinesterase Activities with Caged Cholinergic Ligands — LING PENG AND MAURICE GOELDNER — 265

16. Caged Substrates for Measuring Enzymatic Activity *in Vivo:* Photoactivated Caged Glucose 6-Phosphate — ROBERT R. SWEZEY AND DAVID EPEL — 278

17. Investigation of Charge Translocation by Ion Pumps and Carriers Using Caged Substrates — K. FENDLER, K. HARTUNG, G. NAGEL, AND E. BAMBERG — 289

18. Studies of Molecular Motors Using Caged Compounds — JODY A. DANTZIG, HIDEO HIGUCHI, AND YALE E. GOLDMAN — 307

19. Application of Caged Fluorescein-Labeled Tubulin to Studies of Microtubule Dynamics and Transport of Tubulin Molecules in Axons — TAKESHI FUNAKOSHI AND NOBUTAKA HIROKAWA — 348

20. Two-Photon Activation of Caged Calcium with Submicron, Submillisecond Resolution — EDWARD B. BROWN AND WATT W. WEBB — 356

21. Caged Inositol 1,4,5-Trisphosphate for Studying Release of Ca^{2+} from Intracellular Stores — NICK CALLAMARAS AND IAN PARKER — 380

22. Characterization and Application of Photogeneration of Calcium Mobilizers cADP-Ribose and Nicotinic Acid Adenine Dinucleotide Phosphate from Caged Analogs — KYLE R. GEE AND HON CHEUNG LEE — 403

23. Applications of Caged Compounds of Hydrolysis-Resistant Analogs of cAMP and cGMP	U. BENJAMIN KAUPP, CLAUDIA DZEJA, STEPHAN FRINGS, JÜRGEN BENDIG, AND VOLKER HAGEN	415
24. Caged Probes for Studying Cellular Physiology: Application of o-Nitromandelyloxycarbonyl (Nmoc) Caging Method to Glutamate and a Ca^{2+}-ATPase Inhibitor	FRANCIS M. ROSSI, MICHAEL MARGULIS, ROBERT E. HOESCH, CHA-MIN TANG, AND JOSEPH P. Y. KAO	431
25. Development and Application of Caged Ligands for Neurotransmitter Receptors in Transient Kinetic and Neuronal Circuit Mapping Studies	GEORGE P. HESS AND CHRISTOF GREWER	443
26. Caged Plant Growth Regulators	ANDREW C. ALLAN, JANE L. WARD, MICHAEL H. BEALE, AND ANTHONY J. TREWAVAS	474
27. Use of Caged Compounds in Studies of the Kinetics of DNA Repair	R. A. MELDRUM, R. S. CHITTOCK, AND C. W. WHARTON	483

AUTHOR INDEX 497

SUBJECT INDEX 515

Contributors to Volume 291

Article numbers are in parentheses following the names of contributors.
Affiliations listed are current.

ANDREW C. ALLAN (26), *Department of Plant Genetics, Weizmann Institute of Science, Rehovot, Israel 76100*

CHRISTOPH ALLIN (12), *Lehrstuhl für Biophysik, Fakultät Biologie, 44780 Bochum, Germany*

E. BAMBERG (17), *Max Planck Institute for Biophysics, D-60596 Frankfurt, Germany, and Johann Wolfgang Goethe-Universität Frankfurt am Main Biozentrum, D-60439 Frankfurt am Main, Germany*

HAGAN BAYLEY (7), *Department of Medical Biochemistry and Genetics, Texas A&M Health Science Center, College Station, Texas 77843*

MICHAEL H. BEALE (26), *IACR-Long Ashton Research Station, Department of Agricultural Sciences, University of Bristol, Bristol BS18 9AF, United Kingdom*

JÜRGEN BENDIG (23), *Institut für Organische und Bioorganische Chemie, Humboldt-Universität Berlin, D-10115 Berlin, Germany*

EDWARD B. BROWN (20), *Department of Physics, Cornell University, Ithaca, New York 14853*

CHRISTOPH BURMESTER (14), *Abteilung Physikalische Biochemie, Max Planck Institute for Molecular Physiology, 44139 Dortmund, Germany*

NICK CALLAMARAS (21), *Laboratory of Cellular and Molecular Neurobiology, Department of Psychobiology, University of California, Irvine, California 92697*

BARRY K. CARPENTER (2), *Department of Chemistry, Section of Biochemistry, Molecular and Cell Biology, Cornell University, Ithaca, New York 14853*

VALENTIN CEPUS (12), *Lehrstuhl für Biophysik, Fakultät Biologie, 44780 Bochum, Germany*

CHUNG-YU CHANG (7), *Worcester Foundation for Biomedical Research, Shrewsbury, Massachusetts 01545*

R. S. CHITTOCK (13, 27), *School of Biochemistry, University of Birmingham, Birmingham B15 2TT, United Kingdom*

JODY A. DANTZIG (18), *Pennsylvania Muscle Institute, University of Pennsylvania School of Medicine, Philadelphia, Pennsylvania 19104*

CLAUDIA DZEJA (23), *Institut für Biologische Informationsverarbeitung, Forschungszentrum Jülich, D-52425 Jülich, Germany*

DAVID EPEL (16), *Department of Biological Sciences, Hopkins Marine Station, Stanford University, Pacific Grove, California 93950*

FREDRIC S. FAY* (5), *Department of Physiology, University of Massachusetts, Worcester, Massachusetts 01605*

K. FENDLER (17), *Max Planck Institute for Biophysics, D-60596 Frankfurt, Germany*

STEPHAN FRINGS (23), *Institut für Biologische Informationsverarbeitung, Forschungszentrum Jülich, D-52425 Jülich, Germany*

TAKESHI FUNAKOSHI (19), *Department of Cell Biology and Anatomy, Graduate School of Medicine, University of Tokyo, Hongo, Tokyo 113, Japan*

TOSHIAKI FURUTA (3), *Department of Biomolecular Science, Faculty of Science, Toho University, Chiba 274, Japan*

DANIELA GABRIEL (6), *Max Planck Institute for Biochemistry, 82152 Martinsried, Germany*

KYLE R. GEE (2, 4, 22), *Molecular Probes, Inc., Eugene, Oregon 97402*

* Deceased

KLAUS GERWERT (12), *Lehrstuhl für Biophysik, Fakultät Biologie, 44780 Bochum, Germany*

RICHARD S. GIVENS (1), *Department of Chemistry, University of Kansas, Lawrence, Kansas 66045*

MAURICE GOELDNER (15), *Laboratoire de chimie Bioorganique, Faculté de Pharmacie, Université Louis Pasteur Strasbourg, BP 24-F67401 Illkirch Cedex, France*

YALE E. GOLDMAN (18), *Pennsylvania Muscle Institute, University of Pennsylvania School of Medicine, Philadelphia, Pennsylvania 19104*

H. MAURICE GOODMAN, *Department of Physiology, University of Massachusetts Medical School, Worcester, Massachusetts 01605*

ROGER S. GOODY (14), *Abteilung Physikalische Biochemie, Max Planck Institute for Molecular Physiology, 44139 Dortmund, Germany*

CHRISTOF GREWER (25), *Section of Biochemistry, Molecular and Cell Biology, Cornell University, Ithaca, New York 14853*

VOLKER HAGEN (23), *Forschungsinstitut für Molekulare Pharmakologie, D-10315 Berlin, Germany*

K. HARTUNG (17), *Max Planck Institute for Biophysics, D-60596 Frankfurt, Germany*

MANFRED HEIDECKER (6), *Biomolecular and Cellular Dynamics Group, Max Planck Institute for Biochemistry, 82152 Martinsried, Germany*

GEORGE P. HESS (2, 25), *Section of Biochemistry, Molecular and Cell Biology, Cornell University, Ithaca, New York 14853*

HIDEO HIGUCHI (18), *Department of Metallurgy, Faculty of Engineering, Tohoku University, Sendai 980-77, Japan*

NOBUTAKA HIROKAWA (19), *Department of Cell Biology and Anatomy, Graduate School of Medicine, University of Tokyo, Hongo, Tokyo 113, Japan*

ROBERT E. HOESCH (24), *Medical Biotechnology Center, University of Maryland Biotechnology Institute, and University of Maryland School of Medicine, Baltimore, Maryland 21201*

MITSUO IKEBE (5), *Department of Physiology, University of Massachusetts, Worcester, Massachusetts 01605*

MICHIKO IWAMURA (3), *Department of Biomolecular Science, Faculty of Science, Toho University, Chiba 274, Japan*

ANDREAS H. JUNG (1), *Department of Chemistry, University of Kansas, Lawrence, Kansas 66045*

JOSEPH P. Y. KAO (24), *Department of Physiology, University of Maryland School of Medicine, and Medical Biotechnology Center, University of Maryland Biotechnology Institute, Baltimore, Maryland 21201*

U. BENJAMIN KAUPP (23), *Institut für Biologische Informationsverarbeitung, Forschungszentrum Jülich, D-52425 Jülich, Germany*

EDYTA KRZYMAŃSKA-OLEJNIK (8), *AmberGen, Inc., Boston, Massachusetts 02215*

HON CHEUNG LEE (22), *Department of Physiology, University of Minnesota, Minneapolis, Minnesota 55455*

A. MALLAVARAPU (4), *Biochemistry Department, University of California, San Francisco, California 94143*

MICHAEL MARGULIS (24), *Department of Neurology, University of Maryland School of Medicine, Baltimore, Maryland 21201*

GERARD MARRIOTT (6, 9), *Max Planck Institute for Biochemistry, 82152 Martinsried, Germany*

JAMES A. MCCRAY (10), *Department of Physics, Drexel University, Philadelphia, Pennsylvania 19104, and Division of Neurophysiology, National Institute for Medical Research, London NW7 1AA, England*

R. A. MELDRUM (27), *School of Biochemistry, University of Birmingham, Birmingham B15 2TT, United Kingdom*

W. TODD MILLER (7), *Department of Physiology and Biophysics, SUNY Health Sciences Center, Stony Brook, New York 11794*

T. J. MITCHISON (4), *Department of Cell Biology, Harvard Medical School, Boston, Massachusetts 02115*

G. NAGEL (17), *Johann Wolfgang Goethe-Universitat Frankfurt am Main Biozentrum, D-60439 Frankfurt am Main, Germany*

BRETT NIBLACK (7), *Worcester Foundation for Biomedical Research, Shrewsbury, Massachusetts 01545*

JERZY OLEJNIK (8), *Department of Physics and Molecular Biophysics Laboratory, Boston University, Boston, Massachusetts 02215, and AmberGen, Inc., Boston, Massachusetts 02215*

JOHANNES OTTL (6, 9), *Max Planck Institute for Biochemistry, 82152 Martinsried, Germany*

PENG PAN (7), *Boehringer Ingelheim Pharmaceuticals, Inc., Ridgefield, Connecticut 06877*

CHAN-HO PARK (1), *Research and Development Center, Hansol Institute of Science and Technology, Nomyangju-Si, Kyung Ki-Do, Korea*

IAN PARKER (21), *Laboratory of Cellular and Molecular Neurobiology, Department of Psychobiology, University of California, Irvine, California 92697*

LING PENG (15), *Laboratoire de chimie Bioorganique, Faculté de Pharmacie, Université Louis Pasteur Strasbourg, BP 24-F67401 Illkirch Cedex, France*

GERT RAPP (11), *European Molecular Biology Laboratory, Outstation at DESY, D-22603 Hamburg, Germany*

FRANCIS M. ROSSI (24), *Medical Biotechnology Center, University of Maryland Biotechnology Institute, Baltimore, Maryland 21201*

KENNETH J. ROTHSCHILD (8), *Department of Physics, Boston University, Boston, Massachusetts 02215*

K. E. SAWIN (4), *ICRF, Lincolns Inn, London WC2 3PX, England*

AXEL SCHEIDIG (14), *Abteilung Physikalische Biochemie, Max Planck Institute for Molecular Physiology, 44139 Dortmund, Germany*

R. SREEKUMAR (5), *Department of Physiology, University of Wisconsin, Madison, Wisconsin 53706*

ROBERT R. SWEZEY (16), *Department of Pharmacokinetics and Metabolism, Shaman Pharmaceuticals, Inc., South San Francisco, California 94080*

CHA-MIN TANG (24), *Department of Neurology, University of Maryland School of Medicine, Baltimore, Maryland 21201*

J. A. THERIOT (4), *Whitehead Institute, Cambridge, Massachusetts 02142*

ANTHONY J. TREWAVAS (26), *Institute of Cell and Molecular Biology, University of Edinburgh, Edinburgh EH3 9JH, United Kingdom*

AGNES TROULLIER (12), *Lehrstuhl für Biophysik, Fakultät Biologie, 44780 Bochum, Germany*

CAROLA ULBRICH (12), *Lehrstuhl für Biophysik, Fakultät Biologie, 44780 Bochum, Germany*

JEFFERY W. WALKER (5), *Department of Physiology, University of Wisconsin School of Medicine, Madison, Wisconsin 53706*

JANE L. WARD (26), *IACR-Long Ashton Research Station, Department of Agricultural Sciences, University of Bristol, Bristol BS18 9AF, United Kingdom*

DAVID WARSHAW, *Department of Molecular Physiology and Biophysics, University of Vermont Medical School, Burlington, Vermont 05405*

WATT W. WEBB (20), *Department of Applied and Engineering Physics, Cornell University, Ithaca, New York 14853*

JÖRG F. W. WEBER (1), *Department of Chemistry, University of Kansas, Lawrence, Kansas 66045*

C. W. WHARTON (13, 27), *School of Biochemistry, University of Birmingham, Birmingham B15 2TT, United Kingdom*

Preface

Caged compounds are biomolecules whose activity or function is inactivated by chemical modification with a photolabile group. Excitation of a caged compound with (near-ultraviolet) light results in specific cleavage of a bond within the molecule that rapidly liberates the functional biomolecule of interest. One main advantage of this light-directed perturbation technique centers around its ability to overcome the mixing and diffusion time limits of fast-flow methods. This is achieved by precharging the biological sample with a caged substrate, ligand, or protein conjugate, followed by light-directed activation of the caged compound which triggers the specific reaction under investigation within the excitation volume. In addition, concentration jumps mediated by light-directed activation of caged compounds can be conducted without physically disturbing the preparation. Caged groups described in this volume generate active biomolecules within nanoseconds of absorbing light and, furthermore, by using two-photon excitation of the caged compound, the concentration jump can be confined to a volume having a diameter of approximately 250 nm with little or no irradiation-based damage to the preparation. Light-directed activation of caged compounds and rapid monitoring of the ensuing reaction using photomultiplier or imaging-based techniques have been used to understand the molecular mechanism of several biochemical reactions and processes *in vitro* and *in vivo*. This volume also describes the application of caged compounds to studies that aim to understand the molecular basis of neurotransmission, muscle contraction, and ion channel function.

Topics covered in this book include (a) an up-to-date account of the synthetic chemistry and photochemistry of traditional and newer generations of caged groups, (b) a description of new reagents and methods to prepare monofunctional and bifunctional complexes of a diverse array of caged substrates, ligands, and polypeptides, (c) a state-of-the-art practical and theoretical description of lamp- and laser-based excitation sources for light-directed activation of caged compounds in macro- and microscopic samples, (d) the measurement and interpretation of spectroscopic signals derived from light-directed activation of caged compounds to monitor the kinetics of the reaction being studied, and (e) the application of light-directed activation of caged compounds and caged polypeptides to chemical relaxation studies that aim to understand the molecular mechanism of protein-catalyzed reactions in a diverse array of complex molecular environments.

The primary objective of this volume is to provide newcomers to the field with a central source of valuable practical and theoretical information on caged compounds that will allow them to use light-directed activation of caged compounds to address molecular mechanisms of biochemical reactions or processes within biological samples. It provides the investigator with the most up-to-date and comprehensive account of caged compounds that should aid in the design of experiments to compare and interpret data obtained from kinetic measurements of specific reactions triggered by light-directed activation of caged biomolecules.

This volume is dedicated to the memory of Professor Fredric S. Fay.

GERARD MARRIOTT

METHODS IN ENZYMOLOGY

VOLUME I. Preparation and Assay of Enzymes
Edited by SIDNEY P. COLOWICK AND NATHAN O. KAPLAN

VOLUME II. Preparation and Assay of Enzymes
Edited by SIDNEY P. COLOWICK AND NATHAN O. KAPLAN

VOLUME III. Preparation and Assay of Substrates
Edited by SIDNEY P. COLOWICK AND NATHAN O. KAPLAN

VOLUME IV. Special Techniques for the Enzymologist
Edited by SIDNEY P. COLOWICK AND NATHAN O. KAPLAN

VOLUME V. Preparation and Assay of Enzymes
Edited by SIDNEY P. COLOWICK AND NATHAN O. KAPLAN

VOLUME VI. Preparation and Assay of Enzymes (*Continued*)
Preparation and Assay of Substrates
Special Techniques
Edited by SIDNEY P. COLOWICK AND NATHAN O. KAPLAN

VOLUME VII. Cumulative Subject Index
Edited by SIDNEY P. COLOWICK AND NATHAN O. KAPLAN

VOLUME VIII. Complex Carbohydrates
Edited by ELIZABETH F. NEUFELD AND VICTOR GINSBURG

VOLUME IX. Carbohydrate Metabolism
Edited by WILLIS A. WOOD

VOLUME X. Oxidation and Phosphorylation
Edited by RONALD W. ESTABROOK AND MAYNARD E. PULLMAN

VOLUME XI. Enzyme Structure
Edited by C. H. W. HIRS

VOLUME XII. Nucleic Acids (Parts A and B)
Edited by LAWRENCE GROSSMAN AND KIVIE MOLDAVE

VOLUME XIII. Citric Acid Cycle
Edited by J. M. LOWENSTEIN

VOLUME XIV. Lipids
Edited by J. M. LOWENSTEIN

VOLUME XV. Steroids and Terpenoids
Edited by RAYMOND B. CLAYTON

VOLUME XVI. Fast Reactions
Edited by KENNETH KUSTIN

VOLUME XVII. Metabolism of Amino Acids and Amines (Parts A and B)
Edited by HERBERT TABOR AND CELIA WHITE TABOR

VOLUME XVIII. Vitamins and Coenzymes (Parts A, B, and C)
Edited by DONALD B. MCCORMICK AND LEMUEL D. WRIGHT

VOLUME XIX. Proteolytic Enzymes
Edited by GERTRUDE E. PERLMANN AND LASZLO LORAND

VOLUME XX. Nucleic Acids and Protein Synthesis (Part C)
Edited by KIVIE MOLDAVE AND LAWRENCE GROSSMAN

VOLUME XXI. Nucleic Acids (Part D)
Edited by LAWRENCE GROSSMAN AND KIVIE MOLDAVE

VOLUME XXII. Enzyme Purification and Related Techniques
Edited by WILLIAM B. JAKOBY

VOLUME XXIII. Photosynthesis (Part A)
Edited by ANTHONY SAN PIETRO

VOLUME XXIV. Photosynthesis and Nitrogen Fixation (Part B)
Edited by ANTHONY SAN PIETRO

VOLUME XXV. Enzyme Structure (Part B)
Edited by C. H. W. HIRS AND SERGE N. TIMASHEFF

VOLUME XXVI. Enzyme Structure (Part C)
Edited by C. H. W. HIRS AND SERGE N. TIMASHEFF

VOLUME XXVII. Enzyme Structure (Part D)
Edited by C. H. W. HIRS AND SERGE N. TIMASHEFF

VOLUME XXVIII. Complex Carbohydrates (Part B)
Edited by VICTOR GINSBURG

VOLUME XXIX. Nucleic Acids and Protein Synthesis (Part E)
Edited by LAWRENCE GROSSMAN AND KIVIE MOLDAVE

VOLUME XXX. Nucleic Acids and Protein Synthesis (Part F)
Edited by KIVIE MOLDAVE AND LAWRENCE GROSSMAN

VOLUME XXXI. Biomembranes (Part A)
Edited by SIDNEY FLEISCHER AND LESTER PACKER

VOLUME XXXII. Biomembranes (Part B)
Edited by SIDNEY FLEISCHER AND LESTER PACKER

VOLUME XXXIII. Cumulative Subject Index Volumes I–XXX
Edited by MARTHA G. DENNIS AND EDWARD A. DENNIS

VOLUME XXXIV. Affinity Techniques (Enzyme Purification: Part B)
Edited by WILLIAM B. JAKOBY AND MEIR WILCHEK

VOLUME XXXV. Lipids (Part B)
Edited by JOHN M. LOWENSTEIN

VOLUME XXXVI. Hormone Action (Part A: Steroid Hormones)
Edited by BERT W. O'MALLEY AND JOEL G. HARDMAN

VOLUME XXXVII. Hormone Action (Part B: Peptide Hormones)
Edited by BERT W. O'MALLEY AND JOEL G. HARDMAN

VOLUME XXXVIII. Hormone Action (Part C: Cyclic Nucleotides)
Edited by JOEL G. HARDMAN AND BERT W. O'MALLEY

VOLUME XXXIX. Hormone Action (Part D: Isolated Cells, Tissues, and Organ Systems)
Edited by JOEL G. HARDMAN AND BERT W. O'MALLEY

VOLUME XL. Hormone Action (Part E: Nuclear Structure and Function)
Edited by BERT W. O'MALLEY AND JOEL G. HARDMAN

VOLUME XLI. Carbohydrate Metabolism (Part B)
Edited by W. A. WOOD

VOLUME XLII. Carbohydrate Metabolism (Part C)
Edited by W. A. WOOD

VOLUME XLIII. Antibiotics
Edited by JOHN H. HASH

VOLUME XLIV. Immobilized Enzymes
Edited by KLAUS MOSBACH

VOLUME XLV. Proteolytic Enzymes (Part B)
Edited by LASZLO LORAND

VOLUME XLVI. Affinity Labeling
Edited by WILLIAM B. JAKOBY AND MEIR WILCHEK

VOLUME XLVII. Enzyme Structure (Part E)
Edited by C. H. W. HIRS AND SERGE N. TIMASHEFF

VOLUME XLVIII. Enzyme Structure (Part F)
Edited by C. H. W. HIRS AND SERGE N. TIMASHEFF

VOLUME XLIX. Enzyme Structure (Part G)
Edited by C. H. W. HIRS AND SERGE N. TIMASHEFF

VOLUME L. Complex Carbohydrates (Part C)
Edited by VICTOR GINSBURG

VOLUME LI. Purine and Pyrimidine Nucleotide Metabolism
Edited by PATRICIA A. HOFFEE AND MARY ELLEN JONES

VOLUME LII. Biomembranes (Part C: Biological Oxidations)
Edited by SIDNEY FLEISCHER AND LESTER PACKER

VOLUME LIII. Biomembranes (Part D: Biological Oxidations)
Edited by SIDNEY FLEISCHER AND LESTER PACKER

VOLUME LIV. Biomembranes (Part E: Biological Oxidations)
Edited by SIDNEY FLEISCHER AND LESTER PACKER

VOLUME LV. Biomembranes (Part F: Bioenergetics)
Edited by SIDNEY FLEISCHER AND LESTER PACKER

VOLUME LVI. Biomembranes (Part G: Bioenergetics)
Edited by SIDNEY FLEISCHER AND LESTER PACKER

VOLUME LVII. Bioluminescence and Chemiluminescence
Edited by MARLENE A. DELUCA

VOLUME LVIII. Cell Culture
Edited by WILLIAM B. JAKOBY AND IRA PASTAN

VOLUME LIX. Nucleic Acids and Protein Synthesis (Part G)
Edited by KIVIE MOLDAVE AND LAWRENCE GROSSMAN

VOLUME LX. Nucleic Acids and Protein Synthesis (Part H)
Edited by KIVIE MOLDAVE AND LAWRENCE GROSSMAN

VOLUME 61. Enzyme Structure (Part H)
Edited by C. H. W. HIRS AND SERGE N. TIMASHEFF

VOLUME 62. Vitamins and Coenzymes (Part D)
Edited by DONALD B. MCCORMICK AND LEMUEL D. WRIGHT

VOLUME 63. Enzyme Kinetics and Mechanism (Part A: Initial Rate and Inhibitor Methods)
Edited by DANIEL L. PURICH

VOLUME 64. Enzyme Kinetics and Mechanism (Part B: Isotopic Probes and Complex Enzyme Systems)
Edited by DANIEL L. PURICH

VOLUME 65. Nucleic Acids (Part I)
Edited by LAWRENCE GROSSMAN AND KIVIE MOLDAVE

VOLUME 66. Vitamins and Coenzymes (Part E)
Edited by DONALD B. MCCORMICK AND LEMUEL D. WRIGHT

VOLUME 67. Vitamins and Coenzymes (Part F)
Edited by DONALD B. MCCORMICK AND LEMUEL D. WRIGHT

VOLUME 68. Recombinant DNA
Edited by RAY WU

VOLUME 69. Photosynthesis and Nitrogen Fixation (Part C)
Edited by ANTHONY SAN PIETRO

VOLUME 70. Immunochemical Techniques (Part A)
Edited by HELEN VAN VUNAKIS AND JOHN J. LANGONE

VOLUME 71. Lipids (Part C)
Edited by JOHN M. LOWENSTEIN

VOLUME 72. Lipids (Part D)
Edited by JOHN M. LOWENSTEIN

VOLUME 73. Immunochemical Techniques (Part B)
Edited by JOHN J. LANGONE AND HELEN VAN VUNAKIS

VOLUME 74. Immunochemical Techniques (Part C)
Edited by JOHN J. LANGONE AND HELEN VAN VUNAKIS

VOLUME 75. Cumulative Subject Index Volumes XXXI, XXXII, XXXIV–LX
Edited by EDWARD A. DENNIS AND MARTHA G. DENNIS

VOLUME 76. Hemoglobins
Edited by ERALDO ANTONINI, LUIGI ROSSI-BERNARDI, AND EMILIA CHIANCONE

VOLUME 77. Detoxication and Drug Metabolism
Edited by WILLIAM B. JAKOBY

VOLUME 78. Interferons (Part A)
Edited by SIDNEY PESTKA

VOLUME 79. Interferons (Part B)
Edited by SIDNEY PESTKA

VOLUME 80. Proteolytic Enzymes (Part C)
Edited by LASZLO LORAND

VOLUME 81. Biomembranes (Part H: Visual Pigments and Purple Membranes, I)
Edited by LESTER PACKER

VOLUME 82. Structural and Contractile Proteins (Part A: Extracellular Matrix)
Edited by LEON W. CUNNINGHAM AND DIXIE W. FREDERIKSEN

VOLUME 83. Complex Carbohydrates (Part D)
Edited by VICTOR GINSBURG

VOLUME 84. Immunochemical Techniques (Part D: Selected Immunoassays)
Edited by JOHN J. LANGONE AND HELEN VAN VUNAKIS

VOLUME 85. Structural and Contractile Proteins (Part B: The Contractile Apparatus and the Cytoskeleton)
Edited by DIXIE W. FREDERIKSEN AND LEON W. CUNNINGHAM

VOLUME 86. Prostaglandins and Arachidonate Metabolites
Edited by WILLIAM E. M. LANDS AND WILLIAM L. SMITH

VOLUME 87. Enzyme Kinetics and Mechanism (Part C: Intermediates, Stereochemistry, and Rate Studies)
Edited by DANIEL L. PURICH

VOLUME 88. Biomembranes (Part I: Visual Pigments and Purple Membranes, II)
Edited by LESTER PACKER

VOLUME 89. Carbohydrate Metabolism (Part D)
Edited by WILLIS A. WOOD

VOLUME 90. Carbohydrate Metabolism (Part E)
Edited by WILLIS A. WOOD

VOLUME 91. Enzyme Structure (Part I)
Edited by C. H. W. HIRS AND SERGE N. TIMASHEFF

VOLUME 92. Immunochemical Techniques (Part E: Monoclonal Antibodies and General Immunoassay Methods)
Edited by JOHN J. LANGONE AND HELEN VAN VUNAKIS

VOLUME 93. Immunochemical Techniques (Part F: Conventional Antibodies, Fc Receptors, and Cytotoxicity)
Edited by JOHN J. LANGONE AND HELEN VAN VUNAKIS

VOLUME 94. Polyamines
Edited by HERBERT TABOR AND CELIA WHITE TABOR

VOLUME 95. Cumulative Subject Index Volumes 61–74, 76–80
Edited by EDWARD A. DENNIS AND MARTHA G. DENNIS

VOLUME 96. Biomembranes [Part J: Membrane Biogenesis: Assembly and Targeting (General Methods; Eukaryotes)]
Edited by SIDNEY FLEISCHER AND BECCA FLEISCHER

VOLUME 97. Biomembranes [Part K: Membrane Biogenesis: Assembly and Targeting (Prokaryotes, Mitochondria, and Chloroplasts)]
Edited by SIDNEY FLEISCHER AND BECCA FLEISCHER

VOLUME 98. Biomembranes (Part L: Membrane Biogenesis: Processing and Recycling)
Edited by SIDNEY FLEISCHER AND BECCA FLEISCHER

VOLUME 99. Hormone Action (Part F: Protein Kinases)
Edited by JACKIE D. CORBIN AND JOEL G. HARDMAN

VOLUME 100. Recombinant DNA (Part B)
Edited by RAY WU, LAWRENCE GROSSMAN, AND KIVIE MOLDAVE

VOLUME 101. Recombinant DNA (Part C)
Edited by RAY WU, LAWRENCE GROSSMAN, AND KIVIE MOLDAVE

VOLUME 102. Hormone Action (Part G: Calmodulin and Calcium-Binding Proteins)
Edited by ANTHONY R. MEANS AND BERT W. O'MALLEY

VOLUME 103. Hormone Action (Part H: Neuroendocrine Peptides)
Edited by P. MICHAEL CONN

VOLUME 104. Enzyme Purification and Related Techniques (Part C)
Edited by WILLIAM B. JAKOBY

VOLUME 105. Oxygen Radicals in Biological Systems
Edited by LESTER PACKER

VOLUME 106. Posttranslational Modifications (Part A)
Edited by FINN WOLD AND KIVIE MOLDAVE

VOLUME 107. Posttranslational Modifications (Part B)
Edited by FINN WOLD AND KIVIE MOLDAVE

VOLUME 108. Immunochemical Techniques (Part G: Separation and Characterization of Lymphoid Cells)
Edited by GIOVANNI DI SABATO, JOHN J. LANGONE, AND HELEN VAN VUNAKIS

VOLUME 109. Hormone Action (Part I: Peptide Hormones)
Edited by LUTZ BIRNBAUMER AND BERT W. O'MALLEY

VOLUME 110. Steroids and Isoprenoids (Part A)
Edited by JOHN H. LAW AND HANS C. RILLING

VOLUME 111. Steroids and Isoprenoids (Part B)
Edited by JOHN H. LAW AND HANS C. RILLING

VOLUME 112. Drug and Enzyme Targeting (Part A)
Edited by KENNETH J. WIDDER AND RALPH GREEN

VOLUME 113. Glutamate, Glutamine, Glutathione, and Related Compounds
Edited by ALTON MEISTER

VOLUME 114. Diffraction Methods for Biological Macromolecules (Part A)
Edited by HAROLD W. WYCKOFF, C. H. W. HIRS, AND SERGE N. TIMASHEFF

VOLUME 115. Diffraction Methods for Biological Macromolecules (Part B)
Edited by HAROLD W. WYCKOFF, C. H. W. HIRS, AND SERGE N. TIMASHEFF

VOLUME 116. Immunochemical Techniques (Part H: Effectors and Mediators of Lymphoid Cell Functions)
Edited by GIOVANNI DI SABATO, JOHN J. LANGONE, AND HELEN VAN VUNAKIS

VOLUME 117. Enzyme Structure (Part J)
Edited by C. H. W. HIRS AND SERGE N. TIMASHEFF

VOLUME 118. Plant Molecular Biology
Edited by ARTHUR WEISSBACH AND HERBERT WEISSBACH

VOLUME 119. Interferons (Part C)
Edited by SIDNEY PESTKA

VOLUME 120. Cumulative Subject Index Volumes 81–94, 96–101

VOLUME 121. Immunochemical Techniques (Part I: Hybridoma Technology and Monoclonal Antibodies)
Edited by JOHN J. LANGONE AND HELEN VAN VUNAKIS

VOLUME 122. Vitamins and Coenzymes (Part G)
Edited by FRANK CHYTIL AND DONALD B. MCCORMICK

VOLUME 123. Vitamins and Coenzymes (Part H)
Edited by FRANK CHYTIL AND DONALD B. MCCORMICK

VOLUME 124. Hormone Action (Part J: Neuroendocrine Peptides)
Edited by P. MICHAEL CONN

VOLUME 125. Biomembranes (Part M: Transport in Bacteria, Mitochondria, and Chloroplasts: General Approaches and Transport Systems)
Edited by SIDNEY FLEISCHER AND BECCA FLEISCHER

VOLUME 126. Biomembranes (Part N: Transport in Bacteria, Mitochondria, and Chloroplasts: Protonmotive Force)
Edited by SIDNEY FLEISCHER AND BECCA FLEISCHER

VOLUME 127. Biomembranes (Part O: Protons and Water: Structure and Translocation)
Edited by LESTER PACKER

VOLUME 128. Plasma Lipoproteins (Part A: Preparation, Structure, and Molecular Biology)
Edited by JERE P. SEGREST AND JOHN J. ALBERS

VOLUME 129. Plasma Lipoproteins (Part B: Characterization, Cell Biology, and Metabolism)
Edited by JOHN J. ALBERS AND JERE P. SEGREST

VOLUME 130. Enzyme Structure (Part K)
Edited by C. H. W. HIRS AND SERGE N. TIMASHEFF

VOLUME 131. Enzyme Structure (Part L)
Edited by C. H. W. HIRS AND SERGE N. TIMASHEFF

VOLUME 132. Immunochemical Techniques (Part J: Phagocytosis and Cell-Mediated Cytotoxicity)
Edited by GIOVANNI DI SABATO AND JOHANNES EVERSE

VOLUME 133. Bioluminescence and Chemiluminescence (Part B)
Edited by MARLENE DELUCA AND WILLIAM D. MCELROY

VOLUME 134. Structural and Contractile Proteins (Part C: The Contractile Apparatus and the Cytoskeleton)
Edited by RICHARD B. VALLEE

VOLUME 135. Immobilized Enzymes and Cells (Part B)
Edited by KLAUS MOSBACH

VOLUME 136. Immobilized Enzymes and Cells (Part C)
Edited by KLAUS MOSBACH

VOLUME 137. Immobilized Enzymes and Cells (Part D)
Edited by KLAUS MOSBACH

VOLUME 138. Complex Carbohydrates (Part E)
Edited by VICTOR GINSBURG

VOLUME 139. Cellular Regulators (Part A: Calcium- and Calmodulin-Binding Proteins)
Edited by ANTHONY R. MEANS AND P. MICHAEL CONN

VOLUME 140. Cumulative Subject Index Volumes 102–119, 121–134

VOLUME 141. Cellular Regulators (Part B: Calcium and Lipids)
Edited by P. MICHAEL CONN AND ANTHONY R. MEANS

VOLUME 142. Metabolism of Aromatic Amino Acids and Amines
Edited by SEYMOUR KAUFMAN

VOLUME 143. Sulfur and Sulfur Amino Acids
Edited by WILLIAM B. JAKOBY AND OWEN GRIFFITH

VOLUME 144. Structural and Contractile Proteins (Part D: Extracellular Matrix)
Edited by LEON W. CUNNINGHAM

VOLUME 145. Structural and Contractile Proteins (Part E: Extracellular Matrix)
Edited by LEON W. CUNNINGHAM

VOLUME 146. Peptide Growth Factors (Part A)
Edited by DAVID BARNES AND DAVID A. SIRBASKU

VOLUME 147. Peptide Growth Factors (Part B)
Edited by DAVID BARNES AND DAVID A. SIRBASKU

VOLUME 148. Plant Cell Membranes
Edited by LESTER PACKER AND ROLAND DOUCE

VOLUME 149. Drug and Enzyme Targeting (Part B)
Edited by RALPH GREEN AND KENNETH J. WIDDER

VOLUME 150. Immunochemical Techniques (Part K: *In Vitro* Models of B and T Cell Functions and Lymphoid Cell Receptors)
Edited by GIOVANNI DI SABATO

VOLUME 151. Molecular Genetics of Mammalian Cells
Edited by MICHAEL M. GOTTESMAN

VOLUME 152. Guide to Molecular Cloning Techniques
Edited by SHELBY L. BERGER AND ALAN R. KIMMEL

VOLUME 153. Recombinant DNA (Part D)
Edited by RAY WU AND LAWRENCE GROSSMAN

VOLUME 154. Recombinant DNA (Part E)
Edited by RAY WU AND LAWRENCE GROSSMAN

VOLUME 155. Recombinant DNA (Part F)
Edited by RAY WU

VOLUME 156. Biomembranes (Part P: ATP-Driven Pumps and Related Transport: The Na,K-Pump)
Edited by SIDNEY FLEISCHER AND BECCA FLEISCHER

VOLUME 157. Biomembranes (Part Q: ATP-Driven Pumps and Related Transport: Calcium, Proton, and Potassium Pumps)
Edited by SIDNEY FLEISCHER AND BECCA FLEISCHER

VOLUME 158. Metalloproteins (Part A)
Edited by JAMES F. RIORDAN AND BERT L. VALLEE

VOLUME 159. Initiation and Termination of Cyclic Nucleotide Action
Edited by JACKIE D. CORBIN AND ROGER A. JOHNSON

VOLUME 160. Biomass (Part A: Cellulose and Hemicellulose)
Edited by WILLIS A. WOOD AND SCOTT T. KELLOGG

VOLUME 161. Biomass (Part B: Lignin, Pectin, and Chitin)
Edited by WILLIS A. WOOD AND SCOTT T. KELLOGG

VOLUME 162. Immunochemical Techniques (Part L: Chemotaxis and Inflammation)
Edited by GIOVANNI DI SABATO

VOLUME 163. Immunochemical Techniques (Part M: Chemotaxis and Inflammation)
Edited by GIOVANNI DI SABATO

VOLUME 164. Ribosomes
Edited by HARRY F. NOLLER, JR., AND KIVIE MOLDAVE

VOLUME 165. Microbial Toxins: Tools for Enzymology
Edited by SIDNEY HARSHMAN

VOLUME 166. Branched-Chain Amino Acids
Edited by ROBERT HARRIS AND JOHN R. SOKATCH

VOLUME 167. Cyanobacteria
Edited by LESTER PACKER AND ALEXANDER N. GLAZER

VOLUME 168. Hormone Action (Part K: Neuroendocrine Peptides)
Edited by P. MICHAEL CONN

VOLUME 169. Platelets: Receptors, Adhesion, Secretion (Part A)
Edited by JACEK HAWIGER

VOLUME 170. Nucleosomes
Edited by PAUL M. WASSARMAN AND ROGER D. KORNBERG

VOLUME 171. Biomembranes (Part R: Transport Theory: Cells and Model Membranes)
Edited by SIDNEY FLEISCHER AND BECCA FLEISCHER

VOLUME 172. Biomembranes (Part S: Transport: Membrane Isolation and Characterization)
Edited by SIDNEY FLEISCHER AND BECCA FLEISCHER

VOLUME 173. Biomembranes [Part T: Cellular and Subcellular Transport: Eukaryotic (Nonepithelial) Cells]
Edited by SIDNEY FLEISCHER AND BECCA FLEISCHER

VOLUME 174. Biomembranes [Part U: Cellular and Subcellular Transport: Eukaryotic (Nonepithelial) Cells]
Edited by SIDNEY FLEISCHER AND BECCA FLEISCHER

VOLUME 175. Cumulative Subject Index Volumes 135–139, 141–167

VOLUME 176. Nuclear Magnetic Resonance (Part A: Spectral Techniques and Dynamics)
Edited by NORMAN J. OPPENHEIMER AND THOMAS L. JAMES

VOLUME 177. Nuclear Magnetic Resonance (Part B: Structure and Mechanism)
Edited by NORMAN J. OPPENHEIMER AND THOMAS L. JAMES

VOLUME 178. Antibodies, Antigens, and Molecular Mimicry
Edited by JOHN J. LANGONE

VOLUME 179. Complex Carbohydrates (Part F)
Edited by VICTOR GINSBURG

VOLUME 180. RNA Processing (Part A: General Methods)
Edited by JAMES E. DAHLBERG AND JOHN N. ABELSON

VOLUME 181. RNA Processing (Part B: Specific Methods)
Edited by JAMES E. DAHLBERG AND JOHN N. ABELSON

VOLUME 182. Guide to Protein Purification
Edited by MURRAY P. DEUTSCHER

VOLUME 183. Molecular Evolution: Computer Analysis of Protein and Nucleic Acid Sequences
Edited by RUSSELL F. DOOLITTLE

VOLUME 184. Avidin–Biotin Technology
Edited by MEIR WILCHEK AND EDWARD A. BAYER

VOLUME 185. Gene Expression Technology
Edited by DAVID V. GOEDDEL

VOLUME 186. Oxygen Radicals in Biological Systems (Part B: Oxygen Radicals and Antioxidants)
Edited by LESTER PACKER AND ALEXANDER N. GLAZER

VOLUME 187. Arachidonate Related Lipid Mediators
Edited by ROBERT C. MURPHY AND FRANK A. FITZPATRICK

VOLUME 188. Hydrocarbons and Methylotrophy
Edited by MARY E. LIDSTROM

VOLUME 189. Retinoids (Part A: Molecular and Metabolic Aspects)
Edited by LESTER PACKER

VOLUME 190. Retinoids (Part B: Cell Differentiation and Clinical Applications)
Edited by LESTER PACKER

VOLUME 191. Biomembranes (Part V: Cellular and Subcellular Transport: Epithelial Cells)
Edited by SIDNEY FLEISCHER AND BECCA FLEISCHER

VOLUME 192. Biomembranes (Part W: Cellular and Subcellular Transport: Epithelial Cells)
Edited by SIDNEY FLEISCHER AND BECCA FLEISCHER

VOLUME 193. Mass Spectrometry
Edited by JAMES A. MCCLOSKEY

VOLUME 194. Guide to Yeast Genetics and Molecular Biology
Edited by CHRISTINE GUTHRIE AND GERALD R. FINK

VOLUME 195. Adenylyl Cyclase, G Proteins, and Guanylyl Cyclase
Edited by ROGER A. JOHNSON AND JACKIE D. CORBIN

VOLUME 196. Molecular Motors and the Cytoskeleton
Edited by RICHARD B. VALLEE

VOLUME 197. Phospholipases
Edited by EDWARD A. DENNIS

VOLUME 198. Peptide Growth Factors (Part C)
Edited by DAVID BARNES, J. P. MATHER, AND GORDON H. SATO

VOLUME 199. Cumulative Subject Index Volumes 168–174, 176–194

VOLUME 200. Protein Phosphorylation (Part A: Protein Kinases: Assays, Purification, Antibodies, Functional Analysis, Cloning, and Expression)
Edited by TONY HUNTER AND BARTHOLOMEW M. SEFTON

VOLUME 201. Protein Phosphorylation (Part B: Analysis of Protein Phosphorylation, Protein Kinase Inhibitors, and Protein Phosphatases)
Edited by TONY HUNTER AND BARTHOLOMEW M. SEFTON

VOLUME 202. Molecular Design and Modeling: Concepts and Applications (Part A: Proteins, Peptides, and Enzymes)
Edited by JOHN J. LANGONE

VOLUME 203. Molecular Design and Modeling: Concepts and Applications (Part B: Antibodies and Antigens, Nucleic Acids, Polysaccharides, and Drugs)
Edited by JOHN J. LANGONE

VOLUME 204. Bacterial Genetic Systems
Edited by JEFFREY H. MILLER

VOLUME 205. Metallobiochemistry (Part B: Metallothionein and Related Molecules)
Edited by JAMES F. RIORDAN AND BERT L. VALLEE

VOLUME 206. Cytochrome P450
Edited by MICHAEL R. WATERMAN AND ERIC F. JOHNSON

VOLUME 207. Ion Channels
Edited by BERNARDO RUDY AND LINDA E. IVERSON

VOLUME 208. Protein–DNA Interactions
Edited by ROBERT T. SAUER

VOLUME 209. Phospholipid Biosynthesis
Edited by EDWARD A. DENNIS AND DENNIS E. VANCE

VOLUME 210. Numerical Computer Methods
Edited by LUDWIG BRAND AND MICHAEL L. JOHNSON

VOLUME 211. DNA Structures (Part A: Synthesis and Physical Analysis of DNA)
Edited by DAVID M. J. LILLEY AND JAMES E. DAHLBERG

VOLUME 212. DNA Structures (Part B: Chemical and Electrophoretic Analysis of DNA)
Edited by DAVID M. J. LILLEY AND JAMES E. DAHLBERG

VOLUME 213. Carotenoids (Part A: Chemistry, Separation, Quantitation, and Antioxidation)
Edited by LESTER PACKER

VOLUME 214. Carotenoids (Part B: Metabolism, Genetics, and Biosynthesis)
Edited by LESTER PACKER

VOLUME 215. Platelets: Receptors, Adhesion, Secretion (Part B)
Edited by JACEK J. HAWIGER

VOLUME 216. Recombinant DNA (Part G)
Edited by RAY WU

VOLUME 217. Recombinant DNA (Part H)
Edited by RAY WU

VOLUME 218. Recombinant DNA (Part I)
Edited by RAY WU

VOLUME 219. Reconstitution of Intracellular Transport
Edited by JAMES E. ROTHMAN

VOLUME 220. Membrane Fusion Techniques (Part A)
Edited by NEJAT DÜZGÜNEŞ

VOLUME 221. Membrane Fusion Techniques (Part B)
Edited by NEJAT DÜZGÜNEŞ

VOLUME 222. Proteolytic Enzymes in Coagulation, Fibrinolysis, and Complement Activation (Part A: Mammalian Blood Coagulation Factors and Inhibitors)
Edited by LASZLO LORAND AND KENNETH G. MANN

VOLUME 223. Proteolytic Enzymes in Coagulation, Fibrinolysis, and Complement Activation (Part B: Complement Activation, Fibrinolysis, and Nonmammalian Blood Coagulation Factors)
Edited by LASZLO LORAND AND KENNETH G. MANN

VOLUME 224. Molecular Evolution: Producing the Biochemical Data
Edited by ELIZABETH ANNE ZIMMER, THOMAS J. WHITE, REBECCA L. CANN, AND ALLAN C. WILSON

VOLUME 225. Guide to Techniques in Mouse Development
Edited by PAUL M. WASSARMAN AND MELVIN L. DEPAMPHILIS

VOLUME 226. Metallobiochemistry (Part C: Spectroscopic and Physical Methods for Probing Metal Ion Environments in Metalloenzymes and Metalloproteins)
Edited by JAMES F. RIORDAN AND BERT L. VALLEE

VOLUME 227. Metallobiochemistry (Part D: Physical and Spectroscopic Methods for Probing Metal Ion Environments in Metalloproteins)
Edited by JAMES F. RIORDAN AND BERT L. VALLEE

VOLUME 228. Aqueous Two-Phase Systems
Edited by HARRY WALTER AND GÖTE JOHANSSON

VOLUME 229. Cumulative Subject Index Volumes 195–198, 200–227

VOLUME 230. Guide to Techniques in Glycobiology
Edited by WILLIAM J. LENNARZ AND GERALD W. HART

VOLUME 231. Hemoglobins (Part B: Biochemical and Analytical Methods)
Edited by JOHANNES EVERSE, KIM D. VANDEGRIFF, AND ROBERT M. WINSLOW

VOLUME 232. Hemoglobins (Part C: Biophysical Methods)
Edited by JOHANNES EVERSE, KIM D. VANDEGRIFF, AND ROBERT M. WINSLOW

VOLUME 233. Oxygen Radicals in Biological Systems (Part C)
Edited by LESTER PACKER

VOLUME 234. Oxygen Radicals in Biological Systems (Part D)
Edited by LESTER PACKER

VOLUME 235. Bacterial Pathogenesis (Part A: Identification and Regulation of Virulence Factors)
Edited by VIRGINIA L. CLARK AND PATRIK M. BAVOIL

VOLUME 236. Bacterial Pathogenesis (Part B: Integration of Pathogenic Bacteria with Host Cells)
Edited by VIRGINIA L. CLARK AND PATRIK M. BAVOIL

VOLUME 237. Heterotrimeric G Proteins
Edited by RAVI IYENGAR

VOLUME 238. Heterotrimeric G-Protein Effectors
Edited by RAVI IYENGAR

VOLUME 239. Nuclear Magnetic Resonance (Part C)
Edited by THOMAS L. JAMES AND NORMAN J. OPPENHEIMER

VOLUME 240. Numerical Computer Methods (Part B)
Edited by MICHAEL L. JOHNSON AND LUDWIG BRAND

VOLUME 241. Retroviral Proteases
Edited by LAWRENCE C. KUO AND JULES A. SHAFER

VOLUME 242. Neoglycoconjugates (Part A)
Edited by Y. C. LEE AND REIKO T. LEE

VOLUME 243. Inorganic Microbial Sulfur Metabolism
Edited by HARRY D. PECK, JR., AND JEAN LEGALL

VOLUME 244. Proteolytic Enzymes: Serine and Cysteine Peptidases
Edited by ALAN J. BARRETT

VOLUME 245. Extracellular Matrix Components
Edited by E. RUOSLAHTI AND E. ENGVALL

VOLUME 246. Biochemical Spectroscopy
Edited by KENNETH SAUER

VOLUME 247. Neoglycoconjugates (Part B: Biomedical Applications)
Edited by Y. C. LEE AND REIKO T. LEE

VOLUME 248. Proteolytic Enzymes: Aspartic and Metallo Peptidases
Edited by ALAN J. BARRETT

VOLUME 249. Enzyme Kinetics and Mechanism (Part D: Developments in Enzyme Dynamics)
Edited by DANIEL L. PURICH

VOLUME 250. Lipid Modifications of Proteins
Edited by PATRICK J. CASEY AND JANICE E. BUSS

VOLUME 251. Biothiols (Part A: Monothiols and Dithiols, Protein Thiols, and Thiyl Radicals)
Edited by LESTER PACKER

VOLUME 252. Biothiols (Part B: Glutathione and Thioredoxin; Thiols in Signal Transduction and Gene Regulation)
Edited by LESTER PACKER

VOLUME 253. Adhesion of Microbial Pathogens
Edited by RON J. DOYLE AND ITZHAK OFEK

VOLUME 254. Oncogene Techniques
Edited by PETER K. VOGT AND INDER M. VERMA

VOLUME 255. Small GTPases and Their Regulators (Part A: Ras Family)
Edited by W. E. BALCH, CHANNING J. DER, AND ALAN HALL

VOLUME 256. Small GTPases and Their Regulators (Part B: Rho Family)
Edited by W. E. BALCH, CHANNING J. DER, AND ALAN HALL

VOLUME 257. Small GTPases and Their Regulators (Part C: Proteins Involved in Transport)
Edited by W. E. BALCH, CHANNING J. DER, AND ALAN HALL

VOLUME 258. Redox-Active Amino Acids in Biology
Edited by JUDITH P. KLINMAN

VOLUME 259. Energetics of Biological Macromolecules
Edited by MICHAEL L. JOHNSON AND GARY K. ACKERS

VOLUME 260. Mitochondrial Biogenesis and Genetics (Part A)
Edited by GIUSEPPE M. ATTARDI AND ANNE CHOMYN

VOLUME 261. Nuclear Magnetic Resonance and Nucleic Acids
Edited by THOMAS L. JAMES

VOLUME 262. DNA Replication
Edited by JUDITH L. CAMPBELL

VOLUME 263. Plasma Lipoproteins (Part C: Quantitation)
Edited by WILLIAM A. BRADLEY, SANDRA H. GIANTURCO, AND JERE P. SEGREST

VOLUME 264. Mitochondrial Biogenesis and Genetics (Part B)
Edited by GIUSEPPE M. ATTARDI AND ANNE CHOMYN

VOLUME 265. Cumulative Subject Index Volumes 228, 230–262

VOLUME 266. Computer Methods for Macromolecular Sequence Analysis
Edited by RUSSELL F. DOOLITTLE

VOLUME 267. Combinatorial Chemistry
Edited by JOHN N. ABELSON

VOLUME 268. Nitric Oxide (Part A: Sources and Detection of NO; NO Synthase)
Edited by LESTER PACKER

VOLUME 269. Nitric Oxide (Part B: Physiological and Pathological Processes)
Edited by LESTER PACKER

VOLUME 270. High Resolution Separation and Analysis of Biological Macromolecules (Part A: Fundamentals)
Edited by BARRY L. KARGER AND WILLIAM S. HANCOCK

VOLUME 271. High Resolution Separation and Analysis of Biological Macromolecules (Part B: Applications)
Edited by BARRY L. KARGER AND WILLIAM S. HANCOCK

VOLUME 272. Cytochrome P450 (Part B)
Edited by ERIC F. JOHNSON AND MICHAEL R. WATERMAN

VOLUME 273. RNA Polymerase and Associated Factors (Part A)
Edited by SANKAR ADHYA

VOLUME 274. RNA Polymerase and Associated Factors (Part B)
Edited by SANKAR ADHYA

VOLUME 275. Viral Polymerases and Related Proteins
Edited by LAWRENCE C. KUO, DAVID B. OLSEN, AND STEVEN S. CARROLL

VOLUME 276. Macromolecular Crystallography (Part A)
Edited by CHARLES W. CARTER, JR., AND ROBERT M. SWEET

VOLUME 277. Macromolecular Crystallography (Part B)
Edited by CHARLES W. CARTER, JR., AND ROBERT M. SWEET

VOLUME 278. Fluorescence Spectroscopy
Edited by LUDWIG BRAND AND MICHAEL L. JOHNSON

VOLUME 279. Vitamins and Coenzymes, Part I
Edited by DONALD B. MCCORMICK, JOHN W. SUTTIE, AND CONRAD WAGNER

VOLUME 280. Vitamins and Coenzymes, Part J
Edited by DONALD B. MCCORMICK, JOHN W. SUTTIE, AND CONRAD WAGNER

VOLUME 281. Vitamins and Coenzymes, Part K
Edited by DONALD B. MCCORMICK, JOHN W. SUTTIE, AND CONRAD WAGNER

VOLUME 282. Vitamins and Coenzymes, Part L
Edited by DONALD B. MCCORMICK, JOHN W. SUTTIE, AND CONRAD WAGNER

VOLUME 283. Cell Cycle Control
Edited by WILLIAM G. DUNPHY

VOLUME 284. Lipases (Part A: Biotechnology)
Edited by BYRON RUBIN AND EDWARD A. DENNIS

VOLUME 285. Cumulative Subject Index Volumes 263, 264, 266–289

VOLUME 286. Lipases (Part B: Enzyme Characterization and Utilization)
Edited by BYRON RUBIN AND EDWARD A. DENNIS

VOLUME 287. Chemokines
Edited by RICHARD HORUK

VOLUME 288. Chemokine Receptors
Edited by RICHARD HORUK

VOLUME 289. Solid Phase Peptide Synthesis
Edited by GREGG B. FIELDS

VOLUME 290. Molecular Chaperones
Edited by GEORGE H. LORIMER AND THOMAS BALDWIN

VOLUME 299. Oxidants and Antioxidants (Part A)
Edited by LESTER PACKER

VOLUME 300. Oxidants and Antioxidants (Part B)
Edited by LESTER PACKER

VOLUME 301. Nitric Oxide (Part C: Biological and Antioxidant Properties) (In preparation)
Edited by LESTER PACKER

Fredric Stewart Fay

1943–1997

The scientific community lost one of its most gifted and productive members on March 18, 1997 when Fred Fay suffered a fatal heart attack. Fred was born in New York City and was educated in its public school system. On graduation from the Bronx High School of Science in 1961, he enrolled at Cornell University where he majored in chemistry and was elected to Phi Beta Kappa. His love and talent for science were evident early, not only in the schoolboy experiments conducted in the family apartment, but also in the summers of his undergraduate years when he worked as a lab assistant at the Sloane–Kettering Institute. Fred began his formal scientific career in his junior year at Cornell when he joined the lab of Dr. R. Blake Reeves. With Dr. Reeves Fred first studied the effects of low barometric pressure on the synthesis of a myoglobin-like oxygen-binding pigment in a marine arthropod. This was followed by a study of B_{12} uptake in a protozoan, which led to Fred's first scientific publication [Reeves,

R.B. and Fay, F.S. Cyanocobalamin (vitamin B_{12}) uptake by *Ochromonas malhamensis. Am. J. Physiol.* **210**:1273–1278, 1966]. Reeves had recently arrived at Cornell after completing his doctoral and postdoctoral training in the Physiology Department at Harvard Medical School. As is so often the case, Reeves referred his star pupil to his mentor, Dr. John R. Pappenheimer, and on graduation from Cornell, Fred enrolled in the Medical Sciences Program at Harvard to begin his doctoral studies in physiology.

Pappenheimer had long-standing interests in the regulation of respiration and in oxygen-sensing mechanisms, and at the time was studying the contribution of the ionic composition of cerebrospinal fluid to the regulation of respiration in goats. It was quite natural in this environment for Fred's interest to be tweaked by the question of how tissues sense and respond to changes in oxygen tension. He examined the anomalously high rate of oxygen consumption in the carotid body for clues to its oxygen-sensing function. These studies provided the first accurate data on oxygen consumption by this tissue, which, in fact, consumed oxygen at less than 20% of the previously accepted rate. To complete his thesis project he went on to study contractile responses of the guinea pig ductus arteriosis to increased oxygen availability. This latter research kindled his interest in the physiology of smooth muscle, which was to play a central role in his research career. These early studies also foreshadowed other themes that became prominent in all of his subsequent work. Fred had a need to see what was happening in his experiments. His thesis and the resulting papers beautifully portrayed the morphology of his preparations at the macroscopic and microscopic levels. Studies of both the carotid body and the ductus arteriosus required the custom design and fabrication of suitable apparatus and carried the imprimatur of the inventiveness and skill that characterized all of his later work.

After receiving his PhD in 1969, Fred stayed on at Harvard to continue work on the guinea pig ductus arteriosus. These studies showed all of the elegance and clarity of thought that were to become the hallmark of Fred's science. He found that tension in the ductal wall increased in response to increased partial pressure of oxygen just as well when oxygen was present only in the perfusate within the vessel or when present only in the perifusate bathing the outer adventitial surface that contained nerve endings and mast cells. Because equal but oppositely directed gradients of P_{O_2} produced equivalent contractile responses, he concluded that the smooth muscle itself acted as the oxygen sensor and contracted in a stepwise fashion (Fay, F.S. Guinea pig ductus arteriosus I. The cellular and metabolic basis for oxygen sensitivity. *Am. J. Physiol.* **221**:470–479, 1971). He later showed that the trigger for this response is a unique cytochrome and demonstrated that the ductus could be made to contract by light of the appropriate wavelength if the cytochrome had first been inactivated by carbon monoxide in the dark. Virtually all of the themes that were to characterize his later work were already germinating in these studies: the morphologic studies, particularly associated with the irreversibility of ductal closure and the questions raised with regard to coupling between metabolism and electrical activity and contraction, and the stepwise contractile behavior of "a population of smooth muscle cells of the single-unit type."

Meanwhile, 40 miles to the west, final plans were taking shape to open the University of Massachusetts Medical School. Fred was recruited as an assistant professor of physiology, and was in Worcester to greet the first class of medical students when they arrived in September 1970. He was intimately involved in all of the planning, organizing, and just plain hard work that goes into starting a new department. Much of the credit for whatever success the department now enjoys must go to him. With characteristic energy and drive Fred immediately set up shop in a small lab in the converted tobacco warehouse that was to be the temporary home of the medical school for the next $3\frac{1}{2}$ years. Here he continued his studies on the ductus arteriosus and on the morphologic basis for contractility of smooth muscle in collaboration with Peter Cooke. He also began to lay the groundwork for in-depth studies of smooth muscle physiology.

At that time understanding of how smooth muscle contracts was meager at best. All that was known was derived from studies of multicellular preparations with all of the complexities attributable to heterogeneity of cells and extracellular matrices. Fred began to rectify that situation with characteristic insight and boundless energy. He set about trying to develop a homogeneous population of isolated muscle cells to define the contractile properties of smooth muscle cells themselves. His first attempts were with the intestinal muscles of a large salamander, *Amphiuma* sp., which had unusually large smooth muscle cells. Unfortunately, obtaining a dispersed cell preparation that maintained normal contractility was no simple matter. As he struggled with *Amphiuma,* Bagby and colleagues (Bagby, R., Young, A., Dotson, R., Fisher, B., and McKinnon, K. Contraction of single smooth muscle cells from *Bufo marinus* stomach. *Nature* **234:**351–352, 1971) published their successful preparation of isolated smooth muscle cells from the stomach of *Bufo marinus.* Fred was quick to exploit this preparation and undertook what are now his classic studies of smooth muscle physiology.

If there is any one aspect of Fred's career that stands out, it was his use of the single smooth muscle cell preparation. With it, Fred explored virtually every aspect of smooth muscle physiology and molecular anatomy at the cellular and subcellular level. Fundamental contributions were made concerning the organization of the contractile apparatus and how it is turned on and off in response to various physiologic signals. This unique preparation was not only a model system for smooth muscle, but also a proving ground for Fred's views of cell biology. Working with a medical student, Claudio Delise, Fred showed with scanning electron microscopic images that when a smooth muscle contracts, large "blebs" appear on the plasma membrane, and proposed that the contractile machinery attaches at specialized regions of the plasma membrane. The now classic article describing these findings (Fay, F.S. and Delise, C.M. Contraction of isolated smooth muscle cells—structural changes. *Proc. Natl. Acad. Sci. U.S.A.* **70:**641–645, 1973) was his third most cited paper.

With another medical student, Peter Canaday, he devised an ultrasensitive force transducer. He then developed the techniques for attaching a single smooth muscle cell to the transducer and recorded tensions of a few micrograms when the cell contracted (Fay, F.S. Isometric contractile properties of single isolated smooth muscle cells. *Nature* **265:**553–556, 1977). With this new technology Fred showed that, in contrast to striated muscle or the multifiber smooth muscle preparations used by all previous investigators, the tension of a relaxed smooth muscle cell is virtually independent of its length. These studies highlighted the major contribution of the connective tissue elements to the mechanical properties of smooth muscle tissues and thus confirmed the critical importance of studying isolated smooth muscle cells for the understanding of smooth muscle physiology. In these studies, Fred also made the unique observation that there was a delay between the excitatory stimulus and the onset of force generation. He suggested that the delay was due either (1) to a delay in coupling of the changes in membrane potential to the change in intracellular Ca^{2+}; or, (2) that Ca^{2+} is not rate limiting and that Ca^{2+}-dependent smooth muscle myosin regulatory processes are slow.

The importance of being able to measure intracellular calcium concentrations ($[Ca^{2+}]_i$) was evident, and when he became eligible for his first sabbatical leave, Fred decided to work with Dr. Stuart Taylor at the Mayo Medical School to learn how to measure intracellular Ca^{2+} using aequorin as a calcium indicator (Fay, F.S., Shlevin, H.H., Granger, W.C., and Taylor, S.R. Aequorin luminescence during activation of single isolated smooth muscle cells. *Nature* **280:**506–508, 1979). He microinjected aequorin into individual smooth muscle cells and discovered that on stimulation $[Ca^{2+}]_i$ rises rapidly and remains above basal levels. Therefore, Ca^{2+} delivery to the regulatory proteins is not rate limiting. Rather, it is the myosin light-chain kinase regulatory pathway that is relatively slow, thus generating the observed delay between stimulus and contraction. This finding inevitably generated the next question: What Ca^{2+} handling processes might govern the decay of $[Ca^{2+}]_i$? Its answer would have to wait for the development of new techniques, and he would revisit this concept twenty years later.

Fred loved to hypothesize about molecular mechanisms. He loved to look at a data set and turn it around and around in his mind to uncover every facet. He would challenge his colleagues or fellows to come up with every possibility that might explain the data and then rigorously weigh every potential explanation until only the one or two most plausible survived. He was not shy in going out on a limb to propose unique mechanisms. A good example of this is seen in his second most cited paper (Scheid, C.R., Honeyman, T.W., and Fay, F.S. Mechanism of β-adrenergic relaxation of smooth muscle. *Nature* **277:**32–36, 1979). Fred and his co-workers proposed that β-adrenergic relaxation of smooth muscle cells was a result of cAMP-dependent stimulation of the Na^+,K^+-ATPase pump, which, in lowering intracellular sodium would, in turn, stimulate the Na^+/Ca^{2+} exchanger, with the final outcome being a reduction in $[Ca^{2+}]_i$. It was not until 14 years later, and after his lab had developed the digital imaging microscope (DIM), that Fred was able to expand on this idea. Fred reasoned that if β-adrenergic relaxation involved both Na^+,K^+-ATPase activity coupled to the Na^+/Ca^{2+} exchanger, these two entities might be physically colocalized within the smooth muscle cell. By pushing fluorescence detection coupled to the digital imaging microscope to its technologic maximum, Fred and his colleagues also pushed back the limits of spatial resolution. Using immunohistochemical methods, they demonstrated that the Na^+,K^+-ATPase and the Na^+/Ca^{2+} exchanger were indeed colocalized in the plasma membrane and were restricted to regions that were distinct from those membrane regions that were specialized for mechanical force transmission (Moore, E.D.W., Etter, E.F., Phillipson, K.D., Carrington, W.A., Fogarty, K.E., Lifshitz, L. M., and Fay, F.S. Coupling of the Na^+/Ca^{2+} exchanger, Na^+/K^+ pump, and sarcoplasmic reticulum in smooth muscle. *Nature* **365:**657–660, 1993). With McCarron and Walsh, Fred then went on to show that Na^+/Ca^{2+} exchange regulates cytoplasmic calcium in smooth muscle.

Efforts to understand how smooth muscle contracts had been hampered by the lack of any technology that would allow Fred to look inside the cells and see what was happening molecule by molecule. He therefore set out to devise optical instruments to do just that. The development of the DIM was critical to the next steps in Fred's scientific career. He assembled an extraordinary team of physicists, mathematicians, and computer scientists who together created powerful and innovative optical technology that placed the Imaging Group at the University of Massachusetts at the forefront of biomedical imaging at the cellular and subcellular levels. Since the late 1980s, the digital imaging microscope has evolved to keep pace with the complexity of the scientific problems to which it has been applied. For example, the studies of the distribution of the Na^+,K^+-ATPase and the Na^+/Ca^{2+} exchanger required three-dimensional images of immunolabeled fixed cells. A single data set of multiple focal planes within a cell was taken over a period of several minutes. Such a leisurely time course clearly was inadequate to tackle the rapid dynamic changes that occur in living cells.

In the early 1990s a high-speed instrument was developed that used laser illumination, high-speed piezo electric focus, and a conventional scientific CCD camera partially masked so that different portions of its imaging area could be used to take images a few milliseconds apart. This instrument was used to measure changes in the membrane potential of individual mitochondria that were in motion at speeds of up to 1 micron per second in neuroblastoma cells (Loew, L.M., Tuft, R.A., Carrington, W.A., and Fay, F.S. Imaging in five dimensions: time-dependent membrane potentials in individual mitochondria. *Biophys. J.* **65**:2396–2407, 1993) and was fast enough to permit the study of Ca^{2+} gradients in the first 15 milliseconds of systole in cardiac myocytes (Isenberg, G., Etter, E.F., Wendt-Galliteri, M., Schiefer, A., Carrington, W.A., Tuft, R.A., and Fay, F.S. Intrasarcomere $[Ca^{2+}]$ gradients in ventricular myocytes revealed by high-speed digital imaging microscopy. *Proc. Natl. Acad. Sci. U.S.A.* **93**:5413–5418, 1996). Even greater speed has been achieved with a high-speed CCD camera developed for the Strategic Defense Department, and made possible Fred's last major work in which he studied the relationship between calcium "sparks," which are localized quantal releases of Ca^{2+}, and spontaneous transient outward potassium currents [Kirber, M.T., Etter, E.F., Singer, J.J., Walsh, J.V., Jr., and Fay, F.S. Simultaneous 3D imaging of Ca^{2+} sparks and ionic currents in single smooth muscle cells. *Biophys. J.* **72**:(2)part 2 A295, 1997].

Fred's involvement in his science was far reaching. Although a skilled machinist was available, if a custom part for the microscope had to be made quickly, Fred would use the milling machine himself to build the part. He, as the biologist, did not simply delegate the application of complex physical and mathematical concepts to his team of experts, but remained an interactive contributor to the incorporation of these ideas into workable instruments. Inherent in the workings of the digital imaging microscope was the ability to account for fluorescence from out-of-focus optical planes. This was no small feat. Fred thoroughly understood the physics and mathematics behind these sophisticated deblurring techniques. His most trusted reference in this regard was the "Feynmann Lectures."

Fred was among the earliest investigators to avail themselves of Fura-2 as a Ca^{2+} indicator. He quickly marshaled the resources to obtain the necessary equipment to make ratiometric fluorescent measurements in his isolated smooth muscle cells. Before long he had the requisite technology incorporated into the DIM. These technical advances led to Fred's most cited paper, now an official citation classic (Williams, D.A., Fogart, K.E., Tsien, R.Y., and Fay, F.S. Calcium gradients in single smooth muscle cells revealed by the digital imaging microscope using Fura-2. *Nature* **318**:558–561, 1985). However, Fred was not satisfied with global measurements of Ca^{2+}. He developed the ability to monitor intracellular calcium with both high spatial and temporal resolution. With this technique, Fred and his fellows demonstrated that Ca^{2+} gradients existed within the cell with higher $[Ca^{2+}]_i$ within the nucleus and at the sarcolemma, presumably due to Ca^{2+} stores within the sarcoplasmic reticulum. Fred demonstrated that cytoplasmic Ca^{2+} was about 140 nM under resting conditions and about 700–800 nM under activated conditions. These data indicated the exquisite sensitivity of the contractile apparatus for Ca^{2+}. The specialization of the membrane for calcium handling suggested that there may be significant differences between measurements of global cellular calcium compared to those close to sites of calcium release and storage. This is most evident in Fred's most recent work in which he utilized membrane-bound calcium indicators (Etter, E.F., Minta, A., Poenie, M., and Fay, F.S. Near-membrane $[Ca^{2+}]$ transients resolved using the Ca^{2+} indicator FFP18. *Proc. Natl. Acad. Sci. U.S.A.* **93:**5368–5373, 1996), which were capable of reporting local calcium changes adjacent to the plasma membrane.

Fred contributed not only to our understanding of the contractile machinery, but also of the regulatory processes that govern smooth muscle contraction. A major question in this field was whether activation in smooth muscle contained a calcium-dependent actin filament–linked regulatory system analogous to that in striated muscle. In a novel experiment, Fred and co-workers microinjected the catalytic subunit of myosin light-chain kinase into single smooth muscle cells and observed that, in the absence of changes in intracellular calcium, phosphorylation of the regulatory light chain on smooth muscle myosin was sufficient to elicit a contraction. These data refuted the necessity for an actin-linked regulatory system but did not rule out a possible modulatory role for such a system (Itoh, T., Ikebe, M., Kargacin, G.J., Hartshorne, D.J., Kemp, B.E., and Fay, F.S. Effects of modulators of myosin light-chain kinase activity in single smooth muscle cells. *Nature* **338:**164–167, 1989).

Fred was a renaissance scientist whose interests were not limited to smooth muscle. He had an ongoing interest in chemotaxis and the relationship between changes in intracellular calcium and cellular motility (Brundage, R.A., Fogart, K.E., Tuft, R.A., and Fay, F.S. Calcium gradients underlying polarization and chemotaxis of eosinophils. *Science* **254:**703–706, 1991). His lab also examined a variety of topics that explored the relation of molecular distribution to cell function. Studies either initiated in his lab or conducted collaboratively with other scientists at the University of Massachusetts or around the world included explorations of the association of hexokinase with mitochondria, the distribution of precursor messenger RNA in the nucleus, the codistribution of poly(A) RNA with microfilaments, the localization of a yeast splicing factor in subnuclear domains, the migrations of MAP kinase, and intracellular trafficking of growth factor receptors.

Fred was a superb scholar and teacher. He was not satisfied with proving to himself that he understood the physical principles behind the optical techniques that he incorporated into development of optical instruments. Fred believed that the mark of knowing whether he really understood a concept was to then try and teach it to someone else. Therefore, he designed a successful graduate course in "Optical Methods," which served as a great forum for presenting established ideas and developing new ones. Fred also was an active participant in the Medical Physiology course given each year to first-year medical students. Over the years he taught a wide range of physiological topics, inevitably coming up with new insights and approaches to solidly established findings. Fred always had the patience to help his fellows learn the art of public presentations. He would videotape his fellow's job interview seminars and spend hours going over them point by point until he was satisfied. His standards were high and he brought out the best in all who worked with him.

Fred will be remembered by his students and colleagues in Worcester and on every continent for more than his science, for he was a very special human being. His home, his heart, and his mind were open for all to share, and his wit and energy brought joy to all who were privileged to know him. He worked hard and played hard. He touched many lives, and left them all better for the contact. He took great pleasure and pride in his wife Madeleine, his children, Andrew, Nicholas, and Isabel, and his grandchildren, Sarah, David, and Julia. He also derived great joy from the scientific offspring who worked in his lab. The list includes seven medical students, five graduate students, twenty-three postdoctoral fellows, and six senior scholars who spent their sabbatical leaves with him. Many of his former students and fellows have gone on to build impressive careers for themselves, and their science will always bear Fred's indelible imprint. Fred died far too soon, but we can take some comfort in the knowledge that he packed a great deal into his all too short lifetime. In all, he published more than 125 papers and reviews in the most highly critical and respected journals. His creativity resulted in the award of eight patents, and his technological innovations in optical methods paved the way for him and future generations to "boldly go where no one has gone before."

H. MAURICE GOODMAN
DEPARTMENT OF PHYSIOLOGY
UNIVERSITY OF MASSACHUSETTS
  MEDICAL SCHOOL
WORCESTER, MASSACHUSETTS

DAVID WARSHAW
DEPARTMENT OF MOLECULAR
  PHYSIOLOGY AND BIOPHYSICS
UNIVERSITY OF VERMONT MEDICAL
  SCHOOL
BURLINGTON, VERMONT

[1] New Photoprotecting Groups: Desyl and p-Hydroxyphenacyl Phosphate and Carboxylate Esters

By RICHARD S. GIVENS, JÖRG F. W. WEBER, ANDREAS H. JUNG, and CHAN-HO PARK

The well-established photochemical release of bioorganic substrates represents one of the most ingenious methods for inducing an *in vitro*, controlled biochemical or physiological response yet devised. The major advantages of employing "photoprotecting" or "caged"[1] substrates over conventional methods for substrate release are threefold: (1) the photoprotected or caged substrate can be dispersed throughout a biological preparation without eliciting the substrate's normal stimulus; (2) the release of the substrate can be rigorously controlled, i.e., both the concentration and the spatial distribution can be carefully managed; and (3) the temporal release can be varied over a range of first-order rate constants for release from second to nanosecond domains. First-order rate constants for the release of this magnitude are fast enough to pursue kinetic analyses of practically any subsequent biochemical response.

The necessary properties[2] of a successful caged or phototrigger group for common biological substrates such as ATP, L-glutamate, or γ-aminobutyric acid (GABA) have been suggested and include the following: (1) The photoprotected substrate must be soluble in aqueous buffered media. (2) The cage or phototrigger must be stable to hydrolysis, especially at high ionic strengths. (3) The photochemistry should be efficient, preferably with a quantum efficiency >0.1, and the activating radiation should be >300 nm to avoid competition for the light by biological media. (4) The photoproduct(s) of the protecting group must be biologically benign. In addition to these four necessary requirements, desirable properties for the phototrigger include the following: (5) The photoproducts should not absorb at the same wavelength as the activating radiation for the phototrigger to avoid a light filtering effect by the product at high conversions. (6) The photorelease reaction should be a primary photochemical step, i.e., direct bond cleavage

---

[1] "Caged compound" is a term coined in 1978 for photolabile derivatives of natural substrates such as ATP. The first such derivatives were reported by Kaplan, Forbush, and Hoffman.[4] More recently, the term "phototrigger" has become increasingly more acceptable as a better description of the photoactive protecting group.

[2] General rules for protecting groups in chemical synthesis can also be found in V. N. R. Pillai, *Synthesis* 1–26 (1980) and *Org. Photochem.* **9**, 225–343 (1988).

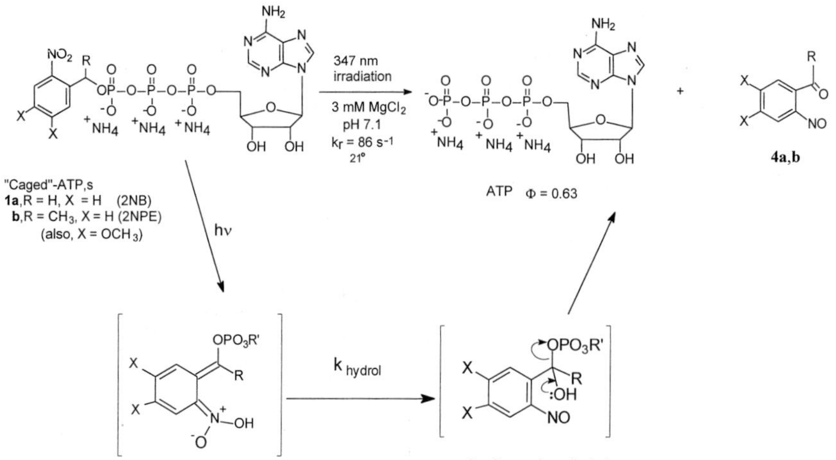

SCHEME 1

to the substrate from the reactive excited state, thus avoiding any long-lived intermediates prior to release of the substrate.

These six criteria seldom are met in their entirety, but they do serve as excellent guidelines for the design and development of new phototriggers.

Over the past twenty years,[3] the emphasis in the field of cage or phototrigger ligands has centered on derivatives of the *o*-nitrobenzyl group. The two basic archetypical chromophores are the 2-nitrobenzyl (2NB, **1a**) and 2-nitrophenethyl (2NPE, **1b**) ligands and their derivatives. This robust chromophore undergoes an initial intramolecular redox reaction on excitation at >300 nm (Scheme 1). An *aci*-nitro intermediate **2** is formed, which subsequently undergoes further molecular reorganization to form a hemiketal or acetal **3**. It is the hydrolysis of **2** that leads to the release of the substrate. The original *o*-nitrobenzyl cage group has thus been ejected as an *o*-nitrosoacetophenone or benzaldehyde (**4a,b**; Scheme 1). For the *o*-nitrobenzyl cages, the photorelease occurs at a modest rate, i.e., $k_r = 1-10^3$ sec^{-1}, and is not very sensitive to minor changes in temperature or the media. Good quantum efficiencies (~0.1–0.6) and absorptivity at wavelengths longer than 300 nm are the distinctive advantages that favor this ligand.

An impressive number of applications of "caged" reactions have appeared that almost exclusively employ a variant of the 2-nitrobenzyl photo-

---

[3] R. S. Givens and L. W. Kueper III, *Chem. Rev.* **93**, 55–66 (1993).

chemistry outlined in Scheme 1.[4–7] Several of these applications have been directed toward understanding fundamental biochemical processes including, for example, the mechanism of release of inorganic phosphate in skeletal muscle,[8] the action of actomyosin in muscle contraction induced by ATP,[9] the role of cAMP in the relaxation of distal muscle,[10] and the action of Ca^{2+}-ATPase in the sarcoplasmic reticulum in active calcium transport during ATP hydrolysis.[11] Other studies have dealt with refining, modifying, and improving the photochemistry of the *o*-nitrobenzyl cage or with optimizing or modifying the reaction conditions.[7,12] Very little has appeared, however, that reveals discovery of new photoactive caging ligands.

Nevertheless, there are several disadvantages that limit the application and usefulness of 2NB/2NPE photochemistry.[12] Among the most problematic are (1) the highly absorbing reactive nitroso ketone or aldehyde **4** that is formed as the side product, (2) the difficulty in synthesis of some of the caged derivatives, (3) the instability of the caged derivative to premature decomposition or to biochemically induced consumption prior to photolysis, and (4) competitive energy migration or electron transfer away from the excited 2-nitroaryl group prior to C–O bond lysis. For 2NB/2NPE, additional disadvantages include the reaction of the *o*-nitrosobenzaldehyde or ketone **4a,b** with peptides and proteins, possibly interfering with the biological processes under examination or with normal spectroscopic monitoring protocols. The relatively modest microscopic rate for release of the nucleotide (i.e., $k_r \sim 1–10^3$ sec^{-1}) is limited by the rate-controlling hydrolysis of the *aci*-nitro intermediate **2** (Scheme 1).[12] The slow release of nucleotides, for example, prevented the complete kinetic analysis of the mechanism of ATP on the contraction of myocardial muscle. Niggli and Lederer[13] found

---

[4] J. H. Kaplan, G. Forbush III, and J. F. Hoffman, *Biochemistry* **17**, 1920–1935 (1978). The recently adopted term "phototrigger" imparts a concept of a very rapid release of the substrate that is now attainable with the newer photoprotecting groups.

[5] See reviews: (a) H. A. Lester and J. M. Nerbonne, *Annu. Rev. Biophys. Bioeng.* **11**, 151 (1982); (b) J. M. Nerbonne, *Optical Methods Cell Physiol.* **40**, Ch. 24, 418–445 (1986); and (c) A. M. Gurney and H. A. Lester, *Physiol. Rev.* **67**, 583–617 (1987).

[6] J. A. McCray and D. R. Trentham, *Annu. Rev. Biophys. Chem.* **18**, 239–270 (1989).

[7] J. E. T. Corrie and D. R. Trentham, in "Bioorganic Photochemistry" (H. Morrison, ed.), Wiley, New York (1993).

[8] J. W. Walker, Z. Lu, and R. L. Moss, *J. Biol. Chem* **267**, 2459–2466 (1992).

[9] E. M. Ostap and D. T. Thomas, *Biophys. J.* **59**, 1235–1241 (1991).

[10] R. F. Willenbucher, Y. N. Xie, V. E. Eysselein, and W. J. Snape Jr., *Am. J. Physiol. (Gastrointest. Liver Physiol.* **14**) **262**, G159–G164 (1992).

[11] S. M. Lewis and D. D. Thomas, *Biochemistry* **30**, 8331–8339 (1991).

[12] (a) J. W. Walker, G. P. Reid, J. A. McCray, and D. R. Trentham, *J. Am. Chem. Soc.* **110**, 7170–7177 (1988); and (b) J. W. Walker, G. P. Reid, J. A. McCray, and D. R. Trentham, *Methods in Enzymol.* **172**, 288–301 (1989).

[13] E. Niggli and W. J. Lederer, *Biophys. J.* **59**, 1123–1135 (1991).

that ADP formation by enzyme-induced hydrolysis of ATP was more rapid than the rate of ATP release by photolysis of the caged nucleotide. The presence of ADP, both from enzymatic formation and as an impurity in the caged ATP sample, severely compromised the study.

To date, however, the many advantageous properties of 2-nitrobenzyl cages have far outweighed the disadvantages, resulting in a plethora of major applications of 2NB/2NPE chemistry to the study of biochemical mechanisms. Nevertheless, future improvements are necessary to meet the increasing demand for faster release rates and a wider selection of excitation wavelengths.

New Phototriggers for Phosphates and Nucleotides:
Desyl and *p*-Hydroxyphenacyl "Cages"

Our studies of desyl and *p*-hydroxyphenacyl phototriggers provide the first general alternative to the 2-nitrobenzyl cage approach. In our initial investigations, we synthesized a series of desyl phosphates[14–17] (e.g., **5a**, Scheme 2) and demonstrated that the release of the phosphate ligands occurs efficiently ($\phi = 0.3$) with rate constants for release estimated at $>10^7$ sec^{-1} [Eq. (1)].[16] Both the quantum efficiencies and the product yields

5a, X = H
b, X = OCH$_3$

were independent of the solvent (benzene, CH$_3$CN, CH$_3$OH, or H$_2$O). Furthermore, the ligands on the phosphate group had no effect on the photochemistry.

To demonstrate the application of this phototrigger for the release of a nucleotide, desyl-cAMP **7a** was synthesized (Scheme 2) and irradiated

---

[14] R. S. Givens and B. Matuszewski, *J. Am. Chem. Soc.* **106**, 6860–6861 (1984).
[15] R. S. Givens and L. W. Kueper III, *Chem. Rev.* **93**, 55–66 (1993).
[16] R. S. Givens, P. S. Athey, L. W. Kueper III, B. Matuszewski, and J.-y. Xue, *J. Am. Chem. Soc.* **114**, 8708–8709 (1992).
[17] R. S. Givens, P. S. Athey, L. W. Kueper III, B. Matuszewski, J.-y. Xue, and T. Fister, *J. Am. Chem. Soc.* **115**, 6001–6012 (1993).

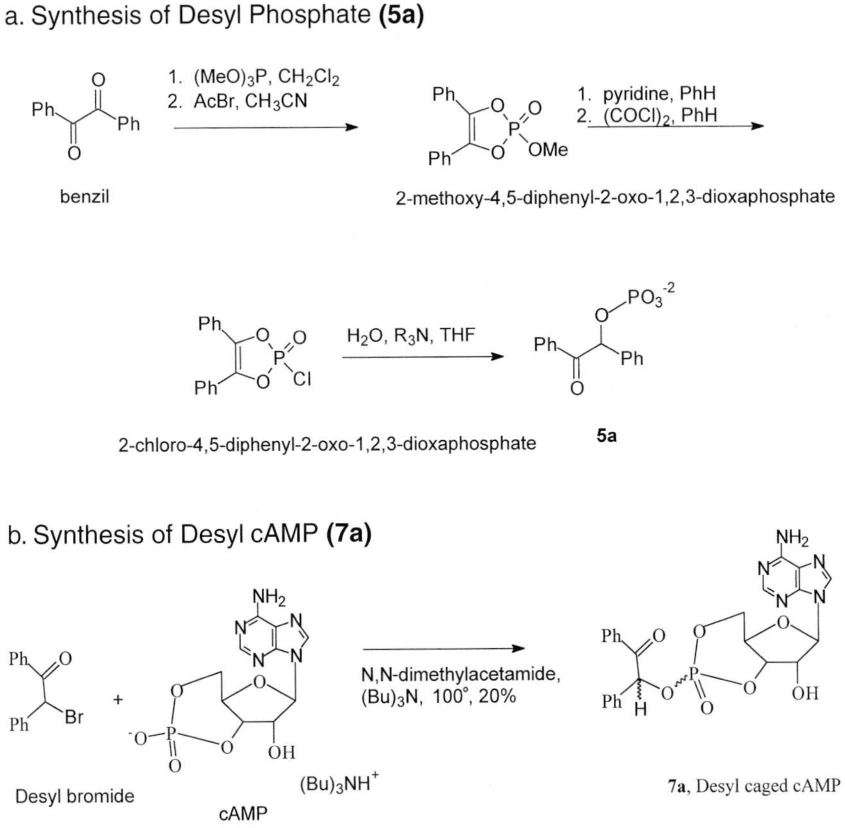

[1] NEW PHOTOPROTECTING GROUPS

a. Synthesis of Desyl Phosphate (5a)

benzil → [1. (MeO)₃P, CH₂Cl₂; 2. AcBr, CH₃CN] → 2-methoxy-4,5-diphenyl-2-oxo-1,2,3-dioxaphosphate → [1. pyridine, PhH; 2. (COCl)₂, PhH] → 2-chloro-4,5-diphenyl-2-oxo-1,2,3-dioxaphosphate → [H₂O, R₃N, THF] → **5a**

b. Synthesis of Desyl cAMP (7a)

Desyl bromide + cAMP (Bu)₃NH⁺ → [N,N-dimethylacetamide, (Bu)₃N, 100°, 20%] → **7a**, Desyl caged cAMP

SCHEME 2

at either 300 or 350 nm.[16,17] The results, shown in Eq. (2), illustrate the highly efficient, rapid release by desyl as the phototrigger. The release of cAMP occurs with a first-order rate constant of 10^8 sec^{-1} and is accompanied by the formation of 2-phenylbenzo[*b*]furan (**6a**). This and all other desyl photoreactions that we have examined are triplet mediated, quenchable

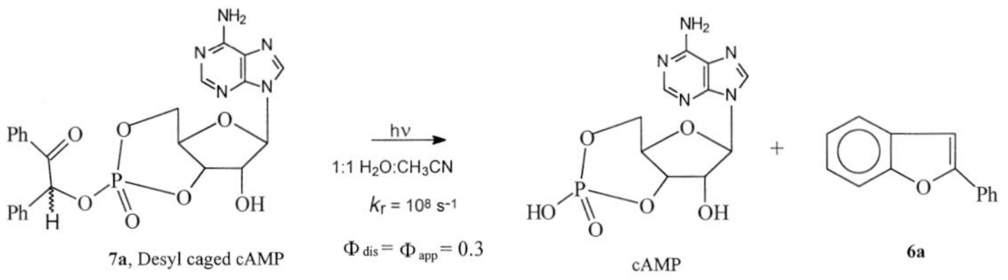

7a, Desyl caged cAMP → [hν, 1:1 H₂O:CH₃CN, $k_r = 10^8$ s^{-1}, $\Phi_{dis} = \Phi_{app} = 0.3$] → cAMP + **6a** (2)

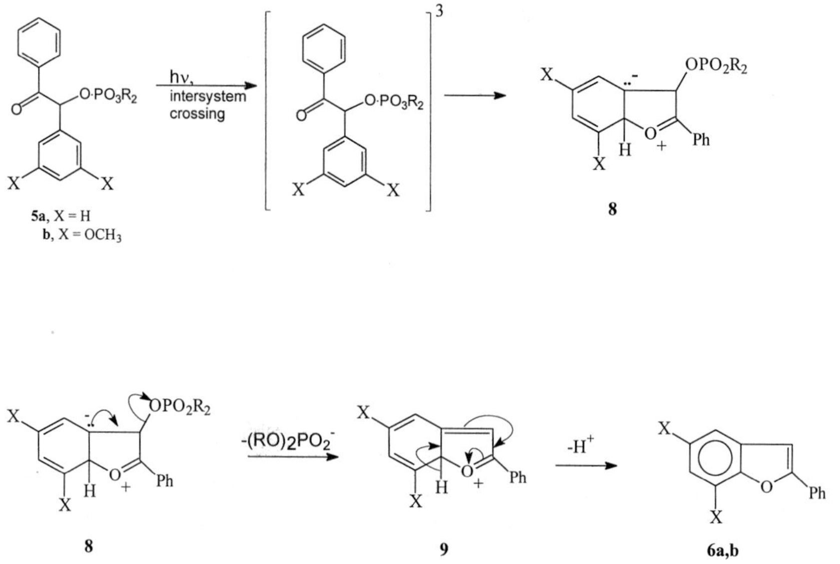

SCHEME 3

by low triplet energy quenchers such as naphthalene (E_T = 62 kcal/mol). Direct observation of the triplet by laser flash photolysis[18] provides definitive evidence for the short-lived triplet (λ_{max} = 460 nm, τ_{trip} = 12 nsec) that decays essentially instantaneously to the 2-phenylbenzofuran **6a** (λ_{max} = 300 nm, k_r = 10^8 sec^{-1}). Based on these preliminary observations, a reasonable sequence of events leading to the formation of the benzofuran begins with the electrophilic n,π^* triplet attacking the benzyl group to generate **8** and then the intermediate protonated dihydrobenzofuran **9**, i.e., simultaneous or sequential loss of the phosphate and the proton generate the furan and the released substrate (Scheme 3). Such an attack on a m,m'-dimethoxybenzyl derivative should be especially favorable due to the electron-donating contributions of the two *ortho, para*-directing methoxy groups (see Scheme 3).

Although this mechanism involves the generation of a reactive intermediate, thus violating the requirement of primary photochemical processes, the lifetime of the intermediate is short enough that the release process is still five orders of magnitude faster than the corresponding *o*-nitrobenzyl derivative. Therefore, benzoin phototriggers may have important applications when absorption at longer wavelengths (>350 nm) and rapid release (>10^8 sec^{-1}) are required.

[18] J. Wirz, R. Givens, and B. Hellrung, unpublished results, 1997.

Subsequently, Corrie and Trentham[19] explored the photochemistry of these and the closely related m,m'-dimethoxybenzoin phosphates as a cage ligand.[20] They have reported the development of caged analog **5c** [Eq. (3)]

$$\text{5c, } m,m'\text{-dimethoxydesyl-ATP} \xrightarrow{h\nu} \text{ATP} + \text{6b} \quad (3)$$

to replace the 2NB/2NPE groups. Their results parallel our findings[16,17] and thus indicate that the desyl group deserves further attention as a cage ligand for nucleotides. In both cases, the photoproducts of the cage ligand appear to be biochemically benign as biological substrates.

Several limitations on the use of benzoin derivatives as phototriggers are apparent from these studies. Among the most severe are (1) the introduction of a new chiral center, which results in diastereomeric mixtures whenever the caged substrate contains one or more chiral centers; (2) the generation of a highly absorbing, fluorescent photoproduct **6a,b**; (3) the instability of the phototrigger for selected derivatives of nucleotides[20]; and (4) the marginal aqueous solubility of the reactant phototrigger and its photoproduct. Each of these limitations can, in principle, be addressed for specific applications. For example, a single enantiomeric benzoin[21] can be synthesized or the phototrigger mixture of diastereomers can be resolved. Analogs containing sulfonic or carboxylic acid functions could be developed to enhance the aqueous solubility of the phototrigger and its photoproduct. These modifications would have a negligible effect on the photochemistry.

A more direct approach to overcoming all four limitations, however, can be anticipated from an examination of the structural similarity of the desyl group to another well-known α-keto derivative, the p-methoxyphenacyl chromophore, first introduced by Sheehan, Wilson, and co-workers.[22] By simply removing the aryl substituent on **5** and substituting a p-hydroxy moiety for Sheehan's p-methoxy group on the phenacyl ring, the aqueous solubility of the resulting phototrigger **10e** should improve dramatically, and the new phototrigger is absent the complications of a chiral center.

---

[19] J. E. T. Corrie, G. P. Reid, D. R. Trentham, M. A. Mazid, and M. B. Hursthouse, *J. Chem. Soc. Perkin Trans.* **I**, 1015–1019 (1992).

[20] J. E. T. Corrie and D. R. Trentham, *J. Chem. Soc. Perkin Trans. I*, 2409–2417 (1992).

[21] M. C. Pirrung and S. W. Shuey, *J. Org. Chem.* **59**, 3890–3897 (1994).

[22] (a) J. C. Sheehan, R. M. Wilson, and A. W. Oxford, *J. Am. Chem. Soc.* **93**, 7222–7228 (1971); and (b) J. C. Sheehan and J. Umezawa, *J. Org. Chem.* **38**, 3771–3774 (1973).

Furthermore, based on the earlier studies by Anderson and Reese,[23] introduction of a *p*-hydroxy and *p*-methoxy substituent (i.e., **10a** or **10d**) promotes a new, competing photorearrangement process yielding a substituted phenylacetate **11** in addition to the normally encountered acetophenone reduction product **12** [Eq. (4)].

$$\underset{\substack{\text{10a, Y = OCH}_3;\ \text{X = Cl}\\ \text{b, Y = OCH}_3,\ \text{X = O}_2\text{CR}\\ \text{c, Y = OCH}_3,\ \text{X = O}_2\text{P(OR)}_2\\ \text{d, Y = OH;\ X = Cl}\\ \text{e, Y = OH,\ X = O}_2\text{P(OR)}_2}}{\text{Ar-CO-CH}_2\text{X}} \xrightarrow[\substack{\text{R'OH}\\ \text{R' = H, CH}_3,\text{ or C(CH}_3)_3}]{h\nu} \underset{\substack{\text{11a, Y = OCH}_3\\ \text{b, Y = OH}}}{\text{Ar-CH}_2\text{-C(O)OR'}} + \underset{\substack{\text{12a, Y = OCH}_3\\ \text{b, Y = OH}}}{\text{Ar-CO-CH}_3} + \text{X} \qquad (4)$$

Parallel investigations of the structurally analogous *p*-methoxyphenacyl phosphates have been reported by Epstein and Garrossian,[24] Baldwin *et al.*,[25] Pirrung,[21] Futura *et al.*,[26] and by us,[14–17,27] that have extended the applications of α-keto derivatives as phototriggers. However, it was not until recently that the first applications of *p*-hydroxyphenacyl phototriggers were reported for phosphate or for a nucleotide, i.e., *p*-hydroxyphenacyl-ATP.[27] Our synthesis of these two examples is outlined in Scheme 4 (structures **13–17**), and the experimental details are provided in the Methods section. Additional information on the synthesis and photochemistry, as well as information on several analogs, appeared in our original articles.[27] In addition to the expected improved solubility and the lack of a diastereomeric mixture of caged derivatives for *p*-hydroxyphenacyl-ATP (**17**), we were pleased to discover that the combined effects on the photochemistry by the *p*-hydroxy substituent and the solvent change to H_2O or buffers were to promote the rearrangement reaction to *p*-hydroxyphenylacetic acid [**11b**, Eq. (5)] to the exclusion of all other possible photoproducts from the phototrigger.

A particularly attractive feature is the fact that the phenylacetic acid does not compete for incident radiation >300 nm, thus permitting higher, in fact complete, conversion of the caged phosphate to ATP [see Eq. (5)]. While this may not be a particular advantage in most physiological stud-

---

[23] J. C. Anderson and C. B. Reese, *Tetrahedron Lett.*, 1–4 (1962).
[24] W. W. Epstein and M. Garrossian, *J. Chem. Soc., Chem. Commun.*, 532–533 (1987).
[25] J. E. Baldwin, A. W. McConnaughie, M. G. Moloney, A. J. Pratt, and S. B. Shim, *Tetrahedron* **46**, 6879–6884 (1990).
[26] T. Futura, H. Torigai, M. Sugimoto, and M. Iwamura, *J. Org. Chem.* **60**, 3953–3956 (1995).
[27] (a) R. S. Givens and C.-H. Park, *Tetrahedron Lett.* **37**, 6259–6262 (1996); and (b) C.-H. Park and R. S. Givens, *J. Am. Chem. Soc.* **119**, 2453–2463 (1997). (Methods adapted in part with permission from the American Chemical Society.)

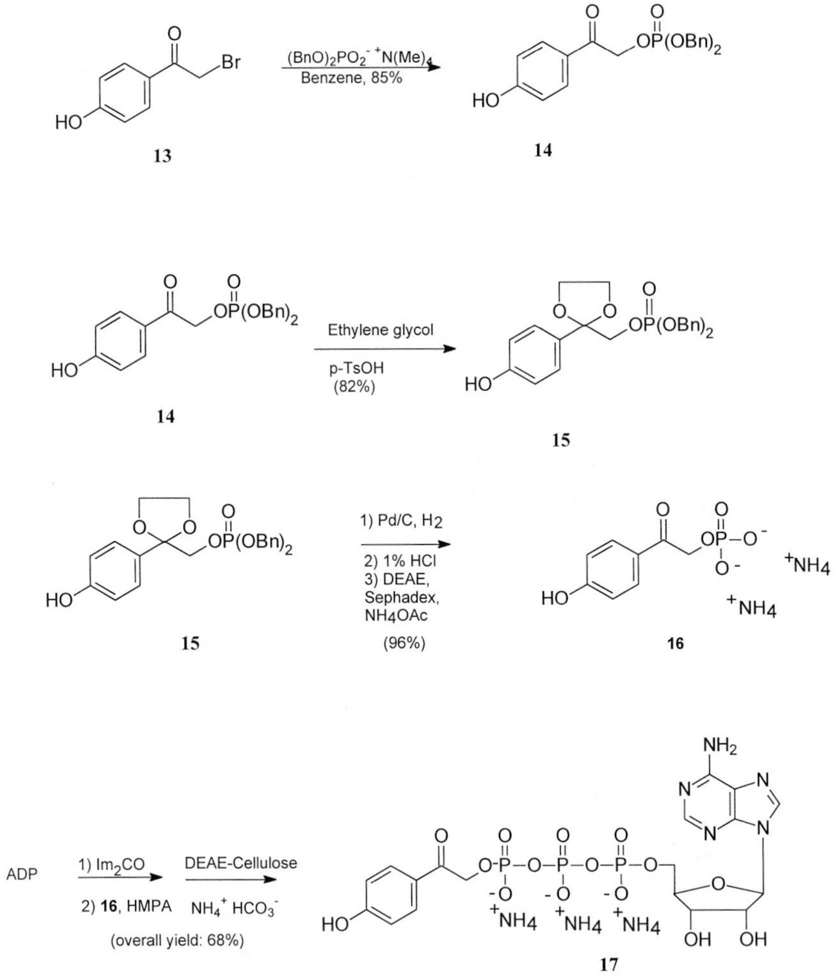

SCHEME 4

ies, it may be important for applications when caged derivatives are introduced as a medicinal agent. Such applications remain to be developed, however.

The rearrangement is proposed to pass through a dienedione intermediate **18** by direct displacement of the phosphate group from the electron-rich, $\pi^* \rightarrow \sigma^*$ electronic excited state, releasing ATP in a primary photochemical process (Scheme 5). Although many of the mechanistic details remain to be determined, the similarity of the phototrigger chemistry to recent electrochemical studies on the radical anion-induced fragmentations of substituted α-phenoxy- and carboxyacetophenones[28] and the correlation of the rate constants for these fragmentations in laser flash photolysis studies[29] lend credence to the process depicted in Scheme 5. To date, however, the dienedione **18** has not been detected. Additional support is the close analogy of the photorearrangement to the anchimerically assisted 1,2-aryl migration during the solvolytic reactions of β-phenethyl halides and tosylates.[30]

Applications of Desyl and p-Hydroxyphenacyl Phototriggers to Excitatory Amino Acids, Peptides, and Carboxylate Esters

Extension of the α-ketophosphate phototriggers to carboxylate release has also been accomplished with both the desyl[31] and the p-hydroxyphenacyl[32] amino acids. For simple amino acids, the applications of these phototriggers has been limited to two examples, the excitatory γ-aminobutyrate GABA and γ-O-substituted L-glutamate derivatives (e.g., **22a,b** and **24a,b**). All of the α-amino-substituted carboxylic acid phototriggers proved to be unstable to aqueous environments, hydrolyzing to the free amino acid and the α-hydroxy phototrigger. The general synthesis for the amino acid deriva-

---

[28] (a) C. P. Andrieux, J.-M. Savéant, A. Tallec, R. Tardivel, and C. Tardy, *J. Am. Chem. Soc.* **118**, 9788–9789 (1996); and (b) M. L. Andersen, W. Long, and D. D. M. Wayner, *J. Am. Chem. Soc.* **119**, 6590–6595 (1997).
[29] N. Mathivanan, L. J. Johnston, and D. D. Wayner, *J. Chem. Phys.* **99**, 8190–8195 (1995).
[30] D. J. Cram, *J. Am. Chem. Soc.* **71**, 3863 (1952).
[31] K. R. Gee, L. W. Kueper III, J. Barnes, G. Dudley, and R. S. Givens, *J. Org. Chem.* **61**, 1228–1233 (1996).
[32] R. S. Givens, A. Jung, C-H. Park, J. Weber, and W. Bartlett, *J. Am. Chem. Soc.* **119**, 8369–8370 (1997).

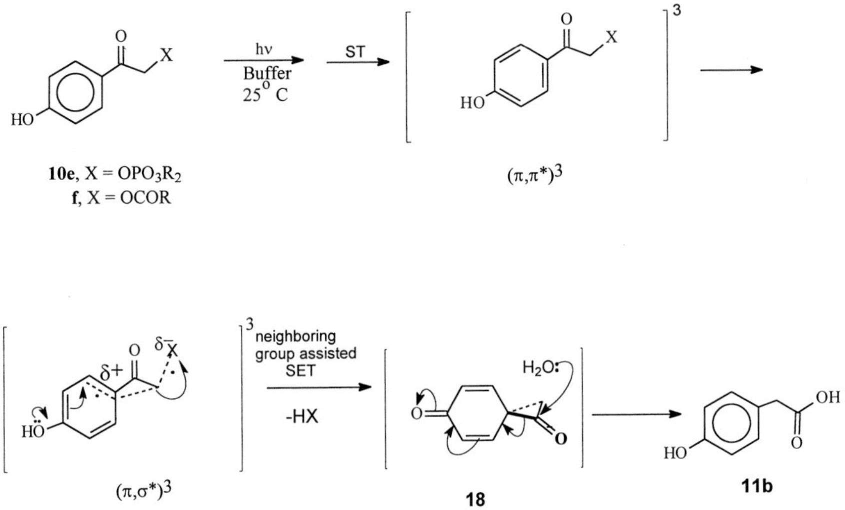

SCHEME 5

tives of the two phototriggers is outlined in Scheme 6 (structures **19–26**) and provided in detail for **22a,b** and **24a,b** in the Methods section. Yields of the photoprotected carboxylates are generally quite good.

The photochemistry of the carboxylate analogs is likewise uncomplicated by side reactions (Schemes 7 and 8). Quantum efficiencies are high, ranging from ~0.1 to 0.3, and the photoreaction is quenchable by Na$^+$ 2-naphthalenesulfonate, thus indicating triplet state intermediacy. The fate of the phototriggers for the carboxylate series is identical to that observed for the phosphate series. This observation provides further evidence that the photochemistry of the phototriggers is independent of the leaving group, at least within the series of phosphates and carboxylate leaving groups. Preliminary evidence shows that the range of leaving groups can be expanded to include sulfonate esters and several other conventional leaving groups.[33]

In an attempt to include peptides and higher homologs of amino acids for applications of the *p*-hydroxyphenacyl phototriggers, the dipeptide Ala-Ala **28** was derivatized by the same method employed for the amino acids and its photochemistry examined in detail (Scheme 8). In contrast to the α-amino acid derivatives noted earlier, this caged dipeptide was stable to aqueous and buffered media,[34] probably a result of the lowered reactivity

---

[33] L. Chimilio, A. H. Jung, J. F. W. Weber, and R. S. Givens, Unpublished results, 1997–1998.
[34] *p*-Hydroxyphenacyl Ala-Ala **27** does slowly hydrolyze in Tris buffer with a half-life of 213 min.[32]

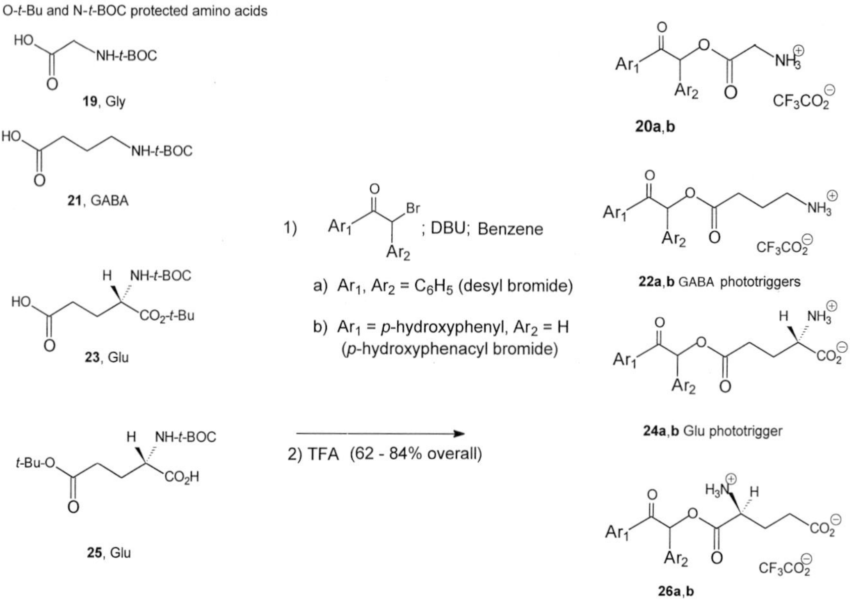

SCHEME 6

O-t-Bu and N-t-BOC protected amino acids

19, Gly
21, GABA
23, Glu
25, Glu

1) Ar$_1$-C(O)-CH(Br)-Ar$_2$; DBU; Benzene

a) Ar$_1$, Ar$_2$ = C$_6$H$_5$ (desyl bromide)
b) Ar$_1$ = p-hydroxyphenyl, Ar$_2$ = H (p-hydroxyphenacyl bromide)

2) TFA (62 - 84% overall)

20a,b
22a,b GABA phototriggers
24a,b Glu phototrigger
26a,b

of the ester to hydrolysis. The lower basicity of the α-amino group as a peptide linkage renders it less susceptible to protonation at pH ~7. For the simpler α-amino acid derivatives, the protonated amino group enhances the nucleophilic attack by H$_2$O, thus leading to hydrolysis even in buffered media.

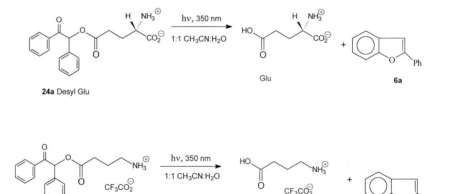

24a Desyl Glu → hν, 350 nm, 1:1 CH$_3$CN:H$_2$O → Glu + **6a**

22a Desyl GABA → hν, 350 nm, 1:1 CH$_3$CN:H$_2$O → GABA + **6a**

SCHEME 7

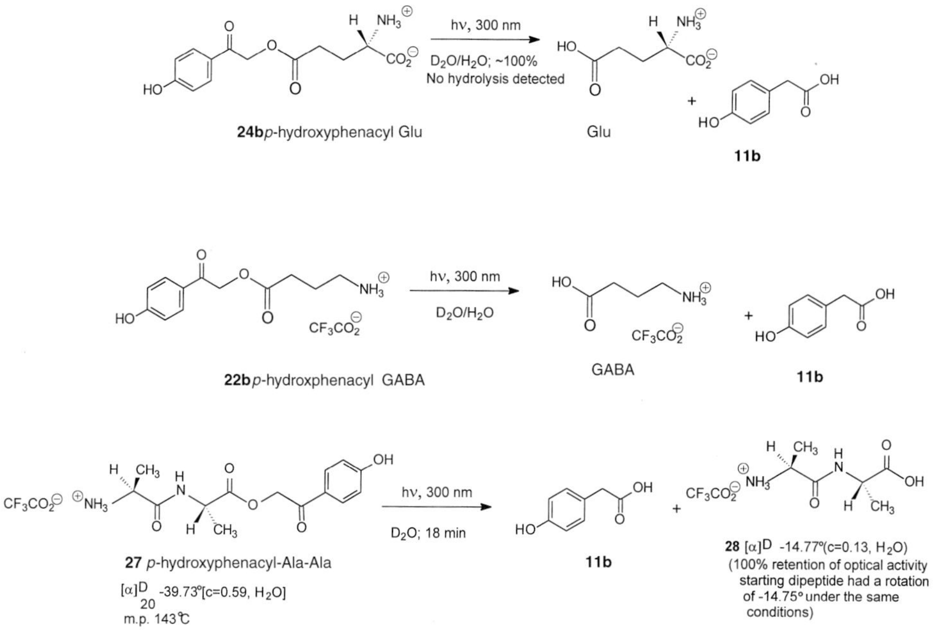

SCHEME 8

Hydrolysis thus avoided, the photochemistry of the caged dipeptide **27** was examined and shown to be in accord with the γ-amino acid derivatives as illustrated in Scheme 7. This reaction was also shown to be a triplet process, occurring at high efficiency, and capable of 100% conversion to photoproducts. Figure 1 illustrates the complete conversion to product for **27**. Determination of the other physical properties of the dipeptide, e.g., melting point (mp), infrared data (IR), and $[\alpha]_D$, further established that the original stereochemistry was not compromised during the photochemical release or the chemical manipulations.[32]

Applications of Desyl and p-Hydroxyphenacyl Phototriggers

Because these new phototriggers have only been known for a short time, there are only a few reported applications. Most notable are the examples of release of the excitatory amino acid L-glutamate, work that has been carried out in the laboratories of Katz and co-workers at the Howard Hughes Medical Institute at Duke University.[31,32]

Katz has found that irradiation of hippocampal pyramidal neurons bathed in either 1 mM desyl glutamate (**24a**) or 50 μM p-hydroxyphenacyl

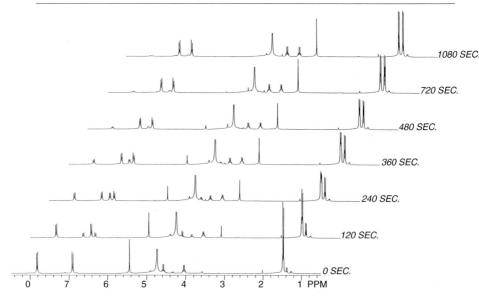

FIG. 1. ^{1}H NMR spectra of the photorelease of Ala-Ala (**28**) from p-hydroxyphenacyl-Ala-Ala (**27**) in D$_2$O at 300 nm. The sample of **27** in D$_2$O was irradiated in a 5-mm-diameter NMR tube at 300 nm in a Rayonet reactor (Southern New England Ultraviolet Company, Hamden, CT). The NMR spectra were obtained before irradiation (lowest spectrum) and after every 2 min of irradiation until the reaction was completed (~12 min). The change in chemical shifts in the aromatic region (δ 6.5–8.0 ppm), the appearance of the benzyl methylene at δ 3.6 ppm, and the disappearance of the phenacyl methylene at δ 5.45 ppm clearly illustrate the conversion of the p-hydroxyphenacyl group to a p-hydroxybenzyl moiety.

glutamate (**24b**) in patch-clamp experiments will generate an action potential when the duration of the light flash releases sufficient L-glutamate to directly stimulate the glutamate receptors (Figs. 2 and 3). The release of glutamate was faster than the duration of the 8- to 10-msec flashes.

In voltage-clamp experiments (Fig. 2), the release of glutamate results in a large inward current causing the cell to escape clamp and fire the action potential. For desyl glutamate (**24a**), the excitation wavelength was 351–364 nm, whereas p-hydroxyphenacyl glutamate (**24b**) responded better to 270-nm irradiation. Finally, in these experiments the desyl glutamate **24a** showed a background intrinsic activity, causing as much as a 300% increase in the holding current. In contrast, the p-hydroxyphenacyl glutamate (**24b**) showed no detectable background activity.

As shown in Fig. 3, cells bathed in 50 μM p-hydroxyphenacyl glutamate (**24b**) and exposed to flashes of different duration will evoke responses in current-clamp recordings in proportion to the length of the exposure until

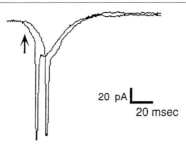

FIG. 2. In a voltage-clamp experiment, desyl L-glutamate (**24a**; 1 mM in artificial cerebral spinal fluid (ASCF) bathing solution) is irradiated near a recording neuron. The resulting large inward current causes the cell to escape clamp and fire an action potential. The response to uncaged glutamate begins as soon as the laser flash (8 msec) begins. [Reprinted in part with permission from K. R. Gee, L. W. Kueper III, J. Barnes, G. Dudley, and R. S. Givens, *J. Org. Chem.* **61,** 1228–1233 (1996). Copyright © 1996 American Chemical Society.]

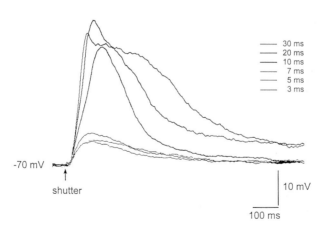

FIG. 3. In a current-clamp recording, *p*-hydroxyphenacyl L-glutamate (**24b**; 50 μM) is irradiated with 270-nm flashes of varying duration. Even in the shortest flash (3 msec), a large voltage change is observed. These recordings were done in the presence of TTX to block regenerative sodium action potentials. (Reprinted in part with permission from R. S. Givens, A. Jung, C-h. Park, J. Weber, and W. Bartlett *J. Am. Chem. Soc.* **119,** 8369–8380 (1997). Copyright © 1997 American Chemical Society.)

the threshold is reached. Exposures of longer duration than 10 msec generate sufficient glutamate to cause a regenerative Ca^{2+} spike, as indicated by the large jump in the response. Figure 4 shows similar photorelease studies but in voltage-clamp recordings. Currents as large as 1500 pA are observed. These studies were all conducted in the presence of tetrodotoxin (TTX) to block regenerative sodium action potentials.

Summary

This chapter introduces two new phototriggers for caging nucleotides, amino acids, peptides, and related functional group derivatives, potentially including higher-order homologs. These new phototriggers have several advantages that favor them over the conventional o-nitrobenzyl derivatives,

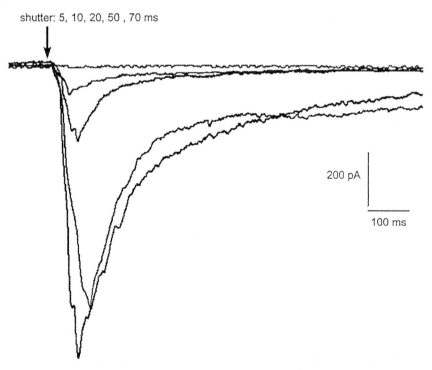

FIG. 4. Voltage-clamp experiments from flash excitation of 50 μM of p-hydroxyphenacyl L-glutamate (**24b**) at 270 nm gave increasingly larger currents, peaking at 1500 pA. Even at very short shutter openings of ~5 msec, a 100-pA inward current was obtained. These experiments were also carried out in the presence of tetrodotoxin (TTX) to prevent regenerative sodium action potentials produced. (Reprinted in part with permission from R. S. Givens, A. Jung, C-h. Park, J. Weber, and W. Bartlett *J. Am. Chem. Soc.* **119,** 8369–8380 (1997). Copyright © 1997 American Chemical Society.)

including a more rapid release of the substrate by a primary photochemical fragmentation process; adequate to excellent aqueous solubility; and stable, benign photoproducts. Added attractive features for the p-hydroxyphenacyl phototrigger include its rearrangement to a phenylacetic acid, thus shifting the chromophore absorption to a shorter wavelength, eliminating its interference with the incident radiation. Finally, the absence of a chiral center eliminates diastereomeric mixtures during synthesis.

It is anticipated that these features will result in added demand for use of α-keto phototriggers for many new applications in biochemistry and physiological studies.

Methods

General Experimental Methods

All nuclear magnetic resonance (NMR) spectra are reported in ppm (δ) from either tetramethylsilane (^{1}H and ^{13}C) or 85% phosphoric acid (^{31}P). Melting points are uncorrected. Triethylamine is dried by refluxing over potassium hydroxide and distilling under argon, collecting the middle fraction, and stored in the dark. Pyridine and N,N-dimethylformamide are dried over phosphorus pentoxide and distilled under reduced pressure, collecting the middle fraction, and stored over sodium hydroxide and molecular sieves, respectively. Benzene is distilled from sodium/benzophenone ketyl prior to use. Technical-grade diethyl ether (Et$_2$O), methanol, dichloromethane (CH$_2$Cl$_2$), and tetrahydrofuran are distilled from calcium hydride.

Synthesis of Desyl Dihydrogen Phosphate, Sodium Salt (5a)[16,17]

2,2,2-Trimethoxy-4,5-diphenyl-2,2-dihydro-1,3,2λ^5-dioxaphosphole. Benzil (35 g, 0.17 mol) is added to a stirred solution of freshly distilled trimethyl phosphite (24 g, 0.19 mol) in 20 ml of dichloromethane at 0° over a 2-hr period. After an additional hour of stirring at 0°, the solvent is removed by a Rotavapor (Büchi Labortechnik AG, Basel, Switzerland), leaving a viscous yellow oil. The oil is used without further purification. The crude yield is nearly quantitative as determined by ^{1}H NMR. ^{1}H NMR (CDCl$_3$) δ 3.73 (d, J = 13.2 Hz, 9H), 7.26 (m, 6H), 7.53 (d, J = 1.8 Hz, 4H); ^{13}C NMR (CDCl$_3$) 55.6 (d, J = 10.5 Hz, CH$_3$), 126.5, 127.8, 128.3, 130.8 [d, J = 12.5 Hz, COP(OCH$_3$)$_3$], 133.9; ^{31}P NMR (CDCl$_3$, 85% H$_3$PO$_4$) δ 33.52 (m, J = 13.2); IR (neat) 3200–2800, 1600, 1450, 1175 (POCH$_3$), 1070, 750 cm^{-1}.

2-Methoxy-4,5-diphenyl-2-oxo-1,3,2λ^5-dioxaphosphate.[16,17] The trimethoxydioxaphosphole is dissolved in 80 ml of dry acetonitrile. In one portion, 3.0 ml (3.9 g, 0.032 mol) of freshly distilled acetyl bromide is added to the stirred solution. Because the reaction is exothermic, the remaining 11.0 ml

(16.7 g, 0.137 mol) of acetyl bromide is added slowly over a 1-hr period at a rate that keeps the solution at 50°. The reaction is followed by the chemical shift and the growth of the methoxy doublet in the ^{1}H NMR. On completing the reaction, the solvent is evaporated to yield a thick yellow oil that is not purified further. The crude yield is >95% based on ^{1}H NMR. ^{1}H NMR (CDCl$_3$) δ 3.94 (d, J = 12.3 Hz, 3H), 7.3 (m, 6H), 7.4 (m, 4H); ^{31}P NMR (CDCl$_3$, 85% H$_3$PO$_4$) δ 11.65 (m, J = 11.3 Hz); IR (neat) 3100, 1600, 1450, 1300, 1250 cm^{-1}.

4,5-Diphenyl-2-oxido-2-oxo-1,3,2λ^5-dioxaphosphole.[16,17] A solution of the cyclophosphate synthesized above (2-methoxy-4,5-diphenyl-2-oxo-1,3,2-dioxaphosphate, 48 g, 0.16 mol) and pyridine (36 g, 0.53 mol) in 90 ml of dry benzene is heated at gentle reflux with stirring for 7 hr, cooled to 20°, the solvent decanted and the precipitate filtered, and thoroughly washed with benzene (2 × 50 ml) under argon. The fine crystalline salt is dried at 25° at 0.3 torr for 12 hr and stored under dry argon. The isolated yield is 95% (57 g, 0.15 mol). mp 149–151°. ^{1}H NMR (DMSO) δ 4.4 (s, 3H), 7.25 (m, 6H), 7.4 (m, 2H), 8.0 (t, J = 6.8 Hz, 2H), 8.5 (t, 1H), 9.0 (d, J = 5.8 Hz, 2H); ^{13}C NMR (DMSO) δ 47.8, 126.22, 127.6, 127.73, 128.32, 131.32 (d, J = 9.25 Hz), 134.54 144.92, 145.52; ^{31}P NMR (DMSO) δ 18.8 (s); IR (KBr) 3500, 3100, 1650, 1505, 1455, 1270, 1140, 1080, 960, 840, 790, 770, 710 cm^{-1}.

Desyl Dihydrogen Phosphate, Aniline Salt (5a, Aniline Salt).[16,17] A solution of 5.0 g (13.6 mmol) of N-methylpyridinium salt of 4,5-diphenyl-2-oxido-2-oxo-1,3,2-dioxaphosphole is treated with phosgene as previously described. After workup, the oil is taken up in 30 ml of tetrahydrofuran and cooled to 0°, to which 0.50 g (28 mmol) of water in 10 ml of tetrahydrofuran (THF) is added dropwise with stirring. During the addition, the color changes from orange to yellow. After 30 min, the solvent is removed to yield a yellow oil that is dissolved in 30–40 ml of ethanol, to which 1.9 g (20 mmol) of aniline is added. The resulting anilinium salt precipitates within a few minutes. The isolated yield of the aniline salt of desyl dihydrogen phosphate is 4.80 g (13.0 mmol, 92% yield). The crude salt is recrystallized from ethanol; mp(decomposition) 131–140°. IR (KBr) 3400–2200, 1700, 1600, 1580, 1500, 1450, 1220, 1190, 1080, 980 cm^{-1}; analysis for C$_{20}$H$_{20}$NO$_5$P calculated: C, 62.34; H, 5.23; N, 3.63. Found: C, 61.58; H, 5.59; N, 3.01.

Desyl Dihydrogen Phosphate, Sodium Salt (5a, Sodium Salt).[16,17] The desyl phosphoric aniline salt (0.52 g, 1.38 mmol) is dissolved by heating in 30 ml of deionized water. The solution is allowed to cool and is then passed through a sodium ion-exchange column (1.5 × 22 cm). Approximately 125 ml of the eluant is collected in a tared flask and the water removed by lyophilization. The sodium salt of **5a**, a white powder, is isolated in 87%

yield (0.42 g, 1.2 mmol). mp/dec 180–183°. ^{1}H NMR (D$_2$O) δ 6.46 (d, J = 8.5 Hz, 1H), 7.2–7.5 (m, 8H), 7.84 (d, J = 7.8 Hz, 2H); ^{13}C NMR (D$_2$O) δ 81.32 (d, J = 4.22 Hz), 130.61, 131.67, 131.72, 131.96, 136.82, 137.10, 138.43 (d, J = 4.6 Hz), 202.01; IR (KBr) 3480, 3020, 1680, 1580, 1480, 1430, 1240, 1200, 1180, 1060 cm^{-1}; UV [H$_2$O, $\lambda_{max}(\varepsilon)$], 251 (11,000), 323 (400). Analysis calculated for C$_{14}$H$_{11}$O$_5$PN: C, 50.02; H, 3.30. Found: C, 51.08; H, 4.00.

Synthesis of Desyl (−)-Adenosine Cyclic 3′,5′-Phosphate (7a)[16,17]

Tri-n-butylammonium Salt of (−)-Adenosine Cyclic 3′,5′-Phosphate. To 1.0 g (2.9 mol) of (−)-adenosine cyclic 3′,5′-phosphate (cAMP) in 250 ml of deionized water is added 1.1 equivalents of tri-*n*-butylamine. The water is removed by lyophilization, yielding a white salt (1.5 g, 2.9 mmol, 100% conversion); dec ≈ 120°. ^{1}H NMR (D$_2$O) δ 0.79 (t, J = 7.3 Hz, 9H), 1.24 (m, J = 7.3 Hz, 6H), 1.52 (m, J = 3.4 Hz, 6H), 2.98 (m, 6H), 4.18 (m, 2H), 4.35 (m, 1H), 4.43 (m, 1H), 4.61 (m, 1H), 6.03 (s, 1H), 8.08 (s, 1H), 8.09 (s, 1H); ^{31}P NMR (D$_2$O) δ −0.91.

Desyl Adenosine Cyclic 3′,5′-Phosphate (7a).[16,17] The procedure of Rameriz[35] is adapted to the synthesis of **7a**. The tri-*n*-butylammonium salt of cAMP (0.5 g, 0.97 mmol) is dissolved in 11 ml of freshly distilled *N,N*-dimethylacetamide and 1.7 ml of tri-*n*-butylamine in a 100 ml round-bottom flask. To this mixture is added dropwise, at 100° over a 10- to 15-min period, a solution of desyl bromide (1.3 g, 4.75 mmol) in 3 ml of *N,N*-dimethylacetamide. Monitoring of the reaction by thin layer chromatography (TLC) in a 90% CHCl$_3$ and 10% methanol solvent system reveals the appearance of the phosphate triester with an R_f of 0.4. After 45–60 min at 100°, the reaction mixture is cooled to room temperature. The solvent is removed by distillation with a Kugelrohr distillation apparatus (Aldrich Chemical Company, Milwaukee, WI). A ^{31}P NMR (CDCl$_3$) of the resulting dark orange oil indicates the reaction is 20–25% complete.

The oil is washed three times with hexane, then dissolved in a minimal amount of chloroform and separated by flash chromatography on a silica column (20 × 3 cm). Elution using 100% chloroform removes the unreacted desyl bromide and nonpolar derivatives of desyl bromide (this is easily noted by the yellow band that readily elutes from the column). The solvent system is then adjusted to 5% methanol and 95% chloroform and the cyclic triester eluted in fractions 20–25 (fraction size ≈ 25 ml). Appropriate fractions are combined and the solvent is removed by evaporation under reduced pressure (Rotavapor). The major impurity found in the triester is tri-*n*-butylamine, which is detected by ^{1}H NMR. Final purification is achieved by dissolving the residue in chloroform and eluting it off a flash

---

[35] F. Rameriz and J. F. Marecek, *Synthesis* 449 (1985).

silica column as previously described, using 100% chloroform to first elute the amine, followed by elution of the triester with 5% methanol in chloroform. Detecting the eluted amine is difficult; therefore, removal of the residual tri-*n*-butylamine was achieved by repeating the above chromatography. Typically, a 100-mg sample is isolated (20% yield): ^{1}H NMR (dimethyl sulfoxide, DMSO) δ 4.0 (m, 1H), 4.25 (m, 2H), 4.4–4.8 (m, 5H), 5.35 (m, 1H), 5.55 (m, 1H), 6.05 (s, 1H), 6.07 (s, 1H), 6.35 (broad doublets, 2 OH), 7.05 (overlapping doublets, 2H), 7.25–7.7 (m, 16H), 7.8–7.96 (m, 6H), 8.2 (s, 1H), 8.27 (s, 1H); ^{31}P NMR (DMSO) δ −5.5 and −6.3; IR (KBr) 1700, 1650, 1600, 1300, 1000, 925, 840, 700 cm^{-1}; UV [H$_2$O, $\lambda_{max}(\varepsilon)$] 253 (20,000), 320 (500); mass spectrum (FAB$^+$) calculated for C$_{24}$H$_{22}$N$_5$O$_7$P: 524 (M$^+$), 330, 195, 115.

Synthesis of p-Hydroxyphenacyl Phototriggers[27]

Synthesis of 2-Bromo-4'-hydroxyacetophenone (13).[27b] 2-Bromo-4'-hydroxyacetophenone (**13**) was prepared by a modification of the method of Durden and Juorio.[36] To 650 ml of a solution of chloroform and ethyl acetate (1:1, v/v) is added 69.8 g (313 mmol) of cupric bromide and 17.2 g (125 mmol) of 4-hydroxyacetophenone. The mixture was heated at reflux under a stream of N$_2$ to remove HBr for 24 hr and additional CuBr$_2$ (5.58 g; 25 mmol) is added if the TLC indicates unreacted 4-hydroxyacetophenone. The mixture is cooled to room temperature and the solution is decanted, filtered through a pad of silica gel, the pad is washed with 9:1 CH$_2$Cl$_2$:ethyl acetate, and the combined solvent mixture evaporated. The crude product is recrystallized from benzene to give 16.8 g (78.1 mmol, 62%) of 2-bromo-*p*-hydroxyacetophenone[27] (**13**): mp 129–131°; ^{1}H NMR (acetone-*d*$_6$) δ 4.64 (s, 2H), 6.97 (d, *J* = 8.7 Hz, 2H), 7.97 (d, *J* = 8.7 Hz, 2H); ^{13}C NMR (acetone-*d*$_6$) δ 122.22, 132.83, 138.29, 169.31, 196.35.

Sequence for Conversion of 13 to p-Hydroxyphenacyl Phosphate (16)

Synthesis of Tetramethylammonium Dibenzyl Phosphate.[37] A solution of dibenzyl monophosphoric acid (4.56 g, 16.44 mmol) in 20 ml of acetone is placed in a 100-ml two-necked flask equipped with a thermometer and a dropping funnel. A 25% aqueous solution of tetramethylammonium hydroxide (5.9 ml, 1.50 g, 16.44 mmol) is added dropwise through the dropping funnel with efficient stirring in an ice bath at 0°. After the addition is

---

[36] (a) D. S. Durden, A. V. Juorio, and B. A. Davis, *Anal. Chem.* **52,** 1815–1820 (1980); and (b) N. P. Buu-Hoi and D. Lavit, *J. Chem. Soc.,* 18–20 (1955).

[37] (a) M. Kulba and A. Zwierzak, *Ann. Soc. Chim. Polonorum* **48,** 1603 (1974); (b) A. Zwierzak and M. Kulba, *Tetrahedron* **22,** 3163–3170 (1971); and (c) E. P. Serebryakov, L. M. Suslova, and V. F. Kucherov, *Tetrahedron* **34,** 345–351 (1978).

complete the resulting solution [neutral to slightly alkaline (pH 7.0–7.5)] is stirred for 2 hr. Evaporation of acetone and H_2O on a Rotavapor gives a concentrated aqueous solution of the salt. The residual H_2O is removed by azeotropic distillation with benzene to yield 5.66 g (98%) of tetramethylammonium dibenzyl phosphate: mp 62–70°; ^{1}H NMR (D_2O) δ 3.15 (s, 12H), 4.84 (d, J = 7.4 Hz, 4H), 7.38 (m, 10H); ^{13}C NMR (D_2O) δ 57.83, 70.31 (d, J = 20.5 Hz), 130.50, 130.96, 131.37, 139.86 (d, J = 27.8 Hz); ^{31}P NMR (D_2O) δ 0.66.

p-Hydroxyphenacyl Dibenzyl Phosphate (14).[27] A mixture of tetramethylammonium dibenzyl phosphate (160 g, 4.70 mmol), *p*-hydroxyphenacyl bromide (**13**; 1.00 g, 4.70 mmol), and 20 ml of benzene is placed in a 50-ml round-bottom flask and refluxed with efficient stirring for 10 hr. The solution is cooled to room temperature and extracted with the H_2O/ethyl acetate solution. After removal of the solvent, the product is recrystallized from CH_2Cl_2 : ethyl acetate (1 : 1) or is purified by a silica gel flash column chromatography (CH_2Cl_2 : ethyl acetate; 60 : 40) to yield 1.63 g (85%) of *p*-hydroxyphenacyl dibenzyl phosphate (**14**). Mp 116.5–118°; ^{1}H NMR (DMSO-d_6) δ 5.13 (d, J = 7.8 Hz, 4H), 5.36 (d, J = 10.7 Hz, 2H), 6.87 (d, J = 7.6 Hz, 2H), 7.38 (m, 10H), 7.82 (d, J = 7.6 Hz, 2H), 10.51 (s, 1H); ^{13}C NMR (DMSO-d_6) δ 68.65 (d, J = 22.3 Hz), 68.73 (d, J = 25.8 Hz), 115.39, 125.14, 127.81, 128.30, 128.41, 130.34, 136.01 (d, J = 29.8 Hz), 162.65, 190.80 (d, J = 16.5 Hz); ^{31}P NMR (DMSO-d_6) δ 0.56; IR (CHCl$_3$) 3212, 1684, 1600, 1456, 1373, 1281, 1115, 984 cm^{-1}. Exact mass calculated for $C_{22}H_{21}O_6P$: 412.1154. Found: 412.1163. Analysis calculated for $C_{22}H_{21}O_6P$: C, 64.08; H, 5.13. Found: C, 64.46; H, 5.18.

p-Hydroxyphenacyl Dibenzyl Phosphate Ethylene Ketal (15).[27] The ketal is prepared by modification of the method of Amos and Ziegler.[38] To the solution of previously dried *p*-hydroxyphenacyl dibenzyl phosphate (**14**; 3.45 g, 8.4 mmol) in 100 ml of dry benzene, a catalytic amount of *p*-toluenesulfonic acid (*p*-TsOH) (79.9 mg, 0.42 mmol) is added, followed by the addition of excess amount of ethylene glycol (1.93 ml, 2.15 g, 34.6 mmol). The solution is refluxed for 5 hr with a Dean–Stark apparatus to remove the water that is generated (CaSO$_4$ is placed in the receiving arm to further trap H_2O). When the reaction is complete, NaHCO$_3$ (0.71 g, 8.4 mmol) is added to neutralize the mixture. The benzene is removed *in vacuo*, ethylene glycol is removed by extraction with H_2O/ethyl acetate to yield a white solid. Further purification is done by recrystallization from Et$_2$O to give 3.19 g (82.1%) of *p*-hydroxyphenacyl dibenzyl phosphate ethylene ketal (**15**): mp 97–98°; ^{1}H NMR (DMSO-d_6) δ 3.80 (m, 2H), 4.00 (m,

---

[38] (a) A. A. Amos and P. Ziegler, *Chem. Ber.* **37**, 345 (1959); and (b) R. E. Ireland and D. M. Walba, *Org. Syn. Coll.* **6**, 567 (1988).

2H), 4.03 (d, J = 6.7 Hz, 2H), 4.95 (d, J = 7.7 Hz, 4H), 6.74 (d, J = 8.6 Hz, 2H), 7.27 (d, J = 8.6 Hz, 2H), 7.33–7.38 (m, 10H), 9.53 (s, 1H); ^{13}C NMR (DMSO-d_6) δ 64.91, 68.34 (d, J = 22.7 Hz), 69.03 (d, J = 24.9 Hz), 107.06 (d, J = 33.4 Hz), 114.71, 127.21, 127.71, 128.28, 128.40, 129.11, 135.96 (d, J = 28.3 Hz), 157.60; ^{31}P NMR (DMSO-d_6) δ −0.02; IR (CHCl$_3$) 3280, 1610, 1512, 1455, 1273, 1168, 1002 cm^{-1}. Exact mass calculated for $C_{24}H_{25}O_7P$: 456.1416. Found: 456.1400. Analysis Calculated for $C_{24}H_{25}O_7P$: C, 63.16; H, 5.52. Found: C, 63.00; H, 5.49.

p-Hydroxyphenacyl Diammonium Phosphate (16).[27] The hydrogenolysis is performed according to the procedure of Heathcock.[39] To the ethylene ketal of p-hydroxyphenacyl dibenzyl phosphate (15; 1.86 g, 4.1 mmol) dissolved in 30 ml of methanol is added 186 mg of 10% Pd/C. The solution is hydrogenated under 10 psi pressure of H$_2$ with stirring for 2 hr followed by the addition of 1% HCl (1 ml). After filtration, the filtrate is evaporated *in vacuo.* The resulting viscous liquid is loaded on 5 g of DEAE-Sephadex (A-50–120; Sigma, St. Louis, MO) pretreated with 10 mM of ammonium acetate. The column is eluted with stepping gradient ammonium acetate solution (200 ml each of 0.0 to 0.15 M solution with 0.05 M increments, then 800 ml of 0.20 M solution to obtain the product). The diammonium salt (16) is eluted at 0.2 M ammonium acetate. Excess ammonium acetate and water are lyophilized off, and the resulting solid is recrystallized from H$_2$O/methanol (1:1) to give 1.05 g (96.2%) of p-hydroxyphenacyl diammonium phosphate (16). Mp 177–192° (dec); ^{1}H NMR (D$_2$O) δ 5.08 (d, J = 5.6 Hz, 2H), 6.92 (d, J = 8.8 Hz, 2H), 7.89 (d, J = 8.8 Hz, 2H); ^{13}C NMR (D$_2$O) δ 69.37 (d, J = 14.2 Hz), 118.46, 128.59, 133.42, 165.03, 200.50 (d, J = 33.3 Hz); ^{31}P NMR (D$_2$O) δ 1.18, IR (KBr) 3350, 3088, 2950, 1679, 1594, 1570, 1440, 1268, 1090 cm^{-1}; UV-vis (CH$_3$CN/H$_2$O) $\lambda_{max}(\varepsilon)$ 220 (8400), 282 (14,000). Analysis Calculated for $C_8H_{15}N_2O_6P$: C, 36.10; H, 5.68; N, 10.52. Found: C, 35.82; H, 5.38; N, 10.33.

Coupling of p-Hydroxyphenacyl Phosphate with ADP

p-Hydroxyphenacyl Adenosine 5'-Triphosphate, Triammonium Salt (17).[27] The coupling reaction is modeled after the method of Hoard and Ott.[40] Dowex 50W-X8 resin (Baker Chemical Co., Phillipsburg, NJ) (10 g) in a sintered glass funnel is treated with 5% hydrochloric acid (20 ml × 2) and washed with water until the filtrate is neutral. The resin is then treated with 20% aqueous pyridine solution and washed with water until the filtrate is neutral. A solution of adenosine 5'-diphosphate (ADP, potassium salt,

---

[39] C. H. Heathcock and R. Ratcliffe, *J. Am. Chem. Soc.* **93**, 1746–1757 (1971).
[40] D. E. Hoard and D. G. Ott, *J. Am. Chem. Soc.* **87**, 1785–1788 (1965).

0.385 g, 0.797 mmol) in 7 ml of water is stirred with the resin for 10 min; the resin is filtered and washed with water (7 ml × 4). Into the filtrate is added tributylamine (0.38 ml, 0.295 g, 1.59 mmol) followed by stirring for 45 min. The solution is evaporated to dryness and then lyophilized to yield a yellow oily residue, which is further dried by sequential cycles of dissolution and evaporation of dry pyridine (3.5 ml × 2) and dry dimethylformamide (DMF, 3.5 ml × 2). The dried residue is redissolved in 3.5 ml of dry DMF, carbonyl diimidazole (0.555 g, 3.42 mmol) is added, and the mixture is stirred under argon for 24 hr at ambient temperature. The reaction is quenched by adding methanol (0.11 ml, 2.8 mmol), the resulting solution is stirred for 3 hr and pumped to dryness. Meanwhile, a solution of p-hydroxyphenacyl diammonium phosphate (**16**; 400 mg, 1.5 mmol) in 13 ml of water is stirred for 45 min with the activated Dowex 50W-X8 resin (10 g, pyridinium form) as described earlier. The resin is filtered off and washed with water (7 ml × 3). Tri-n-octylamine (0.535 g, 0.66 ml, 1.50 mmol) is added to the filtrate, most of the water removed *in vacuo*, and the residue lyophilized to yield a yellow residue, which is subjected to dissolution and evaporation of dry pyridine (5 ml × 2) and dry DMF (5 ml × 2). The gummy residue is redissolved in 3.5 ml of dry DMF and added to the above ADP-imidazole solution. The DMF is pumped off at room temperature. Hexamethylphosphoramide (HMPA, 10 ml) is added to the pale yellow residue, and the solution is stirred under argon for 3 days, for which it is sonicated for 20 min periods every 12 hr. The reaction is quenched by adding 40 ml of water, and is washed with chloroform (35 ml × 4) and hexane (35 ml). The aqueous layer is lyophilized and purified by a diethylaminoethyl (DEAE)-cellulose (170 ml dry volume of Whatman DE 52 conditioned with 10 mM NH$_4$HCO$_3$ solution) column eluted with a stepping gradient of 100 ml of ammonium bicarbonate solution with the concentration from 0.0 to 0.25 M with 0.05 M increments followed by 700 ml of 0.3 M to elute the product. The fractions are monitored by TLC on silica gel (iPrOH : 0.30M aq NH$_4$HCO$_3$ (65 : 35)): R_f (ATP ester) = 0.84; R_f (p-hydroxyphenacyl diammonium phosphate) = 0.58. The fractions containing the desired phototrigger are lyophilized to yield 0.528 g (71.4% purity by mass: yield 68.4%) of p-hydroxyphenacyl adenosine 5′-triphosphate (HATP), triammonium salt (**17**): mp 137° (dec); ^{1}H NMR (D$_2$O) δ 4.21 (m, 1H), 4.27 (s, 2H), 4.37 (t, J = 4.3 Hz, 1H), 4.44 (t, J = 21.5 Hz, 1H), 5.10 (m, 2H), 5.77 (d, J = 5.1 Hz, 1H), 6.52 (d, J = 8.5 Hz, 2H), 7.46 (d, J = 8.8 Hz, 2H), 8.02 (s, 1H), 8.29 (s, 1H); ^{13}C NMR (D$_2$O) δ 68.12 (d, J = 20.7 Hz), 70.87 (d, J = 19.4 Hz), 72.92, 77.77, 86.51 (d, J = 37.1 Hz), 90.20, 117.93, 121.14, 128.20, 133.00, 142.91, 151.02, 152.90, 156.33, 164.12, 197.95 (d, J = 33.2 Hz); ^{31}P NMR (D$_2$O) δ −25.42 (t, J = 48.3 Hz), −14.09

(d, J = 46.1 Hz), −13.74 (d, J = 49.4 Hz); IR (KBr) 3320, 3150, 2991, 2821, 1710, 1679, 1594, 1570, 1453, 1372, 1248, 1112, 1041 cm^{-1}; UV–vis (CH$_3$CN/ H$_2$O) $\lambda_{max}(\varepsilon)$ 260 (30,500), 286 (14,600). Exact mass calculated for C$_{18}$H$_{22}$O$_{15}$P$_3$ (free acid): 641.0482. Found: 641.0460. The *p*-hydroxyphenacyl diammonium phosphate was also recovered.

Synthesis of 2,4′-Dihydroxyacetophenone[41]

A solution of 500 mg (2.33 mmol) of **13** is dissolved in 30 ml benzene: CH$_2$Cl$_2$ (1:1), and 0.1 ml (2.65 mmol) of formic acid is added. The resulting mixture is cooled to 6°. At this temperature a solution of 0.45 ml (3.02 mmol) of 1,8-diazabicyclo[5.4.0]-7-undecene (DBU) in 5 ml benzene is added dropwise over a period of 10 min. The reaction mixture turns light yellow and is stirred for another 45 min at 6°. The mixture is allowed to warm to room temperature and then is stirred until completion (monitoring by TLC, 48 h). The volatiles are removed *in vacuo*, and the resulting 2-formyloxy-4′-hydroxyacetophenone is used without further purification. It is dissolved in 20 ml methanol containing 0.2 g (5 mmol) NaOH and stirred for 4 h. After neutralization with 1*N* HCl, the solution is extracted four times with CH$_2$Cl$_2$ (10 ml each). The organic layers are collected, dried, and evaporated *in vacuo*. The product is purified by silica gel column chromatography (CH$_2$Cl$_2$:ethyl acetate, 9:1) to yield colorless crystals of 2,4′-dihydroxyacetophenone (171 mg, 1.13 mmol, 49%): mp 165–167°; ^{1}H NMR (D$_2$O) δ 7.82 (d, 2H, J = 8.7 Hz), 6.91 (d, 2H, J = 8.7 Hz), 4.89 (s, 2H). ^{13}C NMR (CD$_3$CN) δ 196.81, 161.72, 129.84, 115.09, 64.55. IR (KBr) 1676 cm^{-1}; mass spectrum *m/z* 152 (M$^+$), 136, 121, 93, 65.

Synthesis of Desyl Caged Amino Acids[31,42]

Synthesis of γ-O-Desyl L*-Glutamate* (**24a**).[31,42] A solution of desyl bromide (605 mg, 2.2 mmol), *N-tert*-BOC-L-glutamic acid, α-*tert*-butyl ester **23** (672 mg, 2.2 mmol), and DBU (0.35 ml, 2.3 mmol) is heated at reflux in benzene (25 ml) and stirred overnight to give γ-*O*-desyl *tert*-butyl *N-tert*-BOC L-glutamate in a yield of 1.06 g (96%) as a colorless oil: ^{1}H NMR (CDCl$_3$) δ 7.92 (d, J = 8.1 Hz, 2H), 7.51 (t, J = 7.4 Hz, 1H), 7.47–7.32 (m, 7H), 6.86 (s, 1H), 5.12 (dd, J = 13.6, 7.3 Hz, 1H), 4.5 (br s, 1H), 2.6 (m, 2H), 2.22 (m, 1H), 1.97 (m, 1H), 1.45 (s, 9H), 1.42 (s, 9H).

γ-*O*-Desyl *tert*-butyl *N-tert*-BOC L-glutamate (1.06 g, 2.13 mmol) and trifluoroacetic acid (TFA, 2.0 ml) are stirred at room temperature in dichlo-

---

[41] (a) G. Fodor, O. Kovacs, and T. Mercer, *Acta Chim. Sci. Hung.* **1**, 395–402 (1951); and (b) M. Patzlaff and W. Z. Barz, *Z. Naturforsch.* **33c**, 675–684 (1978).
[42] A. H. Jung, R. S. Givens, and C-H. Park, unpublished results, 1997–1998.

romethane (5 ml) and give, after Sephadex LH-20 purification, γ-O-desyl L-glutamate (**24a**) in a yield of 0.51 g (52%) as a hygroscopic colorless powder; TFA is removed from the product salt during lyophilization: mp 118–120° (dec.); ^{1}H NMR (CDCl$_3$) δ 2.18 (m, 2H), 2.71 (m, 2H), 3.76 (q, 1H), 7.13 (s, 1H) 7.39–7.49 (m, 7H), 7.62 (t, 1H, J = 7.33 Hz), 7.98 (d, 2H, J = 7.35 Hz); ^{13}C NMR (D$_2$O) δ 28.39, 32.654, 56.786, 79.04, 130.95, 131.61, 131.71, 131.81, 131.89, 132.02, 132.43, 132.92, 135.45, 136.35, 137.82, 176.52, 176.68; mass spectrum m/z (relative intensity) 342 (M$^+$ + 1, 95), 230 (45), 212 (95), 195 (64); exact mass calculated = 341.1341; found = 341.1346. UV (CH$_3$CN/H$_2$O) λ$_{max}$ (ε) 252 (9000), 300 (1000), 350 (200). Analysis Calculated for C$_{19}$H$_{19}$NO$_5$ · ½H$_2$O: C, 65.13; H, 5.75; N, 4.00. Found: C, 64.97; H, 5.69; N, 3.88.

Synthesis of O-Desyl γ-Aminobutyric Acid (22a).[31,42] A solution of *N-tert*-BOC γ-aminobutyric acid (**21**; 500 mg, 2.46 mmol), desyl bromide (677 mg, 2.46 mmol), and DBU (396 mg, 2.60 mmol) were heated at reflux in benzene (25 ml) to give *N-tert*-BOC *O*-desyl γ-aminobutyrate in a yield of 0.94 g (96%) as a clear colorless oil: ^{1}H NMR (CDCl$_3$) δ 7.92 (d, J = 7.8 Hz, 2H), 7.4 (m, 8H), 6.86 (s, 1H), 4.7 (br s, 1H), 3.20 (d, J = 5.9 Hz, 2H), 2.53 (m, 2H), 1.87 (dt, J = 6.9, 2.8 Hz, 2H), 1.4 (br s, 9H).

A solution (0.93 g, 2.3 mmol) of the *N-tert*-BOC *O*-desyl γ-aminobutyrate and TFA (8.0 ml, 100 mmol) in chloroform (25 ml) was stirred overnight at room temperature. The volatiles are removed *in vacuo*, and benzene (1 × 15 ml) is evaporated from the residue, leaving a colorless oil. This oil is purified by chromatography on Sephadex LH-20, using water as the eluant. The pure product fractions are combined and lyophilized to give *O*-desyl γ-aminobutyrate (**22a**, *O*-desyl GABA, 0.78 g; 81%) as a colorless hygroscopic powder: mp 82–84°; ^{1}H NMR (D$_2$O) δ 1.82 (m, 2H), 2.38 (m, 2H), 2.84 (t, 2H, J = 7.92 Hz), 6.88 (s, 1H), 6.99 (m, 5H), 7.12 (t, 1H), 7.25 (d, 2H, J = 7.23 Hz), 7.69 (d, 2H, J = 7.23 Hz); ^{13}C NMR (D$_2$O) δ 24.6, 32.9, 41.1, 80.6, 131.1, 131.2, 131.3, 131.7, 132.0, 135.3, 136.1, 136.7, 175.9, 198.8. Analysis calculated for C$_{20}$H$_{20}$NO$_5$F$_3$: C, 58.39; H, 4.90; N, 3.40. Found: C, 58.13, H, 4.75; N, 3.39.

Synthesis of γ-O-(p-Hydroxyphenacyl) Glutamate (24b)[32,43]

Synthesis of Protected γ-O-(p-Hydroxyphenacyl) Glutamate. A suspension of the bromoketone **13** (1.24 g, 5.77 mmol) is stirred under reflux in 20 ml of benzene in order to dissolve the bromoketone, cooled to 6° and a solution of 1.70 g (5.60 mmol) of *N-tert*-BOC-L-glutamic acid, α-*tert*-butyl

---

[43] R. S. Givens, C-H. Park, A. H. Jung, J. F. W. Weber, and W. Bartlet, Full paper in preparation.

ester **23** in 15 ml benzene is added. At this temperature a solution of 0.90 ml (6.02 mmol) of DBU in 10 ml benzene is added dropwise over a period of 15 min. During the addition of the base, the reaction mixture turns light yellow. After the addition, the mixture is allowed to warm to room temperature and is stirred for 17 hr and the solvent removed *in vacuo*. The product is purified by silica gel column chromatography (CH_2Cl_2 : ethyl acetate, 9:1) to yield a yellow solid (1.48 g, 3.38 mmol, 60%). ^{1}H NMR ($CDCl_3$) δ 1.42 (s, 9H), 1.44 (s, 9H), 2.00 (m, 1H), 2.21 (m, 1H), 4.25 (m, 1H), 5.23 (s, 2H), 6.86 (d, 2H, J = 8.4 Hz), 7.76 (d, 2H, J = 8.3 Hz).

Deprotection of Protected γ-O-(p-Hydroxyphenacyl) Glutamate. A solution of the fully protected glutamate (1.23 g; 3.02 mmol) is treated with 10 ml TFA at 4° and stirred for 3 hr at this temperature. Evaporation of the TFA *in vacuo* is followed by the addition of 30 ml ethyl acetate and 30 ml H_2O. The layers are separated and the aqueous layer is evaporated *in vacuo*. Treating the oily residue with 25 ml acetonitrile : H_2O immediately causes a precipitation of colorless cyrstals of **24b** (447 mg, 1.59 mmol, 53%), mp 187–188° (decomposition). ^{1}H NMR (D_2O) δ 2.19 (m, 2H), 2.27 (m, 2H), 3.81 (m, 1H), 5.47 (s, 2H), 6.95 (d, 1H, J = 8.8 Hz), 7.89 (d, 2H, J = 8.8 Hz).

General Procedure for Photolysis

Acetonitrile (high-performance liquid chromatography, HPLC grade) is used without further purification. All water is distilled and passed through a Nanopure deionizing system (Sybron/Barnstead, Inc., Boston, MA). Phosphate buffer for analytical HPLC is prepared using 85% phosphoric acid, tetrabutylammonium phosphate (TBAP, 0.5 mM), and potassium hydroxide to afford a solution of 0.04 M at pH 6. The HPLC system employed consists of two pumps, a controller, an autoinjector fitted with a 10-μL loop, a UV–VIS spectrophotometric detector set at 254 or 280 nm, a recorder, and a C_{18} 5μm 250 × 4.6-nm C_{18} reversed phase column. All analytical HPLC analyses employ either a solvent gradient or an isocratic elution with a flow rate of 1.0–1.5 ml/min. Photolysis is performed on a merry-go-round apparatus equipped with 16 × 300-nm or 4 × 350-nm lamps. The light output for the determination of quantum yields is measured using the potassium ferrioxalate method.[44]

Into a 20 × 180-mm Pyrex tube is placed 5 or 10 ml of an aqueous or alcoholic solution containing the phosphate ester and an appropriate internal standard. The concentration of the phosphate is adjusted to assure complete absorption of the incident radiation, i.e., greater than 3 absorbance

---

[44] C. G. Hatchard and C. A. Parker, *Proc. R. Soc. London* **A-235**, 518–522 (1956).

units at the excitation wavelength. The tube is sealed with a septum, deaerated with argon for at least 20 min at 0°, and photolyzed in the apparatus previously described. A 100-μl aliquot is removed periodically, stored in the dark, and analyzed by HPLC.

Photolysis of Desyl Adenosine Cyclic 3',5'-phosphate (7a) at pH 7.2. Into a 5-ml volumetric flask is placed 42.8 mg (0.032 mmol) of a diastereomeric mixture of desyl adenosine cyclic 3',5'-phosphate (**7a**). The cyclic triester is diluted to 5 ml with 1:1 Tris buffer/dioxane solution. To each of five 10 × 75-mm Pyrex tubes was added 1 ml-aliquot of the stock solution. The solutions are deaerated, irradiated, and prepared for HPLC and ^{31}P NMR analysis as previously described. The apparent pH of the solution before and after photolysis is measured to be 7.2. The quantum efficiencies for the disappearance of desyl adenosine cyclic 3',5'-phosphate (**7a**), Φ_{dis} = 0.39 ± 0.04, the appearance of adenosine cyclic 3',5'-phosphate, Φ_{app} = 0.34 ± 0.03, and the appearance of 2-phenylbenzo[b]furan (**6a**), Φ_{app} = 0.19 ± 0.03.

In a separate experiment, 85.6 mg (0.16 mM) of desyl adenosine cyclic 3',5'-phosphate (**7a**) is dissolved in 10 ml of a 1:1 mixture of dioxane; 0.05 mM tris (in D$_2$O, pH adjusted to 7.0 with NaOH/D$_2$O). Into a series of NMR tubes is placed 1.0 ml of the desyl cAMP solution, deareated with argon, and photolyzed in the merry-go-round apparatus with 4× RPR 3500-Å lamps. Sample tubes are removed periodically and examined by ^{31}P NMR (see Fig. 1). Chemical shifts are reported relative to 85% H$_3$PO$_4$ as an external standard.

Photolysis of Desyl cAMP (7a) at Various pH Values. The cyclic triester **9** (50.3 mg, 0.10 mmol) is dissolved in 3.0 ml of dioxane. Into each of six 10 × 75-mm Pyrex tubes is placed 0.5 ml of the desyl cAMP (**7a**) stock solution, and the volume is brought to 1.0 ml by the addition of 0.5 ml of one of the following buffer solutions: 0.05 M phosphate buffer (pH 7, H$_2$O), 0.05 M phosphate buffer (pH 7, D$_2$O), 0.05 M Tris buffer (pH 7.2, H$_2$O), 0.05 M Tris (pH 7.2, D$_2$O), 0.05 M perchloric acid (pH 2.0, D$_2$O). The pH values of each final solution are determined. The solutions are deaerated with argon for 20 min and irradiated for 10 min with 4× RPR 3500-Å lamps. Prior to HPLC analysis, 0.5 ml from each photolysis mixture is added to 0.25 ml of a solution containing 1,4-diphenyl-1,3-butadiene as an internal standard. Analysis is carried out as described in the General Methods section. The quantum efficiencies for appearance of **6a** are constant at 0.17 ± 0.02 at all pH values. The disappearance of **7a** and appearance of cAMP are also constant at 0.38 ± 0.02 and 0.34 ± 0.02, respectively. These two cannot be measured in the two phosphate buffers, however, due to interferences with the ^{31}P signals from the buffer.

Photolysis of p-Hydroxyphenacyl Diammonium Phosphate (16).[27,35] Into five Pyrex tubes containing 10 ml of 10% CH_3CN in buffer (pH 7.3) are introduced 100 mg (0.38 mmol) of *p*-hydroxyphenacyl diammonium phosphate (**16**) and 0.2 g (1.88 mmol) of anisole as an internal standard. The solution is deaerated and irradiated at 300 nm for 30 min. Aliquots are taken at 5-min intervals. Each aliquot is analyzed by HPLC. The only photoproduct, *p*-hydroxyphenylacetic acid (**16**), is identified by injection on HPLC with an authentic sample. The quantum efficiency for the disappearance of *p*-hydroxyphenacyl diammonium phosphate (**16**) is $\Phi_{dis} = 0.38$ and for the appearance of *p*-hydroxyphenylacetic acid (**11b**, R' = H), $\Phi_{rearr} = 0.12$.

Photolysis of p-Hydroxyphenacyl Adenosine 5'-Triphosphate, Triammonium Salt (17).[27] Into three Pyrex tubes containing 5 ml of Tris buffer (0.05 *M*, pH 7.3) are introduced 20 mg (28.9 µmol) of *p*-hydroxyphenacyl adenosine 5'-triphosphate, triammonium salt (**17**) and 15.3 mg (125 µmol) of benzoic acid as an internal standard. The solution is deaerated and irradiated at 300 nm for 12 min. Aliquots are taken at 3-min intervals and analyzed by HPLC. *p*-Hydroxyphenylacetic acid (**17**) and ATP are identified as photoproducts by injection on HPLC with authentic samples. The quantum efficiency for the disappearance of triammonium salt (**17**) is Φ_{dis} (**17**) = 0.37, and for the appearance of *p*-hydroxyphenylacetic acid (**11b**) is Φ_{rearr} (**11b**) = 0.31, and for ATP, $\Phi_{ATP} = 0.30$.

Photolysis of p-Hydroxyphenacyl Adenosine 5'-Triphosphate, Triammonium Salt (17), in Lactated Ringer's Solution. Into a Pyrex tube containing 5 ml of Ringer's solution (each 100 ml contains 5 g of dextrose hydrous, 600 mg of sodium chloride, 310 mg of sodium lactate, 30 mg of potassium chloride, and 20 mg of calcium chloride, pH 6.5) is introduced 20 mg (28.9 µmol) of *p*-hydroxyphenacyl adenosine 5'-triphosphate, triammonium salt (**17**). The solution is deaerated and irradiated at 300 nm for 12 min. The resulting solution is taken and prepared for HPLC analysis. *p*-Hydroxyphenylacetic acid (**11b**) and ATP are identified as photoproducts by injection on HPLC with authentic samples.

Competitive Quenching Study of p-Hydroxyphenacyl Adenosine 5'-Triphosphate (17) with Sodium 2-Naphthalene Sulfonate. To a 10-ml volumetric flask is added 15 mg (21.7 µmol) of *p*-hydroxyphenacyl adenosine 5'-triphosphate, triammonium salt (**17**), 61 mg (0.5 mmol) of benzoic acid, and Tris buffer (1 *M*, pH 7.3) to the fill line. To four Pyrex tubes is added 1 ml of the above solution. To three of these tubes are added, respectively, 11.5 (50 µmol), 23.0 (100 µmol), and 34.5 mg (150 µmol) of sodium 2-naphthalene sulfonate. The tubes are deaerated for 20 min with argon at room temperature and photolyzed using 4× RPR 350-nm lamps; the reactions are monitored by HPLC. A least-squares analysis is performed on the data obtained as shown in the tabulation.

Parameter[a]	Φ_{dis}(17)	ϕ_{rearr}(11b)	Φ_{ATP}
Sodium naphthalene sulfonate [mM]			
0	0.37 (0.04)	0.31 (0.03)	0.30 (0.03)
50	0.31	0.25	0.25
100	0.27	0.20	0.23
150	0.24	0.16	0.19
Slope $(K_{sv})(M^{-1})$	3.56	6.38	3.61
τ (nsec)	0.54 (0.09)	0.97 (0.10)	0.55 (0.06)
k, 10^8 sec^{-1}	6.90 (1.15)	3.20 (0.33)	5.50 (1.00)

[a] Error shown in parentheses.

Stability Test of p-Hydroxyphenacyl Adenosine 5′-Triphosphate, Triammonium Salt (17), in Various Media. Into six 10-ml test tubes, respectively, containing 5 ml of H_2O, D_2O, Tris buffer (50 mM, pH 7.3), Ringer's solution (I) (each 100 ml contains 600 mg of sodium chloride, 30 mg of potassium chloride, and 20 mg of calcium chloride, pH 6.5), Ringer's solution (II) [each 100 ml contains (I) and 310 mg of sodium lactate, pH 6.5], Ringer's solution (III) [each 100 ml contained (II) and 5 g of dextrose hydrous, pH 6.5] is introduced 5 mg (7.2 μmol) of p-hydroxyphenacyl adenosine 5′-triphosphate, triammonium salt (17). The stability of 17 in six different media was checked by HPLC analysis. In all cases 17 was stable at least 24 hr or longer.

Acknowledgments

We thank the National Science Foundation (Grant NSF/OSR 9255223), the Petroleum Research Fund (Grant 17541-AC4) of the American Chemical Society, and the University of Kansas for financial support. In addition to the co-workers cited in references to our work, we thank the following for assistance in improving the synthetic and photochemical procedures reported herein: Sabine Amslinger, Christiane Burzik, and Russell Herpel.

[2] Synthesis, Photochemistry, and Biological Characterization of Photolabile Protecting Groups for Carboxylic Acids and Neurotransmitters

By KYLE R. GEE, BARRY K. CARPENTER, and GEORGE P. HESS

Introduction

The utility of biologically active compounds whose activity is temporarily masked by strategic attachment of a photolabile, or "caging," group continues to grow.[1-6] The main advantages of reagent application in a biological system by UV photolysis of caged precursors, as opposed to conventional mixing methods, include much better temporal (~1000-fold) and spatial resolution of reagent introduction.[4-10] This becomes important when the processes being studied are very rapid, such as in receptor activation[11] and inhibition.[12,13] The spatial resolution of concentration jumps afforded by the photolysis of caged compounds is also being exploited in complex systems, such as receptor mapping in cells,[7] neuronal mapping in tissue slices,[8,9] and in known circuits of cells controlling pharyngeal pumping in the nematode *Caenorhabditis elegans*.[10]

Attachment of a caging group requires the presence of an appropriate functionality in the compound of interest. Phosphates, carboxylates, phenols, thiols, amines, and amides have all been shown to be suitable sites of attachment of a caging group whose subsequent photolysis liberates the original functionality and thus the native bioeffector molecule.[1] Carboxylates are functional groups common to a large number of natural and unnatural bioeffector molecules, including amino acid neurotransmitters, and thus

---

[1] J. E. T. Corrie and D. R. Trentham, *in* "Bioorganic Photochemistry" (H. J. Morrison, ed.), Vol. 2, p. 243, Wiley, New York, 1993.
[2] S. R. Adams and R. Y. Tsien, *Annu. Rev. Physiol.* **55,** 755 (1993).
[3] J. M. Nerbonne, *Curr. Opin. Neurobiol.* **6,** 379 (1996).
[4] G. P. Hess and C. Grewer, *Methods Enzymol.* **291,** [25] (1998) (this volume).
[5] G. P. Hess, *Biochemistry* **32,** 989 (1993).
[6] G. P. Hess, L. Niu, and R. Wieboldt, *Ann. N.Y. Acad. Sci.* **757,** 23 (1995).
[7] W. Denk, *Proc. Natl. Acad. Sci. U.S.A.* **91,** 6629 (1994).
[8] M. B. Dalva and L. C. Katz, *Science* **265,** 255 (1994).
[9] A. Sawatari and E. M. Callaway, *Nature* **380,** 442 (1996).
[10] H. Li, L. Avery, W. Denk, and G. P. Hess, *Proc. Natl. Acad. Sci. U.S.A.* **94,** 5912 (1996).
[11] N. Matsubara, A. P. Billington, and G. P. Hess, *Biochemistry* **31,** 5507 (1992).
[12] L. Niu and G. P. Hess, *Biochemistry* **32,** 3831 (1993).
[13] L. Niu, L. G. Abood, and G. P. Hess, *Proc. Natl. Acad. Sci. U.S.A.* **92,** 12008 (1995).

general methods for caging carboxylates will prove useful for preparing a wide range of caged bioeffectors.

Caged carboxylates were among the first described in the literature[14-16] on photosensitive protecting groups, although it is only recently that their utility has been exploited for studies of rapid processes in biological systems. The first report of caged versions of biologically relevant carboxylates came from Wilcox et al.,[17] who described making 4,5-dimethoxy-2-nitrobenzyl (DMNB) and 1-(4,5-dimethoxy-2-nitrophenyl)ethyl (DMNPE) esters of the amino acid neurotransmitters aspartate, glutamate, glycine, and γ-aminobutyric acid (GABA). These photosensitive esters were designed to be used in the study of rapid neuronal receptor activation processes, but slow photolysis rates (time constant ~ 1 sec for the production of free amino acid) limited their use for such studies. The synthesis of photolabile o-nitrobenzyl esters of the amino acids serine,[18] alanine, leucine, and lysine[19] was subsequently reported, but again the esters were of limited utility because of suboptimal photolysis parameters.

The use of the α-carboxy-o-nitrobenzyl (CNB) protecting group[20] has resulted in the preparation of caged compounds that are (i) water soluble at neutral pH, (ii) biologically inactive prior to photolysis, (iii) photolyzed in a wavelength region not deleterious to cells or the reaction being investigated ($\lambda > 335$ nm), (iv) photolyzed to the biologically active compound in the useful microsecond time region with sufficient quantum yield to allow kinetic investigations over a wide concentration range of the liberated biologically active compounds, and (v) give rise to photolysis products that are not deleterious to the cells or the reaction being investigated. The compounds caged with the CNB group that meet all these criteria are carbamoylcholine[20] and the neurotransmitters that were caged as CNB esters: the inhibitory neurotransmitter GABA[21] and the excitatory neurotransmitters glutamate[22] and kainic acid.[23] The photolysis of these caged neurotransmitters generated the free amino acids in 20–80 μsec with quan-

---

[14] J. A. Barltrop, P. J. Plant, and P. Shofield, *Chem. Commun.* p. 822 (1966).
[15] A. Patchornik, B. Amit, and R. B. Woodward, *J. Am. Chem. Soc.* **92**, 6333 (1970).
[16] J. C. Sheehan, R. M. Wilson, and A. W. Oxford, *J. Am. Chem. Soc.* **93**, 7222 (1971).
[17] M. Wilcox, R. W. Viola, K. W. Johnson, A. P. Billington, B. K. Carpenter, J. A. McCray, A. P. Guzikowski, and G. P. Hess, *J. Org. Chem.* **55**, 1585 (1990).
[18] M. C. Pirrung and D. S. Nunn, *Bio. Med. Chem. Lett.* **2**, 1489 (1992).
[19] H. Aoshima, D. Tanaka, and A. Kamimura, *Biosci. Biotech. Biochem.* **56**, 1086 (1992).
[20] T. Milburn, N. Matsubara, A. P. Billington, J. B. Udgaonkar, J. W. Walker, B. K. Carpenter, W. W. Webb, J. Marque, W. Denk, J. A. McCray, and G. P. Hess, *Biochemistry* **28**, 49 (1989).
[21] K. R. Gee, R. Wieboldt, and G. P. Hess, *J. Am. Chem. Soc.* **116**, 8366 (1994).
[22] R. Wieboldt, K. R. Gee, L. Niu, D. Ramesh, B. K. Carpenter, and G. P. Hess, *Proc. Natl. Acad. Sci. U.S.A.* **91**, 8752 (1994).
[23] L. Niu, K. R. Gee, K. Schaper, and G. P. Hess, *Biochemistry* **35**, 2030 (1996).

tum yields of 0.15–0.43. Such rapid activation allows these compounds to be used in the study of fast receptor activation,[11,20] inhibition,[12,13,24] and desensitization steps.[20,25,26] However, caging the β-carboxylate group of N-methyl-D-aspartic acid (NMDA) resulted in a compound that was not biologically inert.[27]

Another variant of the nitrobenzyl ester, the 2,2'-dinitrobenzhydryl (DNB) ester of NMDA, photolyzes on the order of 5 μsec with a quantum yield of 0.18.[28] Following up on the favorable photolysis properties of 2-methoxy-5-nitrophenyl (MNP) acetate,[29] the MNP ester of the neurotransmitter glycine was prepared, but it exhibited a limited stability at neutral pH.[30] The MNP ester of β-alanine is rapidly photolyzed to β-alanine, which activates the glycine receptor, and has properties suitable for investigation of the glycine receptor.[31] The MNP esters of both glycine and β-alanine photolyze in less than 2 μsec with quantum yields of 0.2. Benzoinyl, or "desyl," esters of GABA and glutamate were reported with photolysis rate constants on the order of 10^7 sec^{-1} and quantum yields of 0.14–0.3.[32] The suitability of these two compounds for investigations of the GABA and glutamate receptors is not yet known.

This chapter is concerned with the synthesis and purification of examples of carboxylates caged as CNB, 2,2'-dinitrobenzhydryl (DNB), desyl, MNP, and DMNPE esters, as well as the photochemical and biological characterization of representative examples of these caged carboxylates.

Synthesis of Caged Carboxylates

The general strategy for making photosensitive esters involves (A) carbodiimide-mediated coupling of carboxylic acids with photosensitive alcohols, (B) nucleophilic displacement of a halogen anion from an o-nitrobenzyl or desyl halide by a carboxylate anion, or (C) coupling of a carboxylic acid with an α-diazo-o-nitrophenylethyl compound. This last strategy was

---

[24] L. Niu, C. Grewer, and G. P. Hess, *Techn. Protein Chem.* **VII,** 139 (1996).
[25] L. Niu, R. W. Vazques, G. Nagel, T. Friedrich, E. Bamberg, R. E. Oswald, and G. P. Hess, *Proc. Natl. Acad. Sci. U.S.A.* **93,** 12964 (1996).
[26] T. Otis, S. Zhang, and L. O. Trussell, *J. Neurosci.* **16,** 7496 (1996).
[27] K. R. Gee, L. Niu, K. Schaper, and G. P. Hess, *J. Org. Chem.* **60,** 4260 (1995).
[28] K. R. Gee, L. Niu, K. Schaper, and G. P. Hess, submitted (1998).
[29] P. Kuzmic, L. Pavlickova, and M. Soucek, *Coll. Czech. Chem. Commun.* **51,** 1293 (1986).
[30] D. Ramesh, R. Wieboldt, L. Niu, B. K. Carpenter, and G. P. Hess, *Proc. Natl. Acad. Sci. U.S.A.* **90,** 11074 (1993).
[31] L. Niu, R. Wieboldt, D. Ramesh, B. K. Carpenter, and G. P. Hess, *Biochemistry* **35,** 8136 (1996).
[32] K. R. Gee, L. W. Kueper, III, J. Barnes, G. Dudley, and R. S. Givens, *J. Org. Chem.* **61,** 1228 (1996).

initially described by Walker et al.[33] for preparing caged versions of biologically active phosphates such as ATP.

Reagents

All starting materials are available from Aldrich (Milwaukee, WI) as are anhydrous solvents, which are handled under nitrogen or argon using standard syringe techniques. Calcium ionophore A23187 and protected amino acids are available from Sigma (St. Louis, MO); NMDA is from Research Biochemicals, Inc. (Natick, MA).

Chromatography

Flash chromatography is performed essentially according to the method of Still et al.[34] For reversed-phase chromatography, Sephadex LH-20 is employed. The resin is slurried in the solvent noted (usually water) and allowed to swell, followed by elution of the sample under gravity. Product fractions are pooled and the organic solvent (if any) is removed by rotary evaporation, and the remaining aqueous solution is frozen and lyophilized. Analytical thin-layer chromatography (TLC) is performed on aluminum-backed silica gel plates in the indicated solvent systems.

β-O-2,2'-Dinitrobenzhydryl-N-methyl-D-aspartic Acid

The preparation of β-O-DNB-NMDA **(5)** (Scheme 1) requires the protected carbamate N-t-BOC NMDA **(1)**. Initially this carbamate was prepared by acylation of NMDA with *tert*-butyl pyrocarbonate in water/dioxane mediated by sodium hydrogen carbonate.[27] However, this method gives variable yields and is somewhat time-consuming. Pretreatment of NMDA with bistrimethylsilyltrifluoroacetamide (BTTFA) in acetonitrile, followed by *tert*-butyl pyrocarbonate, affords the carbamate reproducibly in high yield in much less time. To a mixture of NMDA (1.0 g, 6.8 mmol) in acetonitrile (anhydrous, 32 ml) is added BTTFA (5.5 ml, 6.8 mmol). The resulting mixture is stirred for 30 min, then di-*tert*-butyl pyrocarbonate (2.2 g, 10 mmol) is added. The resulting slightly hazy mixture is stirred overnight, then quenched by adding 15 ml methanol. The resulting mixture is concentrated *in vacuo* and then dissolved in 2.5% sodium bicarbonate (60 ml). The resulting solution is extracted with ether (3 × 30 ml), then its pH (9) carefully lowered to 1.8 by dropwise addition of 5% HCl. The resulting solution is extracted with ethyl acetate (3× 35 ml). The extract

---

[33] J. W. Walker, G. P. Reid, J. A. McCray, and D. R. Trentham, *J. Am. Chem. Soc.* **110**, 7170 (1988).

[34] W. C. Still, M. Kahn, and A. Mitra, *J. Org. Chem.* **43**, 2923 (1978).

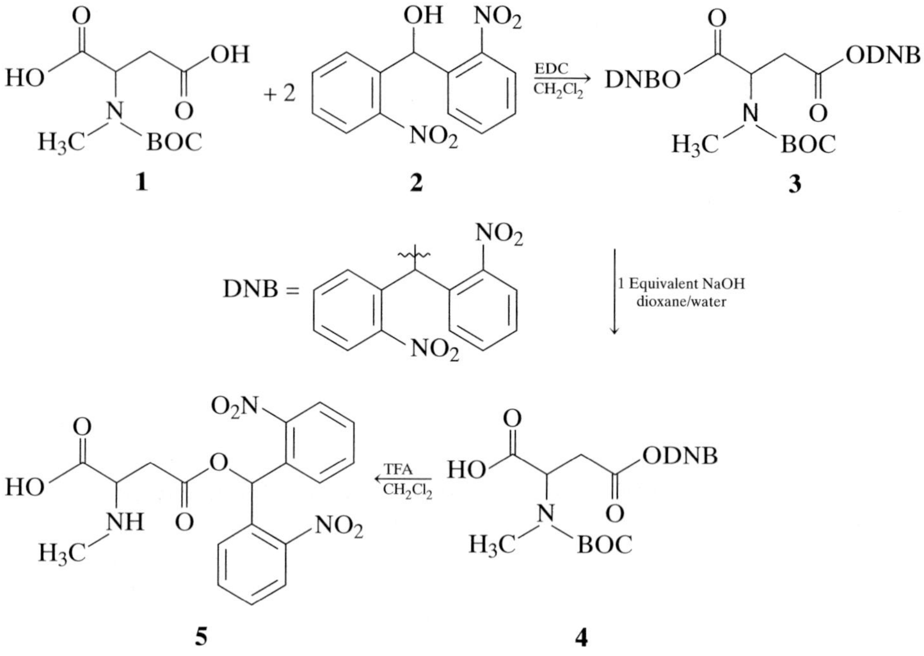

SCHEME 1

is dried over sodium sulfate, then concentrated to *N-t*-BOC NMDA as 1.37 g (82%) of a colorless solid. The carbamate thus obtained is identical to that obtained using the previously described method.[27]

To make the title compound, follow Scheme 1. A mixture of *N-t*- BOC NMDA (**1**, 0.77 g, 3.1 mmol), 2,2′-dinitrobenzhydrol[35] (**2**, 1.70 g, 2.20 mmol), 1-[3-(dimethylamino)propyl]-3-ethylcarbodiimide hydrochloride (EDC) (1.21 g, 6.3 mmol), and catalytic amounts of 4-dimethylaminopyridine (DMAP) and 1-hydroxybenzotriazole (HOBt) in anhydrous dichloromethane (35 ml) are prepared in the cold and allowed to warm to room temperature with stirring. A small amount (10–15%) of monoester is formed, as indicated by TLC analysis of the reaction mixture (R_f 0.35), but the predominant product is the bisester **3** (R_f 0.77), easily distinguished from the starting alcohol (R_f 0.63) on TLC (chloroform/methanol/acetic acid, 50:5:1). After aqueous workup, flash chromatography (using ethyl acetate/hexanes as eluant) gives the starting alcohol **2** as 0.77 g (45%) and the bisester **3** as 0.82 g (35%) of a colorless oil: ^{1}H NMR (CDCl$_3$) δ 8.1 (m, 4H), 7.91 (m,

[35] R. B. Johns and K. R. Markham, *J. Chem. Soc.* p. 3712 (1962).

2H), 7.55 (m, 8H), 7.36 (m, 3H), 7.14 (m, 1H), 4.92 (t, 1H), 3.16 (m, 1H), 2.97 (m, 1H), 2.80 (s, 3H), 1.4 (s, 9H). Analysis calculated for $C_{36}H_{33}N_5O_{14}$: C, 56.92; H, 4.38; N, 9.22. Found: C, 56.84; H, 4.67; N, 8.87. The α-carboxylate **4** is selectively generated by treatment of a solution of **2** (0.80 g, 1.0 mmol) in dioxane with 1.0 equivalent of aqueous sodium hydroxide. The reaction is slow and never reaches completion; if the reaction is pushed by addition of more NaOH, some of the β-carboxylate is generated as well as some nonchromophoric by-products. After 2 days at room temperature, the reaction solution is neutralized with aqueous citric acid and concentrated *in vacuo*, followed by flash chromatography (15% methanol/chloroform) to yield the monoester **4** (0.17 g, 31%) as well as 2,2′-dinitrobenzhydrol **(2)** and bisester **3** (0.29 g, 31%), which is subjected to a second round of alkaline treatment to make more of **4** (80 mg, 30%): ^{1}H NMR (CDCl$_3$) 8.07 (m, 2H), 7.90 (br s, 1H), 7.52 (m, 4H), 7.31 (m, 2H), 7.16 (m, 2H), 4.48 (t, J = 6.5 Hz, 1H), 3.15 (m, 1H), 2.99 (dd, J = 16.4, 7.8 Hz, 1H) 2.82 (s, 3H), 1.34 (s, 9H). Analysis calculated for $C_{23}H_{25}N_3O_{10}$: C, 54.87; H, 5.01; N, 8.35. Found: C, 54.99; H, 5.19; N, 7.79. High-performance liquid chromatography (HPLC) analysis indicates <2% of the β-carboxylate. The amino group in **4** (0.23 g, 0.46 mmol) is deprotected by brief treatment (10 min) with trifluoroacetic acid (TFA) (2 ml) in dichloromethane (4 ml). After concentration the residue is purified by reversed-phase chromatography using water:dioxane (1:1, v/v) as eluant, giving the title compound **(5)** as 192 mg (81%) of a colorless powder: ^{1}H NMR [deuterated dimethyl sulfoxide (DMSO-d_6)] δ 8.12 (td, J = 7.9, 2.7 Hz, 2H), 7.79 (m, 2H), 7.68 (m, 4H), 7.49 (d, J = 8.0 Hz, 1H), 3.46 (t, J = 6.4 Hz, 1H), 2.97 (dd, J = 16.5, 6.4 Hz, 1H), 2.76 (dd, J = 16.5, 6.2 Hz, 1H), 2.45 (s, 3H); ^{19}F (DMSO-d_6) 68.9. Analysis calculated for $C_{18}H_{17}N_3O_8 \cdot 0.5CF_3CO_2H$: C, 50.02; H, 3.88; N, 9.27. Found: C, 50.26; H, 3.74; N, 9.37.

α-Carboxy-2-nitrobenzyl-γ-aminobutyrate

The simplest way to make CNB esters **(8)** is to react carboxylic acids with *tert*-butyl α-bromo-2-nitrophenyl acetate (BBNA) (Scheme 2) in the presence of DBU, followed by TFA treatment to deprotect the caging group carboxylate and any *tert*-butyl carbamate(s) present. The method initially described[22] for making BBNA (radical bromination of *tert*-butyl 2-nitrophenyl acetate) is somewhat time-consuming and involves careful flash chromatographic separation of BBNA from the starting ester, as they elute very close to each other on silica gel. A better route to BBNA involves simple esterification of α-bromo-2-nitrophenylacetic acid (BNPA)[36] with

---

[36] C. Chang, B. Niblack, B. Walker, and H. Bayley, *Chem. Biol.* **2**, 391 (1995).

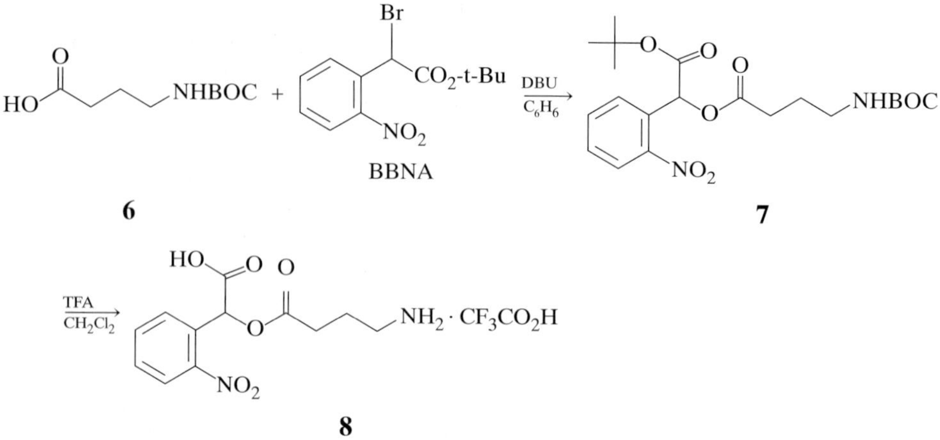

SCHEME 2

N,N'-diisopropyl-*tert*-butylisourea.[37] Although *tert*-butyl ester formation via the isourea route generally affords *tert*-butyl esters in only 40–50% yield, the method is still a significant improvement over that originally described for BBNA when overall ease and speed of synthesis are considered. Thus, to a solution of BNPA (0.50 g, 1.9 mmol) in dichloromethane (15 ml) is added isourea (1.5 g, 7.5 mmol). The resulting mixture is stirred overnight at room temperature, diluted with ethyl acetate (60 ml), and filtered. The filtrate is washed with 5% citric acid (1× 50 ml), 5% sodium bicarbonate (1× 50 ml), and brine (1× 50 ml), dried (sodium sulfate), and concentrated. Flash chromatographic purification of the residue using ethyl acetate/hexanes as eluants gives BBNA as 0.26 g (43%) of a colorless oil that slowly solidifies on standing.

A solution of BBNA (0.26 g, 0.82 mmol), *N-tert*-BOC-γ-aminobutyric acid (**6**, 166 mg, 0.82 mmol) and DBU (0.13 ml, 0.87 mmol) in benzene (10 ml) is refluxed overnight. The DBU hydrobromide is filtered off, and the filtrate concentrated to a pale brown oil and purified by flash chromatography (5% ethyl acetate/chloroform) to give α-*tert*-butylcarboxyl-2-nitrobenzyl *N-tert*-BOC-γ-aminobutyrate (**7**) as 0.36 g (100%) of a clear, pale brown oil: ^{1}H NMR (CDCl$_3$) 8.00 (d, J = 7.6 Hz, 1H), 7.64 (m, 2H), 7.53 (m, 1H), 6.74 (s, 1H), 4.7 (br s, 1H), 3.20 (q, J = 6.3 Hz, 2H), 2.52 (m, 2H), 1.90 (m, J = 6.9 Hz, 2H), 1.43 (s, 9H), 1.40 (s, 9H).

Neat α-*tert*-butylcarboxyl-2-nitrobenzyl *N-tert*-BOC-γ-aminobutyrate (**7**, 0.30 g, 0.68 mmol) is cooled in an ice bath under argon and treated with

---

[37] L. J. Mathias, *Synthesis* p. 561 (1979).

trifluoroacetic acid (2.2 ml, 29 mmol). The resulting solution is incubated overnight at room temperature, and the volatiles are removed to give a pale brown oil. This oil is purified by chromatography on Sephadex LH-20, using water as an eluant. The combined product fractions are lyophilized to give the title compound **8** as its trifluoroacetate salt as 210 mg (77%) of a compact white powder: mp 139–142° (dec); ^{1}H NMR (D_2O) 8.13 (d, J = 8.1 Hz, 1H), 7.81 (t, J = 7.4 Hz, 1H), 7.70 (m, 2H), 6.69 (s, 1H), 3.07 (t, J = 6.4 Hz, 2H), 2.63 (m, 2H), 2.01 (t, J = 7.0 Hz, 2H). Analysis calculated for $C_{14}H_{15}H_2O_8F_3$: C, 42.43; H, 3.82; N, 7.07. Found: C, 42.45; H, 3.97; N, 6.92.

γ-O-Desyl Glutamate

The synthesis of this compound **(12)** has been previously described.[32] Briefly, following Scheme 3, reaction of desyl bromide (**10**, 605 mg, 2.2 mmol), N-t-BOC-glutamic acid, α-*tert*-butyl ester (**9**, 672 mg, 2.2 mmol), and DBU (0.35 ml, 2.3 mmol) in refluxing benzene (25 ml) for 1 hr gives **11** as 1.06 g (96%) of a colorless immobile oil after aqueous workup and flash chromatographic purification (ethyl acetate/chloroform). Reaction of **11** (1.06 g, 2.13 mmol) with TFA (2.0 ml) in dichloromethane (5 ml) for 3 hr gives, after Sephadex LH-20 purification using water, **12** as 0.51 g (52%) of a hygroscopic colorless powder; during lyophilization, the trifluoroacetic acid in the product salt is volatilized: mp 129–132° (dec); ^{1}H NMR (D_2O) 8.07 (d, J = 7.5 Hz, 2H), 7.72 (t, J = 7.4 Hz, 1H), 7.5 (m, 7H), 7.22 (s, 1H),

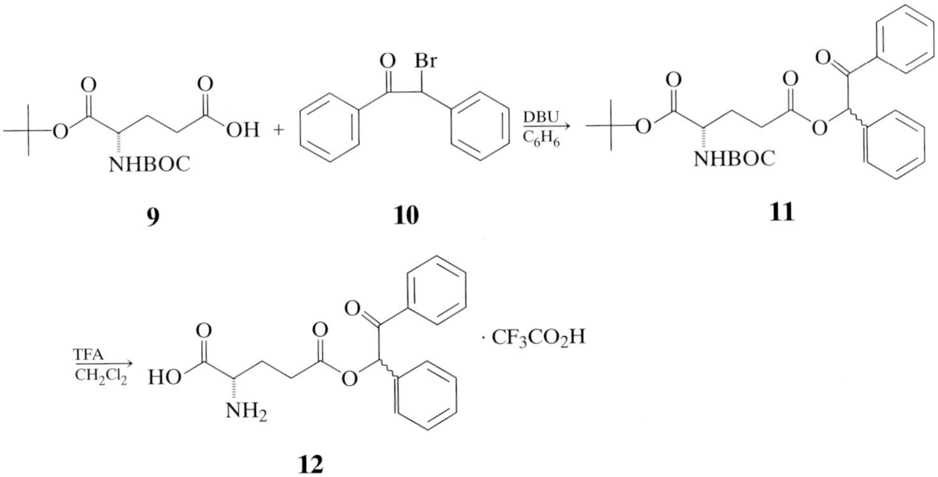

SCHEME 3

3.85 (q, J = 6.6 Hz, 1H), 2.8 (m, 2H), 2.2 (m 2H). Analysis calculated for $C_{19}H_{19}NO_5 \cdot 1/2\ H_2O$: C, 65.13; H, 5.75; N, 4.00. Found: C, 64.97; H, 5.69; N, 3.88.

2-Methoxy-5-nitrophenyl-β-alanine

The synthesis of compound **16** via Scheme 4 has been previously described.[31] Briefly, a solution of *N-t*-BOC β-alanine (**13,** 220 mg, 1.2 mmol), EDC (230 mg, 1.2 mmol), and catalytic amounts of DMAP (10 mg) and HOBt (10 mg) in dichloromethane (7 ml) is cooled to −78° and treated with 2-methoxy-5-nitrophenol (**14,** 170 mg, 1 mmol). The reaction mixture is allowed to warm to room temperature overnight, then washed with water (3× 5 ml), 5% $NaHCO_3$ (1× 5 ml), water (2× 5 ml), 10% citric acid (3× 5 ml), and brine (1× 3 ml), dried (Na_2SO_4), and purified via flash chromatography using hexane/ether as an eluant to give the intermediate carbamate **15** as a white powder in 80% yield: ^{1}H NMR ($CDCl_3$) 8.16 (dd, J = 9.2, 2.7, 1 H), 7.98 (d, J = 2.7, 1H), (d, J = 9.1, 1H), 5.12 (br s, 1H), 3.94 (s, 3H), 3.52 (dt, J = 6.0, 2H), 2.81 (t, J = 6.0, 2H), 1.64 (s, 9H). A solution of **15** (65 mg, 0.19 mmol) is prepared in HCl-saturated ethyl acetate at −78° and allowed to warm to room temperature. The precipitate is filtered, washed with ethyl acetate, and dried *in vacuo* to give the title compound (**16**) as 50 mg (95%) of a white powder: mp 168–175° (dec);

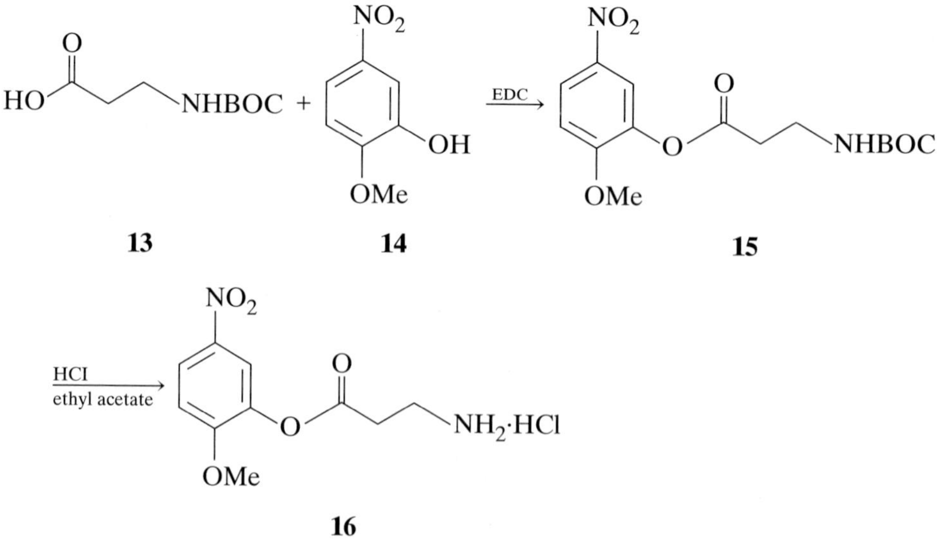

SCHEME 4

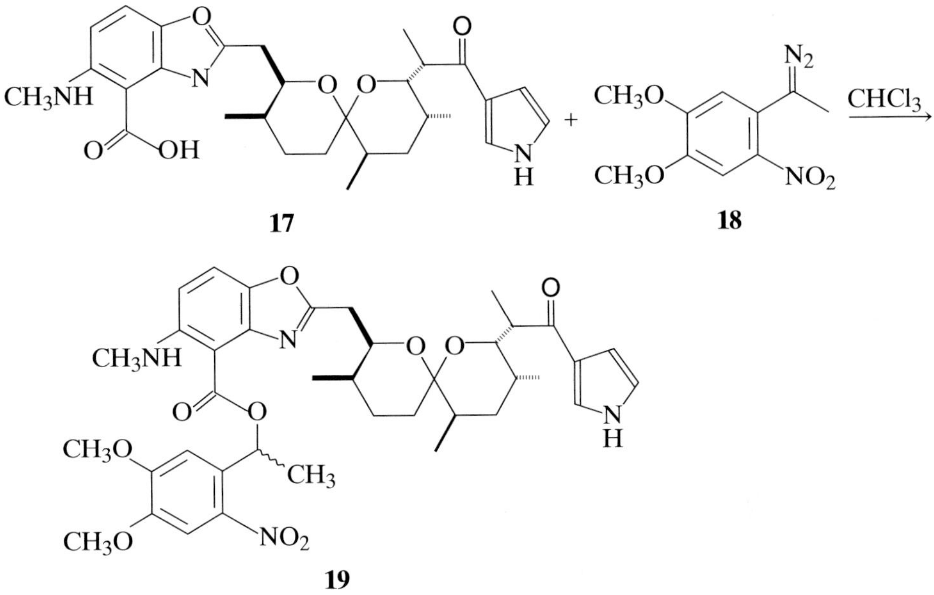

SCHEME 5

^{1}H NMR (D$_2$O) 8.08 (d, J = 8.8 1H), 7.9 (s, 1H), 7.1 (d, 1H), 4.4 (s, 3H), 3.2 (t, J = 5.8, 2H), 2.97 (t, J = 5.8, 2H); HRMS 241.0827 [241.0824 calculated for C$_{10}$H$_{13}$N$_2$O$_5$ (MH$^+$)].

1-(4,5-Dimethoxy-2-nitrophenyl)ethyl A23187

A23187 **(17)** is a calcium ionophore, i.e., a naturally occurring substance that increases intracellular calcium ion concentrations by acting as a calcium carrier across plasma cell membranes. It is often used to increase intracellular calcium concentrations in cell populations suspended in high-calcium media. Attachment of a caging group to its carboxylate moiety abolishes its calcium-channel activity.[38] Ultraviolet photolysis then regenerates free A23187 as judged by intracellular calcium concentration jumps after cells are bathed in solutions of the caged precursor.[39]

Following Scheme 5, to a pale yellow solution of 4,5-dimethoxy-2-nitroacetophenone hydrazone **17** (18 mg, 0.075 mmol) in chloroform (1.5 ml) is added manganese dioxide (70 mg, 0.81 mmol). The resulting mixture is stirred for 15 min in darkness and is then filtered through Celite with a

---

[38] K. R. Gee and P. J. Millard, unpublished results (1995).
[39] A. Ishihara, K. R. Gee, S. Schwartz, K. Jacobsen, and J. Lee, *Biotechniques* **23,** 268 (1997).

chloroform rinse (2× 1 ml). To the combined red–orange diazoethane **(18)** filtrate is added a solution of the ionophore A23187 [**17**, 26.9 mg, 0.0514 mmol, Sigma (St. Louis, MO)] in 1.5 ml chloroform. After the mixture is stirred overnight, the color of the reaction solution is pale yellow and TLC analysis shows no remaining A23187. The reaction solution is diluted with chloroform (5 ml), and glacial acetic acid (5 drops) is added to quench any remaining diazoalkane. Concentration *in vacuo* gives a pale yellow oil, which is purified by flash chromatography on silica gel using 5% methanol/chloroform as an eluant. The title compound **(19)** is obtained as 37 mg (97%) of a pale yellow microcrystalline solid. On TLC the two product diastereomers are resolved using chloroform/methanol/acetic acid (100:5:1) as an eluant, giving two spots at R_f 0.36 and 0.43; the fluorescence of these spots is initially quenched, but the spots fluoresce blue on illumination with a hand-held UV lamp (unmodified A23187 has an R_f of 0.21). Two-dimensional TLC experiments show that after 10 min irradiation at 366 nm with a hand-held UV lamp, both diastereomers of the title compound are photolyzed cleanly into native A23187. For the title compound: mp 108–116° (dec); ^{1}H NMR (CDCl$_3$) δ 9.6, 9.3 (two br s, 1H), 8.2, 8.0 (two s, 1H), 7.6 (m, 3H), 6.9 (m, 2H), 6.6 (t, 1H), 6.25 (dq, 1H), 4.2–2.8 (m, 13H), 1.8–0.7 (m, 26 H). HPLC analysis shows the presence of both product diastereomers, in 93% purity.

Photochemistry of Caged Carboxylates

The mechanism of the photolytic release of carboxylates from *o*-nitrobenzyl esters has been studied in detail,[40,41] as has the photolytic release of carboxylates from desyl esters.[16] We describe general methodology used to determine photochemical properties of caged carboxylates *in vitro*.

Instrumentation

The apparatus used for investigations of the photolysis reaction of caged derivatives of carboxylates is illustrated in Fig. 1.

Kinetic Measurements

Figure 2A illustrates the fast conversion to products observed in the photolysis of MNP-β-alanine **(16)**.[31] The absorbance of the compound released has the wavelength and pH dependence characteristic of 2-methoxy-5-nitrophenol. The photolysis reaction apparently consists of a single expo-

---

[40] H. Schupp, W. K. Wong, and W. Shnabel, *J. Photochem.* **36**, 85 (1987).
[41] Q. Q. Zhu, W. Schnabel and H. Schupp, *J. Photochem.* **39**, 317 (1987).

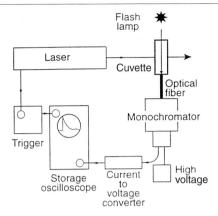

FIG. 1. Laser-flash photolysis apparatus for studying photolysis reactions of caged derivatives of neurotransmitters. Light from a XeCl excimer laser (Lumonics TE-816M, Kanata, Ottawa, Ontario, Canada) is focused to irradiate 40 μl solution in a 2 × 10-mm quartz cuvette.[56] The laser provides 100-nsec pulses of 308 nm light with a pulse energy of 10–50 mJ; flux densities of 108 mJ per mm^2 are produced at the front face of the cuvette. The pulse energy is measured with a Gentec ED200 Joulemeter (Dalton, Canada). A collimated monitor beam is produced by an Oriel (Stratford, CT) flash lamp with a 100-μsec pulse and passed through a Corning WGS360 cutoff filter and then through the quartz photolysis cuvette (Spectrocell Inc., 160 μl, Oreland, PA) oriented with faces perpendicular to the irradiation and monitoring beams, respectively. The monitor beam is at right angles to the laser beam; the path length for the laser and the monitoring beam is 2 and 10 mm, respectively. The monitor beam projects onto the detection apparatus, a photomultiplier–monochromator (single pass) combination (Thorn EM/9635QB, McPherson 275, Fairfield, NJ), for observing the transient absorbance of intermediates produced during photolysis. The detection apparatus is arranged at 90° to the laser. To avoid noise from the discharge of the laser power supply during measurements in the nanosecond time region, the signal emerging from the cuvette is carried by a 250-μm-diameter optical fiber (Fiberguide Industries, Stirling, NJ) from one room housing the laser to the detection system in a second room. This optimizes the detection of the photolysis intermediates and products. A current-to-voltage converter (Thorn EMI Gencom A1, Fairfield, NJ) is used to treat the voltage output, which is filtered at four times the data acquisition rate (Krohn-Hite Model 3396, Avon, MA); signal transients are observed on a storage oscilloscope (Tektronix 549, Beaverton, OR), where the results are displayed and recorded, or are digitized directly into an AST 286 personal computer. Signal capture is optimized by using a variable time-delay trigger (World Precision Instruments A310, Sarasota, FL). The white-light monitoring beam itself produces negligible photolysis of any of the caged compounds we have tested so far. It also does not heat the sample, as determined by monitoring the temperature at which the starting material gives a maximum absorbance with an inserted thermocouple with a 0.1° sensitivity.[56]

nential process with a $t_{1/2}$ value of ~1 μsec. Proof that the photolysis reaction releases β-alanine is ascertained by use of HPLC,[30] cell-flow,[42] and laser-pulse photolysis[4,11,20] experiments with hippocampal neurons containing glycine receptors.[31]

[42] J. B. Udgaonkar and G. P. Hess, *Proc. Natl. Acad. Sci. U.S.A.* **84**, 8758 (1987).

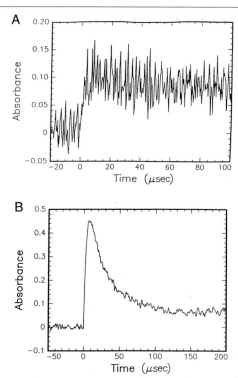

FIG. 2. Rapid absorbance changes produced by photolysis of caged compounds at room temperature by a single pulse of 308-nm light from an XeCl excimer laser (approximately 30–50 mJ). (A) A 2 mM solution of caged MNP-β-alanine **(16)**[31] in 100 mM phosphate buffer, pH 7.1, is photolyzed. The absorbance change is monitored at 405 nm with a single-beam transient spectrophotometer, the signal transfer parameters of which effectively produced 500 kHz filtering of the rapid absorbance jump. (B) A 5 mM solution of CNB-GABA **(8)** in 100 mM phosphate buffer at pH 7.4 is photolyzed.[21] The absorbance change is monitored at 430 nm. (C) Determination of the product quantum yield for photolysis of caged kainate in 100 mM phosphate at pH 6.8 and 22°. A sample cuvette containing 0.5 mM CNB-caged kainate[23] is irradiated, with mixing of the solution between laser pulses of 308 nm with energy output of 50 mJ, and the absorbance, A, is monitored at 430 nm as a function of the number of laser shots, n. The ratio of absorbed photons to targeted molecules (K_E) is 0.87, the path length (l) was 10 mm, and the fraction of the volume irradiated (F) was 0.5. The product quantum yield and molar absorption coefficient are determined, by using Eq. (1), to be 0.34 ± 0.07 and 920 M^{-1} cm^{-1}, respectively. The measurements are made in duplicate, and the standard deviation from the mean is shown.[23]

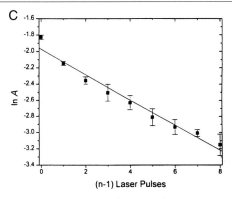

FIG. 2. (*continued*)

When CNB is used to protect the carboxyl function of neurotransmitters, the photolysis reaction is followed by observing transient absorption spectra in the 350- to 500-nm wavelength region characteristic of the *aci*-nitro intermediate in the photolysis reaction.[40,41,43] Experimental evidence indicates that the decay of this intermediate reflects product formation.[33,40,41,43,44]

Figure 2B illustrates the formation and decay of the *aci*-nitro intermediate in the photolysis of CNB-γ-aminobutyric acid **(8)**.[21] The decay phase of the absorbance (Fig. 2B) is consistent with a single exponential process with a $t_{1/2}$ value of 19 μsec. Proof that photolysis results in the release of the inhibitory neurotransmitter of GABA in the photolysis of CNB-GABA is shown by the use of HPLC,[21] cell-flow,[45] and laser-pulse photolysis[4] experiments with cortical neurons containing GABA receptors.[21]

Quantum Yield Determination

The quantum yield of MNP-β-alanine **(16)**[31] is determined with 20-μl aliquots of a solution of 2 mM MNP-β-alanine in 100 mM phosphate buffer, pH 7.1. A quartz cuvette with a 0.2-mm path length is used. The solution is photolyzed by a single 30-mJ light pulse at 308 nm from an excimer laser, and the concentration of the released 2-methoxy-5-nitrophenol, which is assumed to be produced stoichiometrically with β-alanine, is determined spectrophotometrically. A Gentec ED200 Joulemeter (Dalton, Canada) is used to determine the energy absorbed by the sample and by the buffer

---

[43] J. A. McCray and D. R. Trentham, *Proc. Natl. Acad. Sci. U.S.A.* **77,** 7237 (1980).
[44] J. F. Wootton and D. R. Trentham, *NATO Adv. Study Inst. Ser. C* **272,** 277 (1989).
[45] G. P. Hess, J. B. Udgaonkar, and W. L. Olbricht, *Annu. Rev. Biophys. Biophys. Chem.* **16,** 507 (1987).

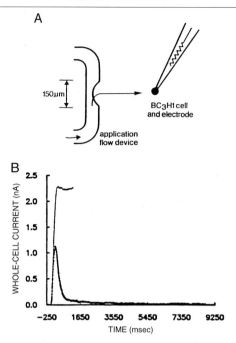

FIG. 3. (A) Schematic drawing (not to scale) of the device used for rapid equilibration of receptors on a cell with neurotransmitters, caged neurotransmitters, and inhibitors.[42,45] The device is made with stainless-steel tubing and is U shaped. Solutions can flow in from one arm of the U and out from a central opening (~150 μm diameter) at the base of the U, as indicated by the arrows. A BC$_3$H1 muscle cell of ~15 μm diameter and containing nicotinic acetylcholine receptors is attached to a whole-cell current-recording electrode.[52] When a valve is opened, the solution containing the biological effector emerges from the central hole of the U tube at a rate of ~1 cm/sec. At the end of the current measurement, which is used to characterize the opening and closing of receptor channels, a valve turns off the flow of solution through the lower arm of the U tube. Simultaneously, another valve opens the connection between the upper arm of the U tube and a suction device. The solution bathing the cell now flows into the U tube at a rate of ~4 cm/sec. This removes the solution bathing the cell and prevents leakage of solutions containing the biological effector from the lower arm of the U tube into the cell compartment. The solution removed by the U tube is automatically replaced by a new solution to bathe the cell. About 3 min of this procedure is sufficient to remove the biological effector from the solution bathing the cell. Typically ~15 measurements can be made with one BC$_3$H1 cell. Leaking of solution from the central tube or one arm of the U tube is prevented by suction applied to the other arm of the U tube. The resulting linear flow rate of the solution is 1–2 cm/sec. The cell suspended from the current-recording electrode is approximately 100 μm away from the hole.[24] (B) An example of an acetylcholine-induced ionic curent in a BC$_3$H1 cell. The cell was placed under whole-cell voltage clamp at a membrane potential of −60 mV. A solution of 200 μM acetylcholine flowed around the cell. The *thin line* parallel to the abscissa was calculated from Eq. (2) and represents the current that would be obtained in the absence of receptor desensitization.[45] (C) Effect of carbamoylcholine concentration on I_A. Carbamoylcholine-induced whole-cell currents through the acetylcholine

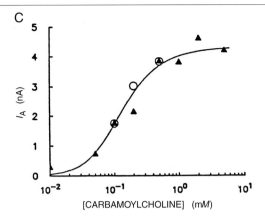

receptors of BC$_3$H1 cells were measured at 23° and −60 mV. The results represent measurements on 10 cells, normalized to each other by standardization of the carbamoylcholine concentration. *Triangles* represent the normalized I_A values. *Circles* represent I_A values calculated from P_o values at three carbamoylcholine concentrations. The *solid line* was drawn using Eq. (2) and values for K_1, ϕ, and $I_M R_M$ of 235 μM, 0.18, and 5.1 nA, respectively, which were obtained from the linear fit of the data using Eq. (2).[42,57]

alone. The quantum product yield estimated to be 0.2 in this experiment is in good agreement with that found for MNP-glycine.[30]

Determination of the product quantum yield can be accomplished without using analytical techniques to isolate the product. The assumption is that the measured optical signal in the photolysis reaction is proportional to the product released. The absorbance of the product A_n measured at the nth laser pulse with mixing of the solution after each pulse is given by Eq. (1):[20]

$$A_n = E_m L c_o \phi K_E \exp[-\phi K_E F(n-1)] \qquad (1)$$

where E_m is the molar extinction coefficient of the absorbance change, L is the length of the light path, and C_o is the initial concentration of the compound that is photolyzed. ϕ is the quantum yield, K_E is the ratio of the number of absorbed photons to the number of target molecules in the laser beam, and F is the fraction of the solution containing the compound photolyzed through which the laser beam passes. It is important to ascertain that the absorbance of the solution at the wavelength used for photolysis does not change during these experiments. The quantum yield and the molar extinction coefficient of the product can be obtained from a semilogarithmic plot of A_n versus $(n-1)$ (Fig. 3C). Such a plot is shown for the photolysis of CNB-caged kainate[23] in which the absorbance measured is due to the *aci*-nitro intermediate observed in the photolysis reaction. A product quantum yield of 0.34 ± 0.07 is obtained from the slope of the line. It is assumed that the concentration of transient intermediate is directly proportional to the concentration of the compound liberated.[33,40,41,43,44]

Biological Characterization of Caged Carboxylates

Crucial in the development of caged biological effectors is the demonstration that the caged compound before photolysis and its photolytic side products are biologically inert in the system under study. To determine whether caged biological effectors are biologically inert, the following question must be asked: Does the caged biological effector initiate, inhibit, or potentiate the reaction to be investigated? Neurotransmitter receptors are reversibly inactivated by their specific neurotransmitter, not in the second time region originally assumed,[46] but in the millisecond time region.[4,26,45,47-50] Techniques were, therefore, developed to investigate receptor-mediated reactions in the millisecond time region.[42,45,47,51] The technique we use to characterize caged neurotransmitters is described next.

Cell-Flow Technique

Instrumentation. The concentration of open receptor channels can be determined by measuring the current passing through the channels at constant transmembrane voltage using the whole-cell current recording technique.[52] A simple device was developed by Krishtal and Pidoplichko[51] to flow neurotransmitter solutions over a single cell suspended in the center of the stream by the recording electrode. A schematic drawing of the cell-flow device is given in Fig. 3A.

The flow rate of the solutions emerging from the flow device is restricted by the stability of the seal between the recording electrode and the cell membrane. At the permissible flow rates, receptors on the cell surface facing the flow device are in contact with the neurotransmitter before the solution reaches the opposite side of the cell and, therefore, desensitize first.

[46] B. Katz and S. Thesleff, *J. Physiol. (Lond.)* **138**, 63 (1957).
[47] G. P. Hess, D. J. Cash, and H. Aoshima, *Nature* **282**, 329 (1979).
[48] B. Sakmann, J. Patlak, and E. Neher, *Nature* **286**, 71 (1980).
[49] L. O. Trussel, G. D. Fischbach, L. I. Thio, and C. Z. Zoramski, *Proc. Natl. Acad. Sci. U.S.A.* **85**, 2834 (1988).
[50] E. L. M. Ochoa, A. Chattopadhyay, and M. G. McNamee, *Cell. Mol. Neurobiol.* **9**, 141 (1989).
[51] O. A. Krishtal and V. I. Pidoplichko, *Neuroscience* **5**, 2325 (1980).
[52] O. P. Hamill, E. Marty, E. Neher, B. Sakmann, and F. J. Sigworth, *Pflugers Arch.* **391**, 85 (1981).

The available hydrodynamic theory[53,54] allows one to correct the observed current for receptor desensitization (inactivation) that occurs while some receptors on the cell surface farther away from the flow device are still equilibrating with the neurotransmitter.[42,45]

A typical cell-flow experiment[42,45] is shown in Fig. 3B. In this experiment a 200 μM solution of acetylcholine flows over a BC_3H1 muscle cell containing nicotinic acetylcholine receptors. The current due to the opening of acetylcholine receptor channels is recorded with a whole-cell current-recording electrode at constant voltage. The current (thick curve) first rises due to the opening of receptor channels, reaches an amplitude of about 1.1 nA, and then falls due to receptor desensitization.

Theory. The thin line in Fig. 3B rising to a value of 2.2 nA is a calculated line and represents the current that would be observed if the receptors did not desensitize. Equation (2) was derived to calculate this current I_A.[42] We define I_A as the amplitude of the current arising from the receptors on the cell surface in the absence of receptor desensitization, at a definite ligand concentration. To take into account the uneven rate at which the ligand solution flows over the cell surface,[42,45] we divide the current time course into constant time intervals (we have used 5 msec) and then correct the current for the desensitization that occurs during each time interval Δt. After the current $(I_{obs})_{\Delta t_i}$ is measured for each of n constant time intervals ($n\Delta t = t_n$), the correct current is given by Eq. (2):

$$(I_A)_{t_n} = (e^{\alpha \Delta t} - 1) \sum_{i=1}^{n} (I_{obs})_{\Delta t_i} + (I_{obs})_{\Delta t_n} \qquad (2)$$

where $(I_{obs})_{\Delta t_i}$ is the observed current during the ith time interval. $(I_A)_{t_n}$ becomes equal to I_A when the value of t_n is equal to or greater than the current rise time, t_r. The current and the current maximum corrected for desensitization are given by the thin line in Fig. 3B. With rise times of 50–220 msec, in measurements made with the same cell and the same ligand concentration, I_A was found to vary by no more than 10%.

Comparison of Cell-Flow Method with Single-Channel Recording Technique and Laser-Pulse Photolysis Technique. The validity of the experimental approach and Eq. (2) (Fig. 3C) used to calculate the concentration of open receptor channels has been determined by two independent tech-

---

[53] V. G. Landau and E. M. Lifshitz, "Fluid Mechanics," Pergamon, Oxford, UK, 1958.
[54] A. G. Levich, "Physicochemical Hydrodynamics," Prentice-Hall, Englewood Cliffs, NJ, 1962.

niques, the single-channel current recording techique,[55] which allows one to determine the fraction of receptor channels that are open[48] at low neurotransmitter concentration, and the laser-pulse photolysis technique using caged neurotransmitters, which has a microsecond time resolution and in which the observed current does not have to be corrected for receptor desensitization.[4,6,11]

Determining Suitability of Caged Neurotransmitters for Biological Experiments Using Cell-Flow Technique

The importance of determining the effects of caged compounds on the activity of the target biological molecule is illustrated with two examples. Both CNB-carbamoylcholine **(20)** and α-methyl-*o*-nitrobenzyl (MNB) carbamoylcholine **(21)** are water soluble and stable in aqueous solutions at neutral pH. On irradiation with light in the 300- to 350-nm region, both compounds release carbamoylcholine with rates and quantum yield that make them suitable for rapid chemical kinetic investigations of nicotinic acetylcholine receptor-mediated reactions. However, the CNB-caged carbamoylcholine is biologically inert and can be used in the investigations of receptor function, whereas the MNB-caged derivative modifies receptor function and cannot be used.

20 R = CO_2H

21 R = CH_3

Another example of the importance of the biological characterization of caged compounds is illustrated by the cell-flow experiments in Fig. 4A and 4B. The experiment in Fig. 4A shows that in the presence of both 300 μM NMDA and 300 μM CNB-NMDA, the observed current amplitude is lower than when CNB-NMDA is not present. This means that the NMDA receptor is inhibited by the caged form CNB-NMDA. It is of interest to note, therefore, that CNB-carbamoylcholine **(20)**,[20] CNB-γ-aminobutyric acid **(8)**,[21] and CNB-glutamate[22] are biologically inert in similar experiments. Figure 4B shows that 10 mM CNB-GABA does not affect the response of a mouse cerebral cortical cell to 250 μM GABA. It is, of course,

---

[55] E. Neher and B. Sakmann, *Nature* **260**, 779 (1976).
[56] A. P. Billington, N. Matsubara, W. W. Webb, and G. P. Hess, *Adv. Protein Chem.* **III**, 417 (1992).
[57] G. P. Hess, *Arch. Phys. Biochem.* **104**, 752 (1996).

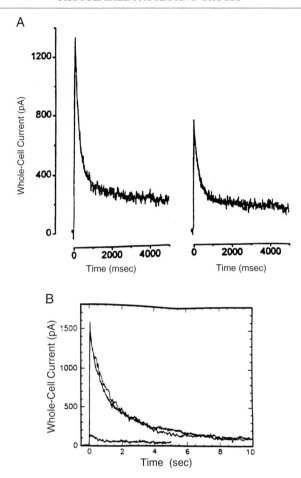

FIG. 4. (A) Whole-cell current induced by NMDA and recorded from individual neonatal rat hippocampal neurons at −60 mV, pH 7.4, and 22°. The transmembrane ion current was generated by (a) flowing 300 μM NMDA over the cell or (b) simultaneously flowing 300 μM NMDA and 300 μM CNB-NMDA in which the β-carboxyl group is caged.[27] (B) Whole-cell current responses of embryonic mouse cerebral cortical cells, pH 7.3, 21–23°, at a transmembrane voltage V_m of − 70 mV. The two curves represent experiments in which the cell was equilibrated with 250 μM GABA in the absence and presence of 10 mM CNB-GABA **(8)**.[21]

also important to ascertain that the side product that is photolytically released together with the desired biological effector is biologically inert.

In the development of any new caged compounds it is important that the reagents meet certain stringent criteria before they are used in the studies for which they have been designed. We have described the synthesis, photochemistry, and biological characterization of photolabile precursors (caged compounds) of carboxylic acids and neurotransmitters. Some of the compounds described can provide new levels of temporal resolution in transient kinetic measurements of function and/or spatial resolution in mapping studies.

Acknowledgments

Work reviewed here was supported by a grant (GM04842) awarded to G.P.H. by the National Institutes of Health Institute of General Medical Sciences.

[3] New Caged Groups: 7-Substituted Coumarinylmethyl Phosphate Esters

By Toshiaki Furuta and Michiko Iwamura

Introduction

The design and synthesis of caged probe molecules that function inside of living cells are challenging problems for organic chemists. As part of a research program directed toward these objectives we have reported the synthesis and utilization of new types of photochemically removable protecting groups that additionally exhibit a strong fluorescence emission.[1-3] Of the four fluorescent, photosensitive protection groups prepared for diethyl phosphate, namely, 1-pyrenylmethyl,[4] 4-(7-methoxycoumarinyl)methyl (MCM),[5] 2-(9,10-dioxo)anthrarylmethyl,[6] and 2-naphthylmethyl, 1-pyrenylmethyl was found to be the best caged group in terms of its photochemical properties. However, its extremely low stability in physiological salt solutions limited its application in studies on living cells, and so we focused our attention on the 4-(7-methoxycoumarinyl)methyl group as a photolabile "cage" group for phosphates.[7] Whether naturally occurring or synthetic, coumarin and many of its related aromatic lactones show interesting photo-

---

[1] M. Iwamura, T. Ishikawa, Y. Koyama, K. Sakuma, and H. Iwamura, *Tetrahedron Lett.* **28,** 679 (1987).
[2] M. Iwamura, K. Tokuda, N. Koga, and H. Iwamura, *Chem. Lett.* 1729 (1987).
[3] M. Iwamura, C. Hodota, and M. Ishibashi, *Synth. Lett.* 35 (1991).
[4] T. Furuta, H. Torigai, T. Osawa, and M. Iwamura, *Chem. Lett.* 1179 (1993).
[5] R. S. Givens and B. Matuszewski, *J. Am. Chem. Soc.* **106,** 6860 (1984).
[6] D. S. Kemp and J. Reczek, *Tetrahedron Lett.* **18,** 1031 (1977).
[7] T. Furuta, H. Torigai, M. Sugimoto, and M. Iwamura, *J. Org. Chem.* **60,** 3953 (1995).

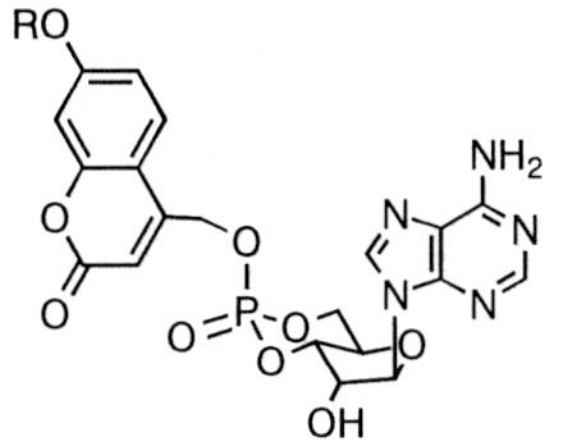

MCM-cAMP : R=CH₃ ACM-cAMP : R=CH₃CO

HCM-cAMP : R=H PCM-cAMP : R=CH₃CH₂CO

FIG. 1. 7-Substituted 4-coumarinylmethyl-caged cAMPs.

biological activities, and a number of these compounds are commercially available and used as fluorescent dyes. Although the 4-coumarinylmethyl group has been used as a fluorescent tag for biological molecules,[5] this highly fluorescent group had not been used as a caging group until our report on a new method for the synthesis of esters of diethyl phosphate and cAMP.[8] The silver(I) oxide promoted condensation method, as we call it, is a one-pot synthesis of phosphate esters from the corresponding phosphate and an alkyl halide in the presence of silver oxide. It is a mild neutral reaction that generates the product in a modest yield along with the silver halide, an easily separable product. Although the product yield is not necessarily high, the main attraction of this reaction is that it is a single-step synthesis that uses commercially available phosphate groups and alkyl halides.

The coumarin-caged cAMP compounds, MCM–cAMP, 4-(7-acetoxycoumarinyl)methyladenosine (ACM)–cAMP, 4-(7-propionyloxycoumarinyl)methyladenosine (PCM)–cAMP and 4-(7-hydroxycoumarinyl)methyladenosine (HCM)–cAMP cyclic 3′,5′-monophosphates (Fig. 1), which differ from each other only by the substituent at the 7-position of the coumarin ring, have been prepared from reactions of the corresponding 4-coumarinylmethyl bromide with cAMP using the silver(I) oxide promoted method. These caged cAMP compounds exhibit similar photocleavage properties and efficiently release cAMP in Ringer's solution on irradiation with near-ultraviolet light. Irradiation of a 0.1 mM solution of any one of the caged cAMPs in Ringer's solution containing 1% (v/v) dimethyl sulfoxide (DMSO) with near-ultraviolet light for 20 min results in the loss of 60–80% of coumarin-caged cAMP and liberation of 40–60% of free cAMP. Another

---

[8] T. Furuta, H. Torigai, T. Osawa, and M. Iwamura, *J. Chem. Soc. Perkin Trans.* **1**, 3139 (1993).

significant feature of these caged cAMPs is their long half-life in the dark in physiological salt solutions compared to those of *o*-nitrobenzyl- or *o*-nitrophenethyl-caged cAMPs, which in our experience readily hydrolyze in aqueous solution.

Higher lipophilicity is an important property of caged nucleotides if they are to be loaded into living cells through the plasma membrane. However, this lipophilicity might also result in significant leakage of the caged cAMP from the cell. To overcome this potential problem, acyloxycoumarin cAMP derivatives were designed and synthesized with the feature that once they enter the cell they are hydrolyzed by intracellular esterases to the more hydrophilic 7-hydyroxycoumarin derivative, HCM–cAMP. The de-esterified caged cAMP compounds are better retained within the cytoplasm, and cells loaded with these caged compounds were found to exhibit a higher cAMP-dependent melanin granule dispersion response compared to other more lipophilic-caged cAMP compounds after irradiation with near-ultraviolet light.[9]

This article summarizes the methods used in our laboratory for the preparation and characterization of 4-coumarinylmethyl-caged phosphate esters and their application in studies of the role of cAMP on the dispersion and aggregation of pigment granules in live cells.

Methods

Materials

(−)-Adenosine cyclic 3′,5′-monophosphate (Acros, Sigma, St. Louis, MO), 4-(bromomethyl)-7-methoxycoumarin (Nakarai, Kyoto, Japan), 4-(bromomethyl)-7-acetoxycoumarin (TCI, Tokyo, Japan), 4-methylumbelliferone (TCI), *p*-toluenesulfonyl hydrazide (Wako, Osaka, Japan), triethylamine (Wako), selenium dioxide (95%, Nakarai), and silver oxide (Wako) are used without further purification. Acetonitrile (dehydrated, Kanto, Tokyo, Japan), dimethyl sulfoxide (Wako), ethanol (99.5%, Wako), and *m*-xylene (Wako) of the highest quality available are used without further purification. Methanol is distilled from $Mg(OCH_3)_2$ under nitrogen. Chromatography is performed using Kieselgel 60 (E. Merck, Darmstadt, Germany) of 0.063- to 0.200-mm-grade for column chromatography and of 0.042- to 0.063-mm-grade silica gel for flash column.

Nuclear magnetic resonance (NMR) is performed using a JNM-GSX 270 spectrometer (Jeol, Tokyo, Japan) with tetramethylsilane as an internal

---

[9] T. Furuta, A. Momotake, M. Sugimoto, M. Hatayama, H. Torigai, and M. Iwamura, *Biochem. Biophys. Res. Commun.* **228**, 193 (1996).

standard. Infrared spectroscopy (IR) is performed using an FT/IR 5000 spectrometer (Jasco, Tokyo, Japan). Ultraviolet (UV) spectra are determined with a U-3210 spectrometer (Hitachi, Tokyo, Japan). High-performance liquid chromatography (HPLC) is performed with a PU980 intelligent HPLC pump equipped with a UV970 intelligent UV/vis detector (Jasco, Tokyo, Japan) using a reversed-phase column, Jasco Crestpac C185S (150 × 4.6 mm) and monitored at 254 nm. The mobile phase is 50% methanol–50% water (v/v). High-resolution mass spectra are obtained using a GC-MATE spectrometer (Jeol, Tokyo, Japan) in FAB mode.

Synthesis of Coumarin-Caged cAMPs

Synthesis of 4-(7-Methoxycoumarinyl)methyladenosine Cyclic 3',5'-Monophosphate

Cyclic AMP (33 mg, 0.10 mmol) and 4-bromomethyl-7-methoxycoumarin (80.7 mg, 0.300 mmol are added to an oven-dried, 30-ml, two-necked, round-bottom flask equipped with a magnetic stirring bar and a rubber septum. The flask is evacuated under vacuum and flushed with argon. Dry acetonitrile (6 ml) and DMSO (1 ml) are added to the reaction flask using a syringe. After the addition of solid silver(I) oxide (44 mg, 0.20 mmol), the resulting black suspension is stirred at 60° for 45 hr under argon. The reaction mixture is filtered through filter paper and the residue washed with chloroform. The combined filtrate is evaporated under reduced pressure. The oily residue is purified by column chromatography (30 g of SiO_2, 4.7% (v/v) methanol–CH_2Cl_2, then 6.2% methanol–CH_2Cl_2) to give 75 mg of a yellow oil. Further purification by flash chromatography (30 g of SiO_2, 4.7% methanol–CH_2Cl_2) gives MCM–cAMP (23 mg, 44%) as a mixture of two stereoisomers [axial (ax)/equatorial (eq) = 2.3/1]; 12.5 mg (29%) of product is obtained from the residual black solid silver salt of cAMP. Separation of the axial and equatorial isomer is achieved by semipreparative reversed-phase HPLC (Cica-Merck, Tokyo, Japan, Lichrosorb RP-18 (7 m) 250 × 250 mm, mobile phase: 60% methanol–40% H_2O, flow rate: 5 ml/min, retention time: axial 33 min, equatorial 36 min). Melting point 260–263° (decomposition); ^{1}H NMR (5% $CD_3OD/CDCL_3$, δ ppm) (axial isomer) 3.90 (3H, s, OCH_3), 4.47 (1H, d, $J = 7.2$ Hz, H'5), 4.70 (1H, dd, $J = 4.6$ and 22.1 Hz, H'4), 4.85 (1H, d, $J = 5$ Hz, H'2), 5.40 (2H, d, $J = 6.5$ Hz, P-O-CH_2), 5.52 (1H, m, H'3), 5.97 (1H, s, H'1), 6.54 (1H, t, $J = 1$ Hz, H3), 6.89–6.91 (2H, m, H5 and H8), 7.74 (1H, dd, $J = 2$ and 6 Hz, H6), 7.92 (1H, s, adenine H2), 8.25 (1H, s, adenine H8); (equatorial isomer) 3.92 (3H, s, OCH_3), 4.48 (1H, d, $J = 6$ Hz, H'5), 4.68 (1H, m, H'4), 4.83 (1 H, d, $J = 5$ Hz, H'2), 5.41 (2H m, P-O-CH_2), 5.51 (1H, m, H'3), 6.04 (1H, s, H'1), 6.49 (1H, t, $J = 1$ Hz, H3), 6.89–6.95 (2H, m, H5 and H8),

7.47 (1H, dd, J = 2 and 6 Hz, H6), 8.03 (1H, s, adenine H2), 8.28 (1H, s, adenine H8); IR (KBr, ν_{max} cm^{-1}) (mixture of two stereoisomers) 1719 (C=O), 1296 (P=O), 1212 (C-OCH$_3$), 1096 (C-OH), 1017 (P-O); UV [50% H$_2$O–dioxane, λ_{max} (ε)] 259 (14,600), 325 (13,300).

Synthesis of 4-(7-Acetoxycoumarinyl)methyladenosine Cyclic 3′,5′-Monophosphate

Silver(I) Oxide Promoted Method. This reaction is performed as described above except it uses 100 mg (0.304 mmol) of cAMP, 267 mg (0.899 mmol) of 7-acetoxy-4-bromomethylcoumarin, 132 mg (0.570 mmol) of silver(I) oxide, 18 ml of acetonitrile, and 3 ml of DMSO and is stirred for 27 hr at 100°. The products are purified by column chromatography (20 g of SiO$_2$, 6.2% methanol–CH$_2$Cl$_2$) and then flash column chromatography (20 g of SiO$_2$, 6.2% methanol–CH$_2$Cl$_2$) to give 22 mg (0.041 mmol, 13.5% yield) of ACM–cAMP as a mixture of two stereoisomers (ax/eq = 3/1).

Diazo-Based Method. Cyclic AMP, 64.3 mg (0.195 mmol), is added to a 20-ml, round-bottomed flask equipped with a magnetic stirring bar and CaCl$_2$ drying tube. Dry DMSO (2 ml) and 4-diazomethyl-7-acetoxycoumarin (52.6 mg, 0.215 mmol) are added to the flask, and the mixture is stirred at room temperature for 4 days. Evaporation and purification by flash column chromatography (10 g of SiO$_2$, 6.2% methanol–CH$_2$Cl$_2$) give 17.4 mg (0.0320 mmol, 16.5% yield) of ACM–cAMP as a mixture of two stereoisomers (ax/eq = 1/1). ^{1}H NMR (5% CD$_3$OD/CDCl$_3$, δ ppm) (axial isomer) 2.35 [3H, s, C(=O)CH$_3$], 4.48 (1H, d, J = 7 Hz, H′5), 4.68 (1H, m, H′4), 4.82 (1H, d, J = 7 Hz, H′2), 5.40 (2H, d, J = 6.5 Hz, P-O-CH$_2$), 5.49 (1H, m, H′3), 5.96 (1H, s, H′1), 6.67 (1H, t, J = 1.5 Hz, H3), 7.07–7.22 (2H, m, H5 and H8), 7.56 (1H, dd, J = 2 and 9 Hz, H6), 7.89 (1H, s, adenine H2), 8.28 (1H, s, adenine H8); (equatorial isomer) 2.36 [3H, s, C(=O)CH$_3$], 4.48 (1H, d, J = 7 Hz, H′5), 4.68 (1H, m, H′4), 4.84 (1H, d, J = 7 Hz, H′2), 5.40 (2H, d, J = 6.5 Hz, P-O-CH$_2$), 5.49 (1H, m, H′3), 6.01 (1H, s, H′1), 6.64 (1H, t, J = 1.5 Hz, H3), 7.07–7.22 (2H, m, H5 and H8), 7.56 (1H, dd, J = 2 and 9 Hz, H6), 7.93 (1H, s, adenine H2), 8.29 (1H, s, adenine H8); IR (KBr, ν_{max} cm^{-1}) (mixture of two stereoisomers) 1721 (C=O), 1267 (P=O), 1203 (C-O), 1135 (C-O), 1067 (C-O), 1015 (P-O); UV [CH$_3$OH, λ_{max} (ε)] 267 (13,400), 313 (7650).

Synthesis of Bromomethyl-7-hydroxycoumarin

4-Bromomethyl-7-acetoxycoumarin, 512.7 mg (1.726 mmol), and 35 ml of wet toluene are added to a 100-ml, round-bottomed flask equipped with a magnetic stirring bar, reflux condenser, and CaCl$_2$ drying tube. The mixture is stirred at 80° for 20 min, 2.8 g of *p*-toluenesulfonic acid on SiO$_2$[10]

---

[10] G. Blay, M. L. Cardona, M. B. Garcia, and J. P. Pedro, *Synthesis* 438 (1989).

is then added to the solution and the mixture stirred at 80° for 5 hr. After cooling to room temperature, the reaction mixture is applied directly to the column and eluted with chloroform. Evaporation of the solvent gives almost pure 4-bromomethyl-7-hydroxycoumarin (420.4 mg, 1.636 mmol, 94.8% yield).

Synthesis of 4-Bromomethyl-7-propionyloxycoumarin

4-Bromomethyl-7-hydroxycoumarin, 318.7 mg (1.240 mmol), and 10 ml of dry acetonitrile are added to a 100-ml, round-bottomed flask equipped with a magnetic stirring bar and $CaCl_2$ drying tube. Triethylamine (0.2 ml, 1.4 mmol) and propionyl chloride (0.2 ml, 2.3 mmol) are added and the mixture stirred at room temperature for 20 min. The mixture is diluted with $CHCl_3$ and then washed sequentially with 0.5 N K_2CO_3, 1N HCl, and saturated NaCl. Drying over anhydrous $MgSO_4$ and evaporation under vacuum give 290.6 mg (0.9715 mmol, 78.3% yield) of 4-bromomethyl-7-propionyloxycoumarin.

Synthesis of 4-(7-Propionyloxycoumarinyl)methyladenosine Cyclic 3',5'-Monophosphate

This reaction is performed as described above except it uses 100 mg (0.304 mmol) of cAMP, 341 mg (1.00 mmol) of 4-bromomethyl-7-propionyloxycoumarin, 132 mg (0.570 mmol) of silver(I) oxide, 18 ml of acetonitrile, and 3 ml of DMSO and is stirred for 16 hr at 120°. The product is purified by column chromatography (30 g of SiO_2, 6.2% methanol–CH_2Cl_2) and then flash column chromatography (20 g of SiO_2, 6.2% methanol–CH_2Cl_2) to give 38 mg (0.0678 mmol, 22.6% yield) of PCM–cAMP as a mixture of two stereoisomers (ax/eq = 3/2). ^{1}H NMR (5% $CD_3OD/CDCl_3$, δ ppm) (axial isomer) 1.27 (3H, t, J = 8 Hz, CH_3), 2.64 (2H, q, J = 8 Hz, CH_2CH_3), 4.50 (1H, d, J = 6 Hz, H'5), 4.67 (1H, m, H'4), 4.86 (1H, d, J = 6 Hz, H'2), 5.41 (2H, d, J = 7 Hz, P-O-CH_2), 5.50 (1H, m, H'3), 5.99 (1H, s, H'1), 6.65 (1H, t, J = 1 Hz, H3), 7.05–7.18 (2H, m, H5 and H8), 7.55 (1H, d, J = 9 Hz, H6), 7.94 (1H, s, adenine H2), 8.26 (1H, s, adenine H8); (equatorial isomer) 1.28 (3H, t, J = 8 Hz, CH_3), 2.64 (2H, q, J = 8 Hz, CH_2CH_3), 4.47 (1H, d, J = 7 Hz, H'5), 4.67 (1H, m, H'4), 4.84 (1H, d, J = 6 Hz, H'2), 5.41 (2H, d, J = 7 Hz, P-O-CH_2), 5.50 (1H, m, H'3), 6.03 (1H, s, H'1), 6.61 (1H, t, J = 1 Hz, H3), 7.05–7.18 (2H, m, H5 and H8), 7.56 (1H, d, J = Hz, H6), 7.97 (1H, s, adenine H2), 8.25 (1H, s, adenine H8); IR (KBr, ν_{max} cm^{-1}) (mixture of two stereoisomers) 1734 (C=O), 1267 (P=O), 1197 (C-O), 1135 (C-O), 1067 (C-O), 1017 (P-O); UV [CH_3OH, λ_{max} (ε)] 265 (13,000), 314 (6600); MS (FAB$^+$) 560 (M+H$^+$), 330, 136, 57; exact mass (FAB$^+$) (M+H$^+$) (calculated, 560.1182; observed, 560.1146).

Synthesis of 4-(7-Hydroxycoumarinyl)methyladenosine Cyclic 3′,5′-Monophosphate

Silver(I) Oxide Promoted Method. This reaction is performed as described above except that it uses 132 mg (0.401 mmol) of cAMP, 298 mg (1.17 mmol) of 7-hydroxy-4-bromomethylcoumarin, 176 mg (0.759 mmol) of silver(I) oxide, 18 ml of acetonitrile, and 3 ml of DMSO, and the reaction is stirred for 24 hr at 80°. The products are purified by column chromatography (20 g of SiO_2, 9% methanol–CH_2Cl_2) and then by flash column chromatography (20 g of SiO_2, 90% methanol–CH_2Cl_2) to give 32 mg (0.063 mmol, 16% yield) of HCM–cAMP as a mixture of two stereoisomers (ax/eq = 2/1).

Diazo-Based Method. This reaction is performed as described above except that it uses 46.6 mg (0.142 mmol) of cAMP and 31.5 mg (0.156 mmol) of 4-diazomethyl-7-hydroxycoumarin. The reaction mixture is stirred for 42 hr at room temperature and purified by flash column chromatography (20 g of SiO_2, 9% methanol–CH_2Cl_2) to give 5.5 mg (0.011 mmol, 7% yield) of HCM–cAMP as a mixture of two stereoisomers (ax/eq = 3/5). ^{1}H NMR (5% $CD_3OD/CDCl_3$, δ ppm) (axial isomer) 4.46 (1H, d, J = 7 Hz, H′5), 4.67 (1H, m, H′4), 4.83 (1H, d, J = 6 Hz, H′2), 5.38 (2H, m, P-O-CH_2), 5.50 (1H, m, H′3), 5.98 (1H, s, H′1), 6.48 (1H, t, J = Hz, H3), 6.82–6.86 (2H, m, H5 and H8), 7.43 (1H, m, H6), 7.96 (1H, s, adenine H2), 8.22 (1H, s, adenine H8); (equatorial isomer) 4.46 (1H, d, J = 7 Hz, H′5), 4.67 (1H, m, H′4), 4.83 (1H, d, J = 6 Hz, H′2), 5.39 (2H, m, P-O-CH_2), 5.50 (1H, m, H′3), 6.03 (1H, s, H′1), 6.44 (1H, t, J = 1 Hz, H3), 7.05–7.18 (2H, m, H5 and H8), 7.43 (1H, m, H6), 8.01 (1H, s, adenine H2), 8.29 (1H, s, adenine H8); IR (KBr, ν_{max} cm^{-1}) (mixture of two stereoisomers) 1715 (C=O), 1272 (P=O), 1207 (C-O), 1141 (C-O), 1067 (C-O), 1006 (P-O); UV [CH_3OH, ν_{max} (ε)] 257 (12,100), 326 (16,800); MS (FAB$^+$) 504 (M+H$^+$), 330, 261, 157, 79.

Results and Discussion

Silver(I) Oxide Promoted Condensation Method

The advantages of this synthesis (Scheme 1) are (1) preparation of the metal or ammonium phosphates is not necessary, (2) the reaction proceeds under neutral conditions, (3) acid or base treatment is not required for the workup or purification step, and (4) choice of incorporated substituents depends only on the availability of the corresponding halide. To optimize the conditions for the reaction between cAMP and benzyl bromide, the effects of solvent, temperature, and reaction time on the product yield were investigated. Unlike the case with phosphoric acid, no alkylated product

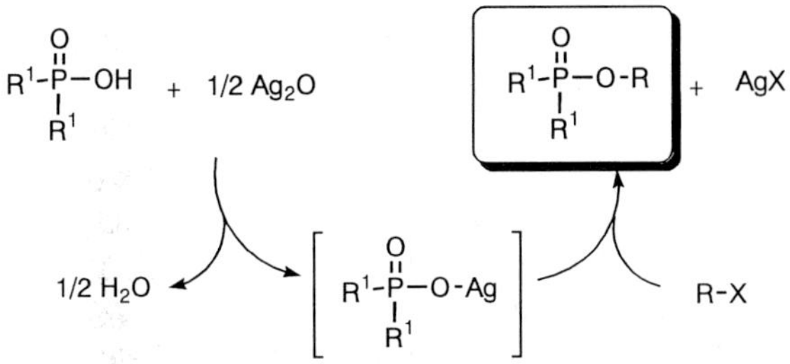

SCHEME 1. Silver(I) oxide promoted condensation reaction.

$$R^1-\underset{R^1}{\overset{O}{\overset{\|}{P}}}-OH + 1/2\,Ag_2O \longrightarrow \left[R^1-\underset{R^1}{\overset{O}{\overset{\|}{P}}}-O-Ag\right] \xrightarrow{R-X} \boxed{R^1-\underset{R^1}{\overset{O}{\overset{\|}{P}}}-O-R} + AgX$$

$$+ 1/2\,H_2O$$

was obtained when reactions were carried out in the absence of acetonitrile. The reaction was found to proceed only when cAMP was dissolved in a little DMSO, with a mixture of acetonitrile and DMSO (6:1, v/v) giving the best result. The highest product yield was obtained at a temperature between 60° and 80°. The duration of a reaction also has an effect on the yield, as shown in Fig. 2. Long reaction times resulted in a decreased product yield, possibly because the cAMP triesters are hydrolyzed by water generated during the course of the reaction.

Thin-layer chromatography (TLC) or HPLC can be used to monitor the progress of the reaction, and it should be terminated when the amount of product has reached its plateau value. The axial isomer always formed more readily than the equatorial isomer (Fig. 2). This result is consistent with earlier reports that showed the equatorial isomer of cAMP benzyl

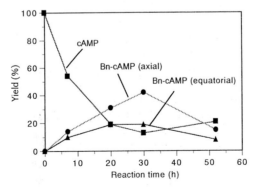

FIG. 2. Product distribution for silver(I) oxide promoted condensation of cAMP and benzyl bromide.

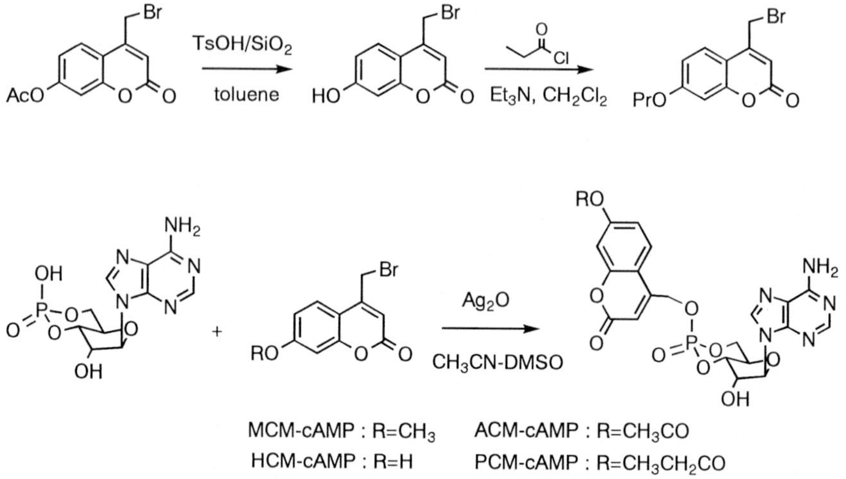

SCHEME 2. Synthesis of coumarin-caged cAMPs by silver(I) oxide promoted method.

ester is hydrolyzed approximately four times faster than the axial isomer,[11] and because of the ground state stereoelectronic effects of six-membered-ring phosphate esters reported by Gorenstein.[12]

Synthesis of Coumarin-Caged cAMPs via Silver(I) Oxide Method

4-(7-Methyoxycoumarinyl)methyl and ACM–cAMP were synthesized (Scheme 2) from the corresponding 7-substituted 4-bromomethylcoumarins, both of which are commercially available, in the presence of silver(I) oxide, with a yield of 44 and 14%, respectively. Both products were obtained as a mixture of axial and equatorial isomers of the underlying dioxaphosphorinan ring. The predominant isomer was axial, and the ratio of the isomers was dependent on the reaction conditions (Fig. 1). The synthesis of PCM–cAMP proceeds with modest yield (22%) using the corresponding bromide that was prepared from commercially available 4-bromomethyl-7-acetoxycoumarin. Although HCM–cAMP was prepared using the same method, with a yield of 16% for one particular run, this synthetic approach was not found to be reproducible. The reason for this result is not yet clear, although the free phenolic hydroxyl substituent on the coumarin ring may affect the condensation reaction.

---

[11] J. Engels and E. J. Schlaeger, *J. Med. Chem.* **20,** 907 (1977).
[12] G. Gorenstein, *Chem. Rev.* **87,** 1047 (1987).

Synthesis of Coumarin-Caged cAMPs via Diazo-Based Method

Ito and Maruyama[13] reported the preparation and characterization of several 7-substituted 4-diazomethylcoumarins. The reaction of cAMP with the 7-substituted (4-diazomethyl)coumarins in DMSO, carried out under conditions similar to those for *o*-nitrophenyldiazoethane,[14] yield the corresponding coumarin-caged cAMPs with a yield of 10–20% The preparation of 7-hydroxy-4-diazomethylcoumarin takes five step sequences with 46% overall yield starting from the commercially available 4-methylumberiferon.[13]

Purification and Structural Determination

Isolation and purification of the caged cAMP from the crude product mixture are usually carried out by column chromatography on silica gel using 4.7% methanol–dichloromethane or 6.2% methanol–dichloromethane as an eluent. Further purification by flash chromatography gave almost pure caged cAMP. Separation of diastereomeric isomers was carried out by semipreparative reversed-phase HPLC using 50% methanol–H_2O as an eluent. Determination of the structure of each isomer was performed mainly by ^1H NMR spectroscopy. Although the coumarinylmethyl-cAMPs are not very soluble in $CDCl_3$, they are soluble in $CDCl_3$ containing 5% CD_3OD. The configurations of the products were assigned on the basis of the ^1H NMR chemical shift differences with respect to the resonance of the H8 proton of the adenine ring. Thus the isomer showing the lower shift was assigned as the axial stereochemistry, in analogy with previous reports.[15] A similar propensity was observed for other cAMP triesters, benzyl (Bn), naphthylmethyl (NM), and 2-nitrobenzyl (NB) esters of cAMP. The ratio of the axial and equatorial isomers in the mixture was determined using the signals of axial and equatorial isomers, which appear as two discrete singlets around 8 ppm. For example, mixtures of the axial and equatorial isomers of MCM–cAMP show two discrete singlets at 8.25 and 8.28 ppm. However, as the axial isomer elutes faster than the equatorial isomer on a reversed-phase column, it was more convenient to determine the yield of each isomer using HPLC. Selected chemical shifts and the retention times are summarized in Table I.

---

[13] K. Ito and J. Maruyama, *Chem. Pharm. Bull.* **31**, 3014 (1983).
[14] J. W. Walker, G. P. Reid, and D. R. Trentham, *Methods in Enzymol.* **172**, 288 (1989).
[15] J. Engels and A. Jager, *Arch. Pharm. (Weinheim)* **315**, 368 (1982).

TABLE I
Selected ¹H NMR Chemical Shifts and HPLC Retention Times

Substrate	Isomer	¹H NMR (ppm)		Retention time (min)[a]
		H8 (Ad)	H2 (Ad)	
MCM–cAMP	ax	8.25	7.92	9.9
	eq	8.28	8.03	12.3
ACM–cAMP	ax	8.28	7.89	8.4
	eq	8.29	7.93	10.1
PCM–cAMP	ax	8.26	7.94	12.2
	eq	8.25	7.97	15.9
HCM–cAMP	ax	8.22	7.96	34[b]
	eq	8.29	8.01	45[b]
Bn–cAMP	ax	8.18	7.82	
	eq	8.31	7.82	
NM–cAMP	ax	8.11	7.89	6.3[c]
	eq	8.26	7.85	12.0[c]
NB–cAMP	ax	8.20	7.88	6.7[d]
	eq	8.33	7.85	9.4[d]

[a] The mobile phase was 50% methanol–50% water (v/v), and the flow rate was 0.6 ml/min unless otherwise noted.
[b] The mobile phase was 35% methanol–65% water (v/v).
[c] The mobile phase was 60% methanol–40% water (v/v), and the flow rate was 0.8 ml/min.
[d] The flow rate was 0.8 ml/min.

Stability of Caged cAMP Compounds in the Dark

The hydrolytic stability of the various cAMP derivatives in 1% DMSO in Ringer's solution stored in the dark is quite variable, and the half-life of each compound is shown in Table II. MCM–cAMP exhibits the highest stability under these conditions, whereas the half-lives of the acyloxycoumarin-caged cAMPs, ACM–cAMP and PCM–cAMP, are far shorter, presumably because the C7-acyloxy group is susceptible to hydrolysis in physiological salt solution and ACM–cAMP is quantitatively converted into HCM–cAMP on prolonged storage in Ringer's solution. Therefore, particular care must be taken in the storage of the acyloxycoumarin-caged cAMPs. We recommend that all compounds be stored in a solid form in a freezer or as 10 mM DMSO stock solutions at $-20°$, where they are stable for several months.

Photochemical and Spectral Properties of Caged cAMP Compounds

Irradiation of MCM–cAMP in 1% DMSO in Ringer's solution gave the parent cAMP and 4-hydroxymethyl-7-methoxycoumarin as photoproducts.[7]

TABLE II
HALF-LIVES OF CAGED cAMPs[a]

Substrate[b]	Half-lives (hr)[c]		Reference
	Axial	Equatorial	
MCM–cAMP	1000	60	7
ACM–cAMP	17.4	15.5	This work
PCM–cAMP	25.6	18.8	This work
HCM–cAMP	28.8	18.9	This work

[a] In 1% DMSO in Ringer's solution.
[b] 10^{-4} M solutions in Ringer's solution containing 1% DMSO.
[c] The exact concentrations were determined by HPLC and analyzed by pseudo-first-order kinetics, using 2-hydroxymethylanthraquinone as an internal standard, which was found to be stable and had no effect on caged cAMPs.

No significant differences were observed in the photochemical reactivity of the axial and equatorial isomers, and therefore it is not necessary to separate the two coumarin-caged cAMP stereoisomers. The mechanism proposed for this photolysis reaction is presented in Scheme 3.

The absorption maxima (λ_{max}) and molar absorptivity (ε) at 340 nm for the different coumarin-caged cAMPs are summarized in Table III. Absorption spectra were measured in methanol because ACM–cAMP and PCM–cAMP are gradually hydrolyzed to HCM–cAMP in Ringer's solution. Introduction of the acyloxy substituent on the C-7 position of the coumarin ring shifts the absorption maximum to a shorter wavelength by about 10 nm and reduces the absorptivity to almost one-half compared to

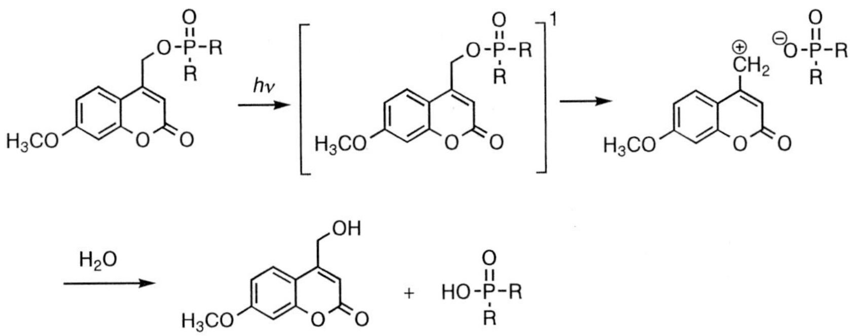

SCHEME 3. Proposed mechanism for photolysis of coumarin-caged phosphate.

TABLE III
PHOTOCHEMICAL AND SPECTRAL PROPERTIES

Substrate[a]	$\lambda_{\max(\varepsilon)}$[b]	ε_{340}[b]	Φ_{dis}[c,d]
MCM–cAMP	325 (13,300), 259 (14,600)	9,600	0.12
ACM–cAMP	313 (7,650), 267 (13,400)	2,600	0.056
PCM–cAMP	313 (6,500), 265 (12,800)	1,900	0.054
HCM–cAMP	326 (16,800), 256 (12,100)	13,100	0.062

[a] All samples were used as a mixture of stereoisomers.
[b] In methanol.
[c] In 1% DMSO in Ringer's solution.
[d] Determined by potassium ferrioxalate actinometry.

those with methoxy or hydroxy substitution. The quantum efficiencies for the disappearance (Φ_{dis}) of the following compounds in Ringer's solution containing 1% DMSO were determined to be 0.12 for MCM–cAMP, 0.056 for ACM–cAMP, 0.054 for PCM–cAMP, and 0.062 for HCM–cAMP, respectively (Table III).

Biological Activity of Caged cAMP Compounds

To evaluate the biological function of the coumarin-based caged cAMPs, the cAMP-dependent granule dispersion response of fish melanophores

TABLE IV
EFFICIENCY OF PHOTOCHEMICAL CONVERSION OF CAGED cAMP AND cAMP-DEPENDENT DISPERSION OF MELANOPHORES

Substrate	Magnitude of dispersion (%)[a]	Conversion of substrate (%)[b]
MCM–cAMP	23 ± 7	60
ACM–cAMP	39 ± 6	23
PCM–cAMP	54 ± 4	9
HCM–cAMP	11 ± 7	64
NPE–cAMP	25 ± 10	32

[a] Dispersing responses of pigment granules after 10 sec irradiation through the objective lense of the fluorescent microscope.
[b] Photlytic consumption of caged cAMPs (100 μM solution in 1% DMSO in Ringer's solution) after 10 sec irradiation through the objective lens of the fluorescence microscope.

isolated from the scales of the dorsal trunk of the wild-type medaka fish (*Oryzias latipes*) was studied. The extent of the dispersion of melanophores is controlled by the intracellular concentration of cAMP, and it is known that cAMP cannot pass through the membrane of live melanophores. The assay is based on an analysis of the distribution of phase-dense granules by phase-contrast microscopy before and after irradiation of caged cAMP loaded cells with near ultraviolet delivered from a mercury arc lamp.[7,9] Image analysis of the extent of granule dispersion after photolysis provides an accurate method to quantify the intracellular concentration of cAMP. To clarify the effect of extracellularly applied caged cAMP, melanophores were pretreated with an α_2-adrenergic agonist (norepinephrine) together with a phosphodiesterase inhibitor, 3-isobutyl-1-methylxanthine (IBMX), which blocks the reaggregation of the dispersing melanin pigments by the intracellular phosphodiesterase (see experimental procedures in Refs. 7 and 9). The results of the dispersing responses for these coumarin-caged cAMPs are shown in Table IV, which also documents the efficiency of these photoactivation reactions.

[4] Caged Fluorescent Probes

By T. J. Mitchison, K. E. Sawin, J. A. Theriot, K. Gee, and A. Mallavarapu

Introduction

Fluorescence microscopy is a powerful tool for observing the distribution of specific molecules in living cells. In many cases, however, it is not possible to infer the dynamic behavior of populations of molecules when they are uniformly labeled with a fluorescent tag. Instead it is necessary to create a local region within the cell where the fluorescence intensity is different from the bulk population and to follow the evolution of this region with time. The classic method for creating a differentially labeled region is fluorescence photobleaching. This method has been used extensively to follow diffusion and directed movements of macromolecule in cells. Photobleaching suffers two main problems: (1) the chemistry of bleaching depends on generation of activated oxygen species that can cause local damage to proteins and membranes, and (2) it can be difficult to accurately track a region of reduced fluorescence within a pool of higher fluorescence under conditions of limited signal.

These considerations led us to develop fluorescence photoactivation as an alternative technology for probing cytoskeleton dynamics. In this approach the protein is tagged with a caged fluorochrome. This probe molecule is nonfluorescent (or fluorescent at a different wavelength) until illuminated with a brief pulse of ultraviolet light. Such illumination leads to photolysis of the caging groups and generation of a fluorescent species. In principle, fluorescence photoactivation can avoid the problem of generation of local oxidative damage inherent to photobleaching and can also produce a more favorable signal-to-noise ratio for imaging. Photoactivation can also produce toxic by-products in the form of the nitrosoaldehyde or nitrosoketone side products from photolysis. However, each molecule of caged fluorochrome releases only one or two molecules of side product, whereas each molecule of fluorochrome in a bleaching experiment probably produces hundreds of activated oxygen molecules before being bleached by reacting with one of them. The main limitation of the photoactivation method has been the development of suitable caged fluorescent probes, and most of this article focuses on our progress in this area.

Requirements for an effective caged fluorescent probe have been discussed.[1,2] Most important are biostability, rapid and efficient photoactivation, good brightness and photostability of the uncaged fluorochrome, practicable synthesis, and friendly protein chemistry. The latter is most difficult to predict from molecular structure—empirically it requires probes that are neither too hydrophobic nor too highly charged, with convenient chemistry for covalent attachment to proteins.

To date, caged fluorochromes have all employed variants of the 2-nitrobenzyl caging group. Two properties have been used to control fluorescence with such groups. In caged fluorescence and rhodamines (Rds), two caging groups act together to pin the xanthene fluorophore in a nonfluorescent lactone tautomer form (Figs. 1 and 3). In caged resorufin (Fig. 2) the fluorophore is held in a nonionizable form, resulting in a blue shift of adsorption and quenching of fluorescence.

The first caged fluorescent probe that was used for a biological experiment was a fluorescein derivative, C2CF-sulfo-N-hydroxysuccinimide (SNHS) (Fig. 1). This probe attached to tubulin led to the discovery of poleward flux in mitotic spindles.[1,3] C2CF is highly hydrophobic, and most proteins other than tubulin tend to aggregate if they are labeled with it. In C2CF-SNHS itself, the sulfosuccinimide group keeps the reagent water

[1] T. J. Mitchison, *J. Cell Biol.* **109**, 637 (1989).
[2] T. J. Mitchison, K. S. Sawin, and J. A. Theriot, in "Handbook of Cell Biology," Vol. 2 (J. E. Celis, ed.), pp. 65–76. Academic Press, New York, 1994.
[3] K. E. Swain and T. J. Mitchison, *J. Cell Biol.* **112**, 941 (1991).

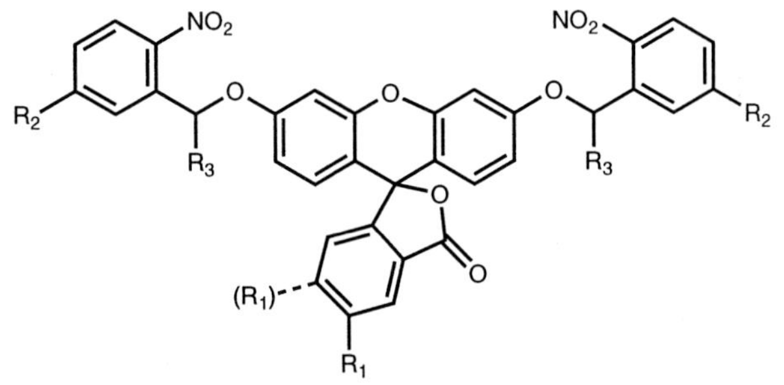

CAGED FLUORESCENT PROBES

FIG. 1. Caged fluoresceins.

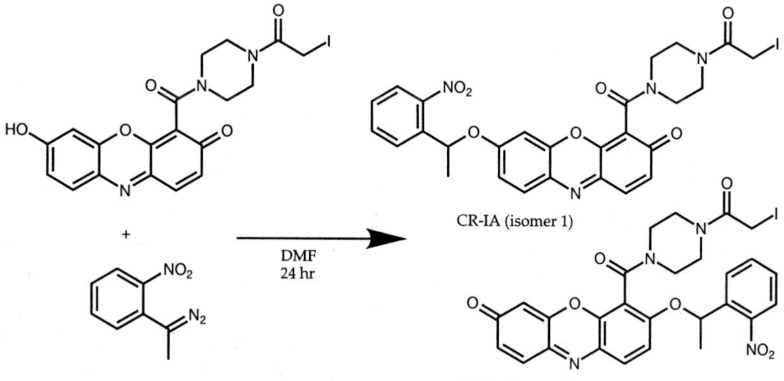

R₁	R₂	R₃	Name
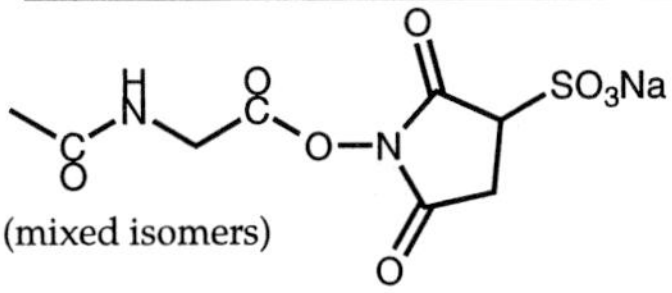 (mixed isomers) — succinimidyl with SO₃Na	H	H	C2CF-SNHS
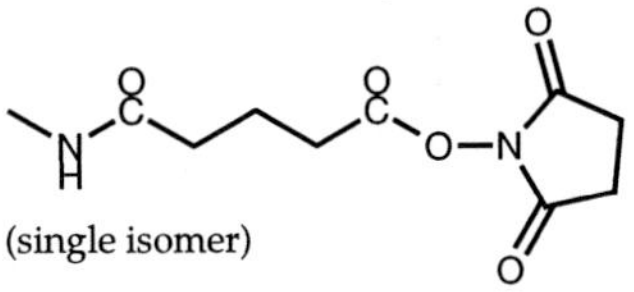 (single isomer) — NHS ester	-O-CH₂-COOH	H	CMNB2AF-NHS
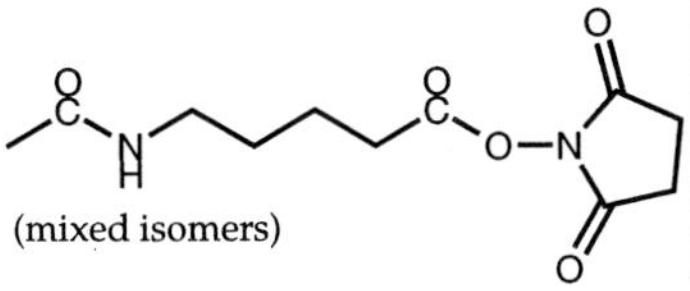 (mixed isomers) — NHS ester	H	-COOH	α-Carboxy-C2CF-NHS

FIG. 2. Structure and synthesis of caged resorufin iodoacetate.

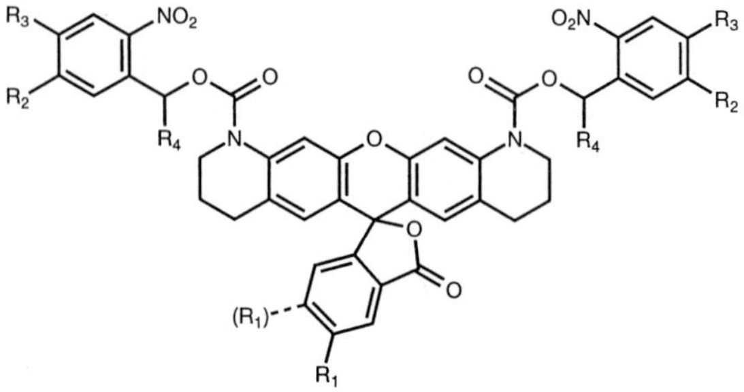

R_1		R_2	R_3	R_4	Name
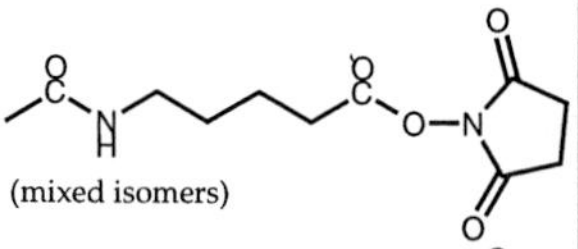 (mixed isomers)		-O-CH$_2$-COOH	-H	-H	CMNB2QRd-NHS
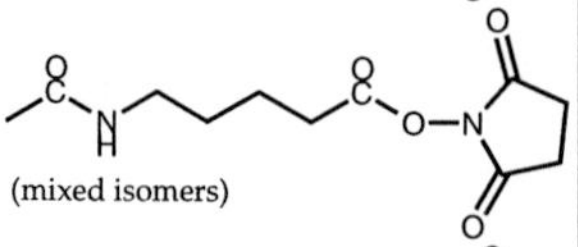 (mixed isomers)		-H	-H	-COOH	α-Carboxy-C2CQRd-NHS
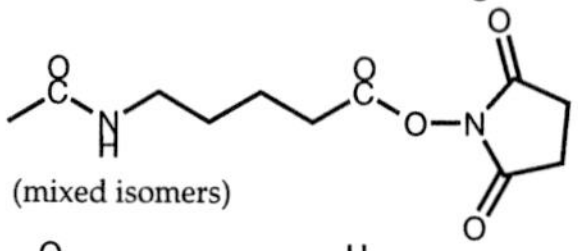 (mixed isomers)		-OCH$_3$	-OCH$_3$	-COOH	α-Carboxy-DM-C2CQRd-NHS
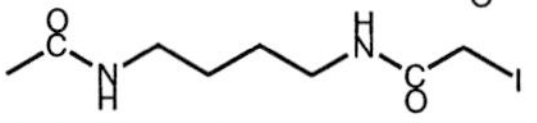 (mixed isomers)		-OCH$_3$	-OCH$_3$	-COOH	α-Carboxy-DM-C2CQRd-IA

FIG. 3. Caged Q-rhodamines.

soluble. As this group hydrolyzes during a labeling reaction, aggregates and precipitates of the reagent usually develop. These problems led us to develop the more water-soluble caged fluorescein CMNB2-AF-N-hydroxysuccinimide (NHS) (Fig. 1), whose applicability is more general. The search for a caged fluorescent probe for actin labeling led to caged resorufin (Fig. 2, see Theriot and Mitchison[4]). Both fluorescein and resorufin suffer rapid photobleaching after activation, which severely limits our ability to image them in cells. This photostability consideration led us to develop the caged rhodamines (Fig. 3). α-Carboxydimethoxy-C2CQRd-NHS and -IA are probably the best caged fluorochromes we currently have in terms of the parameters listed earlier. A practical drawback has been difficulty of synthesis, and in general the area of caged fluorochrome design has been limited

[4] J. A. Theriot and T. J. Mitchison, *Nature* **352**, 126 (1991).

TABLE I
PRACTICAL CONSIDERATIONS FOR DIFFERENT FLUOROPHORES

Caged derivatives of	Excitation/emission maxima (nm) (uncaged)	Photostability (uncaged)	Biostability (caged)	Quantum efficiency of uncaging	Ease of synthesis
Fluorescein	485–495/495–520	Poor	Good	Good	Medium
Q-Rhodamine	540–550/560–580	Good	Good	OK–poor	Harder
Resorufin	580–590/590–610	Poor	Poor	Good	Easy

by the chemistry. The advantages and drawbacks of the different caged fluorochromes are summarized in Tables I and II. Data we have summarized are largely anecdotal, although they coincide with the experience of other laboratories and also theoretical expectations for the different chemistries.

TABLE II
PRACTICAL CONSIDERATIONS FOR DIFFERENT CAGING GROUPS[a]

Caging group	Water solubility	Ease of synthesis	Notes
Nitrobenzyl (as in C2CF-SNHS) [NB]	Poor	Easy	Simplest chemistry
Nitrophenethyl (as in caged resorufin) [NPE]	Poor	Medium	Faster uncaging than nitrobenzyl. By-product less toxic than nitrobenzyl
2-Nitro-5-carboxymethoxynitrobenzyl (as in CMNB2AF) [CMNB]	Good	Medium	Carboxymethoxy group imparts water solubility. Photochemistry similar to nitrobenzyl
α-Carboxynitrobenzyl (as in α-carboxy-C2QRd-NHS) [CNB]	Good	Harder	α-Carboxy imparts water solubility and less toxic by-product. Also increases efficiency of photolysis; promotes uncaging at visible wavelengths
α-Carboxy-4,5-dimethoxynitrobenzyl (as in α-carboxydimethoxy-C2QRd-NHS) [CDMNB]	Good	Harder	As simple α-carboxy. Dimethoxy groups improve efficiency of uncaging at 360 nm

[a] Efficiency of photolysis depends on both the caging group itself and the chemistry that attaches it to the fluorochrome. The phenolic ether linkage in caged fluorescein and caged resorufin is readily cleaved by light with all caging groups. The carbamate linkage in caged rhodamines is more resistant to cleavage. α-Carboxy caging groups improve the efficiency of carbamate photolysis. By-product toxicity refers to the nitrosoaldehyde or ketone formed from the caging group after photolysis. Uncaging at visible wavelengths is an undesirable special property of α-carboxy caging groups whose origin is unclear. For more information, see Haugland.[5] The trivial names for caging groups shown in square brackets indicate the nomenclature used in that reference.

This information is provided in the spirit of helping to guide others in designing biological experiments and hopefully designing new chemistry. For additional information, see Haugland.[5]

Probe Availability and Synthesis

Progress on application of fluorescence photoactivation to biological problems has largely hinged on probe availability and properties. Commercial availability depends on demand and ease of synthesis. Several caged fluorescent probes are currently available from Molecular Probes (Eugene, OR).[5] Where these do not correspond exactly to the probes we have worked with, the considerations listed in Tables I and II may be useful for predicting properties.

Caged fluoresceins (Fig. 1) are fairly easy to synthesize, starting from carboxyfluorescein (usually as a mixture of isomers) or aminofluorescein (available as single isomers). Our synthesis of C2CF was reported in Mitchison[1] and of CMNB2AF in Mitchison et al.[2] The key synthetic step, formation of the bisphenolic ether, used conditions developed by Krafft et al.[6] The main drawback of caged fluoresceins is the lack of photostability of fluorescein itself. Introducing fluorine atoms increases the photostability of fluorescein and decreases its pK_a.[5] Thus caged derivatives of 2',7'-difluorocarboxyfluorescein should in principle make superior probes.

Caged resorufin (Fig. 2) is the easiest probe to synthesize and is also the probe with lowest molecular weight, which should improve its protein chemistry. It is made simply by applying the Trentham diazo caging chemistry described in Walker et al.[7] to commercial resorufin iodoacetate (Boehringer, Mannheim, Germany) and using thin-layer chromatography (TLC) to purify the product.[4] The two isomers can be readily separated, but they seem to have similar properties. If necessary, the 4-carboxyresorufin nucleus can be synthesized by condensing 4-nitrosoresorcinol and 2,6-dihydroxybenzoic acid in the presence of manganese dioxide and reducing the resulting 4-carboxyresasurin with zinc. Caged resorufin has two drawbacks: (1) like fluorescein, resorufin is readily photobleached; because there is a strong Hg line (585 nm) centered on the resorufin adsorption maximum (unlike fluorescein), photobleaching may appear to be very fast indeed; and (2) because the heterocyclic nitrogen atom makes the resorufin ring

---

[5] R. P. Haugland, in "Handbook of Fluorescent Probes and Research Chemicals." Molecular Probes, Eugene, OR, 1996. Pages 447–455, for caged compounds; pp. 22–25 for fluorescein substitutes.

[6] G. A. Krafft, W. R. Sutton, and J. P. Cummings, J. Am. Chem. Soc. **110**, 301 (1988).

[7] J. W. Walker, G. P. Reid, and D. R. Trentham, Methods Enzymol. **172**, 288 (1989).

system electron deficient, it tends to suffer nucleophilic attack and reduction more readily than the other fluorochromes. This tendency is exacerbated in the caged form, where anion formation is blocked (see, e.g., Afanas'eva et al.[8] Under physiological conditions (pH 7.4, 1–5 mM thiol groups) we have found that caged resorufin can undergo both nucleophilic attack and reduction, depending on the thiol used. We have also found that the half-life of caged resorufin in cells is quite short, as little as 20 min when attached to dextran. Attachment to actin prolongs the half-life of the probe, but these chemical considerations make it less attractive for general use. In principle, appropriate substitution of the resorufin ring might ameliorate these problems, but this has not been explored.

Caged rhodamines are the best of the current probes for most applications due to the relative resistance of rhodamines to photobleaching. The synthesis, shown in Fig. 4, is more difficult, and we encourage others to try and improve on our methods. The key synthetic step is the acylation reaction (F in Fig. 4). We used a strong base to deprotonate the amino groups and then added a strong acylating agent (a chloroformate) in a rather uncontrolled reaction with variable and poor yields. The identity of the desired product after purification was confirmed by mass spectrometry. This method was based on Q-rhodamine alkylation conditions described in Arnost et al.[9] All our caged rhodamines are based on the rhodamine derived from 7-hydroxyquinoline, to which we have given the trivial name Q-rhodamines. Q-Rhodamines have very similar optical properties to the popular tetramethyl-rhodamines, including an adsorption maximum near the strong Hg line at 546 nm, high absorption coefficients (>80,000), high quantum efficiencies of fluorescence, and good photostability. Caged rhodamines based on rhodamine 110 (no substitution on the nitrogens) are also useful for making caged derivatives, and the acylation chemistry is somewhat easier. These probes have adsorption maxima near fluorescein (495–515 nm) and are called Rhodamine Green by Molecular Probes.[5] A milder acylation strategy using Steglich's reagent (4,6-diphenylthieno[3,4-d]-1,3,dioxol-2-one 5,5-dioxide) that works well with rhodamine 110 derivatives has been developed (Ottl and Mariott, submitted).

Because caged rhodamines are carbamates rather than phenolic ethers, their quantum efficiency of uncaging is inherently lower than caged fluoresceins. This translates into needing longer pulses of UV light to uncage, with the possibility of damaging the irradiated region of the cell. We have found in practice that only caged rhodamines with α-carboxy caging groups

---

[8] G. B. Afanas'eva, T. S. Viktorova, K. I. Pashkevich, and I. Y. Postovskii, *Chem. Heterocyclic Compounds* **10**, 302 (1974).

[9] M. J. Arnost *et al.*, Polaroid Corp., MA. US Patent number 4,900,686, 1990.

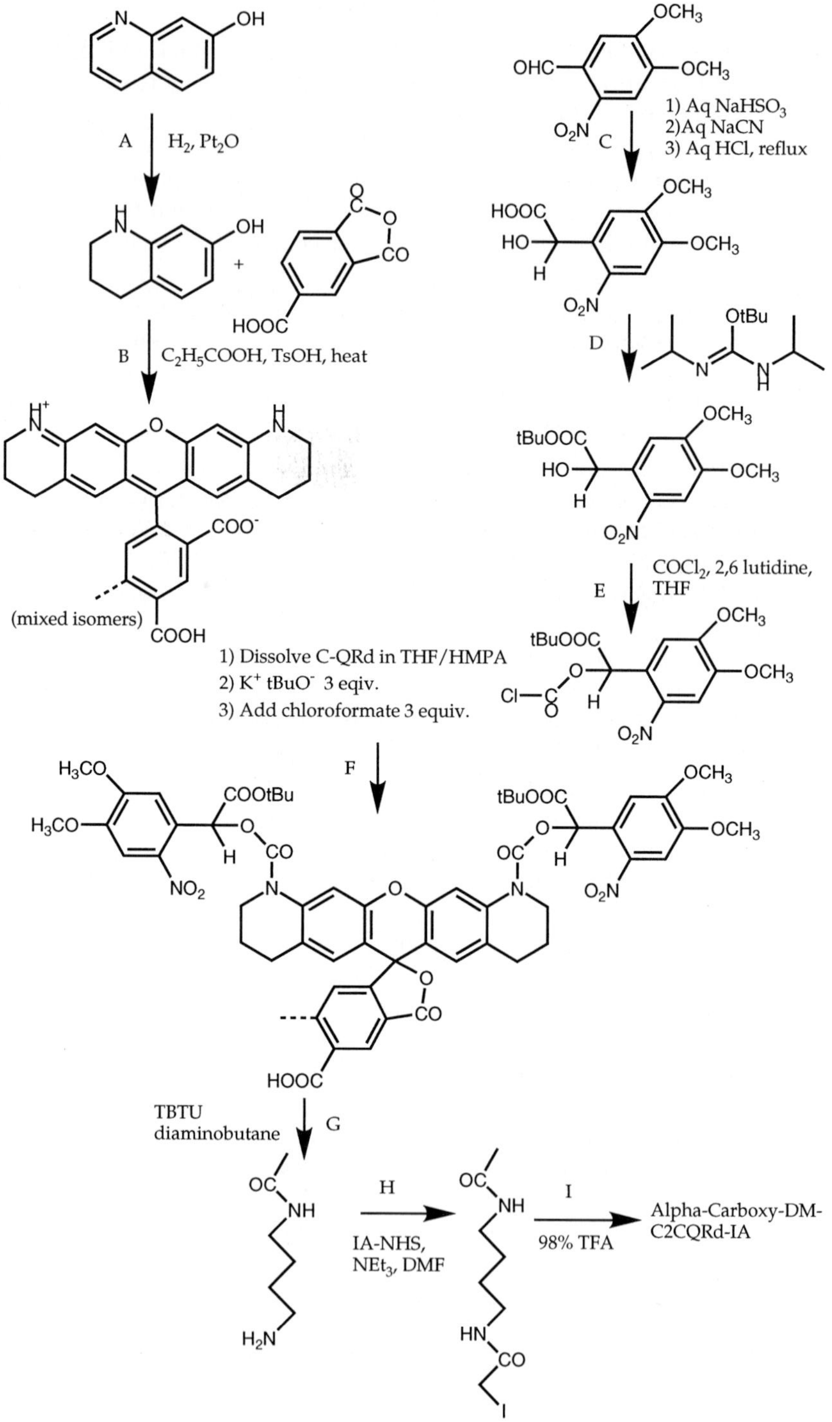

A
H_2, Pt_2O
B
C_2H_5COOH, TsOH, heat
(mixed isomers)
C
1) Aq $NaHSO_3$
2)Aq NaCN
3) Aq HCl, reflux
D
E
$COCl_2$, 2,6 lutidine, THF
1) Dissolve C-QRd in THF/HMPA
2) K^+ $tBuO^-$ 3 eqiv.
3) Add chloroformate 3 equiv.
F
TBTU
diaminobutane
G
H
IA-NHS,
NEt_3, DMF
I
98% TFA
Alpha-Carboxy-DM-C2CQRd-IA

uncaged fast enough to be used as nonperturbing probes of the actin cytoskeleton. The α-carboxy group is known to facilitate the uncaging reaction in other caged compounds.[5] The α-carboxy caging groups also promote water solubility and better protein chemistry, compensating for their more difficult synthesis. The α-carboxy caging groups have one undesirable fea-

---

FIG. 4. Caged Q-rhodamine synthesis. Yields were in the 50–95% range except where indicated. Identification of compounds was based on optical spectroscopy and chemical properties, except for step F, where mass spectrometry was also used to characterize the product. (A) 0.2 M 7-hydroxyquinoline (Eastman) in ethanol was treated with 5 mol% Pt_2O, 2 atm H_2, 24 hr. Solvent evaporated, no purification needed. (B) Equimolar amine and anhydride plus 0.2 equivalents of p-toluenesulfonic acid dissolved in minimal volume of propionic acid. Refluxed 24 hr. Extra propionic acid is added and distilled off twice to remove water. Carboxy-Q-rhodamine purified by two rounds of dissolving in hot aqueous 1.0 M HCl, filtering, cooling, and collecting the precipitate, followed by chromatography on Sephadex LH20 with isocratic elution in methanol. (C) Conditions described in Org. Synth. Coll. **1**, pp. 336–339. Commercial 6-nitro-veratraldehyde contains impurities that do not dissolve in aqueous bisulfite and can be removed by filtration. Product is recrystallized from CH_2Cl_2/hexane. (D) 2-Nitro-4,5-dimethoxynitromandelic acid dissolved in tetrahydrofuran (THF). Two equivalents of the isourea[16] are added. After 2 hr, product is purified by chromatography (SiO_2, 70% hexane, 30% ethyl acetate). (E) Ester is dissolved in THF; 1.5 equivalents of base is added, followed by 1.2 equivalents of phosgene (as a toluene solution). After 10 min the solution is filtered and evaporated. Crude chloroformate is held under vacuum for several hours and used without further purification. (F) Rhodamine is dissolved at 0.1 M in 50% THF, 50% HMPA. K^+-t-BuO$^-$ (1 M in THF) is added with mixing until the solution becomes blue–black (approximately 3 to 4 equivalents). The chloroformate (0.1 M in THF, 3 equivalents) is then added all at once with mixing. The reaction generates many products. The correct one is identified on TLC (SiO_2, 84% benzene, 14% acetone, 2% acetic acid, R_f = 0.3) by its conversion from colorless to red on irradiation with 360-nm light. Purify by chromatography (SiO_2, 94–90% dichloromethane, 4–8% ethyl acetate, 2% acetic acid). Product is further purified by taking up in a minimal volume of hot methanol, cooling to $-20°$ overnight, and collecting the precipitate. Identity of the product is confirmed by mass spectrometry (Isims, strong molecular ion at 1130.5) and visible spectroscopy before and after activation. Yield: Poor, variable. (G). Caged rhodamine in DMF is treated with benzotriazol-1-yltetramethyluronium tetrafluoroborate (TBTU) 2 equivalents, then diaminobutane, 3 equivalents. Crude product is isolated by dissolving in ethyl acetate and washing 3× with water. (H) Product from G in DMF is treated with iodoacetic acid N-hydroxysuccinimide ester, 2 equivalents, and triethylamine, 2 equivalents in DMF. Purify by chromatography (SiO_2, 18% dichloromethane, 80% ethyl acetate, 2% acetic acid). (I) Product of H is deprotected in 98% trifluoroacetic acid (TFA), 2% water, 2 hr. Acid is removed *in vacuo*. Product is dissolved in THF and precipitated with 9 volumes of cold water. For convenience, aliquots of approximately 1 μmol are precipitated in individual microfuge tubes. The product is collected by centrifugation, dried *in vacuo*, and stored in the dark at $-80°$. For protein labeling, an aliquot was resuspended in DMF. The NHS derivatives (Fig. 3) are synthesized by treating the protected, caged-QRd-COOH with NHS and ethyldimethylaminopropylcarbodiimide (EDC) in DMF to give the NHS ester, followed by ε-aminocaproic acid in DMSO/water/triethylamine to give the aminocaproyl derivative, followed by NHS and EDC in DMF to give the final product.

ture we do not understand. They uncage fairly fast in visible light, for example, at 450–520 nm for α-carboxy-4,5 dimethoxy-C2QRd-NHS. α-Carboxy-caged fluorescein (Fig. 1) undergoes similar visible uncaging. This property is peculiar because neither the caging groups nor the lactone tautomer of the fluorochrome should absorb light at these wavelengths. We hypothesize that α-carboxy-caged fluorochromes may contain a significant proportion of the nonlactone tautomer of the fluorochrome in their ground state structure that can absorb light and somehow uncage. This uncaging in visible light has the potential to cause slow activation of the whole field during the observation phase of a photoactivation experiment. In practice, this problem can be largely circumvented by exciting the uncaged fluorochrome with light tightly centered on the adsorption maximum of the uncaged form. For Q-Rds this means isolating the 546-nm Hg line. The level of this light must be controlled anyway to prevent bleaching, and under these conditions further uncaging by the observation light is minimized.

Comparing α-carboxynitrobenzyl caging groups with and without 4,5-dimethoxy substitution (Fig. 3, Table II), we have found that these extra groups significantly increase the rate of photoactivation of caged Q-Rd with 366-nm light. This is presumably due to increased adsorption of photons at 366 nm with the methoxy substituents. The effect is less than the α-carboxy group, but still useful.

Protein Labeling

Most of our experience has been with two proteins, tubulin and actin. Tubulin must be labeled in the polymerized state to protect cysteines and lysines essential for polymerization. We have been successful using only NHS and SNHS esters, which presumably modify lysines. Using polymerized tubulin at pH 8.0–8.6 in glycerol containing buffers and a large excess of reagent, stoichiometries of up to 0.5 caged fluorochrome/tubulin dimer are obtained. These labeled tubulins are somewhat compromised in their polymerization ability, but appear to copolymerize readily with endogenous tubulin. C2CF-SNHS has been used for most work with tubulin, although the other probes work as well. The basic protocol we use to label tubulin on lysines with NHS-ester probes is described in Hyman et al.[10] An updated version of this procedure is described in Desai and Mitchison (*Methods Enzymol.,* in press).

Rabbit muscle actin contains a single reactive cysteine (Cys-374), which has been used as a site of attachment for a variety of fluorescent dyes. We

---

[10] A. A. Hyman, D. Drexel, D. Kellog, S. Salser, K. Sawin, P. Steffen, L. Wordeman, and T. J. Mitchison, *Methods Enzymol.* **196**, 478 (1991).

reacted actin monomers with α-carboxydimethoxy-C2CQRd-IA using a protocol that minimizes the amount of time required for the preparation. We found that this improved yield and minimized hydrolytic activation of the caged rhodamine.

For a typical labeling reaction, 0.5 ml actin (5 mg/ml) is used. Care is taken to minimize oxidation of the cysteine prior to labeling. For this reason, actin monomer, purified according to the method of Pardee and Spudich,[11] is quick frozen and stored in G-buffer (5 mM Tris–Cl, 0.2 mM CaCl$_2$, 0.2 mM ATP, pH 8.0) + 1 mM dithiothreitol (DTT). To remove excess thiol prior to labeling, the actin is thawed and dialyzed against G-buffer + 10 μM DTT for 2 hr at 4°. Because α-carboxydimethoxy-C2CQRd-IA is sparingly soluble in water, we add it at relatively low concentration [20 mM in dimethylformamide (DMF)] at a 5:1 molar ratio to actin. Addition and mixing are performed with rapid vortexing. As with the synthesis of caged compounds, these steps must be performed in safelight. All tubes should be wrapped in an opaque material to occlude light. The solution is then brought to room temperature, and the reaction is allowed to proceed for 30 min to 1 hr. Following this, the sample is put on ice. Free unreacted dye is separated from protein by gel filtration [Pharmacia (Piscataway, NJ) 5 ml Dextran desalting columns] on a column preequilibrated with cold G-buffer + 1 mM DTT. The protein peak is identified by spotting 1 μl of each fraction on paper (nitrocellulose paper or TLC plate) and irradiating the spots for 2–3 min with long-range (360-nm) UV light. Hand-held UV lamps (e.g., ENF-260C, Spectronics Corp., Westbury, NY) work well for this. The fractions (0.5–1 ml) corresponding to the first red/pink peak are pooled. Actin is then polymerized by adjusting the buffer to 75 mM KCl, 50 mM Tris, pH 8.0, 1 mM MgCl$_2$, and 1 mM ATP. The solution is left on ice for 1 hr and is then brought to room temperature for 1 hr. Filaments are pelleted (50,000 rpm, Beckman TLA 100.3 rotor, 1 hr, 4°) and washed once with G-buffer + 1 mM DTT. The pellet should be glassy and translucent. Filaments are resuspended by adding 100 μl G-buffer + 1 mM DTT to the pellet. After a few minutes incubation at 0° to allow swelling, the pellet is broken up and then sonicated briefly. The supernatant is collected and the tube is reextracted with another 100 μl G-buffer + 1 mM DTT to collect any remaining protein. Filaments are depolymerized by dialysis against G-buffer + 1 mM DTT for 24 hr at 4°. For the final 2–3 hr of dialysis, the dialysis solution is replaced by G-Buffer + 0.5 mM glutathione. Aggregates are removed by centrifugation (44,000 Krpm, Beckman TLA100 rotor). The supernatant is collected and quick frozen and stored in liquid N$_2$, typically as 3-μl aliquots.

---

[11] J. D. Pardee and J. A. Spudich, *Methods Cell Biol.* **24**, 271 (1982).

Using actin labeled with α-carboxydimethoxy-C2CQRD-IA by this procedure, and imaging with a cooled CCD camera, we obtain a substantial improvement in signal and signal/noise compared to our earlier caged resorufin work. An example using this probe to analyze the dynamics of actin filaments in *Listeria* tails in a *Xenopus* egg extract is shown in Fig. 5.

Instrumentation

Fluorescence photoactivation imaging at the level of single cells or embryos requires essentially the same apparatus as conventional low-light fluorescence imaging, except that one more light beam must be brought to bear on the specimen for photoactivation. For collection of fluorescence images at low light levels we have used both intensifying cameras and a cooled CCD. In most experiments we collect pairs of images, one showing the uncaged fluorescence signal and the second a phase-contrast image of the cell as a reference image. In some experiments a second fluorescence channel is also collected.[3] One advantage of the cooled CCD camera is its ability to collect high-resolution images at both high and low light levels due to its large dynamic range (e.g., Fig. 5).

To generate the photoactivation beam we use the 366-nm line from a 100-W mercury arc lamp. Others have used the 334–364 lines from an

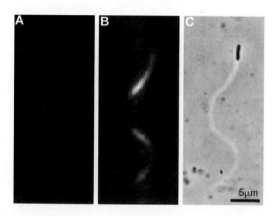

FIG. 5. *Listeria* tail in *Xenopus* extract marked by photoactivating caged rhodamine actin. Rabbit muscle actin was conjugated with α-carboxy-DM-C2QRd-IA as described in the text. This probe was mixed with *Xenopus* egg extract and killed *Listeria* as described in Theriot et al.[17] (A) Image in the rhodamine channel before photoactivation, and (B) an image a few seconds after activation with a bar of light oriented vertical in the image. (C) Corresponding phase-contrast image. Images were collected with a cooled CCD camera. Note the improvement in image resolution compared to a similar experiment reported in Theriot et al.[18] due to the improved probe and use of a CCD camera in place of an ISIT.

argon ion laser.[12] In principle, small, intense areas of illumination can be generated from either source. Laser sources may be easier to focus and their intensity is unlimited. Mercury arc lamps are less expensive, but they have a maximal theoretical intensity at the specimen plane, related to the brightness and size of the arc.

If a mercury source is used to generate the photoactivation beam, a suitable slit or pinhole must be placed in a plane conjugate with the specimen to delimit the illuminated area in the specimen. The simplest photoactivation apparatus simply uses the field diaphragm in the epifluorescence light path as this pinhole. This diaphragm is closed to an appropriate size, a UV excitation filter set is put in the light path (e.g., a DAPI set), and the epifluorescence beam is turned on briefly. Then the diaphragm is opened, a fluorescein or rhodamine filter set is inserted, and observation of the activated zone commences. This setup has been used successfully to follow microtubule flux in newt spindles, with the addition of a slit that can be slid temporarily in place of the field diaphragm during photoactivation.[13]

In general, we prefer an apparatus where the photoactivation and observation beams can be controlled independently by electronic shutters, facilitating reliable positioning of the slit/pinhole and rapid collection of the first images after photoactivation. The two light paths we use are shown in Fig. 6. Either can use a laser in place of the mercury lamp for activation. The choice of light path is a practical consideration, governed by how easy it is to physically access the field diaphragm plane in the epi-illumination tube for a given microscope.

Future Directions

Despite its potential for visualizing dynamic processes in living cells and embryos, fluorescence photoactivation to date has been used in only a handful of laboratories for a limited range of proteins. The main limitation has been probe chemistry and probe availability, as discussed earlier. Expanding the application of the method will hinge primarily on improving the availability of the current probes and developing better ones.

The most obvious avenue for probe improvement is to come up with better conventional caged fluorochromes. Rhodamines are the brightest and most photostable of the current probes. It should be possible to improve on the synthesis methods and perhaps find caging groups that photoactivate as efficiently as the α-carboxy series in UV light but do not photoactivate in visible light. Other fluorescent molecules and other types of caging

---

[12] V. I. Rodionov, S. S. Lim, V. I. Gelfand, and G. G. Borisy, *J. Cell Biol.* **126,** 1455 (1994).
[13] T. J. Mitchison and E. D. Salmon, *J. Cell Biol.* **119,** 569 (1992).

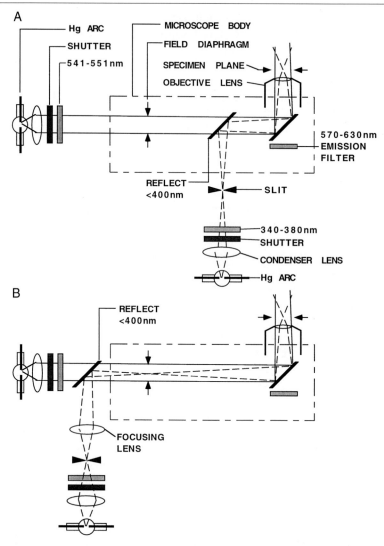

Fig. 6. Schematic diagram of photoactivation light paths. The photoactivation beam is shown by dashed lines, and the fluorescence excitation beam by solid lines. The wavelength values on the excitation and emission filters are appropriate for Q-rhodamine. The schematic depicts an inverted microscope, but the same light paths can be generated on an upright stand. For clarity, components in common to the two light paths are labeled only in (a) and the fluorescence emission light path (which travels through the emission filter) is omitted. (a) In this setup the photoactivation beam is brought into the epifluorescence light path on the objective lens side of the field diaphragm using a dichroic mirror inserted into the epifluorescence light path. The slit is positioned in a plane conjugate to the field diaphragm and the

chemistry should also be considered. It might be possible to make a caged rhodamine by quaternizing one of the nitrogens as opposed to making a carbamate, and such a probe would have higher water solubility. Rhodols, molecules intermediate in structure between fluorescein and rhodamines,[5] are interesting candidates for caging. In principle, a rhodol caged with a single nitrobenzyl ether on the oxygen might be a useful probe with lower molecular weight than the current caged fluoresceins and rhodamines. Cyanine dyes have fluorescence properties superior even to rhodamines as protein-attached probes. It is not obvious how to cage them using nitrobenzyl chemistry, but some other strategy might be feasible. In general, caged fluorochromes are areas where creative chemistry could generate significant payoffs.

In conventional fluorescent imaging, green fluorescent protein (GFP) fusion proteins have risen to prominence because the fluorescent tag can be encoded in the cDNA for the protein. Particularly in usefully mutated forms,[14] GFP appears to be an excellent fluorochrome, with good adsorption coefficients, quantum efficiency, and photostability. Thus it is natural to ask if GFP can be used in any kind of photoactivation experiment. Wild-type GFP is in fact already a caged fluorochrome of sorts. The protein can adopt two conformations with different adsorption maxima, and light adsorption can shift the population of molecules toward the longer wavelength absorbing form.[14] By mutating GFP and selecting appropriate variants, it might be possible to turn it into a very good caged fluorochrome that could replace the small organic probes. An alternative, and perhaps more likely, possibility is that GFP will bring new vigor to the older approach of photobleaching. Since the organic fluorophore is buried within the core of GFP, it is possible that it is less efficient at generating activated oxygen species than conventional organic fluorochromes, and also that when activated oxygens are produced they will react preferentially with GFP itself, rather than nearby proteins. In other words, photobleaching of GFP may be less prone to artifacts due to collateral damage to nearby proteins than is photobleaching with conventional probes. The first GFP

[14] A. B. Cubitt, R. Heim, S. R. Adams, A. E. Boyd, L. A. Gross, and R. Y. Tsien, *Trends Biochem. Sci.* **20**, 448 (1995).

---

specimen, which means it is physically close to the microscope body. This light path has been set up on Zeiss IM35, Zeiss Universal, and Olympus inverted stands. (b) In this setup the photoactivation beam is brought in with a dichroic mirror mounted between the epifluorescence light source and the back of the microscope. It requires no modification of the microscope body, but necessitates an additional lens to focus the image of the slit onto the field diaphragm and specimen plane. This light path has been set up on a Zeiss Axiovert stand.

bleaching experiments are quite encouraging in this light.[15] However, it will be important to test more rigorously the extent of collateral damage in GFP bleaching experiments before adopting this technology more generally.

[15] N. B. Cole, C. L. Smith, N. Sciaky, M. Terasaki, M. Edidin, and J. Lippincott-Schwartz, *Science* **273**, 797 (1996).
[16] L. J. Mathias, *Synthesis* **1979**, 561 (1979).
[17] J. A. Theriot, J. Rosenblatt, D. A. Portnoy, P. J. Goldschmidt-Clermont, and T. J. Mitchison, *Cell* **76**, 505 (1994).
[18] J. A. Theriot, T. J. Mitchison, L. G. Tilney, and D. A. Portnoy, *Nature* **357**, 257 (1992).

[5] Biologically Active Peptides Caged on Tyrosine

By R. Sreekumar, Mitsuo Ikebe, Fredric S. Fay,*
and Jeffery W. Walker

Introduction

Synthetic peptides show considerable promise as selective inhibitors of protein–protein interactions involved in many basic biological processes, including signaling, secretion, and motility.[1] Light-activated caged peptides would be powerful tools[2] for controlling the release of peptides into cells in order to overcome diffusional delays, circumvent microinjection artifacts, and target selected proteins in a time domain that is inaccessible by antisense or gene knockout techniques. In addition, caged peptides would provide a means to spatially map out selected protein activities within living cells.

Peptide probes are usually found by identifying domains involved in critical protein–protein interactions, then narrowing the activity of the domain to a fragment of approximately 5–30 residues by mutagenesis or by screening overlapping peptide sequences. Alternatively, useful peptides are identified by a combinatorial approach involving selection of active peptide sequences out of random libraries.[3] The search process often depends on having a reliable *in vitro* assay of peptide activity that can be

---

* This article is dedicated to the memory of collaborator and friend Fredric S. Fay (deceased), who was a constant source of energy, enthusiasm, and vision.
[1] D. C. Hancock, N. J. O'Reilly, and G. I. Evan, *Mol. Biotech.* **4**, 73 (1995).
[2] S. R. Adams and R. Y. Tsien, *Annu. Rev. Physiol.* **55**, 755 (1993).
[3] P. C. Andrews, D. M. Leonard, W. L. Cody, and T. K. Sawyer, *Methods in Mol. Biol.* **36**, 305 (1994).

used to screen large numbers of candidate peptides rapidly. Once a peptide is identified, development of a caged peptide requires structure–activity information. Various amino acid substitutions need to be evaluated in model peptides to identify positions in the sequence where certain features (e.g., charge, hydrophobicity) are disruptive for target recognition. Alternatively, N or C termini can be modified with tails that interfere with recognition, or a peptide can be conformationally caged by end-to-end or side chain cyclization to render it inactive. The modifying group(s) must be photosensitive so that the peptide can be returned to its original active state on illumination. Each caged peptide must be thoroughly characterized in terms of structure, photochemistry, and biological properties before it can be used with confidence to address specific questions in cellular systems. An ideal peptide probe should display a good selectivity and high (submicromolar) affinity for its target. In the caged form, the peptide should have little or no activity or a greatly reduced affinity and should photolyze efficiently and rapidly to its active form.

The purpose of this article is to describe some practical aspects in the development of caged peptides for use in cell biology. We show here that the relatively straightforward approach of derivatizing individual amino acid side chains on peptides with photolabile groups can lead to useful caged peptides. A general strategy is illustrated for preparing such caged peptides using the tyrosine side chain as an example. Tyrosine was chosen because it is ultraviolet (UV) active and fluorescent, properties that facilitate characterization, and it is a critical side chain in a number of proteins. Successful caging strategies will be described for two model peptides: RS-20, a 20 amino acid calmodulin inhibitory peptide, and LSM1, a 13 amino acid protein kinase autoinhibitory domain peptide (see Scheme 1 for peptide sequences).

Syntheses

Peptides typically have a variety of reactive functional groups, including α-amines, ε-amines, alcohols, carboxylic acids, and sulfhydryls. Selectively directing caging moieties to one site on an unprotected peptide is difficult and often results in mixtures of peptides modified at multiple sites. To avoid this problem, we first synthesize the caged amino acid, then incorporate it into the peptide sequence at the desired position during automated peptide synthesis. This has many advantages: (i) final peptide structures are well defined, (ii) tens of milligrams of peptides are readily prepared, and (iii) more flexibility is available for optimizing peptide properties because a wider variety of amino acid side chains and caging moieties can be consid-

A

RS-20:   NH₃⁺-ARRKW↓QKTGHAVRAIGRLSS-CO₂⁻

LSM-1:   NH₃⁺-LSKDRMKKY↓MARR-CO₂⁻

B

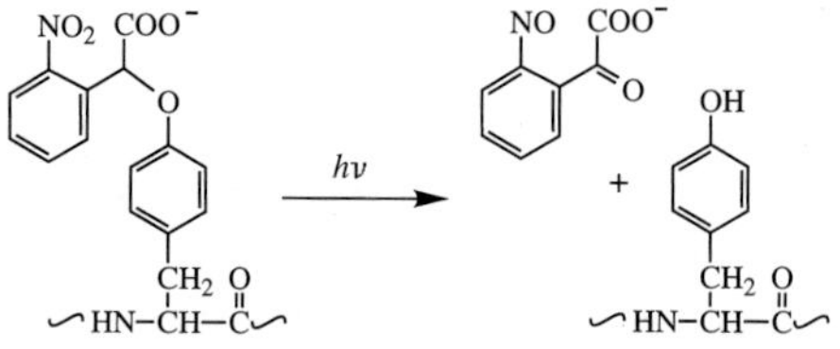

cgY →(hv)→ Y

SCHEME 1. Structures of peptides containing caged tyrosine. (A) Sequence of the calmodulin inhibitor, RS-20, and of the MLCK autoinhibitory peptide, LSM1. Arrows mark the location of caged tyrosines in caged forms of these peptides. (B) Structure and photocleavage reaction for caged tyrosine.

ered. The principal drawback is that chemical synthesis is limited to peptides of up to about 50 amino acid residues in length.

Synthesis of Caged Tyrosine

Step 1: Reactive Caging Moiety: 2-Bromo-α-carboxyl-2-nitrophenyl Methyl Ester (I). The synthetic route for caged tyrosine is illustrated in Scheme 2. The α-carboxyl caging moiety is found to be ideal for phenols,[4] so the preparation of this version of caged tyrosine (cgY[5]) is described in detail. Coupling is carried out with a bromomethyl ester of this moiety, which is prepared in two steps as follows.

To a stirred solution of 1 g 2'-nitrophenylacetic acid (5.5 mmol) in 10 ml benzene is added 3.5 ml anhydrous methanol followed by three drops of concentrated sulfuric acid. The reaction mixture is refluxed for 8 hr with an azeotropic distillation device. Thin-layer chromatography (TLC) in dichloromethane shows quantitative conversion of 2'-nitrophenylacetic

[4] J. W. Walker, H. Martin, F. Schmitt, and R. J. Barsotti, *Biochemistry* **32,** 1338 (1993).
[5] cgY, O-(α-Carboxyl-2-nitrobenzyl)tyrosine incorporated into a peptide (see Scheme 1 for structure). Amino acids are identified by standard single-letter code: A, Ala; C, Cys; D, Asp; E, Glu; F, Phe; G, Gly; H, His; I, Ile; K, Lys; M, Met; N, Asn; P, Pro; Q, Gln; R, Arg; S, Ser; T, Thr; V, Val; W, Trp; Y, Tyr.

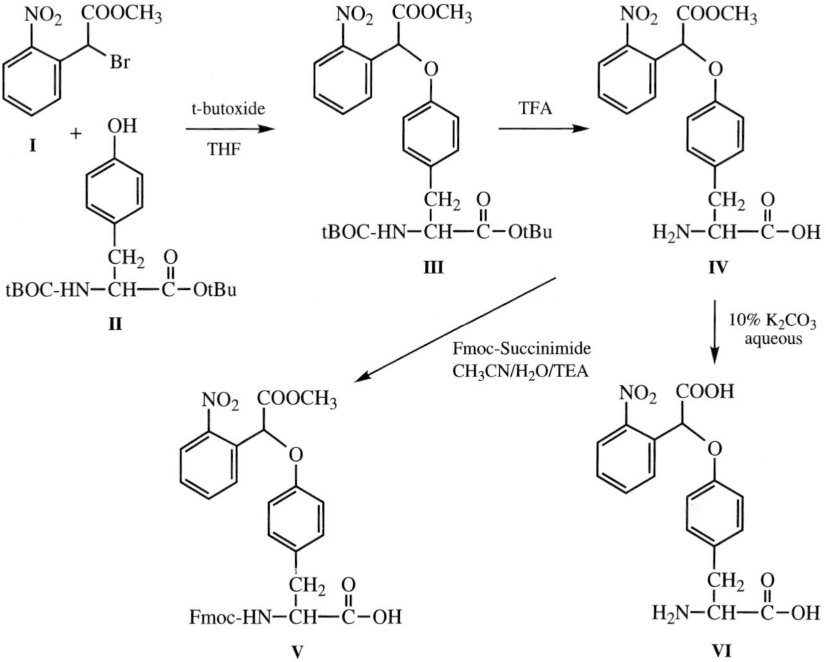

SCHEME 2. Synthetic route for caged tyrosine.

acid ($R_f = 0.01$) to the corresponding methyl ester ($R_f = 0.28$). Purification by silica gel flash chromatography in hexane/ethyl acetate yields 0.95 g of product as a white solid (95% yield). Of this product, 2′-nitrophenyl methyl acetate, 6.2 g (31.8 mmol), is then dissolved in 100 ml carbon tetrachloride in a 500-ml round-bottom flask. While stirring with a magnetic stir bar, 0.31 g benzoyl peroxide (1.28 mmol) is added, followed by 6.22 g N-bromosuccinimide (34.9 mmol). The flask is fitted with a reflux condenser and the reaction mixture is refluxed for 24 hr. Thin-layer chromatography in dichloromethane shows a new spot **I** ($R_f = 0.43$) and some starting material ($R_f = 0.28$). Extending the reaction time does not improve the extent of formation of **I,** but modest improvements are afforded by another 24 hr of reflux after filtering the reaction through Whatman (Clifton, NJ) #1 paper and recharging with fresh benzoyl peroxide and N-bromosuccinimide. In preparation for purification, the reaction mixture is filtered through Whatman #1 paper and rotary evaporated to near dryness. Approximately 3-ml aliquots of the crude mixture are chromatographed on a 2-in. (inside diameter) flash column containing 400 g silica gel in hexane. Compound **I** is eluted in ethyl acetate/hexane (1:18, v/v) with an overall yield of 3.08 g (50%).

Step 2: Protected Tyrosine: α-N-BOC-tyrosine tert-Butyl Ester ***(II)***. To selectively derivatize tyrosine on its phenolic side chain, the α-carboxyl and α-amino groups need to be protected. This is accomplished with a *tert*-butyl ester and a *tert*-butyloxocarbonyl (BOC) moiety, respectively. *tert*-Butyltyrosine is commercially available (Bachem, Torrance, CA) so it is typically used as starting material. The amino group is protected with the commercial reagent BOC-ON (Sigma, St. Louis, MO). This reaction proceeds in high yield, and the product, **II**, is easily purified by flash chromatography.

To 1.78 g tyrosine *tert*-butyl ester (7.5 mmol) dissolved in 30 ml dioxane/water (5:1, v/v) is added 1.11 ml triethylamine (TEA, 8 mmol) and then 2.0 g BOC-ON (8.2 mmol). The reaction mixture is stirred overnight at room temperature under a nitrogen atmosphere. The progress of the reaction is monitored by TLC on silica gel plates containing a fluorescent indicator that localizes UV active compounds as dark spots. Thin-layer chromatography in ethyl acetate/hexane (1:1, v/v) shows three spots: tyrosine-*tert*-butyl ester (R_f = 0.05), **II** (R_f = 0.5), and BOC-ON (R_f = 0.65). The reaction solvent is removed, and the crude product is purified in one batch by flash chromatography using a 2-in. (inside diameter) glass column containing 300 g silica gel in petroleum ether. The column is developed with 400 ml petroleum ether/ether (6:1, v/v), then 400 ml petroleum ether/ether (4:1, v/v), and finally **II** elutes from the column in 300 ml petroleum ether/ether (3:2, v/v) in a yield of 1.74 g (97%). Protected tyrosine is now ready for coupling to the caging moiety.

Step 3: Coupling. An equal molar amount of *tert*-butoxide is mixed with **II** to enhance the reactivity of the phenolic group. Then the reactive photolabile group is added in the form of a bromide, and the reaction proceeds to >70% yield. To a 100-ml round-bottom flask under a nitrogen atmosphere is added 2.5 g protected tyrosine, **II** (7.4 mmol) in 10 ml tetrahydrofurant (THF). Then 7.6 ml 1 *M* potassium *tert*-butoxide in THF is added dropwise while stirring with a magnetic stir bar. The reaction mixture immediately turns dark purple, then brown. Finally, 2.48 g 2-bromo-2'-nitrophenyl methyl acetate, **I** (9 mmol), is added and left to stir overnight at room temperature. The yellow–brown suspension is then centrifuged, and TLC of the supernatant solution in petroleum ether/ether (2:1, v/v) shows the product, **III** (R_f = 0.12), and traces of starting materials, **II** (R_f = 0.01) and **I** (R_f = 0.31). Reaction solvent is removed by rotary evaporation, and the entire reaction mixture is chromatographed using the 2-in. (inside diameter) flash column with 400 g silica gel in hexane. The column is washed with 500 ml dichloromethane/hexane (2:1, v/v), and **III** is eluted with 200 ml dichloromethane with an overall yield of 1.8 g (72%).

Step 4: Deprotection. Removal of α-amino and α-carboxyl protecting groups is accomplished by treatment in neat trifluoroacetic acid (TFA) at room temperature. Lower levels of TFA do not remove the *tert*-butyl ester. The methyl ester on the caging moiety is stable in the presence of neat TFA, and for most applications this methyl ester is maintained until the final step of caged peptide preparation. The progress of deprotection of α-amines and α-carboxyls is readily monitored by analytical reversed-phase high-performance liquid chromatography (HPLC) using a Whatman ODS-3 column (1 × 25 cm) and a solvent composed of acetonitrile/water/TFA (40:60:0.1, v/v/v). Using a flow rate of 1.5 ml/min, **III** elutes at 7 min and fully deprotected caged tyrosine methyl ester, **IV**, elutes at 4 min. Deprotection in neat TFA at room temperature is complete in 4 hr. Caged tyrosine methyl ester, **IV**, is purified in approximately 90% yield by semipreparative reversed-phase HPLC using a 2.5 × 25-cm ODS-3 column and the same solvent system at 3 ml/min.

In some instances, particularly for evaluation of the properties of caged tyrosine itself or for the use of caged tyrosine as a control compound in biological experiments, the methyl ester, **IV**, is hydrolyzed to the free acid, **VI**. The methyl group is removed by treatment with aqueous 10% (w/v) K_2CO_3 at room temperature for 1 hr. Caged tyrosine free acid, **VI**, is purified by reversed-phase HPLC, rotary evaporated to dryness, and stored as a solid at −20°.

Step 5: F-Moc Caged Tyrosine (V). For caged peptide synthesis, *N*-(9-fluorenyl)methoxycarbonyl (F-moc) chemistry is used for selective blocking and deblocking of α-amino groups during peptide elongation. Therefore, caged tyrosine methyl ester is made ready for incorporation into peptides by derivatizing its α-amino groups with an F-moc moiety using a modification of the method of Ten Kortenaar *et al.*[6]

Caged tyrosine methyl ester (**IV**, 1.04 g, 1.95 mmol) is suspended in 7 ml acetonitrile, 4 ml water, and 0.7 ml TEA; 0.65 g F-moc succinimide (1.93 mmol) is added portionwise with stirring. The pH of the homogeneous reaction mixture is initially approximately 8.5 but drops as the reaction proceeds. Small aliquots of TEA are added to maintain the pH around 8 (using standard pH paper). Progress of the reaction is monitored by analytical reversed-phase HPLC in acetonitrile/water/TFA (60:40:0.1, v/v/v), and is complete within approximately 30 min. The reaction mixture is washed with 0.1 *M* HCl (3 × 25 ml), and the organic phase is separated and dried by rotary evaporation. Compound **V** is purified by semipreparative reversed-phase HPLC in acetonitrile/water/TFA 45:55:0.1 (v/v/v) at

---

[6] P. B. W. Ten Kortenaar, B. G. Van Duk, J. M. Peeters, R. J. Raaben, P. J. H. M. Adams, and G. I. Tesser, *Int. J. Peptide Protein Res.* **27**, 398 (1986).

3 ml/min. The yield of purified **V** is 0.58 g (56%). Compound **V** is stable for at least 6 months when stored as a solid at −20°.

Step 6: Synthesis of Peptides Containing Caged Tyrosine. Peptide synthesis is carried out on an Applied Biosystems, Inc., Model 432A (Foster City, CA) solid-phase synthesizer using F-moc chemistry and a Wang resin.[7] Activation of α-carboxyls for peptide bond formation is accomplished with a fivefold excess of 2-(1*H*-benzotriazolyl)-1,1,3,3-tetramethuronium hexafluorophosphate (HBTU), and selective removal of the F-moc group is carried out in 10% piperidine in dimethylformamide (DMF). Side chain protecting groups are *tert*-butyl for S, T, and Y; trityl for Q and N; BOC for K; and 2,2,5,7,8-pentamethylchroman-6-sulfonyl (Pmc) for R. F-moc tyrosine methyl ester, **V** (45 mg, 75 μmol), is packed into a plastic amino acid cartridge that is compatible with the synthesizer. The peptide sequence is determined by the alignment of F-moc amino acid cartridges in the carriage wheel, and F-moc tyrosine methyl ester is placed in its desired position in the sequence. Double coupling is carried out during the step(s) in which the caged amino acid is incorporated. After the peptide synthesis is complete, the peptide column containing the derivatized resin is washed with 20 ml dichloromethane and dried in a nitrogen stream for 15 min. Cleavage is accomplished by transfer of the resin into a 10-ml polypropylene tube followed by the addition of 1 ml of cleavage cocktail with stirring for 2–4 hr. The composition of the cocktail is TFA/thioanisole/ethanedithiol (92.5:5:2.5, v/v/v). Crude peptide is precipitated from this reaction mixture by filtration to remove the resin followed by dropwise addition of filtrate to 10 volumes of ice-cold *tert*-butyl methyl ether. The precipitate is then washed with 3 × 25 ml of the same solvent, dried in a vacuum desiccator for 15 min, suspended in 2 ml water, frozen, and lyophilized to a white powder.

Peptides are purified by analytical reversed-phase HPLC in acetonitrile/water/TFA. To generate the free α-carboxyl moiety on the cage, purified peptide is demethylated by treatment with 10% K_2CO_3, neutralized with 1 *M* HCl, and repurified by reversed-phase HPLC. Each peptide requires different elution conditions for purification depending on composition and sequence, but peptide demethylation nearly always results in a shift to an earlier retention time (e.g., see Fig. 6). Interestingly, in the case of 9cgY-LSM1 (LSM1 with cgY in position 9) the demethylated caged peptide resolves into two HPLC peaks, which appear to be stereoisomers (see below). Typically, the first indications that the caged peptide synthesis has been successful include (i) the peptide contains a UV absorbance peak near 270 nm (Fig. 1) and (ii) the peptide photolyzes smoothly to a new HPLC peak (Fig. 2). Final confirmation of peptide structures relies on

---

[7] S. S. Wang, *J. Am. Chem. Soc.* **95,** 1328 (1973).

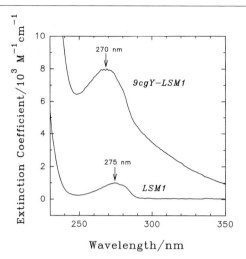

FIG. 1. UV/VIS absorbance spectra for equal concentrations of LSM1 and 9cgY-LSM1 in aqueous 20 mM Tris, pH 7.5.

amino acid analysis and mass spectrometry before and after photolysis. Peptides are double lyophilized to remove traces of TFA and are stored frozen in powder form prior to biological experiments.

Spectral and Photochemical Properties

Extinction Coefficient

The caging moiety and the tyrosine side chain provide useful UV-active chromophores for estimating peptide concentrations. To determine the extinction coefficient for each chromophore, UV/visible scans are taken of the peptides in aqueous solution, then aliquots are quantified by amino acid analysis. Peptide tyrosines have a UV maximum at 275 nm and a ratio of absorbance to moles of peptide (normalized to alanine) equal to 10^3 (Fig. 1). Thus, the extinction coefficient for a peptide containing only tyrosine chromophores (i.e., no F or W) is $\varepsilon_{275\,nm} = 1000\ M^{-1}\ cm^{-1}$ per tyrosine. By a similar analysis, peptides containing α-carboxyl caged tyrosine, cgY, display a UV maximum at 270 nm and an extinction coefficient of $\varepsilon_{270\,nm} = 8000\ M^{-1}\ cm^{-1}$ per caged tyrosine (Fig. 1). Errors in these values are estimated to be ±12%.

Quantum Yield

A convenient method for measuring photolytic quantum yield involves admixture of the compound to be tested with a reference compound whose

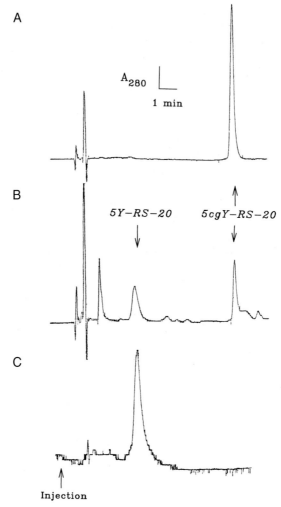

FIG. 2. Analytical reversed-phase HPLC traces of 5cgY-RS-20 before (A) and after (B) 3 min photolysis. (C) Synthetic 5Y-RS-20 injected as a marker.

quantum yield is known.[8] This is valid strictly only if the two compounds have equivalent absorption properties in the wavelength range used for irradiation. Therefore, it is important to mix the compounds at equal concentrations and to ensure that the compounds possess the same photolabile

[8] J. W. Walker, G. P. Reid, and D. R. Trentham, *Methods Enzymol.* **172**, 288 (1989).

protecting group. An HPLC method is then needed that resolves the two compounds from each other and from photolysis by-products.

Caged tyrosine, **VI**, is mixed with caged phenylephrine each at 30 μM in 50 mM Tris, pH 7.5, 1 mM dithiothreitol (DTT). One milliliter of this mixture is placed in a quartz cuvette (1 cm path length) and irradiated with a xenon arc lamp filtered with a UG11 filter to pass near-UV light (300–400 nm). The solution is occasionally mixed to promote uniform illumination, and at 1-min intervals 100-μl aliquots are removed and injected onto the analytical HPLC. The amount of material in each peak is quantified and plotted as a function of irradiation time on a semilogarithmic plot (Fig. 3). The slope of this plot relative to the reference compound is a measure of the quantum yield. To check for competing side reactions during photolysis it is necessary to carry out a direct measure of the amount of product formed. For caged tyrosine, **VI**, and the two peptides described here, a stoichiometric amount of tyrosine or tyrosine peptides is formed, indicating that there are no alternative routes for breakdown of the caged compounds and that the values reported in Fig. 3 are equivalent to the product quantum yield.

Photolysis Kinetics

The rate of photolysis is initially characterized for the caged amino acid rather than the caged peptide because larger quantities are available. Two

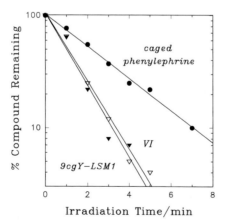

FIG. 3. Steady-steady quantum yields (Φ) determined by mixing caged tyrosine, **VI**, or caged peptide, 9cgY-LSM1, with an equal concentration of reference compound caged phenylephrine (Φ = 0.11).[4] Mixtures were irradiated with a xenon arc lamp (300–400 nm), and aliquots were chromatographed by HPLC as in Fig. 2 to determine the amount remaining. Φ = 0.31 for **VI**; Φ = 0.32 for 9cgY-LSM1.

milliliters of 1 mM caged tyrosine, **VI**, in 25 mM N-[2-hydroxyethyl]piperazine-N'-[2-ethanesulfonic acid] (HEPES), pH 7.0, is placed in a 1 × 1-cm quartz cuvette with all four walls polished. The sample is photolyzed with a xenon flash lamp (Hi Tech Scientific, Salisbury, UK), and then absorbance or fluorecence transients are monitored at 90° to the photolysis beam. The custom-built transient spectrophotometer has been described previously.[9] A similar absorbance transient is observed for both compound **VI** and the caged tyrosine peptide 9cgY-LSM1. The absorbance transient decays biexponentially with a poorly resolved process occurring with a rate contant of about 600 sec^{-1} followed by a slower decay at 50 sec^{-1}. Further characterization involves varying the detecting wavelength with a monochromator, which reveals a peak absorbance at 435 nm consistent with the existence of an *aci*-nitro anion intermediate in the photolytic reaction.[4] This decay process has been equated with the rate of formation of the biologically active species in at least one case,[9] but the mechanism of photolysis is poorly understood for most 2'-nitrobenzyl compounds. For instance, the significance of the two distinct decay processes is unknown. Therefore, it is useful to have additional data concerning the rate of appearance of the photoproduct of interest.

Caged tyrosine is of interest in this regard because tyrosine is an intrinsically fluorescent compound, and addition of the caging moiety greatly reduces its fluorescence (Fig. 4A). Flash photolysis with fluorescence detection allows the rate of formation of tyrosine to be monitored. Fluorescence transients following flash photolysis reveal a step increase in tyrosine fluorescence (excitation at 280 nm, emission at 300 nm) for both caged tyrosine, **VI** (Fig. 4B), and the caged peptide 9cgY-LSM1. The time resolution of these measurements is approximately 10 msec, indicating that the rate of formation of tyrosine is greater than 100 sec^{-1} and too fast to measure with our current instrumentation. Such fast photolysis rates will no doubt enhance the utility of caged tyrosine and caged tyrosine–containing peptides in kinetic studies. For example, photorelease of a tyrosyl side chain in a disordered region of a peptide or protein could be used to initiate protein folding. The large fluorescence increase could then be used as an environmentally sensitive signal to monitor the folding process.

Deciding on Caging Strategy: Structure–Activity Analysis

To prepare an inactive version of a peptide, one must be armed with knowledge of which features of the peptide are important for biological

---

[9] J. W. Walker, G. P. Reid, J. A. McCray, and D. R. Trentham, *J. Am. Chem. Soc.* **110**, 7170 (1988).

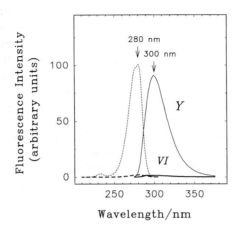

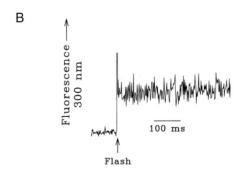

Fig. 4. (A) Fluorescence spectra of 0.1 mM tyrosine (Y, plain lines) or 0.1 mM caged tyrosine (compound **VI**, bold lines). Excitation spectra are illustrated by broken lines and emission spectra by solid lines. (B) Fluorescence increase with a sampling rate of 500 sec^{-1} following pulse photolysis of 1 mM compound **VI** in 25 mM HEPES, pH 7.0 at room temperature.

recognition and activity. Two model peptides are described (Scheme 1). One is the widely studied calmodulin-binding peptide, RS-20, whose sequence is derived from the calmodulin-binding domain of smooth muscle myosin light-chain kinase (MLCK).[10] A wealth of structural information exists

[10] T. J. Lukas, W. H. Burgess, F. G. Prendergast, W. Lau, and D. M. Watterson, *Biochemistry* **25**, 1458 (1986).

about how this peptide interacts with its target protein, calmodulin.[11,12] Critical hydrophobic residues have been identified in positions 5 and 18 of this peptide. Using the two assays described below, we found that substitution of W5 or L18 with glutamate effectively eliminated its interaction with calmodulin. In contrast, substitution into either position with lysine only modestly reduced its binding affinity toward calmodulin. Therefore, because negative charge is more disruptive than positive charge, we synthesized peptides containing negatively charged caged tyrosine in these positions to create caged RS-20 peptides. As described below, only the peptide containing caged tyrosine in position 5 displays large differences in calmodulin binding before and after photolysis. Importantly, the peptide with *both* W5 and L18 substituted with caged tyrosine is *inactive* even after photolysis. This peptide is useful as an inactive caged peptide to control for the effects of UV exposure, generation of photolytic by-products, and nonspecific effects of peptides during physiological experiments. Peptides with the same composition but with a scrambled sequence can also be used for this purpose.

The second peptide of interest is an autoinhibitory domain peptide, which is also from smooth muscle MLCK. This 13 amino acid peptide is thought to have a critical tyrosine in position 9 based on mutagenesis studies of the whole enzyme.[13] Substitution of Y9 with glutamate eliminated MLCK inhibitory activity and indicated that the simple replacement with the negatively charged caged tyrosine would generate a caged LSM1 peptide.

Caged RS-20: Calmodulin Binding

The high-affinity calmodulin-binding peptide RS-20 is caged by replacing W5 with caged tyrosine. Caged forms of this peptide chromatographed as a single peak in reversed-phase HPLC and photolyzed to 5Y-RS-20 (RS-20 with Y in position 5) as expected (Fig. 2). A rapid test of the calmodulin-binding activity of these peptides is performed on a calmodulin affinity column. Five milliliters of calmodulin agarose (Sigma) is poured into a 1×10-cm glass column and equilibrated with 50 mM Tris, pH 7.5, 1 mM CaEGTA, 32 μM Ca^{2+} at 4°; 100 nmol 5cgY-RS-20 (RS-20 with cgY in position 5) is dissolved in 1 ml of eluant and is passed over the column while 14×1.5-ml fractions are collected. Analysis of each fraction by UV absorbance at 270-nm reveals a peak in the column void volume (Fig. 5),

---

[11] M. Ikura, G. M. Clore, A. M. Gronenborn G. Zhu, C. B. Klee, and A. Bax, *Science* **256**, 632 (1992).
[12] A. Crivici and M. Ikura, *Annu. Rev. Biophys. Biomol. Struct.* **24**, 85 (1995).
[13] M. Tanaka, R. Ikebe, M. Matsuura, and M. Ikebe, *EMBO J.* **14**, 2839 (1995).

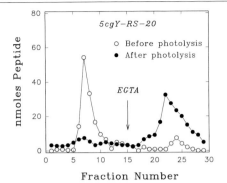

FIG. 5. Calmodulin binding determined by affinity chromatography on calmodulin-agarose. 5cgY-RS-20 did not bind to calmodulin-agarose in 32 μM free Ca^{2+}, 50 mM Tris, pH 7.5, 4° (○). After photolysis, the peptide bound tightly under the same conditions and was eluted with 1 mM EGTA, 50 mM Tris, pH 7.5 (●).

which on reversed-phase HPLC analysis is found to comigrate with 5cgY-RS-20.

An identical aliquot of 100 nmol of 5cgY-RS-20 in eluant is then irradiated for 10 min with the 300–400-nm filtered output of a 75-W xenon lamp. This sample is applied to the calmodulin-agarose column, and fractions are collected and analyzed as before. No peaks are observed until the eluant is changed to 50 mM Tris, pH 7.5, 1 mM ethylene glycol N,N,N',N'-tetraacetic acid (EGTA), no Ca^{2+}, at which time a peak appears (Fig. 5) that comigrates with 5Y-RS-20 on reversed-phase HPLC. Data demonstrate that before photolysis, 5cgY-RS-20 elutes in the column void volume and therefore does not bind appreciably to calmodulin. After photolysis, the peptide photoproduct 5Y-RS-20 binds strongly to calmodulin in a Ca^{2+}-dependent manner (Fig. 5). Calmodulin affinity chromatography provides a rapid means for screening large numbers of caged peptides directed against calmodulin. The capacity of the column is approximately 500–1000 nmol of peptide, and overall recoveries of peptides are greater than 50%.

Caged RS-20: Calmodulin-Dependent Myosin Light-Chain Kinase Activity

For a more quantitative analysis of the interaction between peptides and calmodulin, a calmodulin-dependent MLCK assay is used.[10,14] In this assay, the efficacy of peptide binding is assessed by examining its ability to

---

[14] M. Ikebe, M. Stepinska, B. E. Kemp, A. R. Means, and D. J. Hartshorne, *J. Biol. Chem.* **262**, 13828 (1987).

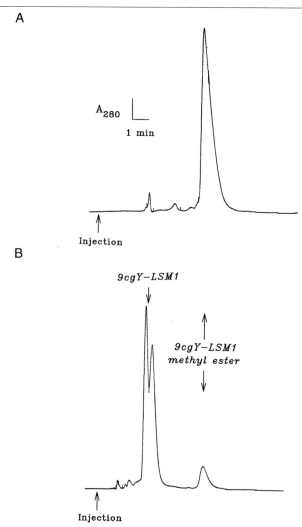

Fig. 6. Reversed-phase HPLC traces showing the shift in retention time and appearance of stereoisomers of 9cgY-LSM1 as a result of demethylation. (A) Purified 9cgY-LSM1. (B) Purified 9cgY-LSM1 treated with 10% aqueous K_2CO_3 for 60 min at room temperature.

reduce calmodulin-dependent activation of MLCK over a range of peptide concentrations. The midpoint of the peptide inhibition curve, $K_{0.5}$, is taken as a measure of relative peptide affinity for calmodulin.

The assay cocktail consists of 100 nM calmodulin, 0.5 μg/ml smooth muscle MLCK, 200 μg/ml myosin regulatory light chain, and a standard buffer containing 30 mM Tris, pH 7.5, 100 mM KCl, 0.1 mM CaCl$_2$, and 1 mM MgCl$_2$. The phosphorylation reaction is started by adding 50 μM [γ-^{32}P]ATP and is continued for 5 min at 25°. Incorporation of ^{32}P into myosin light chain is measured by trapping the protein on a filter and then counting the filter in a liquid scintillation counter. Caged peptides inhibit this calmodulin-dependent activity only weakly with apparent $K_{0.5}$ = 100 μM, whereas uncaged versions of the same peptide inhibit nearly 2 orders of magnitude more potently with an apparent $K_{0.5}$ near 1 μM.

Caged LSM1: Myosin Light-Chain Kinase Catalytic Domain Activity

The inhibitory activity of the MLCK autoinhibitory peptide LSM1 is evaluated according to the inhibition of the kinase activity of the constitutively active MLCK. This unregulated form of MLCK is prepared by trypsin treatment of the whole enzyme as described.[14] The assay conditions are the same as above except 1 mM EGTA is included and no Ca^{2+}/calmodulin is added. LSM1 inhibits MLCK catalytic activity with an apparent $K_{0.5}$ of about 1 μM under the conditions employed (Fig. 7).

The caged LSM1 peptide exhibited the unique property of resolving into two components of roughly equal quantity on the HPLC (Fig. 6). None of the caged forms of RS-20 behaved in this way, and even caged LSM1 chromatographed as a single HPLC peak prior to K$_2$CO$_3$ demethylation (Fig. 6A). Each of these demethylated caged LSM1 peaks photolyzed to LSM1 with a similar photosensitivity, and each component had the expected mass for caged tyrosine LSM1. The two compounds appear to be stereoiso-

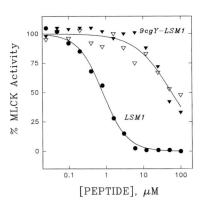

FIG. 7. *In vitro* enzyme assay of a caged peptide before and after photolysis. The two stereoisomers of 9cgY-LSM1 (Fig. 6B) were resolved and tested separately for their ability to inhibit the activity of MLCK catalytic domain. First isomer ($\triangledown$) and second isomer ($\blacktriangledown$) off HPLC column. Both isomers of 9cgY-LSM1 photolyzed to LSM1 ($\bullet$).

mers containing the R and S configurations in the caging group. Titration of the MLCK activity with different concentrations of peptides revealed that both isomers of caged LSM1 were similarly weakly potent as inhibitors of MLCK activity (Fig. 7). Thus, asymmetry in the caging moiety has no obvious effect on the biological activity of this caged peptide. The peptide photoproduct of each isomer of the caged peptide inhibited MLCK approximately 2 orders of magnitude more potently than did the caged forms (Fig. 7).

Summary

We have demonstrated the feasibility of preparing caged peptides by derivatizing a single amino acid side chain in peptides up to 20 amino acids long. Two peptides are illustrated whose activities are reduced by nearly 2 orders of magnitude using this caging approach. The specific strategy described here of derivatizing tyrosine side chains with a charged caging moiety should be generally applicable in the preparation of caged peptides that have a critical tyrosine residue (e.g., LSM1) or that have critical hydrophobic patches (e.g., RS-20). Other amino acid side chains are also accessible via this caging strategy. Derivatives of threonine, serine, lysine, cysteine, glutamate, aspartate, glutamine, and asparagine can be prepared and site specifically inserted into peptides in an analogous manner. The caged peptides synthesized and purified by the methods described here are compatible with biological samples, including living cells,[15,16] and have been used to demonstrate the central importance of calmodulin, MLCK, and, by inference, myosin II in ameboid locomotion in polarized eosinophil cells.[16] Photoactivation of peptides within cells should provide a wealth of new information in future investigations by allowing specific protein activities to be knocked out in an acute and spatially defined way.

Acknowledgments

The authors thank Dr. Robert Carraway for performing peptide synthesis during early stages of this work and for providing helpful advice on synthetic procedures. We thank Dr. Gary Case of the University of Wisconsin Biotechnology Center for advice on peptide synthesis. The assistance of Dr. Masanori Tanaka with the MLCK assays is gratefully acknowledged. Research described here was supported by NIH Grants P01 HL47053, K04 HL03119 and NSF Grant MCB 922082 to J.W.W., and PO1 HL47530 to M.I.

[15] R. M. Drummond, R. Sreekumar, J. W. Walker, R. E. Carraway, M. Ikebe, and F. S. Fay, *Biophys. J.* **70**, A238 (1996).
[16] J. W. Walker, S. H. Gilbert, R. M. Drummond, M. Yamada, R. Sreekumar, R. E. Carraway, M. Ikebe and F. S. Fay, *Proc. Natl. Acad. Sci. U.S.A.* **95**, 1568 (1998).

[6] Light-Directed Activation of Protein Activity from Caged Protein Conjugates

By GERARD MARRIOTT, JOHANNES OTTL, MANFRED HEIDECKER, and DANIELA GABRIEL

Introduction

Light-directed generation of biomolecules from their *caged* precursors has emerged as a powerful technique for rapidly manipulating the concentration of enzyme substrates and protein ligands and for fluorescently marking specific proteins at defined locations in living cells.[1] By monitoring the time dependence of the relaxation of the reaction under study from its light-induced perturbed state to the new equilibrium state, kinetic parameters of the reaction can be determined and used to gain an understanding of the reaction mechanism.[2,3] One of the most attractive features of light-directed activation of caged components is that it allows kinetic parameters to be determined for specific protein-catalyzed reactions within complex biomolecular environments: for example, time-resolved X-ray diffraction of crystals composed of the Ras P21-caged GTP complex has been used to study the dynamics of protein motions in the crystal during the hydrolysis of GTP after its photoactivation from caged GTP;[4,5] ATP has been rapidly generated from caged ATP in skinned muscle fibers to investigate the kinetics of the actomyosin ATPase reaction and its coupling to force production;[6,7] and calcium ions and inositol triphosphate have been generated from their caged precursors in living cells to investigate the role of calcium ion-based signaling during cell motility and after fertilization of *Xenopus* oocytes.[8,9] Rapid mixing techniques are unsuitable in these molecular environments because of incomplete mixing and potential damage to the preparation under study caused by the delivery and rapid flow of solutions.

---

[1] S. Adams and R. Y. Tsien, *Methods Cell Biol.* **55**, 755 (1993).
[2] J. A. McCray and D. R. Trentham, *Annu. Rev. Biophys.* **18**, 239 (1989).
[3] J. A. Theriot and T. J. Mitchison, *Nature* **352**, 126–131 (1991).
[4] I. Schlichting, S. C. Almo, G. Rapp, K. Wilson, K. Petratos, A. Lentfer, A. Wittinghofer, W. Kabsch, E. F. Pai, G. Petsko, and R. S. Goody, *Nature* **345**, 309 (1995).
[5] A. Sheidig, C. Burmester, and R. S. Goody, *Methods Enzymol.* **291**, [14], 1998 (this volume).
[6] Y. E. Goldman, M. G. Hibberd, and D. R. Trentham, *J. Physiol.* **354**, 605 (1984).
[7] J. A. Dantzig and H. Higuchi, and Y. E. Goldman, *Methods Enzymol.* **291**, [18], 1998 (this volume).
[8] I. Parker and I. Ivorram, *J. Physiol.* **461**, 133 (1993).
[9] F. S. Fay, S. H. Gilbert, and R. A. Brundage, *Ciba Found. Symp.* **188**, 121 (1995).

Furthermore, most commercial fast-flow mixing instruments have mixing dead times that exceed 1 msec, whereas light-directed activation of caged compounds based on α-carboxy-2-nitrobenzyl derivatives can be used to establish a concentration jump within a few microseconds.[10,11]

Light-directed perturbations of protein activity are usually achieved by activation of a caged enzyme substrate or caged protein ligand,[12,13] although a complementary approach has been developed in which near-ultraviolet light is used to directly trigger protein activity from an inactive, caged protein conjugate.[14–18] Caged protein conjugates are prepared by chemically modifying amino acid residues that are essential for protein activity using photocleavable labeling reagents. These labeling groups may be removed from the protein on irradiation with near-ultraviolet light with a concomitant recovery of protein activity, often to the prelabeled level. Caged protein conjugates should not be confused with caged fluorophore-labeled protein conjugates, which are fully active proteins that harbor covalently bound, nonfluorescent chromophores (caged fluorophores) that can be rendered fluorescent after irradiation with near-ultraviolet light.[19]

Our interest in developing reagents and labeling procedures to prepare caged proteins has emerged as part of a research program aimed at understanding the molecular basis of cell motility. Microinjected caged proteins, labeled with fluorescent or caged fluorescent dyes for imaging purposes, may be used to directly manipulate the activity of G-actin and actin-binding proteins at defined locations in motile cells, and, in combination with time-resolved fluorescence image microscopy, this technique provides opportunities to investigate the kinetics of G-actin polymerization/sequesterization and the role of specific actin-binding proteins in the regulation of actin filament dynamics and cell motility.

We have developed a number of caged reagents and associated labeling procedures that can be used to prepare caged conjugates of almost any protein by targeting essential lysine or cysteine residues.[15,16,18] Amino group directed caged reagents based on 6-nitroveratryloxycarbonyl chloride

[10] K. R. Gee, R. Wieboldt, and G. P. Hess, *J. Am. Chem. Soc.* **116,** 8366 (1994).
[11] G. P. Hess and C. Grewer, *Methods Enzymol.* **291,** [25], 1998 (this volume).
[12] Y. E. Goldman, M. G. Hibberd, J. A. McCray, and D. R. Trentham, *Nature* **300,** 701 (1982).
[13] G. C. Ellis-Davies and J. H. Kaplan, *Proc. Natl. Acad. Sci. U.S.A.* **91,** 187 (1994).
[14] P. D. Senter, M. J. Tansey, J. M. Lambert, and W. A. Blättler, *Photochem. Photobiol.* **42,** 231 (1985).
[15] G. Marriott, H. Miyata, and K. Kinosita, *Biochem. Int.* **26,** 9943 (1992).
[16] G. Marriott, *Biochemistry* **33,** 9092–9097 (1994).
[17] C. Ghung-Yu, B. Niblack, B. Walker, and H. Bayley, *Chem. Biol.* **2,** 391 (1995).
[18] G. Marriott and M. Heidecker, *Biochemistry* **35,** 3170 (1996).
[19] T. J. Mitchison, K. E. Sawin, J. A. Theriot, K. Gee, and A. Mallavarapu, *Methods Enzymol.* **291,** [4], (1998) (this volume).

(NVOC-Cl) have been used to prepare caged conjugates of G-actin[16] and profilin,[20] and these conjugates can be reactivated at will upon illumination with near-ultraviolet light. A new thiol reactive caged reagent, 4,5-dimethoxy-2-nitrobenzyl bromide (DMNBB), has been used to prepare a caged heavy meromyosin (HMM) conjugate that is unable to couple the energy derived from ATP hydrolysis to the production of molecular forces that move actin filaments,[18] and light-directed activation caged HMM has been used to restore the coupling between these two processes. This property has proven useful in studying specific aspects of the actin-activated, force-generating ATPase reaction of myosin and its subfragments in solution, in the *in vitro* motility assay,[18] and within skinned muscle fibers (Koegler, Ruegg, and Marriott, 1988, unpublished results).

This article reviews experimental procedures used by our group to prepare and characterize caged reagents and caged protein conjugates of cytoskeleton proteins. The labeling methods described in this article are quite general and may form the basis of a labeling procedure to cage almost any protein. Similarly optimized irradiation conditions are presented that may be used for light-directed activation of caged protein conjugates.

Methods

Materials

The following materials are used:
6-Nitroveratryloxycarbonyl chloride (NVOC-Cl), stored under dry silica gel in the dark (Fluka, Ronkonkoma, NY); 4,5-Dimethoxy-2-nitrobenzyl bromide (DMNBB); acetone (Aldrich, Milwaukee, WI).
Tetramethylrhodamine (TMR) phalloidin; Na_2^+-ATP; polylysine 10,000 MW; dimethylaminopyridine; catalase (CAT); glucose oxidase (GO) (Sigma, St. Louis, MO).
5′-Iodoacetamidotetramethylrhodamine (IA-TMR, a gift of Dr. John Corrie, NIMR, Mill Hill, London), also available as mixed isomers from Molecular Probes (Eugene, OR).
AB buffer: 25 mM imidazole, 25 mM KCl, 1.0 mM dithiothreitol (DTT), 4 mM $MgCl_2$, pH 7.0.
AB/GO/CAT/glucose/DTT buffer; glucose oxidase 100 μg/ml; catalase 18 μg/ml; glucose 3 mg/ml.
G-buffer: 2 mM Tris, 0.2 mM ATP, 0.2 mM $CaCl_2$, 1 mM DTT, pH 8.0.

---

[20] D. Gabriel, diploma thesis, Ludwigs Maximilians Universität, München (1995).

Instrumentation

Absorption measurements are performed using the Hewlett-Packard (Palo Alto, CA) 8452A diode array spectrophotometer. Light-scattering measurements are made in an Aminco Bowman Luminescence Spectrometer Series 2 (Sopra, Buttelborn, Germany).

Consideration for Preparation of Caged Proteins by Modification of Lysine Residues

NVOC-Cl was widely employed in the 1970s in peptide and carbohydrate chemistry as a light-sensitive protection group for the amino groups of lysine and amino sugars.[21] The carbamate bond linking the NVOC to the amino group may be cleaved on illumination with near-ultraviolet light, releasing the deprotected peptide or sugar with a respectable yield.[22-23] We have shown that caged proteins can be prepared using NVOC-Cl as a reagent to modify essential lysine residues. The carbamate bond that links NVOC to lysine residues of the protein is also easily cleaved with ultraviolet light, generating the free amino base of the lysine residue, carbon dioxide, and 4,5-dimethoxy-2-nitrosobenzaldehyde, as illustrated in Fig. 1.[16]

In general, derivatization of lysine residues of proteins has a neutral effect on their activity, although a couple of reports in the literature describe how the activity of certain proteins is specifically inhibited on modification with lysine-directed reagents. For example, the polymerization activity of G-actin is blocked when its lysine-61 residue is modified with fluorescein isothiocyanate (FITC),[24] and the phosphodiesterase-activating function of calmodulin is reduced about fivefold after modification of its lysine-75 residue with the N-hydroxysuccinimide ester of biotin.[25] In designing a labeling protocol to prepare caged profilin, we took advantage of the finding that when profilin is labeled with an average of one tetramethylrhodamine succinimide ester group, its ability to bind G-actin is reduced by 30%,[26] and presumably higher labeling ratios would lead to a further attenuation of this interaction. Interestingly, one of the seven lysine residues of profilin

---

[21] V. N. R. Pillai, *Synthesis* **26,** 1 (1980).
[22] B. Amit, U. Zehavi, and A. Patchornik, *J. Org. Chem.* **39,** 192 (1974).
[23] S. P. A. Fodor, J. L. Read, M. C. Pirrung, L. Stryer, A. T. Lu, and D. Solas, *Science* **251,** 767 (1991).
[24] L. D. Burtnick, *Biochim. Biophys. Acta* **791,** 57 (1984).
[25] D. Mann and T. C. Vanaman, *Methods Enzymol.* **139,** 433 (1992).
[26] J. A. Theriot, J. Rosenblatt, D. A. Portnoy, P. J. Goldschmidt-Clermont, and T. J. Mitchison, *Cell* **76,** 505 (1994).

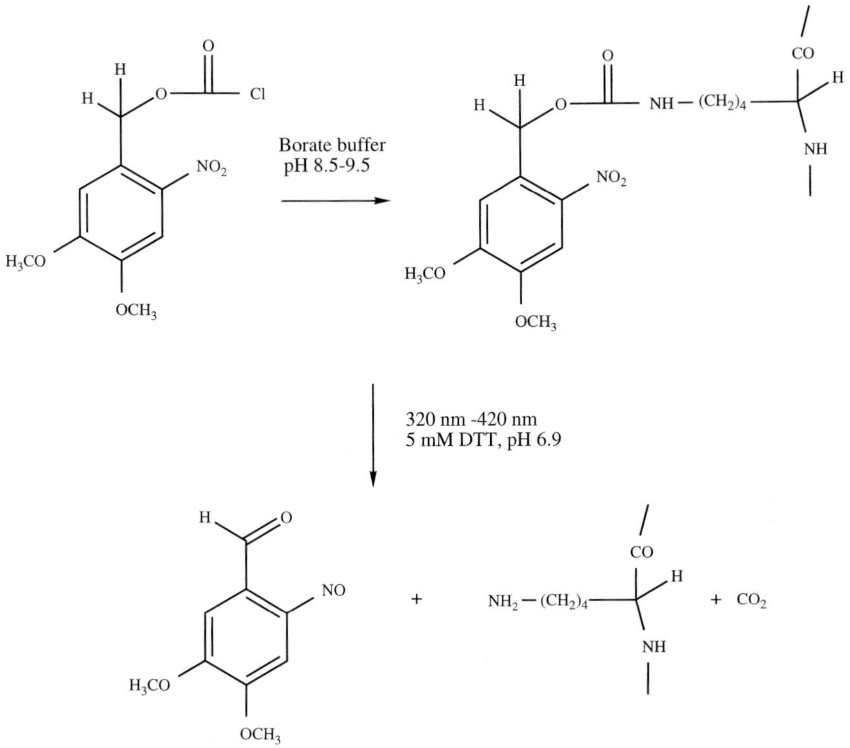

FIG. 1. Proposed reaction mechanism of the NVOC-lysine photodeprotection reaction.

is located in a putative actin-binding domain, and we assume the caged profilin is formed after modification of this residue with NVOC-Cl.

Absorption spectrophotometric analysis of the NVOC-Cl-modified protein conjugate, after centrifugation and removal of excess NVOC-Cl and other breakdown products, can be used to determine the number of NVOC groups attached to the protein. In general the caged protein conjugate should be labeled with the minimum number of NVOC groups necessary to inactivate its activity. To illustrate this point we find that the polymerization activity of caged G-actin containing between three and five NVOC groups is fully restored after irradiation with near-ultraviolet light, whereas very little polymerization activity is recovered after irradiation of a caged G-actin containing an average of eight NVOC groups (Ref. 16 and Marriott, 1997, unpublished observations).

Procedures

Preparation of 4-Dimethylaminopyridine Salt of 6-Nitroveratryloxycarbonyl Chloride

4-Dimethylaminopyridine (DMAP), 11 mg (0.09 mmol), dissolved in 300 µl dry acetonitrile is mixed with 20 mg of NVOC-Cl dissolved in 200 µl dry dioxane. After a 15-min reaction time the yellow–orange precipitate is centrifuged for 10 min at 10,000 g and the supernatant discarded. The precipitate is washed with dioxane and centrifuged again for 10 min at 10,000 g. The solid residue is dried and stored over silica in the dark and used within a month of its preparation.

Yield: 25 mg (0.063 mmol) = 88%
Molecular Weight = 397.83 g/mol; $C_{16}H_{20}N_{30}Cl$; cation MG = 362.38
FAB-MS: m/z 362.3 (M^+)
mp = 80° (decomposition)
^{1}H NMR: ($DCCl_3$, 500 MHz): δ [ppm] = 8.12 (d, 2H), 7.71 (s, 1H), 7.02 (s, 1H), 6.71 (d, 2H), 5.60 (S, 2H), 3.97 (S, 3H), 3.94 (s, 3H), 3.20 (s, 6H).

Preparation and Spectroscopic Characterization of 6-Nitroveratryloxycarbonyl-Labeled Polylysine

A freshly prepared solution, 100 µl, of 10 mM NVOC-Cl in acetone is added rapidly to 2 mg/ml polylysine in 0.1 M sodium borate buffer, pH 9.0, and the reaction is left for 30 min at 20° in the dark. Unreacted and hydrolyzed NVOC-Cl, which may form a precipitate during the reaction, are removed by centrifugation at 10,000 g for 20 min. This supernatant fraction is dialyzed overnight against two changes of 3 liters of 10 mM PIPES (piperazine-N,N'-bis-[2-ethane-sulfonic acid]), pH 6.9. Because polylysine does not contain any aromatic amino acids, it can be used to define the ultraviolet spectral properties of NVOC-lysine in the conjugate. The absorption of NVOC-labeled polylysine at 350 nm is used to calculate its concentration in the conjugate using an extinction coefficient of 5000 M^{-1} cm^{-1}.[27] Using this calculated concentration and the absorption value of NVOC-polylysine at 280 and 290 nm, it is possible to calculate, using the Beer–Lambert equation, the extinction coefficients of NVOC at 290 nm (3400 M^{-1} cm^{-1}) and at 280 nm (3050 M^{-1} cm^{-1}). Knowledge of these two extinction coefficients and those of the protein at 280 or 290 nm may then be used to determine the NVOC/protein labeling ratio for almost any NVOC-labeled protein conjugate.

[27] J. F. Cameron and J. M. J. Fréchet, *J. Am. Chem. Soc.* **113**, 4202 (1991).

Extinction Coefficients Used in This Study

G-Actin (290 nm)[28] = 26,000 M^{-1} cm^{-1}
Profilin I (280 nm)[29] = 19,950 M^{-1} cm^{-1}
Profilin II (280 nm)[29] = 18,700 M^{-1} cm^{-1}
NVOC (350 nm)[27] = 5,000 M^{-1} cm^{-1}
NVOC (290 nm) = 3,400 M^{-1} cm^{-1}
NVCO (280 nm) = 3,050 M^{-1} cm^{-1}
TMR-IA (549 nm)[30] = 96,400 M^{-1} cm^{-1}

Preparation of Caged G-Actin

A freshly prepared solution, 100 μl, of 0.1 M NVOC-Cl solution in acetone is added in rapid succession to 1.95 ml of a sodium borate based G-buffer (2 mM borate, 0.2 mM ATP, 0.2 mM CaCl$_2$, 1 mM DTT, pH 8.5) and 6 ml of 36 μM G-actin, prepared according to the method of Pardee and Spudich,[30a] and the reaction is left for 30 min in the dark. The final concentrations of G-actin and NVOC-Cl are 27 μM and 1.25 mM, respectively, with an acetone content of 1.25%. The addition of NVOC-Cl in this manner reduces the amount of denatured protein that forms by directly mixing acetone and aqueous solution of proteins. After 30 min the sample is centrifuged at 10,000 g for 10 min at 4° to remove insoluble material and then dialyzed overnight against 2 mM PIPES-based G-buffer, pH 6.9, at 4° with two changes of buffer. Magnesium chloride and KCl are added to the G-actin conjugate at 2 mM and 0.1 M, respectively, and after 90 min at 20°, the reaction is centrifuged at 100,000 g for 60 min at 4°. The pellet fraction, which contains polymerization competent G-actin, is discarded, and the supernatant fraction, which contains caged G-actin, is analyzed by absorption spectroscopy. The absorption value at 290 nm of G-actin in the caged G-actin conjugate is determined by subtracting the contribution to the 290-nm absorption from the NVOC group. The G-actin concentration in the conjugate is then calculated using an extinction coefficient at 290 nm of 26,000 M^{-1} cm^{-1}.

Characterization of Caged G-actin

Using the conditions of NVOC conjugation described above, the labeling ratio of NVOC to G-actin normally falls between three and five NVOC groups per G-actin monomer, with a caged G-actin yield of 30 to 60%. The

---

[28] D. J. Gordon, D. Eisenberg, and E. D. Korn, *J. Biol. Chem.* **251,** 4778 (1976).
[29] M. Haugwitz, A. Noegel, D. Rieger, and M. Schleicher, *J. Cell Sci.* **100,** 481 (1994).
[30] J. E. T. Corrie and J. S. Craik, *J. Chem. Soc. Perkin Tran.* **1,** 2967 (1994).
[30a] J. D. Pardee and J. A. Spudich, *Methods Cell Biol.* **24,** 271 (1982).

absorption spectrum of an 11.3 μM solution of caged G-actin containing 3.9 NVOC groups per actin monomer in 2 mM PIPES-based G-buffer, pH 6.9, is shown in Fig. 2A. The absorption of NVOC on G-actin is centered at 350 nm and extends to 420 nm. Consequently, all operations on caged G-actin must be conducted in the dark or at least under subdued light. Caged G-actin is best used within a day or two of its preparation and should be centrifuged at 100,000 g for 60 min before an experiment. The polymerization competence of G-actin and NVOC-labeled G-actin in F-buffer can be shown by sodium dodecyl sulfate–polyacrylamide gel electrophoresis (SDS–PAGE). Solutions of G-actin or caged G-actin, 20 μM, in 2 mM PIPES-based G-buffer are made into F-buffer, and after a 90-min period to allow for polymerization, the F-actin fraction is separated from G-actin by high-speed centrifugation (60 min at 4° in a Beckman airfuge at 22 psi). The amount of protein in the supernatant and pellet fractions is analyzed by SDS–PAGE and densitometry of the Coomassie-stained gel.

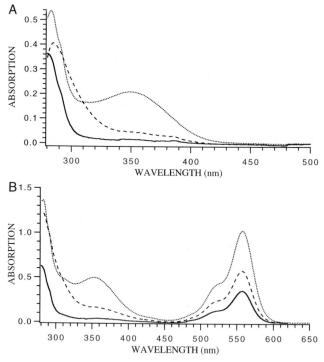

FIG. 2. Absorption spectra of (A) caged G-actin; (B) 25 μM TMR-labeled caged G-actin (solid lines, caged; dotted lines, irradiated F-actin pellet resuspended in 500 μl F-buffer; dashed lines, irradiated supernatant).

We normally find that more than 95% of the control G-actin can polymerize, whereas less than 5% of the caged G-actin is found in the pellet fraction.

Preparation of Tetramethylrhodamine-Labeled Caged G-Actin

In principle, fluorescence image microscopy of fluorescently labeled caged G-actin microinjected into living cells can be used to show that caged G-actin incorporates into the actin cytoskeleton only after it has been irradiated with near-ultraviolet light. The following method is used to prepare a caged G-actin fluorescently labeled at cysteine-374 residue with TMRIA. G-Actin at a concentration of 100 μM is dialyzed against 2 mM Tris-based G-buffer, pH 8.0, in the absence of DTT, clarified by centrifugation, and treated at 4° in an overnight reaction with a 2-mole excess of 5'-IATMR, added from a stock solution of 20 mM in dimethylformamide (DMF) that is stored at $-80°$. Free dye is removed from the conjugate by dialysis against three changes of 2 mM Tris-based G-buffer, pH 8.5, containing 1 mM DTT, and then centrifuged at 100,000 g for 60 min at 4°. To remove polymerization-incompetent G-actin in this preparation, MgCl$_2$ and KCl are added to 2 and 50 mM, respectively, to make F-buffer, and after 90 min at 20° the solution is centrifuged at 100,000 g for 60 min at 4°. The F-actin pellet is resuspended in 2 mM borate-based G-buffer and dialyzed overnight against 1 liter of 2 mM borate-based G-buffer. The fluorescently labeled G-actin in the supernatant is centrifuged once more at 100,000 g for 60 min at 4° and the concentration of G-actin established using the Bradford assay.[31] 5'-IATMR-labeled G-actin at 25 μM is treated with NVOC-Cl, added from a fresh 0.1 M acetone stock solution, to achieve a final concentration of 1.0 mM. After a 30-min reaction time at 20° the double-labeled conjugate is centrifuged at 10,000 g for 20 min at 4° to remove insoluble material, and the supernatant is dialyzed overnight against two changes of 2 liters of 2 mm PIPES-based G-buffer, pH 6.9, containing 1 mM DTT, and then centrifuged at 100,000 g for 60 min at 4°. To initiate polymerization, MgCl$_2$ and KCl are added to the supernatant to final concentrations of 2 and 50 mM, respectively, and after 90 min at room temperature the reaction is centrifuged at 100,000 g for 60 min at 4°; the TMR-labeled caged G-actin conjugate is present in the supernatant fraction. The yield of TMR-labeled caged G-actin is generally higher if G-actin is first reacted with the fluorophore and then caged with NVOC-Cl. The absorption spectrum of a 25 μM solution of TMR-labeled caged actin is shown in Fig. 2B. The concentration of TMR in the conjugate of 10 μM is calculated using an extinction coefficient of TMR of 96,400 M^{-1} cm^{-1} at 549 nm, and the G-actin concentration of 25 μM is determined using the Bradford assay.[31]

[31] M. M. Bradford, *Anal. Biochem.* **72,** 248 (1976).

Photoactivation of Caged Protein Conjugates in Solution

Prior to photoactivation, the caged G-actin conjugate is dialyzed into PIPES-containing F-buffer containing 5 mM DTT, pH 6.9, and centrifuged at 100,000 g for 60 min to remove any denatured protein or any polymerization competent G-actin that may form during storage. Samples of caged G-actin, 500 μl, are irradiated in a quartz cuvette (4 × 10 mm) with the near-ultraviolet light output of a 100-W high-pressure mercury arc lamp selected with an interference filter that transmits light between 340 and 400 nm (Dr. Rapp, Opto-Elektronik, Hamburg, Germany). The intense infra-red irradiation of the mercury arc lamp is blocked using two copper sulfate glass filters and a 10-mm-thick distilled water filter. To ensure a uniform irradiation of the caged protein solution in the cuvette, especially during a prolonged irradiation treatment, the near-ultraviolet light beam is defocused at the center of the cuvette, which is occasionally agitated.

Fluorescence and Photoactivation Microscopy

Simultaneous fluorescence microscope-based imaging of actin filaments and photoactivation of caged HMM in an *in vitro* motility assay is performed essentially as described by Marriott and Heidecker[18] and Mitchison and co-workers[19,32] and is illustrated in Fig. 3. The near-ultraviolet photolysis beam is selected with a broad-band near-ultraviolet interference filter (340–400 nm) at the source of a 100-W mercury arc lamp that is aligned perpendicular to the body of an Axiovert 35 microscope (Zeiss, Oberkochen). A dichroic mirror that reflects light between 340 and 420 nm, mounted onto the normal mercury arc lamp port of the microscope, directs the photolysis beam into the microscope. A second 100-W mercury arc lamp, aligned parallel to the microscope body, is used to excite rhodamine/CY3–based fluorescent dyes in the image field. A custom-made bandpass filter that selects the 546-nm line (Chromatech, Brattleboro, VT) is placed in front of the lamp, and this beam passes through the external dichroic mirror and into the microscope. The fluorescence excitation filter in the microscope filter assembly is removed in order to transmit the two coaligned excitation beams. The dichroic mirror and emission filter are custom made for the CY3 fluorophore (Keio filter set; Chromatech, Brattleboro, VT); the CY3 dichroic filter reflects the photolysis and fluorescence excitation beams to the microscope objective, while the emission filter blocks the ultraviolet light and transmits the CY3 or rhodamine emission between 565 and 590 nm. TMR-phalloidin–labeled actin filaments in the image field are recorded with an intensified charge-coupled device camera (C2400, Hamamatsu Pho-

---

[32] T. J. Mitchison, *J. Cell Biol.* **109,** 637 (1989).

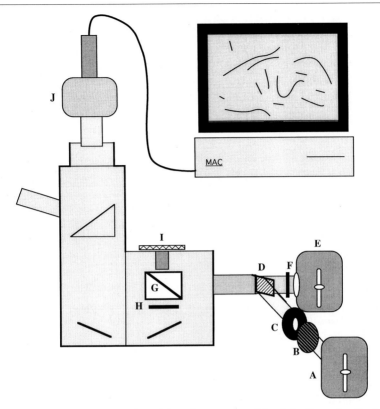

FIG. 3. Schematic representation of the microscope used for light-directed activation of caged proteins with simultaneous imaging of tetramethylrhodamine fluorescence. A, 100-W mercury arc lamp for irradiation of caged proteins; B, 340- to 400-nm interference filter; C, manually controlled shutter with adjustable iris; D, dichroic mirror mounted in an assembly attached to the lamp port of the microscope used to reflect light between 340–420 nm into the microscope; E, 100-W mercury arc lamp used for excitation of fluorescence dyes; F, 546-nm interference filter; G, dichroic mirror from CY3 custom filter set used to reflect 340- to 400-nm and 546-nm light; H, emission filter from CY3 custom filter set used to collect light between 565 and 590 nm and to block near-ultraviolet light; I, *in vitro* motility chamber; J, intensified charged-coupled device camera.

tonics, Herrsching, Germany), and the images are simultaneously saved on a video tape and in the memory of a frame grabber at 0.5-sec intervals (Scion Inc., Frederick, PA) under the control of a Macintosh IIfx computer.

Functional Analysis of Caged G-Actin and Decaged G-Actin

The activity of G-actin is measured by its ability to polymerize in F-buffer. A 25 μM solution (500 μl) of caged G-actin in 2 mM PIPES-based F-buffer, pH 6.9, containing 5 mM DTT is irradiated in a microcuvette (4 × 10 mm) with near-ultraviolet light as described in the previous section. Aliquots of the protein sample, 50 μl, are removed at defined exposure times and following a 90-min period at 20° in the dark to allow for polymerization; the samples are centrifuged in an airfuge at 22 psi for 60 min at 4° to pellet F-actin. The supernatant is removed and the pellet is resuspended in 50 μl of PIPES-based F-buffer. The amount of actin in the supernatant and pellet fractions is determined by SDS–PAGE as described previously. Approximately 50% of polymerization-competent G-actin is generated from caged G-actin following 2 min of irradiation with near-ultraviolet light, and this yield increases to almost 90% after 12 min of irradiation (Fig. 4). Proof that the caged groups have been removed from the polymerization-competent, decaged G-actin fraction is shown by absorption spectrophotometric analysis of the pellet and supernatant fractions of the high-speed centrifugation (Fig. 2A). The ratio of the absorbance at 290 and 350 nm, an indicator of the amount of decaged protein, is much higher in the pellet fraction than in the preirradiated sample, whereas the absorption of the supernatant fraction is dominated by the soluble photoproducts, as SDS–PAGE shows less than 5% of the protein remains in this fraction after 12 min of irradiation (Fig. 4). Similar results are obtained with the TMR-labeled caged G-actin conjugate, although the yield of polymerization-competent TMR-labeled G-actin obtained after 12 min of irradiation with ultraviolet light is lower than that found for the nonfluorescent-caged G-actin (Fig. 2B). The *in vitro* motility assay is used to show that F-actin filaments generated by the photoactivation of caged G-actin are functional and indistinguishable from normal actin filaments in terms of the precentage

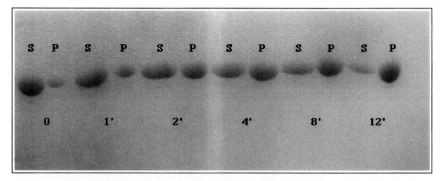

Fig. 4. SDS–PAGE and Coomassie-stained gel of the 100,000-g supernatant and pellet fractions of 20 μM caged G-actin in 2 mM PIPES-based F-buffer as a function of the exposure time to near-ultraviolet light.

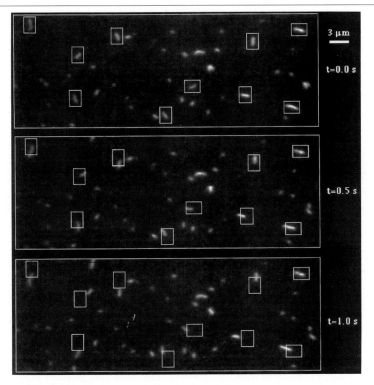

FIG. 5. *In vitro* motility assay F-actin filaments labeled with TMR-phalloidin. The actin was prepared by light-directed activation of a 500-μl sample of caged G-actin as described in the text. Scale bar is 3 μm. The filaments were observed to slide on the HMM surface for several minutes. The three images captured by the frame grabber show the movement of filaments over a 1-sec interval.

of filaments that move on the HMM surface with a maximum sliding velocity of 3 μm/sec (Fig. 5).

Preparation of Profilin

Recombinant profilin from *Dictyosteium discoideum* is isolated from an *Escherichia coli* strain transfected with the expression vector JM83 pIMS profilin I or profilin II.[33] Induction of the JM83 pIMS profilin I and profilin II is achieved by addition of 0.5 mM IPTG (iso proplylthio-galactopyranoside) for 2 hr at 37°. The cells are centrifuged (45 min, 10,000 g) and washed with TE buffer (10 mM Tris–HCl, 2 mM EDTA, pH 8.0). After a second

[33] L. Eichinger, A. A. Noegel, A. A., and M. Schleicher, *J. Cell Biol.* **112,** 665 (1991).

centrifugation (45 min, 10,000 g) the pellet is resuspended in TEDABP buffer [10 mM Tris–HCl, 1 mM EGTA, 1 mM DTT, 0.02% NaN$_3$, 1 mM benzamidine, 0.5 mM phenylmethylsulfonyl fluoride (PMSF), pH 8.0] and freeze-thawed twice in liquid nitrogen. After lysis with 0.4 mg/ml lysozyme, the cells are homogenized on ice and sonicated seven to eight times. SDS–PAGE analysis is used to show that the majority of the profilin is solubilized in TEDABP containing 2 and 7 M urea, respectively. The two urea wash fractions are pooled and dialyzed against TEDABP and are then applied to a TEDABP equilibrated polyproline-Sepharose column prepared according to the method of Rozycki et al.[34] Affinity-bound profilin is eluted with TEDABP buffer in two steps using 2 and 7 M urea. Profilin I and II are found primarily in the 7 M urea fraction. This fraction is dialyzed against several changes of 50 mM Na$_4$B$_2$O$_4$, 1 mM EDTA, pH 8.5, prior to labeling with the DMAP salt of NVOC-Cl.

Preparation of 6-Nitroveratryloxycarbonyl-Labeled Profilin

Purified profilin, 4 ml (40 μM), in 50 mM borate buffer, 1 mM EDTA, pH 8.5, is mixed with a fourfold molar excess of solid NVOC-Cl–DMAP, a water-soluble and activated form of NVOC-Cl.

After a 1-hr reaction at 4°, the slightly yellow reaction mixture is centrifuged (10,000 g, 30 min) and the white–yellow pellet is discarded, and the supernatant is dialyzed four times at 4° against 2 liters of 2 mM Tris–HCl, 0.5 mM DTT, 0.1 mM CaCl$_2$, pH 8.0. The solution is then clarified by centrifugation (10,000 g, 30 min, 4°), and the protein and NVOC concentration in the caged profilin conjugate are determined by absorption spectroscopy using the extinction coefficients listed earlier. The best preparations of caged profilin contain two to three NVOC groups per profilin molecule.

Functional Analysis of Caged Profilin

The activity of profilin is determined by its ability to bind to G-actin in physiological salt with a concomitant inhibition of G-actin polymerization. The time-dependent increase in a light-scattering signal that accompanies the transition of G-actin to F-actin may be used to monitor the kinetics of the polymerization reaction. The scattering signal is measured in an SLM-AMINCO fluorimeter (Rüttelborn, Germany) through crossed polarizers setting the excitation and emission wavelengths at 350 nm with a 4-nm bandpass. The 350-nm probing light is highly attenuated compared to the amount of light delivered by the 100-W mercury arc lamp and is not expected to result in the photoactivation of any significant amount of caged protein. A 4.7 μM

---

[34] M. Rozycki, C. E. Schutt, and U. Lindberg, *Methods Enzymol.* **196,** 100 (1991).

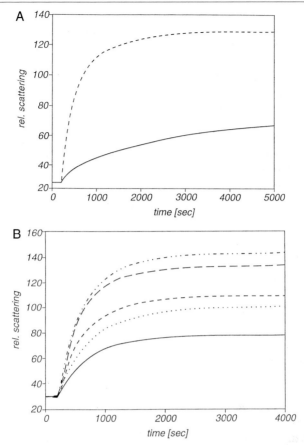

FIG. 6. Polymerization of G-actin monitored by light scattering of F-actin filaments. The polymerization was induced by the addition of MgCl$_2$ and KCl to G-actin in G-buffer, and the scattered light was monitored through cross-polarizers using 350 nm for both excitation and emission wavelengths. (A) The upper curve is the polymerization profile for 4.7 μM G-actin, and the lower curve is the profile for 4.7 μM G-actin in the presence of 12 μM profilin. (B) The curves (from top to bottom) represent the polymerization of 4.7 μM G-actin in the presence of caged profilin that has been irradiated with near-ultraviolet light as described in the Methods section for 0, 2.5, 5, and 10 min, respectively.

solution of G-actin from *Dictyostelium discoideum*, prepared according to Prassler *et al.*,[34a] in 2 mM PIPES-based G-buffer is treated with MgCl$_2$ to 1 mM and KCl to 50 mM, respectively, to start polymerization, and the light-scattering signal is measured for 100 min (Fig. 6A). Normal profilin

---

[34a] J. Prassler, S. Stocker, M. Heidecker, G. Marriott, and G. Gerisch, *Mol. Cell Biol.* **8**, 83 (1997).

at 12 μM binds to G-actin and greatly decreases the rate and extent of G-actin polymerization compared to G-actin alone (Fig. 6A). However, caged profilin at 14 μM has almost no effect on the kinetics or extent of polymerization because of its inability to bind G-actin (Fig. 6B, second curve from top). To show that profilin is activated from caged profilin after exposure to ultraviolet light, 100-μl samples of caged profilin subjected to irradiation for 2.5, 5, and 10 min as described earlier are added to separate solutions of G-actin to achieve a full profilin concentration of 12 μM and their effect on polymerization, initiated by the addition of $MgCl_2$ and KCl to 2 and 50 mM, respectively, is measured by light scattering, as shown in Fig. 6B. Increased exposure of the caged profilin conjugate to intense ultraviolet light leads to a corresponding increase in the inhibition of G-actin polymerization, which evidently results from the binding of functional, photoactivated profilin to G-actin.

Preparation of Caged Proteins by Modification of Cysteine Residues

A second labeling strategy to prepare caged protein conjugates is based on the targeted modification of essential cysteine residues of the protein using photolabile alkyl halides. For example, each of the two heads of HMM harbor an ATPase domain with two highly reactive cysteines (Cys-707 and Cys-697) that can be specifically derivatized with a variety of haloacetyl-alkylating reagents. Modification of either cysteine residue results in considerable perturbations of the actin-activated ATPase of myosin or its subfragments.[35,36] However, it is important to recognize that at a functional level, the activity of myosin is to couple the energy derived from the actin-activated ATPase to the production of molecular forces that move actin filaments. In this context the activity of HMM is better characterized using the *in vitro* motility assay[37] rather than assays that measure only its ATPase activity. Thus HMM labeled at Cys-707 with thiol reactive groups exhibits normal or even elevated Mg^{2+}, or Ca^{2+}-activated ATPase, but these conjugates do not support the sliding of F-actin filaments in the *in vitro* motility assay.[18,38] We reasoned that if the thiol group of Cys-707, or Cys-697, could be alkylated using a photolabile protection group for mercaptans,[15] then this conjugate would not be able to couple the energy derived from hydrolysis of ATP to force production and the movement of actin filaments, until the caged groups are removed from the conjugate with near-ultraviolet light. This type of caged HMM can be prepared by reacting

[35] S. A. Mulhern and E. Eisenberg, *Biochemistry* **17**, 4419 (1978).
[36] E. Reisler, *Methods Enzymol.* **85**, 84 (1982).
[37] S. J. Kron and J. A. Spudich, *Proc. Natl. Acad. Sci. U.S.A.* **83**, 6272 (1985).
[38] D. D. Root and E. Reisler, *Biophys. J.* **63**, 730 (1992).

bromomethyl-2-nitro-4,5-dimethoxyenzene (DMNBB: Fig. 7) with HMM under neutral to slightly basic conditions, where it specifically alkylates cysteine-707. The 4,5-dimethoxy-2-nitrobenzyl bromide group has a broad near-ultraviolet absorption band that extends into the blue region of the spectrum, and its high molar absorptivity ($E_{350\,nm}$ of 5000 M^{-1} cm^{-1}) is a useful feature if the caged protein is to be operated on as a monolayer or in optically thin samples such as cells and single muscle fibers.

Preparation of Caged Heavy Meromyosin

Heavy meromyosin is prepared from rabbit muscle according to Marriott and Heidecker[18] using a purification procedure that has been optimized to preserve the functional activity of myosin. Actin is labeled with tetra-

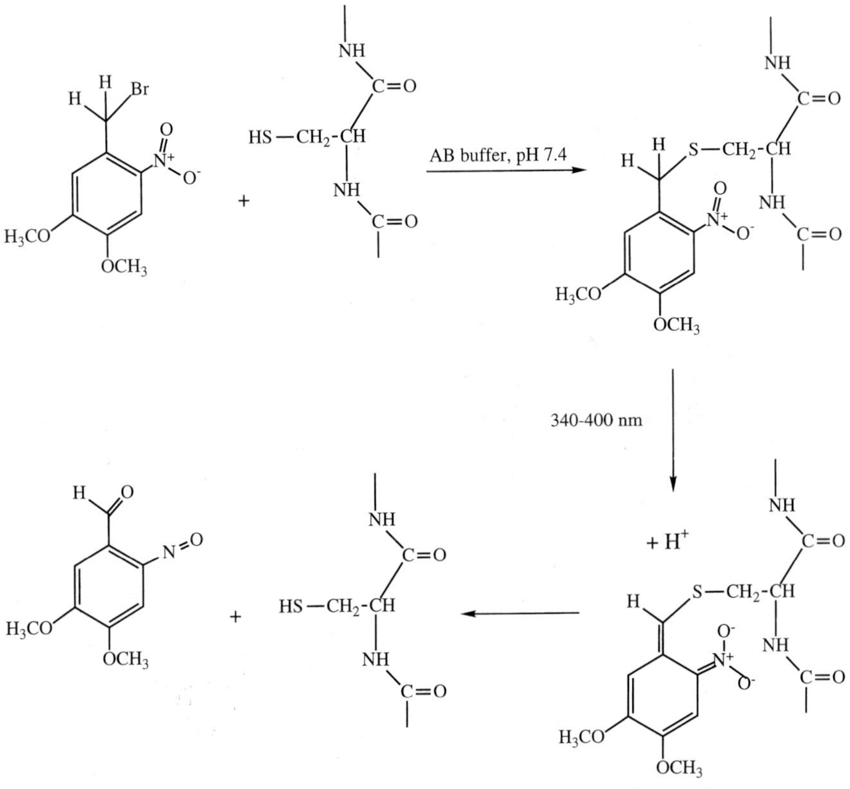

FIG. 7. Structure of DMNBB and proposed mechanism of the protection and light-directed cleavage of the thioether bond linking DMNBB to the thiol of Cys-707 of HMM.

methylrhodamine phalloidin according to the method of Heidecker et al.[39] All labeling reactions that use DMNBB are performed in a darkened room or protected from room light with aluminum foil, and because DMNBB is a very potent alkylating reagent it must be handled with due care and should be dispensed and weighed only in small quantities in a well-ventilated hood. DMNBB-labeled HMM is prepared by treating HMM at 18.6 μM in AB in the absence of DTT with DMNBB (added from a fresh 9 mM stock solution in DMF) to a final concentration of 186 μM for 75 min at room temperature. After an overnight dialysis at 4° against 2 × 1 liter of AB buffer with 10 mM DTT and clarification by centrifugation at 100,000 g for 20 min at 4°, the absorption spectrum of the protein conjugate is recorded and the labeling ratio calculated using an extinction coefficient of DMNBB at 350 nm of 5000 M^{-1} cm^{-1}, and the HMM concentration is determined using the Bradford assay.[31]

These labeling conditions normally result in the modification of two 4,5-dimethoxy-2-nitrobenzyl groups per HMM molecule, or one caged group per ATPase. Inclusion of ATP to 1 mM does not influence the labeling ratio, and DMNBB groups are not incorporated into HMM after a prereaction with NEM (N-ethylmaleimide) according to the method of Reisler[36] (data not shown), which would suggest the DMNBB labeling site is restricted to the SH-1 thiol. The ATPase activity of HMM and caged HMM is measured using the malachite green assay for the detection of phosphate according to Kodama et al.[40] The Ca^{2+}-activated ATPase of caged HMM is increased more than fivefold compared to unmodified HMM (data not shown). This activating effect is often used as an indicator for the modification of the SH-1 thiol group of myosin and its subfragments.[36]

Irradiation of caged HMM at 18 μM in AB buffer/10 mM DTT with near-ultraviolet light (340–400 nm) leads to a dose-dependent decrease in the intensity of the $S_0 - S_1$ transition. After dialysis, a comparison of the absorption spectrum of a nonirradiated caged HMM with the caged HMM conjugate that had been exposed to irradiation for 30 min shows that the ultraviolet light treatment removes most, if not all, of the protection groups. The structureless, residual near-ultraviolet absorption spectrum of the deprotected conjugate probably results from secondary photochemical reactions of the photoproduct 2-nitrosobenzaldehyde with HMM during the long exposure time. However, in the in vitro motility assay, full activation of HMM requires, at most, several hundred milliseconds of illumination with near-ultraviolet light, during which time these secondary photochemical reactions should be negligible.

[39] M. Heidecker, Y. Yan-Marriott, and G. Marriott, *Biochemistry* **34,** 11017 (1995).
[40] T. Kodama, K. Fukui, and K. Kometani, *J. Biochem.* **99,** 1465 (1986).

Transient Kinetics Investigations of Photoactivation of Caged Heavy Meromyosin

The kinetics of the photoactivation reaction of caged HMM is determined by time-resolved absorption spectroscopy after irradiation of caged HMM with a pulse of near-ultraviolet light, using an instrument similar to that described by Uhl et al.[41] Transient absorption spectra of the *aci*-nitro intermediate of DMNBB recorded at defined times (0.1 to 50 msec) after irradiation with a 50-nsec laser-generated pulse of 380-nm light are shown in Fig. 8A. Less than 1% of the DMNBB in caged HMM is photolyzed per irradiation pulse, as measured by the change in the value of the steady-state absorption of the caged group at 350 nm. The peak of the DMNBB *aci*-nitro transient absorption spectrum occurs at 440 nm and is already present 100 μsec after the photolysis pulse. The value of the transient absorption at 440 nm decays in a first-order reaction with a rate constant of 45.6 sec^{-1} at pH 7.4 (Fig. 8B). Because this decay defines the rate-limiting step in the photocleavage reaction,[2] it is reasonable to assume that this will also be the rate at which the actin-activated, force-generating ATPase activity of HMM is generated by light-directed activation of caged HMM.

Preparation and Photoactivation of Caged Heavy Meromyosin in In Vitro Motility Assay

The *in vitro* assay is performed according to Marriott and Heidecker.[18] The *in vitro* chamber is composed of a meticulously cleaned bottom coverslip (50 × 25 mm) coated with nitrocellulose and separated from an upper acid-washed coverslip (24 × 24 mm) by two strips of paper greased with silicon on both sides. Heavy meromyosin at 50 μg/ml in AB buffer is introduced from one side of the chamber for 90 sec and then again from the opposite side for 90 sec. The chamber is washed free of HMM by three changes of AB buffer without DTT and then incubated for 30 min in the dark with AB buffer containing 45 μM DMNBB, added from a fresh DMF stock solution. After this reaction time, excess DMNBB is washed out of the chamber with three changes of AB buffer containing 10 mM DTT. Actin filaments labeled with tetramethylrhodamine phalloidin are then added to a concentration of 20 nM, and after 90 sec unbound filaments are washed out with AB/GO/CAT/DTT buffer and 1 mM ATP is added to the chamber in AB/GO/CAT/glucose/DTT buffer to initiate sliding of filaments by fully functional HMM.

The labeling conditions described above produce a monolayer of caged HMM that does not support motility of F-actin filaments in the *in vitro*

[41] R. Uhl, B. Meyer, and H. Desel, *J. Biochem. Biophys. Methods* **10**, 35 (1984).

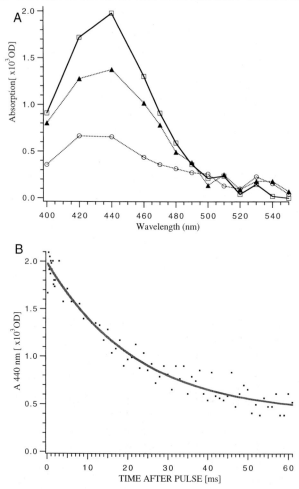

Fig. 8. (A) Selected time-resolved absorption spectra of a 21.4 μM solution of DMNBB-labeled HMM in AB buffer containing 5 mM DTT following irradiation with a 10-nsec pulse of 380-nm light; 0.1 msec (top), 10 msec (middle), and 50 msec (bottom). (B) Time-resolved absorption decay of sample described in A recorded at 440 nm.

motility assay in the presence of 1 mM ATP (Fig. 9), even though its Ca^{2+}-activated. ATPase activity is elevated compared to unmodified HMM. Irradiation of the caged HMM-coated surface with one or two 500-msec pulses of 340–400 nm light delivered from the mercury arc lamp leads to the movement of more than 90% of the actin filaments with a sliding velocity of up to 4 μm/sec (Fig. 9), which is similar to that measured for normal HMM. After the photolysis pulse, shorter actin filaments of 2 μm or smaller run smoothly, whereas longer filaments sometimes sever in the first few seconds and then run smoothly. Actin filaments in the image field adjacent to the photolyzed image field do not move, which allows hundreds of flash experiments to be made on a single chamber. Irradiation of a surface of caged HMM with a polarized pulse of near-ultraviolet light, to photoselect those DMNBB molecules attached to HMM whose absorption dipole moment lies in the direction of the electric vector of the photolysis beam, results in filaments moving in all directions and is not restricted to the direction of the electric vector of the photolysis light. These data suggest indirectly that the head domains of the surface-bound HMM have a considerable degree of rotational freedom around their point of attachment.

The millisecond photogeneration kinetics of actin-activated ATPase activity from caged HMM, together with the high efficiency of light-directed cleavage of the thioether bond, makes caged myosin a unique probe to study the mechanism of muscle contraction using *in vitro* model systems and

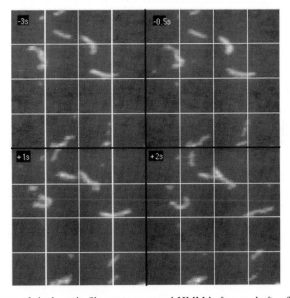

FIG. 9. Images of single actin filaments on caged HMM before and after flash photolysis of the image field. Images of the fluorescent actin filaments on caged HMM taken from −3 to −0.5 sec do not show any actin sliding motion. Irradiation of the image field for 0.5 sec initiates the movement of most actin filaments in the field. The velocity of the filaments is 2–4 μm/sec. Actin filaments in the image field adjacent to the irradiated area do not move (data not shown).

in permeabilized muscle fibers. Interestingly, generation of actin-activated ATPase activity from caged HMM has similar activation kinetics to caged ATP[2] but requires a fraction of the irradiation energy and liberates far less photoproduct and, unlike caged ATP, caged myosin does not suffer the complication of being a competitive inhibitor of myosin.[42]

Concluding Comments

We have shown that caged proteins can be prepared using both lysine- and cysteine-directed caged reagent labeling approaches. The technique will be useful to cage the activity of a large variety of proteins, although one limitation of the lysine-directed labeling strategy is the requirement that wild-type proteins must have a reactive lysine residue in the active site. The cysteine-directed labeling approach has an important advantage in this regard in that standard site-directed mutagenesis techniques can be used to engineer fully active proteins that harbor one or more cysteine residue(s) in the active site. Modification of such critically located cysteine residues with DMNBB or other thiol-directed caged reagents will no doubt improve the ease with which the activity of proteins can be caged.[17] Light-directed activation of protein activity may be used to investigate the function of diverse proteins within living cells. Our interests center on the role of actin-binding proteins, motor proteins, and the small GTPase signaling molecules in the process of cell motility. Light-directed activation of caged conjugates of these proteins may be used to trigger the activity of the protein of interest several hours after their microinjection, a time period necessary for the cell to recover from injuries associated with the impalement. Along this line of research, we have initiated programs to develop and apply caged conjugates of regulatory actin-binding proteins, other myosin isoforms, and constitutively active forms of small GTPases, which will be used to study spatial and kinetic aspects of protein signaling pathways that regulate actin filament dynamics and cell motility.

---

[42] J. Sleep, C. Herrmann, T. Barman, and F. Travers, *Biochemistry* **33,** 6038 (1993).

[7] Caged Peptides and Proteins by Targeted Chemical Modification

By HAGAN BAYLEY, CHUNG-YU CHANG, W. TODD MILLER, BRETT NIBLACK, and PENG PAN

Introduction

Considerable attention has been directed at the preparation of small caged molecules, as witnessed by the material in this volume and as reviewed elsewhere.[1-4] In contrast, only recently has momentum gathered around attempts to produce caged macromolecules for experimental biology. The primary focus has been on caged peptides and proteins, although other macromolecules or supramolecular assemblies have been examined, e.g., caged oligonucleotides[5] and photomodulatable lipid bilayers.[6,7] The caged peptides and proteins described here are 2-nitrobenzyl derivatives that are irreversibly activated on irradiation. A degree of reversible control of enzyme activity can be provided by the introduction of photoisomerizable groups into proteins.[8] However, perfect on–off switching is a formidable problem and has not yet been achieved.

Previous Approaches

Caged proteins have been made by the random introduction of photocleavable protecting groups through chemical modification of reactive amino acid side chains.[9-13] However, because of its predictability and repro-

---

[1] H. Morrison, ed., "Biorganic Photochemistry, Vol. 2: Biological Applications of Photochemical Switches," Wiley-Interscience, New York, 1993.
[2] S. R. Adams and R. Y. Tsien, *Annu. Rev. Physiol.* **55**, 755 (1993).
[3] R. S. Givens and L. W. Kueper, *Chem. Rev.* **93**, 55 (1993).
[4] J. M. Nerbonne, *Curr. Opin. Neurobiol.* **6**, 379 (1996).
[5] P. Ordoukhanian and J.-S. Taylor, *J. Am. Chem. Soc.* **117**, 9570 (1995).
[6] A. Ulman, "Ultrathin Organic Films," Academic Press, San Diego, CA, 1991.
[7] J. L. Thomas and D. A. Tirrell, *Acc. Chem. Res.* **25**, 336 (1992).
[8] I. Willner and R. Shai, *Angew. Chem. Int. Ed. Engl.* **35**, 367 (1996).
[9] G. Marriott, H. Miyata, and K. Kinosita, *Biochem. Int.* **26**, 943 (1992).
[10] G. Marriott, *Biochemistry* **33**, 9092 (1994).
[11] S. Thompson, J. A. Spoors, M.-C. Fawcett, and C. H. Self, *Biochem. Biophys. Res. Commun.* **201**, 1213 (1994).
[12] R. Golan, U. Zehavi, M. Naim, A. Patchornik, and P. Smirnoff, *Biochim. Biophys. Acta* **1293**, 238 (1996).
[13] C. H. Self and S. Thompson, *Nature Med.* **2**, 817 (1996).

ducibility, targeted modification is likely to be more useful. Indeed, targeted modification will be essential for caging single domains in multifunctional proteins. In one sophisticated form of targeted modification, caged proteins have been obtained by the incorporation of unnatural amino acids at specific sites during *in vitro* translation.[14] A second rational approach to the photoregulation of enzymes involves the use of photolabile active-site directed reagents. For example, the protease thrombin can be caged as an *o*-hydroxycinnamoyl acyl enzyme.[15,16] Noncovalent photosensitive reagents that block the active site of a protein until irradiated might also be used. A third approach is targeted chemical modification of single cysteine residues,[17,18] which can be introduced by genetic engineering,[17] and this is described here. Proteins might also be inactivated by site-specific crosslinking to another macromolecule with a photocleavable bifunctional reagent. For example, light-activated immunotoxins have been made by cross-linking.[19] However, in this example, the sites of attachment were not defined. Surprisingly, less work still has been done on caged peptides as compared to caged proteins, although many amino acids with side chains blocked with photolabile protecting groups are available for peptide synthesis.[20] Techniques for caging unprotected peptides in solution are given here.

Caging Peptides and Proteins on Sulfur

Two related approaches for caging peptides and proteins are described here. They are the attachment of photocleavable protecting groups to the sulfur atoms of (1) cysteinyl residues or (2) thiophosphoryl groups. In general, the chemistry has been evaluated with small peptides and then extended to proteins. Nevertheless, it is stressed that caged peptides are themselves potentially useful reagents. For example, when uncaged, peptides might prevent protein–protein interactions or be released from the blockade of the active site of an enzyme.

The highlights of the approaches are summarized as follows. Caged

---

[14] D. Mendel, J. A. Ellman, and P. G. Schultz, *J. Am. Chem. Soc.* **113,** 2758 (1991).

[15] B. L. Stoddard, J. Bruhnke, N. Porter, D. Ringe, and G. A. Petsko, *Biochemistry* **29,** 4871 (1990).

[16] B. L. Stoddard, J. Bruhnke, P. Koenigs, N. Porter, D. Ringe, and G. A. Petsko, *Biochemistry* **29,** 8042 (1990).

[17] C.-Y. Chang, B. Niblack, B. Walker, and H. Bayley, *Chem. Biol.* **2,** 391 (1995).

[18] G. Marriott and M. Heidecker, *Biochemistry* **35,** 3170 (1996).

[19] V. S. Goldmacher, P. D. Senter, J. M. Lambert, and W. A. Blättler, *Bioconjug. Chem.* **3,** 104 (1992).

[20] R. P. Haugland, "Handbook of Fluorescent Probes and Research Chemicals," 6th Ed. Molecular Probes, Eugene, OR, 1996.

cysteinyl or thiophosphoryl peptides have been made by the chemical modification of unprotected peptides in aqueous solution. This circumvents procedures that may seem trivial to the chemist but are generally unappealing to a cell biologist or molecular biophysicist. Mass spectrometry has been used to analyze the caging and uncaging of peptides. Notably, characteristic peaks are generated from the 2-nitrobenzyl group by the laser of matrix-assisted laser desorption/ionization time-of-flight (MALDI-TOF) instruments. Proteins have also been caged on cysteinyl residues. Single-cysteine mutants are now available for many interesting proteins. Cysteine-scanning mutagenesis can be combined with targeted chemical modification to determine which mutants will be suitable for caging. In some cases, it has been possible to generate sufficient protein for this analysis by *in vitro* translation. In our original approach, the modification reagent differed from the reagent ultimately used for caging. However, if the caging reagent is available in abundance, there is no reason it cannot be used directly in conjunction with scanning mutagenesis. Caging reactions can be monitored by examining shifts in the electrophoretic mobility of proteins in extended sodium dodecyl sulfate (SDS)–polyacrylamide gels. Natural cysteines can also be used as targets for caging reactions. In this case, it can simplify matters to remove irrelevant cysteines by mutagenesis. Because many proteins are members of large families, it is likely that a caging reaction developed for one member of the family will be applicable to others, minimizing the need for extensive structure–function analyses. Thiophosphoryl peptides are readily caged. Indeed, thiophosphoryl groups can be caged in the presence of free cysteinyl residues (Fig. 1). Many cell-signaling proteins are regulated by phosphorylation. Therefore, if this approach can be extended to proteins, it will obviate the need for lengthy searches for suitable cysteine mutants.

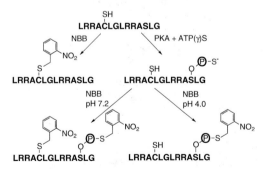

FIG. 1. Selective modification of thiophosphoryl groups in the presence of cysteinyl residues. At low pH, modification of thiophosphorylserine by 2-nitrobenzyl bromide (NBB) in a model peptide can be achieved in the presence of a free sulfhydryl group. [From P. Pan and H. Bayley, *FEBS Lett.* **405,** 81 (1997).]

TABLE I
YIELDS FOR CAGING OF PEPTIDES, AND YIELDS AND QUANTUM EFFICIENCIES FOR
PHOTOLYSIS OF CAGED PEPTIDES[a]

Peptide	Reagent	% yield of Caging	% yield of Photolysis[c]	ϕ
C-Kemptide[b]	NBB	95	70	0.62
C-Kemptide	DMNBB	95	67	0.15
C-Kemptide	BNPA	95	62	0.21
Thiophosphorylkemptide	NBB	95	70	0.23
Thiophosphorylkemptide	DMNBB	95	55	0.06
Thiophosphorylkemptide	BNPA	10	—	—

[a] From P. Pan and H. Bayley, *FEBS Lett.* **405**, 81 (1997).
[b] Kemptide is Leu-Arg-Arg-Ala-Ser-Leu-Gly.
[c] Photolysis was at pH 5.8.

Summary of Findings

Two major considerations have dictated the choice of sulfhydryl-directed 2-nitrobenzyl reagents for derivatizing peptides and proteins. First is the ability to obtain a high yield of caged product. The possibility of carrying out the caging reaction in aqueous solution is especially important for proteins. Second, the caged reagent should be efficiently photolyzed at wavelengths that do not damage biological molecules. Therefore, a caging group should have a high product quantum yield (ϕ) in the near-ultraviolet (UV) or visible (e.g., $\phi \geq 0.1$) at neutral pH, with a useful extinction coefficient (ε) (e.g., ≥ 1000 M^{-1} cm^{-1}). With regard to these criteria, the 2-nitrobenzyl groups has proved satisfactory in many applications with small molecules, and therefore it was our first choice for peptides and proteins.

Three 2-nitrobenzyl reagents have been explored for caging cysteine-containing peptides: 2-nitrobenzyl bromide (NBB), 2-bromo-2-(2-nitrophenyl)acetic acid (BNPA), and 4,5-dimethoxy-2-nitrobenzyl bromide (DMNBB).[21] All were effective (Table I). 2-Bromo-2-(2-nitrophenyl)acetic acid was synthesized as a highly water-soluble reagent,[17] but the model peptide we used was released less readily when caged with BNPA than with NBB. When DMNBB was used to produce a caged reagent with a higher extinction in the near-UV [e.g., 2-nitrobenzyl (NB), $\varepsilon_{312} \sim 1300$; 4,5-dimethoxy-2-nitrobenzyl (DMNB), $\varepsilon_{312} \sim 4500$], a compensating reduction in product quantum yield was observed. Therefore, the DMNB-caged peptide was photolyzed at about the same rate as the NB-caged peptide.

[21] P. Pan and H. Bayley, *FEBS Lett.* **405**, 81 (1997).

Where yet longer wavelengths (≥360 nm) are required to prevent tissue damage and NB absorption is still weaker, DMNB peptides could still be useful. Poorly soluble reagents such as NBB and DMNBB must be added from a concentrated stock solution in organic solvent. Nevertheless, the final concentration of organic solvent can be as low as 2%, which is tolerated by peptides and most proteins.

Interestingly, all three reagents have also been used to cage proteins. In the case of a pore-forming bacterial exotoxin, α-hemolysin (αHL), BNPA was highly effective, yielding a caged cysteine mutant [α-hemolysin mutant Arg-104→Cys(αHL-R104C)-α-carboxynitrobenzyl(CNB)] with no residual activity in a very sensitive assay.[17] Uncaging occurred with ~60% efficiency. Interestingly, when NBB was reacted with αHL-R104C, only 75% inactivation was obtained, and no further decrease in activity occurred after prolonged incubation (unpublished data, 1994). The negative charge at residue 104 of α-hemolysin contributed by the CNB caging group must play an important role in abolishing pore-forming activity. In contrast, NBB has proven to be the most effective reagent for caging the catalytic (C) subunit of cAMP-dependent protein kinase (PKA).[22] Compared with NBB, BNPA and DMNBB were somewhat less effective at inactivating a single-cysteine version of the C subunit, and the caged proteins obtained with these reagents were markedly less well activated on irradiation.[22] In another example, caged heavy meromyosin was best produced by selective derivatization of one of two natural cysteines with DMNBB.[18] In this case, an alternative reagent, 1-(bromomethyl)-2-nitro-4-benzoic acid,[9] gave a caged protein that was not efficiently activated by light.

The efficacy of a reagent depends, then, on the nature of the target protein. This situation might be rationalized in several ways. For example, the rate of caging might be affected by the microenvironment of the cysteinyl residue (although in the cases cited, the reactions were pushed to completion), the activity of the protein might be significant after complete modification with one reagent but not with another, or the efficiency of the uncaging reaction (product quantum yield: ϕ) might also be affected by the microenvironment on the protein, and in this case by quenching of the excited state, e.g., by nearby tryptophans. In short, with a particular protein, it is advisable to test all three 2-nitrobenzyl reagents described here, as well as other reagents that are becoming available. Of course, the appropriate control incubations must also be done to ensure that the protein is not damaged by the modification or reactivation conditions. Finally, we have reported the caging of thiophosphorylserine peptides with NBB and DMNBB. In this case, BNPA appeared to react too slowly to be useful.[21]

---

[22] C.-Y. Chang, T. Fernandez, R. Panchal, and H. Bayley, in preparation (1998).

Goals

Caged peptides and proteins should prove useful in several areas of experimental biology, including cell biology and molecular biophysics. The capacity to control the time, dose, and site of activation of caged reagents is a tremendous advantage. For example, in cell biology, the ability to load cells with caged signal transduction proteins by microinjection and trigger their activity at will should be invaluable.[23] Such studies could be carried out with cells in culture or with living tissues, such as early embryos for studies of development. In biophysics, supramolecular assembly or enzymatic catalysis might be triggered with light and monitored with techniques ranging from single-crystal X-ray diffraction to circular dichroism or infrared spectroscopy. The criteria for reagents for the various applications can differ greatly. For example, in cell biology, it would be best to produce a caged peptide or protein with no residual activity at all. Then, the photolytic release of even 5% of the theoretical maximum activity within a cell would be very useful. In contrast, in many biophysics experiments, 5% residual activity might be tolerable. However, a far greater extent of uncaging would be needed in most circumstances (e.g., 95%). The rate of release after a short light flash might also be more important here than in a cell biology experiment, with the important exception of fast synaptic transmission.[4,24] This rate depends on the breakdown of photogenerated intermediates and should not be confused with the steady-state rate of production of uncaged products during continuous photolysis, which is a function of lamp intensity. The 2-nitrobenzyl group is generally lost fairly slowly after excitation, in milliseconds to seconds.[1,25,26] In favorable circumstances, release in the microsecond range occurs.[25] Protecting groups that are lost much more quickly are under development.[27] Finally, in biotherapeutics, caged peptides and proteins might be activated in specific tissues in controlled doses, at defined times.[28] Design criteria here must include the use of the longest wavelengths possible to prevent tissue damage and to allow tissue penetration by light.

In our laboratory, we are especially interested in caged peptides and proteins for signal transduction research. We have developed reagents for

---

[23] S.-H. Wang and G. J. Augustine, *Neuron* **15**, 755 (1995).
[24] G. P. Hess, L. Niu, and R. Wieboldt, *Ann. N.Y. Acad. Sci.* **757**, 23 (1995).
[25] R. Wieboldt, K. R. Gee, L. Niu, D. Ramesh, B. K. Carpenter, and G. P. Hess, *Proc. Natl. Acad. Sci. U.S.A.* **91**, 8752 (1994).
[26] A. Barth, K. Hauser, W. Mäntele, J. E. T. Corrie, and D. R. Trentham, *J. Am. Chem. Soc.* **117**, 10311 (1995).
[27] C.-H. Park and R. S. Givens, *J. Am. Chem. Soc.* **119**, 2453 (1997).
[28] H. Bayley, F. Gasparro, and R. Edelson, *Trends Pharmacol. Sci.* **8**, 138 (1987).

the controlled permeabilization of cells, including protease-activated[29] and metal-regulated α-hemolysins,[30] as well as a caged hemolysin.[17] Caged cysteinyl and thiophosphoryl model peptides have been tested.[21] Therefore, we are now in a position to make caged peptides to control protein–protein interactions (e.g., those involving proteins that bind to serine phosphate, such as 14-3-3 proteins[31]) and to control enzyme activity (e.g., caged kinase pseudosubstrates[32]). A caged cAMP-dependent protein kinase catalytic subunit has been developed,[22] and we hope to apply the technology we have used to additional protein kinases, GTP (guanosine triphosphate)-binding proteins, transcription factors, and other signaling molecules. We are also extending the approach to thiophosphoryl proteins. The activities of many proteins are regulated by phosphorylation, which can often be mimicked by thiophosphorylation. Furthermore, thiophosphorylated proteins are often relatively resistant to cellular phosphatases, so an uncaged thiophosphoryl protein would persist in a cell. Because the requirements for phosphorylation are quite specific, we can be certain that caging will block the functional effects of thiophosphorylation and that uncaging will restore them.

Materials

Kemptide (Leu-Arg-Arg-Ala-Ser-Leu-Gly) is purchased from Peninsula Laboratories (San Carlos, CA). C-Kemptide (Leu-Arg-Arg-Ala-Cys-Leu-Gly) is synthesized at the Worcester Foundation Protein Chemistry Facility. Kemptide CS-dimer (Leu-Arg-Arg-Ala-Cys-Leu-Gly-Leu-Arg-Arg-Ala-Ser-Leu-Gly) was synthesized by Quality Controlled Biochemicals (Hopkinton, MA) and further purified by high-performance liquid chromatography (HPLC). Recombinant α-hemolysin R104C was purified at the Worcester Foundation for Biomedical Research by Stephen Cheley and Christopher Shustak after expression in *Staphylococcus aureus*. α-HL-R104C is also obtained by *in vitro* transcription and translation in an *Escherichia coli* S30 system.[17] A plasmid encoding the mutant C subunit of PKA, C subunit mutants Cys-343→Ser(C343S), was kindly provided by Dr. Susan Taylor (University of California, San Diego). Protein kinase inhibitor peptide (PKIP) Affi-Gel 10 is prepared as described.[33] Adenosine 5'-O-(3-thiotriphosphate) (ATPγS) is purchased from Calbiochem-Novabiochem

[29] R. G. Panchal, E. Cusack, S. Cheley, and H. Bayley, *Nature Biotechnol.* **14,** 852 (1996).
[30] M. J. Russo, H. Bayley, and M. Toner, *Nature Biotechnol.* **15,** 278 (1997).
[31] A. J. Muslin, J. W. Tanner, P. M. Allen, and A. S. Shaw, *Cell* **84,** 889 (1996).
[32] R. B. Pearson and B. E. Kemp, *Methods Enzymol.* **200,** 62 (1991).
[33] S. R. Olsen and M. D. Uhler, *J. Biol. Chem.* **264,** 18662 (1989).

(La Jolla, CA). 2-Bromo-2-(2-nitrophenyl)acetic acid was synthesized[17] and is now available from Molecular Probes (Eugene, OR). 2-Nitrobenzyl bromide is purchased from Fluka (Ronkonkoma, NY), and DMNBB from Aldrich (Milwaukee, WI). 4-Acetamido-4'-[(iodoacetyl)amino]stilbene 2,2'-disulfonate (IASD) and 1-(2-nitrophenyl)ethyl (NPE) phosphate (diammonium salt) are from Molecular Probes. Inorganic phosphate is measured with a kit from Sigma (St. Louis, MO).

Mass Spectrometry

Mass spectrometry is done with a PerSeptive Voyager matrix-assisted laser desorption/ionization mass spectrometer (MALDI-MS) equipped with a 337-nm nitrogen laser (PerSeptive Biosystems, Framingham, MA). Analyte samples in ~25% (v/v) acetonitrile in 0.1% aqueous trifluoroacetic acid (TFA) or in a low-salt buffer solution (0.6 μl) are spotted onto the sample plate. The matrix, α-cyano-4-hydroxycinnamic acid, is then added as a saturated solution in 25% (v/v) acetonitrile in 0.1% aqueous TFA (0.6 μl) and the sample is dried at room temperature before analysis at a scan rate of 5 sec^{-1}. When 2-nitrobenzyl peptides are examined, the following phenomena prove useful for structure confirmation[21]: (1) a peak at MH$^+$−16, presumably due to loss of oxygen from the -NO$_2$ group[34]; (2) a peak at MH$^+$−135 corresponding to loss of the 2-nitrobenzyl group; and (3) in the case of caged thiophosphoryl groups, a peak at MH$^+$−151 corresponding to loss of the 2-nitrobenzyl group and a further 16 mass units, presumably due to sulfur–oxygen exchange. These additional peaks arise through photolysis with the 337-nm laser, as they are not observed when electrospray ionization mass spectrometry (ESI-MS) is carried out. Sulfur–oxygen exchange does not occur when caged thiophosphoryl peptides are photolyzed in solution. Finally, when fragmentation for mass mapping is performed, e.g., with trypsin, the fragment carrying the 2-nitrobenzyl group also undergoes the same photochemistry in the MALDI-MS, which aids in the identification of the modified residue. Mass spectra that illustrate these phenomena are included below.

Caged Cysteinyl Peptides

C-Kemptide (Leu-Arg-Arg-Ala-Cys-Leu-Gly, 0.2 mM) is reacted with NBB, DMNBB, or BNPA in 100 mM Tris-HCl, pH 7.2, at 25° in the dark for 1 hr. Stock solutions (50 mM) of NBB, DMNBB, and BNPA are

---

[34] H. A. Morrison, in "The Chemistry of the Nitro and Nitroso Groups" (H. Feuer, ed.) pp. 165–213. Interscience, New York, 1969.

prepared in, respectively, 95% ethanol, dimethylformamide, and 100 mM Tris-HCl, pH 7.2. Final concentrations in the reaction mixture are 2 mM NBB, 1 mM DMNBB, and 2 mM BNPA. The caged C-kemptides are purified by HPLC with a Vydac C_{18} column (Vydac, Hesperia, CA) [buffer A: 0.1% TFA in water; buffer B: 0.08% TFA in acetonitrile/water (7:3)] (Fig. 2A). The conversion is ~95% in all three cases. When examined by MALDI-MS, the purified peptides give the expected MH$^+$ peaks as well as the laser-generated fragments described earlier (Fig. 2B). C-Kemptide caged by BNPA appears as two peaks (both m/z 966) on HPLC, corresponding to the diastereomers of α-carboxy-2-nitrobenzyl C-kemptide.

Thiophosphorylation of Model Peptides

Kemptide (Leu-Arg-Arg-Ala-Ser-Leu-Gly, 1 mg, 1.3 μmol) or kemptide CS-dimer (Leu-Arg-Arg-Ala-Cys-Leu-Gly-Leu-Arg-Arg-Ala-Ser-Leu-Gly, 1 mg, 0.65 μmol) is dissolved in 100 mM 3-(N-morphilino)propanesulfonic acid (MOPS), pH 7.0, containing 100 mM KCl and 10 mM MgCl$_2$ (300 μl). ATPγS in H$_2$O (300 μl) is added in fivefold molar excess over peptide. Thiophosphorylation is initiated by the addition of C subunit of PKA (2 μl; final concentration of 1.0 μg/ml), followed by incubation at room temperature for 2 hr. Thiophosphorylated peptides are purified by HPLC, with the Vydac C_{18} column, and then analyzed by MALDI-MS. MH$^+$ ions are observed at m/z 869 for thiophosphorylkemptide and m/z 1638 for the kemptide CS-dimer. Purified thiophosphorylated peptides are lyophilized and stored at $-20°$.

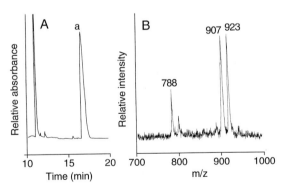

FIG. 2. HPLC purification and MALDI-MS analysis of C-kemptide after reaction with NBB. (A) HPLC chromatogram of a reaction mixture consisting of 20 μM C-kemptide and 0.2 mM NBB after incubation for 60 min in 100 mM Tris–HCl, pH 7.2, at 25°. Peak a is the caged peptide. (B) MALDI-MS of purified NB-C-kemptide from (A). The MH$^+$ peak is at m/z 923. Peaks at MH$^+$−16 and MH$^+$−135 are also seen.

Caged Thiophosphoryl Peptides

Caged thiophosphorylkemptides are synthesized by reaction with NBB and DMNBB under the same conditions used for caging C-kemptide. Both derivatives are formed in ~95% yield, purified by HPLC (Fig. 3A), and analyzed by MALDI-MS, which again gives the expected MH^+ peaks as well as the laser-generated fragments described earlier (Fig. 3B). No reaction occurs between thiophosphorylkemptide and BNPA at neutral or basic pH. When thiophosphorylkemptide is incubated with 2 mM BNPA at pH 4.0, 37°, for 4 hr, a yield of ~10% of the diastereomers of the caged thiophosphorylkemptide is obtained. Because of the low yields, under relatively harsh reaction conditions, these derivatives were not further investigated.

Selective Reaction at Thiophosphoryl Group

Kemptide CS-dimer (0.2 mM) or thiophosphorylkemptide CS-dimer (0.2 mM) and NBB (2 mM) are allowed to react in 100 mM Tris–HCl, pH 7.2, at 25° in the dark for 100 min. Kemptide CS-dimer caged on the cysteinyl residue (MALDI-MS, m/z 1677) and thiophosphorylkemptide CS-dimer caged on both the cysteinyl residue and thiophosphoryl serine (MALDI-MS, m/z 1909) are obtained.

The reactivities of cysteinyl and thiophosphoryl groups toward NBB differ at low pH and permit the caging of thiophosphorylserine in the

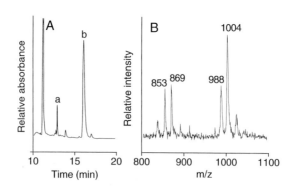

FIG. 3. HPLC purification and MALDI-MS analysis of thiophosphorylkemptide after reaction with NBB. (A) HPLC chromatogram of a reaction mixture containing 20 μM thiophosphorylkemptide and 0.2 mM NBB after incubation for 60 min in 100 mM Tris–HCl, pH 7.2, at 25°. Peak a is unreacted peptide and peak b the caged peptide. (B) MALDI-MS of purified NB-thiophosphorylkemptide from (A). The MH^+ peak is at m/z 1004. Peaks at MH^+-16, MH^+-135, and MH^+-151 are also seen.

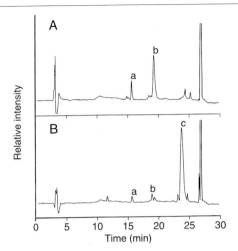

FIG. 4. HPLC chromatograms of reaction mixtures containing 20 μM thiophosphoryl CS-dimer and 0.2 mM NBB incubated at pH 4.0 and pH 7.2. (A) Incubation in 100 mM sodium acetate, pH 4.0, at 25° for 100 min. (B) Incubation in 100 mM Tris–HCl, pH 7.2, at 25° for 100 min. a, thiophosphoryl CS-dimer; b, thiophosphoryl CS-dimer caged on thiophosphate; c, thiophosphoryl CS-dimer caged on both cysteine and thiophosphate.

presence of free cysteine in the same peptide. Presumably, the cysteinyl residue (p$K_a \simeq 9.5$) is protonated under these conditions, whereas the thiophosphoryl group is a reactive monoanion (p$K_a \simeq 5.5$).[35] Selective caging of thiophosphorylkemptide CS-dimer (0.2 mM) with NBB (2 mM) in 80% yield is achieved in 100 mM sodium acetate buffer, pH 4.0, at 25° in the dark for 100 min (Fig. 4). MALDI-MS (m/z 1774) of the HPLC purified product is consistent with only one 2-nitrobenzyl group per peptide. The thiophosphoryl CS-dimer, caged with NBB at pH 4.0, is digested with trypsin, and the fragments are analyzed directly by MALDI-MS (Fig. 5). The peak at m/z 734 is consistent with the fragment RAS^P*LG, where S^P* is the caged thiophosphoryl residue, and the peak at m/z 600 with its uncaged product.

Caging at the sulfur of thiophosphoryl groups should be applicable to *native* proteins provided that they are not irreversibly denatured or precipitated at the low pH values required for selective modification. Shifts in the pK_a values of cysteinyl and thiophosphoryl groups induced by microenvironments on a protein may facilitate or complicate selective modification, and the reaction conditions will have to be adjusted for each case.

[35] E. K. Jaffe and M. Cohn, *Biochemistry* **17,** 652 (1978).

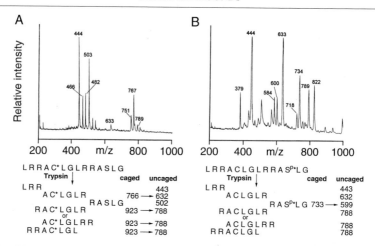

FIG. 5. MALDI-MS analysis of trypsin digests. (A) NB-caged CS-dimer (nonthiophosphorylated) and (B) NB-caged thiophosphoryl CS-dimer (made at pH 4.0). The peptides were digested with trypsin (1 μg/ml) in 25 mM NH$_4$HCO$_3$ at pH 8.9 for 10 min and then subjected to MALDI-MS analysis. Identified tryptic fragments, including those uncaged by the laser, are tabulated as unprotonated species. [From P. Pan and H. Bayley, *FEBS Lett.* **405**, 81 (1997).]

α-Hemolysin R104C Caged on Cys-104

αHL-R104C (0.36 mg/ml; stored in 10 mM sodium acetate, 150 mM NaCl, pH 5.2) is dialyzed against 10 mM Tris–HCl, pH 8.5, containing 5mM 2-mercaptoethanol. Dialyzed R104C (100 μl), 10 mM dithiothreitol (DTT) in water (20 μl), and 1.0 M Tris-HCl, pH 8.5 (40 μl) are incubated for 5 min at room temperature, prior to the addition of 100 mM BNPA in 100 mM sodium phosphate, pH 8.5 (40 μl). After 3 hr at room temperature, 1.0 M DTT (5 μl) is added, and excess reagents are removed from the protein by gel filtration on Bio-Gel P-2 (Bio-Rad, Richmond, CA) eluted with 10 mM Tris–HCl, pH 8.5, containing 50 mM NaCl. About 80% conversion to caged αHL-R104C is achieved as judged by sodium dodecyl sulfate–polyacrylamide gel electrophoresis (SDS–PAGE). The remaining 20% of the protein is converted to two inactive products that migrate near unmodified αHL-R104C.

Caging of *in Vitro* Translated Protein

In many cases, sufficient protein for analysis can be obtained by caging the product of an *in vitro* translation and, often, the unfractionated translation mix can be used. An *Escherichia coli* S30 transcription/translation system can produce 10–50 μg protein/ml. In calculating the molar excess of caging reagent to be used, thiols present in the translation mix should be taken into account. Modification of *in vitro*-translated protein is a particulary useful approach for scanning the properties of a collection of cysteine mutants, before choosing one for detailed analysis or increased production.

αHL-R104C (30 μl of S30 translation mix) is reduced for 5 min at room temperature by the addition of 10 mM DTT (15 μl in water), 1.0 M Tris–HCl, pH 8.5 (30 μl), and water (45 μl). BNPA (100 mM) is added in 100 mM sodium phosphate, pH 8.5 (30 μl) and the reaction is incubated at room temperature in the dark for 9 hr. The progress of a modification reaction can be followed by gel-shift electrophoresis (Fig. 6). Excess BNPA and its low-mass sulfhydryl adducts are then removed by repeated cycles of dilution with 100 mM Tris–HCl, pH 8.5, and concentration by ultrafiltration (Amicon, Microcon 3, Millipore Corp., Bedford, MA). The final volume (135 μl) is readjusted to 150 μl with 100 mM Tris–HCl containing 10 mM DTT so that the final concentration of DTT is 1 mM. The sample is stored at $-20°$ before use.

Catalytic Subunit of cAMP-Dependent Protein Kinase Caged at Cys-199

The Cα subunit of mouse PKA contains two cysteines: the crucial Cys-199 and the dispensable Cys-343. We avoid possible complications with Cys-343, such as intermolecular disulfide bond formation, by using the mutant C343S kindly provided by the laboratory of Dr. Susan Taylor (Uni-

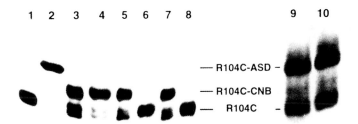

FIG. 6. Photolysis of αHL-R104C-CNB examined by SDS–PAGE. ^{35}S-labeled R104C was treated as described in the text and subjected to electrophoresis in a 12% SDS–polyacrylamide gel. An autoradiogram of a region between the 29.5 and 45.5-kDa markers is shown. Lane 1, R104C; lane 2, R104C treated with IASD; lane 3, R104C treated with BNPA; lane 4, R104C treated with BNPA and then further treated with IASD; lane 5, BNPA adduct of R104C after removal of excess reagent; lane 6, as lane 5 but photolyzed at pH 6.0; lane 7, BNPA adduct of R104C after removal of excess reagent; lane 8, as lane 7 but photolyzed at pH 8.5; lane 9, as lane 6, but the photolysis products were further treated with IASD; lane 10, as lane 8, but the photolysis products were further treated with IASD. [From C.-Y. Chang, B. Niblack, B. Walker, and H. Bayley, *Chem. Biol.* **2**, 391 (1995).]

versity of California, San Diego). Caging the C subunit turns out to be more difficult than the caging of αHL-R104C. Numerous sets of conditions were tested, as documented elsewhere.[22] In this case, NBB gives the best results. BNPA and DMNBB give similar inactivation, but poorer reactivation on photolysis.

C343S (4.3 μM in 100 mM Tris–HCl, pH 8.5, 1 mM EDTA) is incubated with NBB (2 mM) at 25° for 45 min. NBB is added from a 50 mM stock in ethanol, resulting in a final ethanol concentration of 4%, which is tolerated by the enzyme. Excess NBB is removed by gel filtration. Immediately before use, Sephadex G-25 (Sigma) is equilibrated in 20 mM Tris–HCl, pH 8.5, 200 mM KCl, degassed, and packed into a small gel filtration column (1 ml). The reaction mix containing the modified C subunit (50 to 150 μl) is applied and eluted with the same buffer, and fractions (50 μl) containing C343S-NB are collected. The specific catalytic activity of the C subunit is measured before and after reaction with NBB.[36] The residual specific activity is ~8%. As judged by SDS–PAGE, the extent of modification is ~95%. The residual activity can be reduced to ~4% by a second modification with N-ethylmaleimide. The reaction mix is adjusted to pH 7.4, and N-ethylmaleimide is added to a final concentration of 10 mM and allowed to react at 25° for 20 min.

Unmodified C343S can be removed with PKIP coupled to Affi-Gel 10. To treat NBB-modified C343S, the matrix is equilibrated with 2× PKIP binding buffer [40 mM Tris–HCl, pH 7.4, 4 mM MgCl$_2$, 2 mM ATP, 1 mM phenylmethylsulfonyl fluoride, 2 mM EDTA, 2 mM DTT, and 0.2% Nonidet P-40 (NP-40)]. The supernatant is removed and the gel gently mixed with an equal volume of eluant from the Sephadex G-25 column at 4° for 2 hr. The PKIP Affi-Gel is then separated by centrifugation. The supernatant contains the C343S-NB (86% recovery). The activity of C343S-NB is similar before and after the PKIP Affi-Gel treatment, suggesting that the residual activity after NBB treatment is derived from the modified C subunit and not from traces of unmodified protein.

Photolysis of Caged Peptides and Proteins

We estimated the quantum yields for formation of uncaged product from both peptides and proteins by comparison with the rate of photolytic phosphate release from 1-(2-nitrophenyl)ethyl phosphate (NPE phosphate, "caged phosphate"), for which $\phi = 0.54$ as determined by the ferrioxalate actinometer method.[37,38] NPE phosphate is a convenient reagent with which

[36] R. Roskoski, *Methods Enzymol.* **99**, 3 (1983).
[37] J. H. Kaplan, B. Forbush, and J. F. Hoffman, *Biochemistry* **17**, 1929 (1978).
[38] J. W. Walker, G. P. Reid, J. A. McCray, and D. R. Trentham, *J. Am. Chem. Soc.* **110**, 7170 (1988).

to make this comparison, although a reading of the original literature[37,38] suggests that the ϕ value has not been determined with certainty and is perhaps wavelength dependent. Further, the pH at which ϕ was determined and the pH dependence of ϕ are not stated. Our experiments suggest that the pH dependence is weak. Therefore, at the very least, the relative ϕ values obtained with NPE phosphate are of value.

For a dilute solution, the rate of photolysis of a reagent is (see, e.g., Bayley[39]):

$$v \simeq 2.303\phi I_0 \varepsilon c$$

where v is the rate of photolysis of the reagent in M sec^{-1}, ϕ is the product quantum yield, I_0 is the incident light intensity in millimoles of photons sec^{-1} cm^{-2}, ε is the extinction coefficient in M^{-1} cm^{-1}, and c is the concentration in M. The half-time of photolysis follows:

$$\tau \simeq 0.3/\phi I_0 \varepsilon$$

Therefore, for low light absorption, the half-time of the reaction is inversely proportional to the quantum yield, light intensity, and extinction coefficient, as intuition would suggest. It is not a function of the concentration of the reagent.

A quantum yield can therefore be obtained by using

$$\phi_2 = \phi_1 \varepsilon_1 \tau_1 / \varepsilon_2 \tau_2$$

where ϕ_1 is a known product quantum yield such as that of NPE phosphate.

This equation applies for a single wavelength (i.e., it is fine for a laser), so it may be necessary to correct for the number of photons absorbed over the entire range of wavelengths emitted by a lamp and passed by a filter. It is especially useful for caged proteins, which are generally used as relatively dilute solutions.

If the absorbance is high, the equation does not apply. Obviously, in this case all the incident light is absorbed by the sample and the value of ε is irrelevant. Here, the initial rate of photolysis will be a function of ϕ and independent of the concentration of reactant:

$$v_1/v_2 = \phi_1/\phi_2$$

Photolysis Conditions

Caged peptide (300 μl, 1.6–1.8 mM) in 100 mM sodium acetate, pH 5.8, is placed in the well of a 96-well microplate on ice. Irradiation is

---

[39] H. Bayley, "Photogenerated Reagents in Biochemistry and Molecular Biology," Elsevier, Amsterdam, 1983.

carried out with a 30-W UV lamp, which is set 1.5 cm above the plate (2200 μW cm^{-2}, peak emission 312 nm, Cole-Parmer, Chicago, IL). A glass filter, providing a window at 290–380 nm (Oriel Corp., Stratford, CT), is used. Irradiated material is removed at various time points and analyzed by HPLC. Values of ϕ are calculated by using the equation for solutions of high absorbance (Table I). Photolysis is considerably less efficient at neutral pH: NB-C kemptide ($\phi_{\text{pH 5.8}} = 0.62$; $\phi_{\text{pH 7.2}} = 0.14$), NB-thiophosphoryl-kemptide ($\phi_{\text{pH 5.8}} = 0.23$; $\phi_{\text{pH 7.2}} = 0.04$).

αHL-R104C-CNB in 100 mM Tris–HCl (pH 6.0 or 8.5) containing 1 mM DTT (60 μl) is placed in a well of a 96-well microtiter plate and irradiated on ice through a 285-nm cutoff filter (Oriel) at 3.5 cm from a Foto UV 300 illuminator (Fotodyne, Hartland, WI). To assay for the unmasking of Cys-104, samples are treated with IASD and analyzed by SDS–PAGE (Fig. 6). For the sample at pH 8.5, irradiated αHL-R104C-CNB (5 μl) is diluted with 100 mM Tris–HCl, pH 8.5 (3 μl), and reacted with 100 mM IASD in water (2 μl) for 1 hr at room temperature. For the sample at pH 6.0, irradiated αHL-R104C-CNB (5 μl) is diluted with 1.0 M Tris–HCl, pH 8.5 (2 μl), and water (1 μl) and treated in the same way.

C343S-NB collected from Sephadex G-25 is either left at pH 8.5 or adjusted to pH 6.0 with 0.5 M sodium phosphate, pH 5.5. Solutions are made 8.5 mM in DTT before photolysis. Samples (40 μl, 0.5 μM protein) are placed in wells of a 96-well microtiter plate, which is put on ice. Photolysis is performed through the 285-nm cutoff filter with a 30-W UV lamp, which is set 2 cm above the plate (Cole-Parmer). At various time points after the initiation of photolysis, samples are taken for kinase assay (5 μl) and IASD modification (32 μl). The IASD is added to 20 mM, and the sample is incubated for 20 min before analysis by SDS–PAGE (Fig. 7). The quantum efficiency is calculated using the equation for solutions of low absorbance ($\phi_{\text{pH 6.0}} = 0.84$; $\phi_{\text{pH 8.5}} = 0.14$).

No clear rules for estimating ϕ emerge from our studies or those of others. There is no escaping the fact that for each new caged reagent the efficiency of photolysis and indeed the identity of the photolysis product

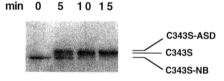

FIG. 7. Photolysis of C343S-NB. C343S-NB was prepared as described and photolyzed at pH 6.0 for 0, 5, 10, and 15 min. The photolyzed samples were reacted with IASD to form C343S-ASD. [From C.-Y. Chang, T. Fernandez, R. Panchal, and H. Bayley, *in preparation* (1998).]

must be determined experimentally. Obviously, a high value for ϕ is worthless if the desired product is not formed. Reagents with $\phi \geq 0.05$ should be useful for most applications, but again the photosensitivity of each experimental system must be examined. Furthermore, ϕ should be determined under conditions relevant to the application. The efficiency of photolysis of 2-nitrobenzyl reagents is pH dependent and is generally worse at neutral pH than at mildly acidic pH. In our hands, the sensitivity of ϕ to pH in the physiological range varies with the nature of the linkage (e.g., cysteine versus thiophosphate[21]) and in proteins with the site of attachment.[17] Multisubunit proteins may present an additional problem. In the case of αHL-R104C-CNB, uncaged protein is released in ~60% yield on irradiation, but only 10–15% of the pore-forming activity is regained. The αHL pore is a heptamer, and it is likely that improperly photolyzed subunits have a negative effect on activity.

Future Goals

Our immediate goals include applying the reagents we have so far developed to cellular studies. Caged peptides should present little difficulty in this regard. However, caged proteins must be concentrated for injection into cells and, if possible, stored in this form. For example, it is unlikely that glycerol can be used as a cryoprotectant. With regard to caged peptides, one of the most useful groups of reagents will be molecules caged at thiophosphoryltyrosine. Although there are proteins that bind to phosphoserine (e.g., 14-3-3 protein), phosphotyrosine interactions, e.g., with SH2 domains, are of paramount importance in cell signaling. In preliminary experiments with Arg-Arg-Leu-Ile-Glu-Asp-Ala-Ile-Tyr-Ala-Ala-Arg-Gly thiophosphorylated on Tyr with the insulin receptor kinase domain, the thiophosphoryl peptide reacted efficiently with NBB, but photolysis of the caged peptide was very inefficient ($\phi < 0.05$). The production of peptides thiophosphorylated at tyrosine has also been a problem: receptor and nonreceptor tyrosine kinases do not catalyze this reaction efficiently. In many cases, we were unable to detect thiophosphorylation of peptides with ATPγS using tyrosine kinases (e.g., Abl, Src, and epidermal growth factor (EGF) and insulin receptors) under conditions that work well when ATP itself is used. A promising approach is to use divalent metal ions other than Mg^{2+} in the reaction (Grace et al.[40] and W. T. Miller, unpublished, 1997). Another area of attention is to test reagents other than 2-nitrobenzyl halides for caging cysteinyl and thiophosphoryl peptides and proteins. Especially promising is the p-hydroxyphenacyl group.[27] We have tested p-hy-

[40] M. R. Grace, C. T. Walsh, and P. A. Cole, *Biochemistry* **36,** 1874 (1997).

droxyphenacyl bromide (a gift of Dr. Richard Givens, University of Kansas, Lawrence) on cysteinyl peptides with encouraging results. This reagent should provide caged peptides and proteins that are rapidly uncaged after a light flash,[27] with ϕ values that are less sensitive to pH (unpublished data, 1996), and which contain no additional chiral centers (cf. BNPA). Furthermore, the nitrosoketones and nitroaldehydes generated from 2-nitrobenzyl caging groups are highly reactive and should be scavenged, e.g., by thiols.[37,38] The p-hydroxyphenacyl group is converted to the harmless 4-hydroxyphenylacetic acid after hydrolysis of a spiroketone intermediate. Because the intermediate has the potential to react with nucleophilic groups on biological macromolecules, further analysis of the uncaging of p-hydroxyphenacyl compounds is required in a biological context.

Long-term goals include the development of better methods for getting caged peptides and proteins into cells and the synthesis of reversible photoactivatible reagents (i.e., reagents with on–off switches). A possible alternative to microinjection is to piggyback a caged protein on another protein capable of translocation across cell membranes. An attractive prospect is VP22, which is a herpes virus protein with the remarkable ability to enter living cells even when fused to quite large polypeptides such as green fluorescent protein.[41] VP-22 could be fused with an appropriate cysteine mutant. The molecule would be caged *in vitro* before incubation with target cells. We have already developed pores that can be switched on and off by divalent metal ions, either during assembly or as the full assembled pore.[30,42,43] We hope to extend this approach to photomodulated proteins by targeted modification with photoisomerizable groups, such as azobenzenes or spiropyrans.[1,8] Another interesting approach to caged proteins might be to split the protein of interest with an intervening sequence and prevent protein splicing by caging of a conserved cysteine in the splice junction.[44] Control of protein splicing has already been achieved by introducing a caged amino acid during *in vitro* translation.[45] A alternative way of generating cysteines at defined sites is by *in vitro* chemical ligation of synthetic peptides.[44,46]

---

[41] G. Elliott and P. O'Hare, *Cell* **88,** 223 (1997).
[42] B. Walker, J. Kasianowicz, M. Krishnasastry, and H. Bayley, *Protein Eng.* **7,** 655 (1994).
[43] B. Walker, O. Braha, S. Cheley, and H. Bayley, *Chem. Biol.* **2,** 99 (1995).
[44] Y. Shao and S. B. H. Kent, *Chem. Biol.* **4,** 187 (1997).
[45] S. N. Cook, W. E. Jack, X. Xiong, L. E. Danley, J. A. Ellman, P. G. Schultz, and C. J. Noren, *Angew. Chem. Int. Ed. Engl.* **34,** 1629 (1995).
[46] P. E. Dawson, T. W. Muir, I. Clark-Lewis, and S. B. H. Kent, *Science* **266,** 776 (1994).

Acknowledgments

We thank John Leszyk at the W.M. Keck Protein Chemistry Facility at the Worcester Foundation for peptide synthesis and help with HPLC and mass spectrometry. We are also grateful to Stephen Cheley, Rekha Panchal, Christopher Shustak, and Barbara Walker for advice and materials. This work was supported by grants from the NIH, NS26760 (to H.B.) and CA 58530 (to W.T.M.). P.P. was supported by the NIH postdoctoral training grant T32 NS07366. The MALDI-TOF mass spectrometer was purchased through an instrumentation grant from the NSF, BIR-9512226 (to H.B.).

[8] Photocleavable Affinity Tags for Isolation and Detection of Biomolecules

By JERZY OLEJNIK, EDYTA KRZYMAŃSKA-OLEJNIK, and KENNETH J. ROTHSCHILD

Introduction

The incorporation of affinity tags into biomolecules is often used in molecular biology and medicine for isolation and detection. Isolation usually involves the attachment or incorporation of an affinity label followed by separation through the interaction of the label with an affinity media. Direct detection involves monitoring the tag through radioactivity, spectroscopically (fluorescence, mass spectrometry), or through its secondary interaction with another molecule (colorimetry, luminescence). However, conventional methods are limited by the difficulty of removing (releasing) the affinity label from the target biomolecule. In particular, it is often desirable to obtain an unlabeled biomolecule after it has been isolated and/or detected by a label-affinity interaction.

The use of avidin–biotin technology has proved to be an important tool in numerous areas of biochemistry, molecular biology, and medicine,[1–3] including detection of proteins by nonradioactive immunoassays,[4,5] cytochemical staining,[6] cell separation,[7] isolation of nucleic acids, detection of

---

[1] M. Wilchek and E. A. Bayer, *Methods Enzymol.* **184**, 1990.
[2] M. Wilchek and E. A. Bayer, in "Immobilised Macromolecules" (U. B. Sleytr, ed.), pp. 51–60, Springer, London, UK, 1993.
[3] M. D. Savage, G. Mattson, S. Desai, G. W. Nielander, S. Morgensen, and E. J. Conklin, "Avidin-Biotin Chemistry: A Handbook." Pierce Chemical Company, Rockford, IL, 1992.
[4] J. L. Guesdon, *J. Immunol. Methods* **150**, 33–49 (1992).
[5] T. Sano, C. L. Smith, and C. R. Cantor, *Science* **258**, 120–122 (1992).
[6] A. A. Rogalski and S. J. Singer, *J. Cell Biol.* **101**, 785–801 (1985).
[7] J. Wormmeester, F. Stiekema, and C. DeGroot, *Methods Enzymol.* **184**, 314–319 (1990).

specific DNA/RNA sequences by hybridization,[8] and probing conformational changes in ion channels.[9] The technique derives its usefulness from the extremely high affinity of the avidin–biotin interaction ($K = 10^{15}\ M^{-1}$) and the ability to biotinylate a wide range of target biomolecules such as antibodies, nucleic acids, and lipids.

The first step in the isolation of a target molecule is its biotinylation or the biotinylation of an agent that binds the target molecule (e.g., an antibody or hybridization probe). This biotinylation can be achieved by reaction of an appropriate reagent that is most often directed at either amino or sulfhydryl groups on the target. The biotinylated molecule or target complex is then separated from other molecules in a heterogeneous mixture using affinity media that rely on the avidin–biotin interaction.[1] Once the target molecule is bound through the avidin–biotin interaction, it is often important to recover it in an unmodified and biologically functional form.

In the case of DNA, a biotin moiety can be introduced (i) during its solid support synthesis using biotinyl phosphoramidite;[10–14] (ii) enzymatically with biotinylated nucleoside triphosphate analogs, such as biotin-11-dUTP;[15] or (iii) through postsynthetic modifications with suitable biotinylation reagents such as biotin-NHS ester, biotin hydrazide,[8,16] or photobiotin.[17] In the case of proteins, a biotin moiety can be introduced by direct biotinylated or through *in vitro* translation using, for example, biotinylated lysyl-tRNA.[18]

Existing methods for the release of the target include (i) dissociation of the biotin–avidin complex (6–8 M guanidinium hydrochloride),[3] (ii) utilization of biotin analogs with decreased affinity toward avidin (e.g., iminobiotin),[19] and (iii) the use of biotin derivatives with chemically cleav-

---

[8] J. L. McInnes and R. H. Symons, *in* "Nucleic Acid Probes" (R. H. Symons ed.), pp. 33–80, CRC Press, Inc., Boca Raton, FL, 1989.

[9] S. L. Slatin, X. Q. Qiu, K. S. Jakes, and A. Finkelstein, *Nature* **371**, 158–161 (1994).

[10] R. T. Pon, *Tetrahedron Lett.* **32**, 1715–1718 (1990).

[11] U. Pieles, B. S. Sproat, and G. M. Lamm, *Nucleic Acids Res.* **18**, 4355–4360 (1990).

[12] K. Misiura, I. Durrant, M. R. Evans, and M. J. Gait, *Nucleic Acids Res.* **18**, 4345–4354 (1990).

[13] A. M. Alves, D. Holland, and M. D. Edge, *Tetrahedron Lett.* **30**, 3089–3092 (1989).

[14] A. J. Cocuzza, *Tetrahedron Lett.* **30**, 6287–6290 (1989).

[15] P. R. Langer, A. A. Waldrop, and D. C. Ward, *Proc. Natl. Acad. Sci. U.S.A* **78**, 6633–6637 (1981).

[16] P. Tijssen, "Laboratory Techniques in Biochemistry and Molecular Biology: Hybridization with Nucleic Acid Probes." Elsevier, New York, 1993.

[17] A. C. Forster, J. L. McInnes, D. C. Skingle, and R. H. Symons, *Nucleic Acids Res.* **13**, 745–761 (1985).

[18] T. V. Kurzchalia, M. Wiedmann, H. Breter, W. Zimmermann, E. Bauschke, and T. A. Rapoport, *Eur. J. Biochem.* **172**, 663–668 (1988).

[19] K. Hoffman, S. W. Wood, C. C. Brinton, J. A. Montibeller, and F. M. Finn, *Proc. Natl. Acad. Sci. U.S.A* **77**, 4666–4668 (1980).

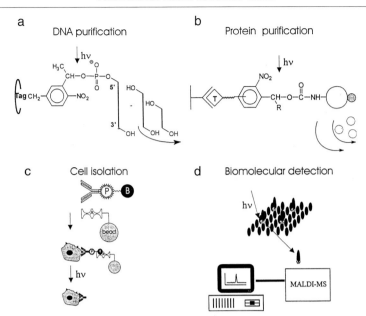

FIG. 1. Selected applications of photocleavable tags in biotechnology: (a) removal of failure sequences from synthetic DNA, (b) isolation of protein from heterologous mixture, (c) isolation of cells displaying specific surface antigen, and (d) spatially addressable photorelease and detection by mass spectrometry.

able spacer arms.[20–26] However, none of these methods allows release of the target molecule or target complex in an unaltered form because the biotin or part of the cleavable spacer arm remains attached. In addition, they all require harsh reaction conditions that can denature the target biomolecule.

As described in this article, the use of photocleavable tags (PC tags) in the isolation/detection of biomolecules eliminates many of the difficulties associated with conventional affinity labels by providing a rapid and effective method of removing the label in a single step. As illustrated in Fig. 1, PC tags offer advantages in numerous applications involving the isolation of proteins, nucleic acids, lipids, cells, and biomolecular assemblies. More

[20] M. Shimkus, J. Levy, and T. Herman, *Proc. Natl. Acad. Sci. U.S.A.* **82**, 2593–2597 (1985).
[21] T. Herman, E. Lefever, and M. Shimkus, *Anal. Biochem.* **156**, 48–55 (1986).
[22] B. A. Dawson, T. Herman, and J. Lough, *J. Biol. Chem.* **264**, 12830–12837 (1989).
[23] B. A. Dawson, T. Herman, A. L. Haas, and J. Lough, *J. Cell. Biochem.* **46**, 166–173 (1991).
[24] G. A. Soukup, R. L. Cerny, and L. J. Maher III, *Bioconj. Chem.* **6**, 135–138 (1995).
[25] B. D. Gildea, J. M. Coull, and H. Koster, *Tetrahedron Lett.* **31**, 7095–7098 (1990).
[26] E. Leikauf, F. Barnekow, and H. Koster, *Tetrahedon* **51**, 3793–3802 (1995).

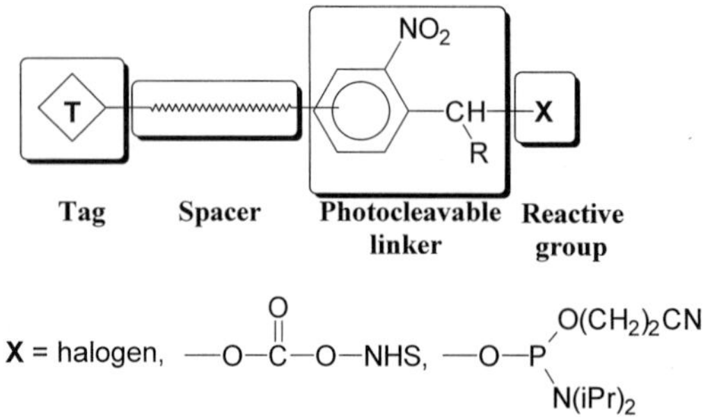

FIG. 2. General design of photocleavable tags based on α-substituted (R = alkyl, aryl) 2-nitrobenzyl moiety.

advanced applications include the selective release of biomolecules from a two-dimensional array and the construction of biomolecular assemblies at the nanometer scale.

Design of Photocleavable Affinity Tags

The general design of photocleavable tags is shown in Fig. 2. These reagents consist of an affinity tag **(T)** (biotin or specific reactive group), which is capable of strong specific interaction or chemical reaction with an appropriate support. The tag is linked through a variable-length spacer arm to the photoreactive moiety (photocleavable linker), in this case 2-nitrobenzyl derivative. The photoreactive moiety can be derivatized with a variety of reactive groups **(X)**, which provide for selective reactivity toward a particular chemical group on the target biomolecule. Examples of the reactive groups include halogen (sulfhydryl reactive), *N*-hydroxysuccinimide (NHS)-carbonate (amine reactive), and *N,N*-diisopropyl-2-cyanoethyl phosphoramidite (reactive toward hydroxyl groups).

We have introduced a new class of biotin derivatives, termed photocleavable biotins (PC-biotins), which eliminate many of the limitations of conventional biotins.[27,28] These reagents consist of a biotinyl moiety linked through a spacer arm (5-aminocaproic acid) to an α-substituted 2-nitrobenzyl nucleus that bears a reactive group. Reactivity of these compounds depends on the nature of the reactive group and can be directed toward either primary amino groups (PC-biotin NHS carbonate; compound **8,**

[27] J. Olejnik, S. Sonar, E. Krzymanska-Olejnik, and K. J. Rothschild, *Proc. Natl. Acad. Sci. U.S.A.* **92,** 7590–7594 (1995).

[28] J. Olejnik, E. Krzymanska-Olejnik, and K. J. Rothschild, *Nucleic Acids Res.* **24,** 361–366 (1996).

Scheme 1) or hydroxyl group (PC-biotin phosphoramidite; compound **11**, Scheme 2) on the target biomolecule. The spacer is designed to provide sufficient distance between the target molecule and the biotinyl moiety in order to allow for effective binding to streptavidin. The α-substituted 2-nitrobenzyl moiety was selected because its derivatives, such as esters and amides, are known to exhibit highly efficient and fast photoreactions.[29–31] The NHS esters and carbonates react selectively under mild conditions (pH 8) with primary amino groups to give amides and carbamates, respectively.[32,33]

PC-Biotin phosphoramidite (compound **11**, Scheme 2) was designed for direct use in any automated DNA/RNA synthesizer employing standard phosphoramidite chemistry. PC-biotin phosphoramidite contains a biotinyl moiety that bears a 4,4'-dimethoxytrityl group on an N-1 nitrogen atom. This group enables monitoring of coupling efficiency by trityl cation assay. In this case, the 2-nitrobenzyl group is derivatized with N,N'-diisopropyl-2-cyanoethyl phosphoramidite—a reagent compatible with state of the art automated nucleic acid synthesis. The selective reaction of PC-biotin phosphoramidite with the free 5'-OH group of a full-length oligonucleotide chain results in the introduction of a phosphodiester group linked to a photocleavable biotin moiety.

Materials and Methods

All chemicals are purchased from Aldrich (Milwaukee, WI), unless otherwise indicated. ^{1}H NMR spectra are recorded on a Varian XL-400 or Varian Unity Plus (Palo Alto, CA) spectrometer at 400 MHz in deuterated DMSO or CDCl$_3$ solution with chemical shifts (δ, ppm) reported relative to a tetramethylsilane internal standard. ^{31}P nuclear magnetic resonance (NMR) spectra are recorded in CDCl$_3$ on a JEOL JNM-GSX270 spectrometer (Peabody, MA) at 109.36 MHz with chemical shifts (δ, ppm) reported relative to an 85% H$_3$PO$_4$ external standard. Emission spectra are measured on a SLM 48000 fluorimeter (Rochester, NY) using 380 nm excitation. Infrared spectra are recorded on a Nicolet 740 FTIR spectrometer (Madi-

---

[29] J. E. T. Corrie, *J. Chem. Soc. Perkin Trans.* **I**, 2161–2166 (1993).
[30] T. Milburn, N. Matsubara, A. P. Billington, J. B. Udgaonkar, J. W. Walker, B. K. Carpenter, W. W. Webb, J. Marque, W. Denk, J. A. McCray, and G. P. Hess, *Biochemistry* **28**, 49–55 (1989).
[31] J. W. Walker, G. P. Reid, J. A. McCray, and D. R. Trentham, *J. Am. Chem. Soc.* **110**, 7170–7177 (1988).
[32] M. Wilchek and E. A. Bayer, *Methods Enzymol.* **184**, 123–138 (1990).
[33] A. K. Ghosh, T. T. Duong, S. P. McKee, and W. J. Thompson, *Tetrahedron Lett.* **33**, 2781–2784 (1992).

son, WI). Oligonucleotide synthesis is performed on an Applied Biosystem DNA/RNA synthesizer (Model 392; Foster City, CA). Samples are irradiated with a Blak Ray XX-15 UV lamp (Ultraviolet Products Inc., San Gabriel, CA) at a distance of 15 cm (emission peak 365 nm; 300-nm cutoff; 1.1 mW intensity at 31 cm). UV-VIS spectra are recorded on a Shimadzu (Columbia, MD) 2101PC spectrophotometer. High-performance liquid chromatography (HPLC) analysis is performed on a Waters (Milford, MA) system consisting of a U6K injector, 600 Controller, Novapak C_{18} (3.9 × 150 mm) column, and a 996 photodiode array detector. For oligonucleotide analysis, the following conditions are used: buffer A, 0.1N triethylamine acetate, pH 6.0; buffer B, acetonitrile. Elution is performed using a linear gradient (8–45%) of buffer B in buffer A over 45 min at a 1-ml/min flow. Preparative purifications are achieved on a Waters Novapak C_{18} RCM cartridge (8 × 100 mm) using conditions as specified above, except for flow rate, which is increased to 2 ml/min. Fractions are then analyzed, pooled, and freeze-dried. For Leu-Enk experiments, buffer A is 0.1% trifluoroacetic acid (TFA) in water, buffer B is 0.1% TFA in 3:1 (v/v) acetonitrile/water. Elution is performed using a linear gradient (30–45%) of B over 10 min followed by 45% of B isocratic for 10 min. Reagents and conjugates should be protected from bright sunshine but are found to be stable in artificial light.

Synthesis of Photocleavable Biotin Derivatives

Photocleavable Biotin N-hydroxysuccinimide Carbonate: Scheme 1. 5-Methyl-2-nitrobenzoic acid (**1**, Scheme 1) (5 g, 27.6 mmol) is added in small portions to thionyl chloride (16.4 g, 138 mmol). After 10 hr of stirring at room temperature, excess thionyl chloride is evaporated to give the acid chloride **(2)**. Magnesium turnings (1.07 g, 44.2 mmol), absolute ethanol (6 ml), chlorobenzene (8 ml), and 0.1 ml of dry CCL_4 are refluxed for 2 hr. Diethyl malonate (4.82 g, 30 mmol in 10 ml of chlorobenzene) is added, followed by the addition of the acid chloride **(2)** (5.49 g, 27.5 mmol in 10 ml of chlorobenzene). The reaction mixture is stirred for 1 hr, acidified with 20 ml of 2 N sulfuric acid, extracted with chloroform (3 × 20 ml), dried, and evaporated to dryness. The residue is dissolved in acetic acid (8.25 ml); 5.4 ml of H_2O and 1 ml of concentrated H_2SO_4 are added and the mixture is refluxed for 6 hr. The reaction mixture is neutralized with aqueous Na_2CO_3 and extracted with $CHCl_3$ (3 × 10 ml). Chloroform extracts are combined, dried, and after removing the solvents *in vacuo,* the residue is crystallized from 95% ethanol to give 5-methyl-2-nitroacetophenone **(3)** (4.46 g, 81%).[34] Compound **(3)** (3.51 g, 19.6 mmol), N-bromosuccin-

---

[34] L. M. Long and H. D. Troutman, *J. Am. Chem. Soc.* **71**, 2473–2475 (1949).

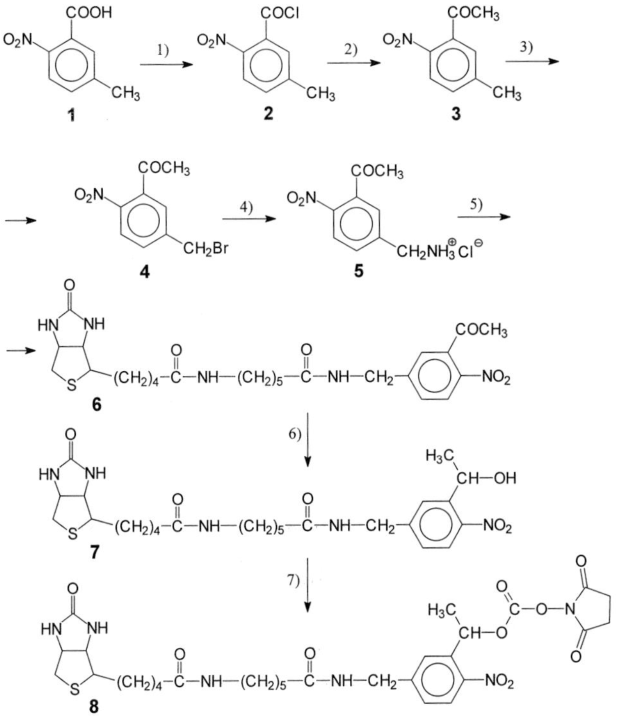

SCHEME 1. (1) $SOCl_2$; (2) (i)$Mg[CH(COOEt)_2]_2$, (ii) H^+/H_2O, Δ; (3) NBS/peroxide; (4) (i) hexamethylenetetramine, (ii) HCl/ethanol; (5) biotinamidocaproic acid/DCC/Et_3N; (6) $NaBH_4$; (7) N,N'-disuccinimidyl carbonate/Et_3N.

imide (3.66 g, 20.6 mmol), and benzoyl peroxide (46 mg, 0.01 equivalents) are refluxed in 20 ml of CCl_4 for 5 hr. The precipitate is filtered off, filtrate evaporated, and the residue crystallized from CCl_4 to give 5-bromomethyl-2-nitroacetophenone (4) (3.64 g, 72%).[35,36] Compound (4) (2.0 g, 7.75 mmol) is added to a solution of hexamethylenetetramine (1.14 g, 8.13 mmol) in 15 ml of chlorobenzene. The mixture is stirred overnight, and the resulting precipitate of 5-aminomethyl-2-nitroacetophenone hydrochloride (5) is filtered off and washed with chlorobenzene (10 ml) and diethyl ether (20 ml). The precipitate (2.93 g, 7.36 mmol) is suspended in 39 ml of chlorobenzene and 19.5 ml of methanol. To this suspension 4.9 ml of concentrated HCl is added and the mixture is stirred overnight. Diethyl ether, 70 ml, is

[35] P. D. Senter, M. J. Tansey, J. M. Lambert, and W. A. Blättler, *Photochem. Photobiol.* **42**, 231–237 (1988).
[36] T. Doppler, H. Schmid, and H. J. Hansen, *Helv. Chim. Acta* **62**, 291–303 (1979).

then added, and the resulting precipitate is washed with 2 × 50 ml of ether and dried under vacuum. Dimethylformamide (10 ml) is added followed by the addition of a 5-biotinamidocaproic acid (3.29 g, 1.25 equivalents, in 35 ml of N-methylpyrrolidone), dicyclohexylcarbodiimide (2.28 g, 1.5 equivalents), and triethylamine (1.28 ml, 1.25 equivalents). The solution is stirred overnight at room temperature, the precipitate is filtered off, and the filtrate is added to 700 ml of diethyl ether. The resulting precipitate is dried and purified on a silica gel column using step gradient (5–20%) of methanol in $CHCl_3$ to give 5-(5-biotinamidocaproamidomethyl)-2-nitroacetophenone (**6**) (2.27 g, 58%).

Intermediate **6** (1 g, 1.87 mmol) is dissolved in 15 ml of 70% ethanol, and after cooling the solution to 0°, sodium borohydride (141 mg, 4 equivalents) is added. The solution is stirred at 0° for 30 min and at room temperature for an additional 2 hr. The reaction is quenched by the addition of 1 ml acetone, neutralized with 0.1 N HCl, and concentrated to about 5 ml. The aqueous layer is decanted, and the residue is washed with water (3 × 5 ml) and dried to give 5-(5-biotinamidocaproamidomethyl)-1-(2-nitrophenyl)ethanol (**7**) (0.71 g, 71%).

Compound **7** (1.07 g, 2 mmol) is dissolved in 10 ml dimethylformamide. To this solution N,N'-disuccinimidyl carbonate (Fluka, Ronkonkoma, NY) (1 g, 1.5 equivalents) is added followed by triethylamine (0.81 ml, 3 equivalents). After 5 hr of stirring at room temperature, solvents are evaporated to dryness, and the residue is applied to a silica gel column and purified using a step gradient of methanol in chloroform to give 1.04 g (69%) of 5-(5-biotinamidocaproamidomethyl)-1-(2-nitrophenyl)ethyl-N-hydroxysuccinimidyl carbonate (PC-biotin-NHS) ester (**8**), which is characterized as follows: mp 113–114° (uncorrected); CI-MS (M^+ = 676.5); UV–VIS (in phosphate buffer, pH 7.4): λ_1 = 2.04 nm, ε_1 = 19190 M^{-1} cm^{-1}; λ_2 = 272 nm, ε_2 = 6350 M^{-1} cm^{-1}; ^{1}H NMR (δ; ppm): 8.48 (t, 1H), 8.05–8.03 (d, 1H), 7.75–7.71 (t, 1H), 7.66 (s, 1H), 7.46–7.45 (d, 1H), 6.44 (s, 1H), 6.37 (s, 1H), 6.28–6.27 (m, 1H), 4.39 (m, 2H), 4.30 (m, 1H), 4.12 (m, 1H), 3.57 (d, 2H), 3.09 (m, 1H), 3.01–2.99 (m, 2H), 2.79 (m, 5H), 2.58–2.55 (d, 1H), 2.17–2.15 (m, 2H), 2.04–2.02 (m, 2H), 1.72–1.71 (m, 2H), 1.66–1.43 (m, br, 6H), 1.38–1.36 (m, br, 2H), 1.26–1.25 (m, br, 3H); IR (KBr): $\nu_{C=O}$ 1815 and 1790 cm^{-1}.

Photocleavable Biotin Phosphoramidite: Scheme 2. 5-(6-Biotinamidocaproamidomethyl)-2-nitroacetophenone (**6**) (0.5 g, 0.94 mmol) is dried by coevaporation with anhydrous pyridine (3 × 2 ml) and then dissolved in 5 ml of the latter. To this solution 4,4'-dimethoxytrityl chloride (DMTr-Cl) (0.634 g, 1.87 mmol) is added, followed by 4-dimethylaminopyridine (0.006 g, 0.046 mmol). The reaction mixture is stirred at room temperature for 5 hr and then an additional 0.317 g of DMTr-Cl is added. After 24 hr, the reaction is quenched with methanol (1 ml), poured into 100 ml of 0.1 M

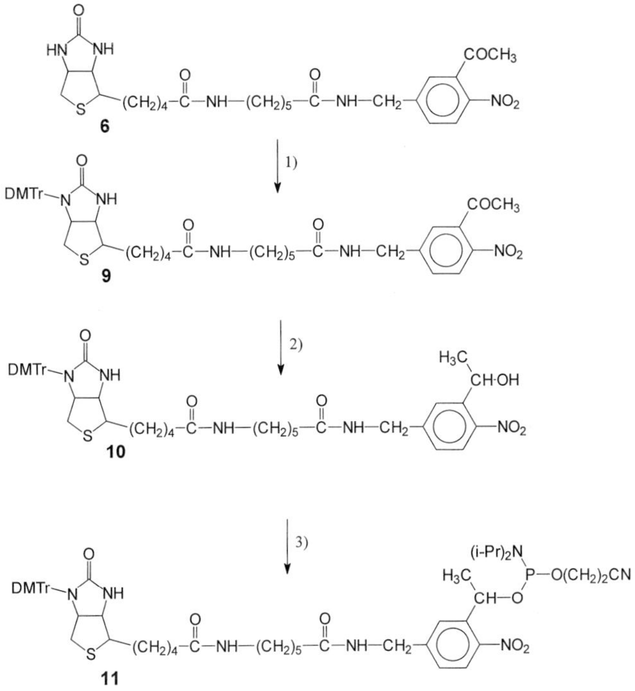

SCHEME 2. (1) DMTr-Cl/DMAP; (2) NaBH₄; (3) 2-cyanoethoxy-*N,N*-diisopropyl-chlorophosphine/DIPEA. [Adapted from J. Olejnik, E. Krzymanska-Olejnik, and K. J. Rothschild, *Nucleic Acids Res.* **24**, 361–366 (1996).]

sodium bicarbonate, and extracted with dichloromethane (3 × 50 ml). Evaporation of combined extracts gives yellow oil, which is further purified on a silica gel column using a step gradient of methanol in dichloromethane/ 0.2% triethylamine. Appropriate fractions are pooled and evaporated to give compound **9** as a white foam (0.73 g, 93% yield).

1-*N*-(4,4'-Dimethoxytrityl)-5-(6-biotinamidocaproamidomethyl)-2-nitroacetophenone (compound **9**, Scheme 2) (0.85 g, 1.016 mmol) is dissolved in 7 ml of ethanol, and sodium borohydride (0.028 g, 0.74 mmol) is added with stirring. After 1 hr the reaction is quenched with 4 ml of acetone and evaporated under reduced pressure to give a yellow oil, which is redissolved in 10 ml of methanol, and the solution is added to 120 ml of water. Precipitate is isolated by centrifugation (7000 rpm, 45 min) and dried *in vacuo* over KOH to give compound **10** (0.7 g 82%).

1-*N*-(4,4'-Dimethoxytrityl)-5-(6-biotinamidocaproamidomethyl)-1-(2-nitrophenyl)ethanol (compound **10**) (0.186 g, 0.22 mmol) is placed in an oven-dried flask with a magnetic stirring bar, sealed with a septum, and dried for at least 6 hr *in vacuo*. Anhydrous acetonitrile (0.003% water) (1 ml) is added through septum under argon. Subsequently, N,N-diidopropylethylamine (0.15 ml, 0.88 mmol) is added followed by 2-cyanoethoxy-N,N-diisopropylchlorophosphine (0.052 g, 0.22 mmol). After 1 hr another 0.5 equivalent of the phosphine is added. After an additional 2 hr at room temperature the reaction mixture is treated with 0.3 ml of ethyl acetate, followed by a saturated saline solution (10 ml), and extracted with dichloromethane (3 × 10 ml). The organic layer is washed with water, dried over sodium sulfate, evaporated under reduced pressure, and purified on a silica gel column using a step gradient (0–3%) of triethylamine in acetonitrile. Appropriate fractions are pooled and evaporated to give compound **11** as a white foam (0.144 g, 62% yield). Thin-layer chromatography (TLC), $CH_3CN:(C_2H_5)_3N$, 95:5, v/v; $R_f = 0.48$. ^1H NMR (ppm): 7.79–7.32 (m, 1H), 7.65–7.61 (m, 1H), 7.25–7.19 (m,5H), 7.13–7.05(m, 4H), 6.91–6.85 (m, 1H), 6.75–6.73 (m, 4H), 5.75–5.66 (br s, 1H), 5.54–5.43 (m, 1H), 5.22s, 5.12d (1H), 4.38–4.26 (m, 3H), 4.23–4.11 (m, 2H), 3.88–3.77 (m, 1H), 3.73 (s, 1H), 3.66–3.54 (m, 2H), 3.46–3.37 (m, 1H), 3.29–3.21 (m, 1H), 3.10–3.02 (m, 2H), 2.65–2.60 (m, 1H), 2.54–2.44 (m, 1H), 2.40–2.32 (m, 1H), 2.26–2.20 (dd, 1H), 2.10–2.11 (app. t, 1H), 2.08–2.01 (m, 2H), 1.62–1.58 (m, 6H), 1.55–1.45 (m, 4H), 1.39–1.35 (t, 2H), 1.31–1.27 (t, 2H), 1.16–1.07 (m, 9H), 0.87–0.83 (dd, 3H). ^{31}P NMR (ppm): 146.7, 147.9. Analysis calculated for $C_{52}H_{72}N_7O_9S_1P_1$: C 62.32%, H 7.24%, N 9.78%; found: C 62.03%, H 6.79%, N 9.20%.

Preparation of Photocleavable-Biotin Conjugates and Photocleavage Reaction

Reaction of Photocleavable-Biotin-N-hydroxysuccinimide with Leucine-Enkephalin and Photocleavage. Leucine-enkephalin (Leu-Enk) (Sigma Chemical Co., St. Louis, MO) (200 μl, 15.5 μmol/ml in 0.1 N $NaHCO_3$, pH 8.0) and PC-Biotin-NHS ester (200 μl, 17 μmol/ml in dimethylformamide) are mixed and stirred overnight at room temperature and used without further purification. PC-Biotin-Leu-Enk (1.93 μmol/ml in 25 mM phosphate buffer, pH 7.4) is irradiated for 10 min, with aliquots being withdrawn after 1, 2, 3, 4, 5, and 10 min.

Interaction of Photocleavable-Biotin-Leu-Enk with Immobilized Avidin or Streptavidin. PC-Biotin-Leu-Enk (10 nmol) is added to a suspension of monomeric avidin or streptavidin (tetrameric) agarose beads (15 nmol) and incubated with gentle agitation for 30 min. The suspension is then spin-filtered using the Ultrafree MC filter unit, 0.22 μm (Millipore, Bedford, MA), for 3 min at 5000 rpm, and filtrate containing unbound PC-biotin-Leu Enk is subjected to photolysis as described earlier. The released free Leu-Enk is assayed using fluorescamine (see below). The residue containing PC-biotin-Leu-Enk complexed with avidin or streptavidin agarose beads is resuspended in phosphate buffer (1 ml) and irradiated as described earlier. The released Leu-Enk is assayed using fluorescamine.

Fluorescamine Assay. Fluorescamine solution (10 μl, 10 mg/ml in dry acetone) is added to a solution of Leu-Enk (final volume 3 ml, 10–125 nmol in borate buffer; pH 9.0) and is incubated for 10 min. Fluorescence is measured at 488 nm after excitation at 383 nm. The assay is found to be linear in the concentration range studied. Time dependence for photocleavage of PC-biotin-Leu-Enk is studied by removing small aliquots from the photoreaction at various times of irradiation and subsequent determination of the released Leu-Enk.

5'-Photocleavable-Biotin-Oligonucleotide Synthesis. The PC-biotin phosphoramidite (**11**), 0.1 M solution in anhydrous acetonitrile, is attached to the extra port of the Applied Biosystem 392 DNA/RNA synthesizer. The syntheses are carried out at a 0.2-μmol scale using cyanoethyl phosphoramidites. For the last coupling (introduction of **11**) the coupling time is increased to 120 sec as recommended for conventional biotin phosphoramidite.[37] Typical coupling efficiency (as determined by trityl cation conductance) is between 95 and 97%. Standard detritylation as well as cleavage and deprotection procedures are used. Control 5'-phosphorylated sequences are synthesized using the chemical phosphorylation reagent Phosphalink (Applied Biosystem, Foster City, CA) according to the manufacturer's instructions.[38]

Affinity Purification and Photocleavage. Crude 5'-PC-biotin-oligonucleotide (16 nmol) is added to a suspension of streptavidin-agarose beads (700 μl, 24 nmol) (Sigma, St. Louis, MO), and the suspension is incubated at room temperature for 1 hr. It is then spin-filtered (5 min, 5000 rpm) using a 0.22-μm Ultrafree MC filter (Millipore, Bedford, MA). Beads on the filter are washed with 100 μl of phosphate buffer (pH 7.2) and spin-filtered (three times). Finally, the beads are resuspended in 700 μl of phosphate buffer and irradiated for 5 min. After irradiation the suspension is spin-filtered, the beads are washed with phosphate buffer (3 × 100 μl), and the combined filtrate volume is adjusted to 1 ml and analyzed by UV absorption spectroscopy or HPLC.

Time Dependence of Photocleavage. In order to calculate the time dependence of the photocleavage, HPLC-purified 5'-PC-biotin-(dT)$_7$ (48 nmol) is incubated with 1.5 equivalents of streptavidin-agarose beads for 1 hr. The beads are spin-filtered, washed, resuspended in phosphate buffer (pH 7.2), and irradiated. Aliquots (200 μl each) are withdrawn from 0, 0.25, 0.5, 1, 2, 4, 6, and 10 min of irradiation, spin-filtered, and washed as described earlier. The filtrate volume is adjusted to 1 ml and the absorbance

---

[37] Applied Biosystems, User Bulletin No. 70, 1992.
[38] Applied Biosystems, User Bulletin No. 86, 1994.

at 260 nm measured. A sample of 700 µl of streptavidin, which has not been incubated with oligonucleotide, is spin-filtered and the UV absorption measured, serving as background. A similar measurement is made on a sample of oligonucleotide not incubated with streptavidin (16 nmol, 700 µl phosphate buffer, serving as 100% control). The molar extinction coefficient at 260 nm for the PC-biotin moiety is determined separately (4700 M^{-1} cm^{-1}), and this value is subtracted from the estimated (assuming molar extinction coefficient equal to 12,000 for each dT) molar extinction coefficient of 5'-PC-biotin-(dT)$_7$ (88,700) for photorelease efficiency calculations. In order to determine the time course for photocleavage in solution, (dT)$_7$-5'-PC-biotin (1 OD$_{260}$) is dissolved in 1 ml of phosphate buffer and irradiated at 300–350 nm. Aliquots (10 µl) are withdrawn after 0, 0.25, 0.5, 1, 2, 4, 6, and 10 min of irradiation and injected onto an HPLC column. The percentage conversion is calculated from the ratio of the area of the particular peak [i.e., 5'-PC-biotin-(dT)$_7$ or 5'-p-(dT)$_7$] over the sum of the areas of the component peaks, molar extinction coefficients of the components adjusted as described earlier.

Evaluation of Photocleavable Biotins

Evaluation of Photocleavable-Biotin Peptides

The properties of the PC-biotin-NHS ester **(8)** have been studied using leucine-enkephalin (Leu-Enk), a pentapeptide (Tyr-Gly-Gly-Phe-Leu) as a model substrate. First, PC-biotin-NHS ester is reacted with Leu-Enk to yield PC-biotin-Leu-Enk. High-performance liquid chromatography traces of Leu-Enk (trace a, Fig. 3), the PC-biotin-NHS ester (trace b), and the PC-biotin-Leu-Enk (trace c) are shown in Fig. 3. It can be seen from trace c that the formation of PC-biotin-Leu-Enk is complete because the peaks of Leu-Enk (trace a) and PC-biotin-NHS ester (trace b) are absent. Instead, two new peaks (peaks 1 and 2) are seen, which have identical UV–VIS absorbance spectra (Fig. 3). These two peaks can be attributed to PC-biotin-Leu-Enk, with the separation most likely due to the presence of chiral center in the reagent.

Figure 3 also shows that illumination of PC-biotin-Leu-Enk ($\lambda > 300$ nm) results in almost complete photolysis in less than 5 min. Along with the decrease in intensity of peaks 1 and 2, associated with PC-biotin-Leu-Enk, there is a parallel appearance of a photoproduct, as indicated by the increase in peak 3 (traces d and e, Fig. 3). The identical retention time of this peak (3) and the Leu-Enk control (peak 4) establish this photolysis product as Leu-Enk. Thus, we conclude that Leu-Enk is photoreleased from PC-biotin-Leu-Enk in less than 5 min of illumination in a completely

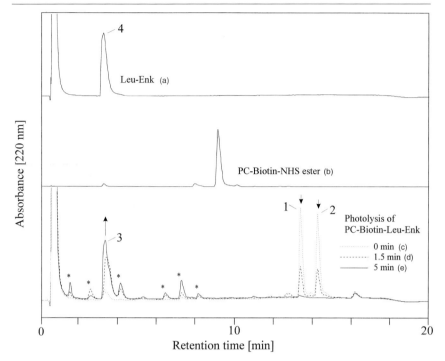

FIG. 3. Reaction of PC-biotin-HS with leucine-enkephalin (Leu-Enk) and its photolysis analyzed by HPLC.[27] Trace a, Leu-Enk; trace b, PC-biotin-NHS ester; traces c, d, and e, PC-biotin-Leu-Enk irradiated for 0, 1.5, and 5 min, respectively.

unaltered form. Other peaks (labeled as asterisks) that appear during photolysis are most likely due to the 2-nitrosoacetophenone derivative and other minor photolysis products that have also been observed in the photolysis of α-substituted 2-nitrobenzyl compounds.[39,40]

The time dependence for photocleavage of PC-biotin-Leu-Enk in solution is also measured using fluorescamine, which reacts only with the free N-terminal amino group of Leu-Enk to form a fluorophore (λ_{ex} = 383 nm, λ_{em} = 480 m)[41] and not with PC-biotin-Leu-Enk, which lacks any free

[39] T. Milburn, N. Matsubara, A. P. Billington, J. B. Udgaonkar, J. W. Walker, B. K. Carpenter, W. W. Webb, J. Marque, W. Denk, J. A McCray, and G. P. Hess, *Biochemistry* **28**, 49–55 (1989).

[40] J. W. Walker, G. P. Reid, J. A. McCray, and D. R. Trentham, *J. Am. Chem. Soc.* **110**, 7170–7177 (1988).

[41] S. Udenfriend, T. Stein, P. Böhlen, W. Dairman, W. Leimbruger, and M. Weigele, *Science* **178**, 871–872 (1972).

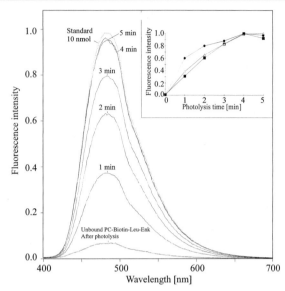

FIG. 4. Photorelease of Leu-Enk from PC-biotin-Leu-Enk assayed by fluorescamine.[27] PC-biotin-Leu-Enk was irradiated in solution (■) or complexed with streptavidin–agarose (●) or monomeric avidin agarose (△).

amino group. Figure 4 (inset) shows the time course for the release of Leu-Enk on photocleavage of PC-biotin-Leu-Enk in solution. No fluorescence is detected before illumination (0 min), confirming that no unreacted Leu-Enk exists in solution. The amount of Leu-Enk released at different times due to photocleavage is also calculated using the fluorescamine assay. Data confirm that the Leu-Enk is almost completely released from PC-biotin-Leu-Enk in solution within 5 min of UV illumination.*

In order to determine if the biotinyl moiety of PC-biotin-Leu-Enk retains its affinity to avidin, we complexed PC-biotin-Leu-Enk with monomeric avidin-coated agarose beads. For this purpose, PC-biotin-Leu-Enk was incubated for 30 min in a suspensin of these beads at a PC-biotin-Leu-Enk/avidin ratio of 1/1.5. The suspension was then spin-filtered for 3 min. The filtrate was analyzed for unbound PC-biotin-Leu-Enk by illuminating it for 6 min, which is sufficient to release Leu-Enk, as shown by the HPLC experiments (Fig. 3). As seen in Fig. 4, only a small amount of Leu-Enk

---

* The small drop in measured fluorescence after 5 min is most likely due to the presence of the PC-biotin photoproduct, which may act as either a quenching agent or an inner filter. Note in the case of PC-biotin-Leu-Enk complexed to beads such a drop is not observed because the PC-biotin photoproduct remains bound to the avidin.

is still present in the filtrate (~6%) after spin-filtering the beads. This indicates that the free PC-biotin-Leu-Enk binds efficiently to both the monomeric avidin and streptavidin-coated agarose beads.

It was also established that Leu-Enk is completely released from PC-biotin-Leu-Enk when complexed with monomeric avidin-coated agarose. For this experiment, avidin-bound PC-biotin-Leu-Enk was photocleaved by illumination of the resuspended agarose beads. As seen in Fig. 4, the fluorescamine-based assay shows that approximately 9.7 nmol of Leu-Enk is released into solution in approximately 5 min. This is close (within the estimated error) to the amount of PC-biotin-Leu-Enk (9.4 nmol) immobilized on monomeric avidin beads. Furthermore, HPLC analysis confirmed that the released photoproduct is Leu-Enk (data not shown). A similar experiment (Fig. 4, inset) using streptavidin-coated agarose beads also established that PC-biotin-Leu-Enk is efficiently bound by streptavidin and that Leu-Enk is completely released on illumination in less than 4 min.

Evaluation of PC-Biotin Oligonucleotides

The heptamer, 5'-PC-biotin-(dT)$_7$, was assembled using PC-biotin phosphoramidite (11) in an automated DNA/RNA synthesizer. The unmodified sequence, 5'-OH-(dT)$_7$, and a 5'-phosphorylated sequence, 5'-p-(dT)$_7$, were prepared using standard procedures (see Materials and Methods). Figure 5 shows the HPLC trace of 5'-PC-biotin-(dT)$_7$ (trace a, Fig. 5). Two main peaks are observed in this trace with a retention time of 23.9 and 24.5 min. These two peaks can be attributed to the two diastereoisomers

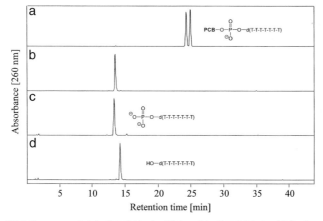

FIG. 5. HPLC traces of (a) 5'-PC-biotin-(dT)$_7$, (b) 5'-PC-biotin-(dT)$_7$ irradiated with near-UV light for 5 min, (c) 5'-p(dT)$_7$, and (d) 5'-OH-(dT)$_7$. [Adapted from J. Olejnik, E. Krzymanska-Olejnik, and K. J. Rothschild, *Nucleic Acids Res.* **24,** 361–366 (1996).]

generated by the introduction of the PC-biotin moiety onto the 5' end of the oligonucleotide.[30] Compared with the unmodified oligonucleotide, 5'-OH-(dT)$_7$ (trace d, Fig. 5, retention time 14.5 min), the PC-biotin-modified oligonucleotide (trace a, Fig. 5) shows an increased retention time, which is typical for biotinylated oligonucleotides.[12,25] We conclude from these data that 5'-PC-biotin moiety is retained during cleavage and deprotection of the oligonucleotide with ammonia [5'-phosphorylated oligonucleotide is not present in the 5'-PC-biotin-(dT)$_7$ sample].

The interaction of the PC-biotin-modified oligonucleotide with streptavidin and the photorelease of the oligonucleotide was evaluated by incubating the 5'-PC-biotin-(dT)$_7$ with streptavidin–agarose beads; separating the beads from the solution by spin-filtering; and irradiating resuspended beads with 300–350 nm light. The effects of irradiating the resuspended beads for 4 min are shown in Fig. 5. The two peaks assigned to 5'-PC-biotin-(dT)$_7$ (trace a, Fig. 5) disappear and a single peak appears (trace b, Fig. 5) with a retention time of ~13 min. The retention time of this peak is almost identical to that of the reference 5'-phosphorylated sequence, i.e., 5'-p-(dT)$_7$ (trace c, Fig. 5). These data conclusively show that irradiation causes cleavage of the PC-biotin moiety and release of 5'-phosphorylated oligonucleotide into solution.

We also measured the time dependence of the photoconversion of 5'-PC-biotin-(dT)$_7$ into 5'-p-(dT)$_7$ in solution. For this purpose, a 5'-PC-biotin-(dT)$_7$ solution was subjected to irradiation with 300–350 nm light and the reaction mixture was analyzed by reversed-phase HPLC after different irradiation times (Fig. 6). It can be seen from the decrease of the intensity of peaks at 23.7 and 24.3 min assigned to 5'-PC-biotin-(dT)$_7$ and the increase of the intensity of single peak at ~13 min assigned to 5'-p-(dT)$_7$ that the photoreaction is complete in approximately 4 min. The appearance of the additional small peaks with a retention time of approximately 33 min can be attributed to formation of the biotinyl-2-nitrosoacetophenone derivative and other minor photoproducts identified previously.[39,40]

The time dependence and efficiency of the photocleavage of 5'-PC-biotin-(dT)$_7$ complexed with streptavidin–agarose beads was determined by measuring the absorbance of the supernatant at 260 nm, which reflects the amount of 5'-p-(dT)$_7$ released into solution. The initial A_{260} value at 0 min (Fig. 6, inset) corresponds to less than 3% of the 5'-PC-biotin-(dT)$_7$ prior to complexation with streptavidin–agarose beads. This result shows that 5'-PC-biotin-(dT)$_7$ is almost quantitatively (97%) complexed with streptavidin–agarose. On irradiation, the photorelease is very rapid (4 min) and reaches 92% of the estimated absorption due to 5'-PC-biotin-(dT)$_7$ prior to complexation (see Materials and Methods). No further increase in absorbance was observed after 6 min of irradiation. In a separate experi-

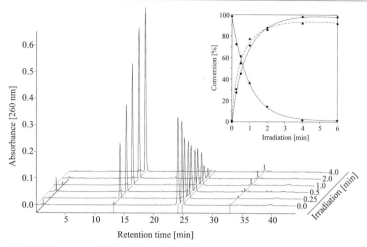

FIG. 6. HPLC traces of 5'-PC-biotin-(dT)$_7$ after increasing times of irradiation.[28] See Materials and Methods for more details. (Inset) Time dependence of photocleavage 5'-PC-biotin-(dT)$_7$→5'-p-(dT)$_7$ reaction in solution (solid lines); (◆) concentration of 5'-PC-biotin-(dT)$_7$; (■) concentration of 5'-p-(dT)$_7$. For comparison, time dependence for photorelease of 5'-p-(dT)$_7$ from 5'-PC-biotin-(dT)$_7$–streptavidin-agarose beads complex is shown (dashed line, ●).

ment, synthetic 5'-p-(dT)$_7$ was incubated with streptavidin–agarose beads. It was found that approximately 8% of 5'-p-(dT)$_7$ binds nonspecifically to the streptavidin–agarose beads (data not shown). Thus, the incomplete release of 5'-p-(dT)$_7$ appears to be due to nonspecific binding and not to incomplete photocleavage.

In order to evaluate the usefulness of PC-biotin phosphoramidite for synthesis and affinity purification/phosphorylation of longer oligonucleotides, two 5'-PC-biotin-labeled sequences, a 50-mer and a 60-mer, were prepared. After deprotection, the crude 5'-PC-biotin oligonucleotides were separately incubated with streptavidin–agarose beads. The beads were then washed, resuspended, and finally irradiated to obtain the full-length phosphorylated oligonucleotides. Figure 7 shows the results of polyacrylamide gel electrophoresis (PAGE) of the crude 50-mer (lane 1) and 60-mer (lane 4) oligonucleotides, and the affinity purified and photocleaved oligonucleotides (50-mer, lane 2; 60-mer, lane 5). Polyacrylamide gel electrophoresis of the supernatant obtained after isolation of the oligonucleotides with streptavidin–agarose beads is also shown (50-mer, lane 3; 60-mer, lane 6). In agreement with earlier studies, the biotinylated oligonucleotides migrate slower, whereas 5'-phosphorylated sequences migrate faster than sequences with 5'-OH.[10] It can be further seen that affinity purification and photocleavage result in a compact band, indicative of high purity and homogeneity,

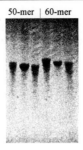

1 2 3 4 5 6

FIG. 7. PAGE (20%) of 50-mer (lanes 1–3) and 60-mer (lanes 4–6) visualized by UV shadowing, each lane representing 0.25 OD_{260} units of material.[28] Lanes 1, 4: crude 5'-PC-biotin-oligonucleotides; lanes 2, 5: PC-biotin-affinity purified and phosphorylated oligonucleotides; lanes 3, 6: filtrate containing sequences not bound to streptavidin.

in contrast to crude material, which exhibits a much broader band with additional material appearing at lower molecular weight. Because the latter material did not bind to streptavidin, it is likely to correspond to the failure sequences.

Concluding Remarks

As described, we have designed, synthesized, and evaluated two photocleavable biotin derivatives that can be used as reagents to label biomolecules. These represent a much broader class of photocleavable affinity tags that are currently under development. Several properties are important for a PC tag to be useful in the isolation of biomolecules. They include: (1) *Reactivity:* The PC tag should react selectively and efficiently with target molecules under mild conditions. (2) *Complexation:* The PC tag should retain a high affinity toward the capture molecule, such as avidin. (3) *Photocleavage:* The intensity and exposure time of light required for complete photocleavage should be minimal. Additional requiremens for PC phosphoramidites include compatibility with standard cycles and procedures for automated DNA/RNA synthesis. The PC-biotin-NHS ester reported here satisfies all of these criteria. It reacts with any substrate having primary amino group under mild conditions, exhibits a high affinity toward (strept)avidin, and allows fast (~4 min) and efficient (~99%) photorelease of the substrate in an unaltered form. Biotin-NHS ester and its analogs have been widely used as biotinylation reagents for a variety of biomolecules, including antibodies,[3] hormones,[1] amino-derivatized nucleic acids,[8] and lipids.[42] Similar protocols should be applicable for the biotinylation of these

---

[42] M. R. Soria, H. Loughrey, A. Ferraretto, A. M. Cannon, G. Acerbis, F. Sudati, G. Bottiroli, and M. Masserini, *J. Liposome Res.* **3**, 543–549 (1993).

molecules with PC-biotin-NHS ester, thereby expanding the range of applications of biotin-avidin technology.

PC-Biotin phosphoramidite also satisfies the above requirements, including automated synthesis. It reacts efficiently (95–97% coupling yield) with the 5′-hydroxyl group of the growing oligonucleotide chain under standard conditions. The PC-biotin moiety introduced onto synthetic oligonucleotide is stable during deprotection and shows high affinity toward streptavidin. The PC-biotin moiety is rapidly (<4 min) and efficiently photocleaved on illumination with near-UV light, resulting in 5′-phosphorylated DNA. Thus, the addition of a 5′-PC-biotin moiety onto oligonucleotides through the use of PC-biotin phosphoramidite provides a *photoremovable* affinity tag for fast and efficient purification and phosphorylation of synthetic DNA/RNA. This is especially important for the removal of failure sequences that are formed as a result of imperfections during the synthesis cycle. Note, however, that this method would not eliminate deletion sequences that are also known to occur.[43,44]

The ability to 5′-phosphorylate an oligonucleotide through photocleavage is another important advantage of this approach. Many applications of synthetic oligonucleotides require phosphorylation on the 5′ end, including gene construction, cloning, oligonucleotide ligation assay (OLA),[45] the ligation chain reaction,[46] and total cDNA sequencing. Typically, 5′-phosphorylation is achieved by either enzymatic or chemical methods. The use of enzymes involves several time-consuming steps and results often in nonquantitative phosphorylation. Chemical phosphorylation is possible during oligonucleotide synthesis using commercially available phosphoramidites such as Phosphalink[38]; however, the resulting product still requires purification.

In addition to PC-biotin-mediated affinity purification and phosphorylation, PC-biotin oligonucleotides should be useful in a number of other applications. For example, they could be used as primers for polymerase chain reaction (PCR), thereby simplifying the streptavidin-mediated affinity purification of PCR products from a reaction mixture containing template DNA, polymerases, and other components. This procedure would yield unmodified amplified fragments suitable for sequencing or cloning. Other possible applications include isolation of DNA/RNA macromolecular complexes[20–24] and controlled photorelease of oligonucleotides for the triggering of DNA–protein interactions and for therapeutic purposes.

---

[43] J. Temsamani, M. Kubert, and S. Agrawal, *Nucleic Acids Res.* **23,** 1841–1844 (1995).
[44] K. L. Fearon, J. T. Stults, B. J. Bergot, L. M. Christensen, and A. M. Raible, *Nucleic Acids Res.* **23,** 2754–2761 (1995).
[45] U. Landegren, R. Kaiser, J. Sanders, and L. Hood, *Science* **241,** 1077–1080 (1988).
[46] F. Barany, *Proc. Natl. Acad. Sci. U.S.A.* **88,** 189–193 (1991).

Photocleavable affinity tags can also be used for spatially addressable photorelease of biomolecules for diagnostic applications.

In general, photocleavable biotin derivatives can overcome one of the most critical limitations of avidin–biotin technology, i.e., the irreversibility of the avidin–biotin interaction, without sacrificing the high affinity of this interaction. For this purpose, it will be important to synthesize other PC-biotin compounds that exhibit selectivity for other functional groups, such as sulfhydryls. In addition, it is desirable to be able to vary the spacer arm length and adjust the wavelength for optimal photocleavage to suit specific applications. Importantly, the synthetic strategy employed here for the synthesis of PC-biotin-NHS ester and PC-biotin phosphoramidite can be readily adapted for the synthesis of other photocleavable affinity tags.[47,48] For example, the specific photoreactive moiety, reactive group, spacer arm, and affinity tag can be replaced easily to create a variety of different reagents using the current synthetic approach.

Photocleavable biotins are expected to be useful in a variety of applications relating to the isolation of biological molecules. Suitable target biomolecules include proteins, nucleic acids, carbohydrates, lipids, and macromolecular assemblies. PC-Biotin should also be useful in the isolation of special cells expressing unique cell-surface antigens, which are targets of PC-biotin-labeled antibodies. In addition to the direct chemical reaction of PC-biotin with biomolecules, they can also be incorporated enzymatically during their synthesis. For example, photocleavable biotins can be introduced into nascent proteins during *in vitro* synthesis by using aminoacyl-tRNAs with suitable amino acids, such as lysine, that will selectively react with PC-biotin-NHS ester. This will be especially useful for the rapid and efficient isolation of nascent proteins. PC-Biotin-labeled nucleotides can also be incorporated into nucleic acids through enzymatic or chemical synthesis, thus presenting new opportunities in gene cloning, PCR diagnostic assays, and *in situ* hybridizations.

Acknowledgments

The authors thank Dr. Stacie Williams for proofreading of the manuscript. This work was supported by a grant from the Army Research Office (ARO) (DAAL03-92-G-0172) to K.J.R. The work was also supported by SBIR grants to AmberGEn, Inc., from the Army Research Office (ARO) DAAH04-96-C-0050 and the National Institutes of Health IR43 GM54920-01.

[47] J. Olejnik, E. Krzymanska-Olejnik, K. J. Rothschild, *Nucleic Acids Res.*, submitted.
[48] J. Olejnik, E. Krzymanska-Olejnik, K. J. Rothschild, *in preparation.*

[9] Synthesis and Applications of Heterobifunctional Photocleavable Cross-Linking Reagents

By GERARD MARRIOTT and JOHANNES OTTL

Introduction

Absorption of near-ultraviolet light by biomolecules with one or more functional moiety protected with a nitrobenzyl-based "caged" group leads to a photoisomerization reaction in which the bond linking the functional group of interest to the benzylic carbon of the caged reagent is cleaved. The light-directed photocleavage reaction generates the biomolecule of interest and a photoproduct in the irradiated volume within a few milliseconds or faster.[1–5] The ability to generate light-directed concentration jumps of protein ligands and substrates provides an opportunity to measure transient kinetic data of specific protein activity in complex molecular environments[6] that include muscle fibers and living cells.[1,2,7] Monofunctional caged reagents have been used to prepare inactive yet photoactivatable enzyme substrates,[2] receptor ligands,[8] and fluorescent probes.[7,9] In addition, caged reagents capable of labeling amino acid and thiol groups have been used to inhibit the activity of proteins (caged proteins) by modifying one or more essential amino acid residues.[10–14] These caged groups can be removed from the inactive protein conjugate with concomitant recovery of its activity after irradiation with near-ultraviolet light.

In addition to these monofunctional caged reagents, heterobifunctional photocleavable reagents have been described to protect the amino or thiol

---

[1] J. A. McCray and D. R. Trentham, *Annu. Rev. Biophys. Chem.* **18**, 239 (1989).
[2] J. H. Kaplan, B. Forbush, III, and J. F. Hoffman, *Biochemistry* **17**, 1929, 1978.
[3] B. Amit, U. Zehavi, and A. Patchornik, *J. Org. Chem.* **39**, 192 (1974).
[4] J. F. Cameron and J. Frechet, *J. Am. Chem. Soc.* **113**, 4303 (1991).
[5] D. Gravel, S. Murray, and G. Ladouceur, *J. Chem. Soc. Chem. Commun.* 1828 (1985).
[6] S. R. Adams and R. Y. Tsien, *Annu. Rev. Physiol.* **55**, 755–784 (1993).
[7] J. A. Theriot and T. J. Mitchison, *Nature* **352**, 126 (1991).
[8] M. Wilcox, R. W. Viola, K. W. Johnson, A. P. Billington, B. K. Carpenter, J. A. McCray, A. P. Cuzikowski, and G. P. Hess, *J. Org. Chem.* **55**, 1585 (1990).
[9] T. J. Mitchison, *J. Cell Biol.* **109**, 637 (1989).
[10] G. Marriott, H. Miyata, and K. Kinosita, *Biochem. Int.* **26**, 9943 (1992).
[11] C.-Y. Chang, B. Niblack, B. Walker, and H. Bayley, *Chem. Biol.* **2**, 391 (1995).
[12] G. Marriott, *Biochemistry* **33**, 9092 (1994).
[13] P. D. Senter, M. Tansey, J. Lambert, and W. Blättler, *Photochem. Photobiol.* **42**, 231 (1985).
[14] J Ottl, D. Gabriel, and G. Marriott, *Bioconjugate Chemistry* **9**, March–April (in press).

group of small molecules[15] and proteins[10,13] as their corresponding photolabile carbamate or thioether, respectively.[10,13,15] The second reactive group of the cross-linking reagent may then be used to attach the caged compound to a second biomolecule; for example, an antibody molecule to target the caged compound to a specific site,[13] a derivatized surface, or a fluorescent dye for imaging purposes or to probe protein–protein interactions.[10,15] On irradiation of the cross-linked protein complex with near ultraviolet light, the bond linking the caged reagent to the protected group of the molecule of interest is cleaved and the two biomolecules dissociate at a diffusion controlled rate. Photocleavable cross-linked complexes have been described in which either the activity of the protein of interest is blocked and activated on irradiation with light,[10,13] or enhanced in the cross-linked complex and reduced to a normal level after irradiation.[12] This article describes heterobifunctional, photocleavable cross-linking reagents and associated labeling techniques that can be used to prepare caged protein conjugates that may be used to achieve a photoactivation or photodeactivation of protein activity.

Materials

Reagents

4-Bromomethylbenzoic acid is purchased from Tokyo Kasei Company (Tokyo, Japan). Fuming nitric acid, N-hydroxysuccinimide, dicyclohexylcarbodiimide, 3,4-dimethoxy-6-nitrobenzaldehyde, di-N-succinimidyl carbonate (DSC), vinylmagnesium bromide, *meta*-chloroperbenzoic acid (MCPBA), 4-dimethylaminopyridine (DMAP), 4,6-diphenylthieno[3,4-d][1,3]dioxol-2-one 5,5-dioxide (TDO) are from Aldrich (Milwaukee, WI). Bovine serum albumin (BSA), polylysine with an average molecular mass of 40,000 Da and 2-iminothiolane are purchased from Sigma (St. Louis, MO). Tetramethylrhodamine iodoacetamide (IATMR), acrylodan, and aminodextran are purchased from Molecular Probes (Eugene, OR) or obtained as a gift from Dr. John Corrie, NIMR, Mill Hill, London. Rhodamine 110 is from Lambda Physik (Göttingen, Germany). All other reagents are of the highest quality available and obtained from Sigma or Aldrich unless stated otherwise. G-Actin is prepared according to Marriott.[12]

Instrumentation

Nuclear magnetic resonance (NMR) spectra are measured on a 500-MHz Bruker Instrument or a 400-MHz instrument (JOEL, JNM-GX 400,

---

[15] J. Olejnik, S. Sonor, E. Kryzymanska-Olejnik, and K. Rothschild, *Proc. Natl. Acad. Sci. U.S.A.* **92,** 7590 (1995).

Tokyo, Japan), mass spectrophotometry on a Finnigan MAT 900 or Finnigan HSQ 30 instrument. Infrared spectra are recorded on a Perkin-Elmer 1760 X FT-IR spectrophotometer or a JASCO A202 instrument of samples suspended in Nujol. Absorption spectra are recorded on a HP 82152 diode array spectrophotometer (Hewlett Packard) or a Shimadzu UV350A double-beam instrument. Fluorescence spectrometry is performed on an SLM-AB2 fluorometer (Sopra, Buttlelborn, Germany) or a Hitachi F-4010 instrument. Light-directed photoactivation of caged compounds is performed essentially as described by Marriott.[12]

Miscellaneous Methods

Reaction of 2-Iminothiolane with Bovine Serum Albumin and Aminodextran

Bovine serum albumin, 4 mg, dissolved in 1 ml of 50 mM borate buffer, pH 8.5, is treated with 40 μl of a freshly prepared solution of 3.5 mg of 2-iminothiolane in 250 μl distilled water. The solution is left at room temperature for 1 hr and then centrifuged for 10 min at 10,000 rpm. The protein conjugate is dialyzed overnight at 4° against two changes of phosphate buffer, pH 7.5, previously saturated with nitrogen gas to prevent oxidation of the thiol groups. After dialysis the thiolated BSA conjugate is centrifuged at 14,000 rpm for 20 min at 4°. A similar labeling procedure is used to prepare the thiolated dextran from aminodextran.

Determination of Free Thiol Content of Bovine Serum Albumin

Of a 10 mM stock solution of acrylodan (the thiol reactive form of the fluorescent dye prodan) dissolved in DMF, 15 μl is added to 300 μl of a 60 μM solution of iminothiolane–BSA at pH 8.0. The reaction is left for 2 hr at room temperature, and the solution centrifuged for 10 min at 10,000 rpm and dialyzed for 2 days against phosphate buffer, pH 7.5, at 4° with three changes of buffer. The protein conjugate is centrifuged for 20 min at 14,000 rpm at 4° before recording its absorption spectrum. The concentration of prodan in the conjugate is calculated from the absorption value at 385 nm using an extinction coefficient of 18,500 M^{-1} cm^{-1}.[16]

Labeling of G-Actin with Tetramethylrhodamine Iodoacetamide

Rabbit muscle G-actin, purified according to Marriott,[12] in G-buffer (2 mM Tris, 0.2 mM CaCl$_2$, 0.2 mM ATP, pH 8.0) containing 0.5 mM

[16] G. Marriott, K. Zechel, and T. M. Jovin, *Biochemistry* **27**, 6214 (1988).

dithiothreitol (DTT) is treated with 13 μl of a 10 mM stock solution of IATMR dissolved in DMF, and the reaction is left for 2 hr at room temperature. The protein is centrifuged for 30 min at 100,000 g at 4° and dialyzed overnight against two changes of G-buffer without DTT at 4°. The actin conjugate is centrifuged at 100,000 g for 1 hr, and the supernatant is made into F-buffer (G-buffer containing 2 mM $MgCl_2$ and 0.1 M KCl) and left for 2 hr at room temperature. Ultracentrifugation and resuspension of the pellet in G-buffer are followed by an overnight dialysis against G-buffer without DTT. The protein solution is clarified by centrifugation (100,000 g for 1 hr), and absorption spectrometry is used to calculate a labeling ratio of 0.4 fluorophore per actin monomer using an extinction coefficient for IA-TMR of 96,500 M^{-1} cm^{-1} at 550 nm.[14]

4-Bromomethyl-3-nitrobenzoic Acid Succinimide Ester: A Simple Heterobifunctional, Photocleavable Cross-Linking Reagent

This reagent, described by Marriott et al.[10] has been used to prepare cross-linked complexes of actin to study properties of the nucleation stage in the polymerization of G-actin,[10] and the stabilization of actin filament ends by cross-linked actin oligomers containing at least three monomers.[17]

Syntheses

4-Bromomethyl-3-nitrobenzoic Acid. In a well-ventilated hood, 5 g of 4-bromomethyl-3-nitrobenzoic acid (BNBA) is added gradually to 50 ml of fuming nitric acid maintained at −11° using a KCl–ice bath. Control of the reaction temperature is critical for both safety and high product yield. After 2 hr the reaction mixture is poured onto crushed ice and the product is recrystallized from dichloromethane–heptane (1:1, v/v). Mass spectrophotometric analysis of the product shows it contains the expected molecular ions of 269 and 271 m/z and a debrominated fragment of 190 m/z.

4-Bromomethyl-3-nitrobenzoic Acid Succinimide Ester. 4-Bromomethyl-3-nitobenzoic acid, 2.5 g, is dissolved in 20 ml of dry acetonitrile followed by 1.05 g of N-hydroxysuccinimide and 2.15 g of dicyclohexylcarbodiimide, and the reaction is left overnight at room temperature. After removal of the dicyclohexylurea by centrifugation, filtration, and evaporation of the solvent, the product is recrystallized from dichloromethane–heptane (1:1), yielding 2.02 g of a white solid. Mass spectrophotometric analysis reveals molecular ions of the product of 356 and 358 m/z and a debrominated fragment of 227 m/z. NMR: δ 2.94 (4H, s, succinimide methylene); δ 4.87 (2H, s, bromoalkane methylene); δ 7.778 (1H, dd, aro-

---

[17] H. Miyata, K. Kinosita, and G. Marriott, *J. Biochem.* **121,** 527 (1997).

matic); δ 8.783 (1H, d, aromatic). This reagent and several longer chain derivatives are now available commercially from the Dojin-do Chemical Co. (http://www.dojindo.co.jp).

4-(2-Hydroxymercaptomethyl)-3-nitrobenzoic Acid. A solution containing 100 mg of BNBA and a fivefold molar excess of 2-mercaptoethanol in 5 ml of tetrahydrofuran is heated to reflux in the dark for 2 days. The product identified by thin-layer chromatography (TLC) on silica gel plates is purified by silica gel chromatography using chloroform–methanol (2:1, v/v). The molar absorption extinction coefficient at the 2-nitrobenzyl group is determined to be 500 M^{-1} cm^{-1} at 350 nm.

Biochemical and Photochemical Properties of 4-Bromomethyl-3-nitrobenzoic Acid Succinimide Ester

4-Bromomethyl-3-nitrobenzoic acid and BNBA-SE (Fig. 1) react with the thiol group of cysteine residues of actin and BSA between pH 7 and 8 to form the corresponding thioether. Relatively harsh reaction conditions are required to prepare the 2-mercaptoethanol derivative of BNBA. Irradiation of a solution of 4-(2-hydroxyethylmercaptylmethyl)-3-nitrobenzoic acid with near-ultraviolet light results in the photocleavage of the thioether bond and release of mercaptoethanol and a nitrosobenzaldehyde derivative. Infra-red spectroscopy of 4-(2-hydroxyethylmercaptylmethyl)-3-nitrobenzoic acid recorded before and after irradiation with near-ultraviolet light shows the disappearance of the strong nitrobenzyl absorption band at 1530 cm^{-1}, and this result is consistent with a bond cleavage reaction mechanism that proceeds via the photoisomerization of the nitrobenzyl group.

Preparation of Photolabile Dimer of Actin

Cross-linking of F-Actin with 4-bromomethyl-3-nitrobenzoic acid succinimide ester. The photocleavable cross-linking reagent BNBA-SE is used to prepare an actin dimer whose nucleating activity in the polymerization reaction of G-actin can be controlled with near-ultraviolet light. Cross-linking of actin oligomers is achieved by incubating a solution of F-actin at 25 μM in thiol-free F-buffer with BNBA-SE at 12.5 μM for 1 hr at 20°. The F-actin fraction is centrifuged at 100,000 g at 4° for 1 hr, and the pellet is resuspended in G-buffer containing 1 mM 2-mercaptoethanol and dialyzed against this buffer for 24 hr at 4°. After a 5-min bath sonication at 4°, the protein is centrifuged again at 100,000 g, the supernatant is set aside, and the pellet is subjected to a second round of depolymerization. The combined supernatant fractions are subjected to gel-exclusion chromatography using Sephadex G-150 (Pharmacia, Freiburg, Germany). Fractions

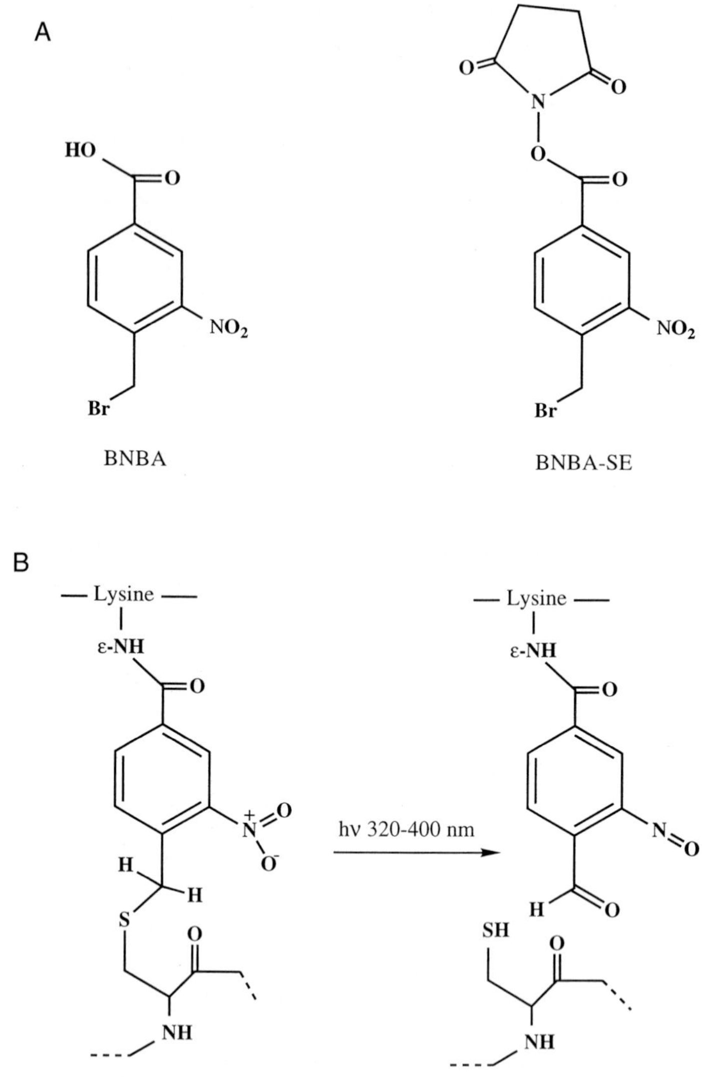

FIG. 1. (A) Molecular structures of BNBA and BNBA-SE. (B) Light-directed photocleavage reaction of cross-linked actin.

containing cross-linked actin dimer are identified by sodium dodecyl sulfate–polyacrylamide gel electrophoresis (SDS–PAGE) and pooled.

Cross-linking of actin monomers in F-actin with BNBA-SE proceeds very efficiently using the reaction conditions described. Competition labeling experiments using acrylodan reveal that the highly reactive cysteine-

374 of one actin monomer is the target of the reactive bromoalkane of the caged reagent (data not shown). A lysine residue from an adjacent actin monomer in the filament most likely reacts with the activated carboxyl group of BNBA-SE. The cross-linking reaction generates a ladder of actin oligomers containing up to 14 or more monomer units.[17] The actin dimer is purified from this ensemble of cross-linked actin monomers using size-exclusion chromatography and is shown to be fully active by measuring its ability to polymerize in F-buffer (Fig. 2B). As expected from earlier studies,

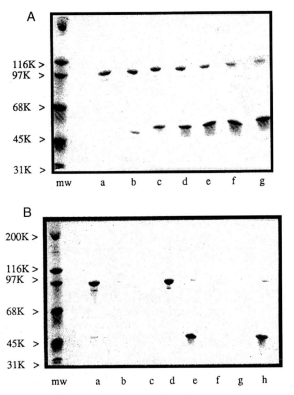

FIG. 2. (A) SDS–PAGE analysis of the light-directed photocleavage of the BNBA-SE-mediated cross link between actin monomers in the purified actin dimer. Numbers of pulses of 355-nm light delivered by the laser to a 0.85 μM sample of the cross-linked actin dimer in F-buffer is (a) 0, (b) 30, (c) 60, (d) 90, (e) 150, (f) 250, and (g) 400. (B) Polymerizability of the BNBA-SE cross-linked actin dimer. (a) Supernatant fraction and (b) pellet fraction of a 0.85 μM solution of the actin dimer in G-buffer after centrifugation at 4° for 60 min at 100,000 g. (c) Supernatant fraction and (d) pellet fraction of a 0.85 μM solution of the actin dimer in F-buffer after centrifugation for 60 min at 100,000 g. (e) Supernatant fraction and (f) pellet fraction of a 0.85 μM solution of the actin dimer in F-buffer containing 5 mM DTT after irradiation with 400 pulses of 355-nm light and centrifugation for 60 min at 100,000 g.

the actin dimer acts as a nucleating species in the polymerization of G-actin.[18,19] The nucleating activity of the dimer is quantified by measuring the rate of elongation of the actin polymerization reaction using the change in the fluorescence emission of the prodan conjugate of G-actin against different concentrations of the actin dimer (Fig. 3A). This nucleation activity is lost after irradiating the dimer with near-ultraviolet light; photocleavage of the cross-link and the concomitant disappearance of actin dimer and the increase in actin monomer are confirmed by SDS–PAGE and densitometric analysis of the Coomassie-stained bands (Fig 2A). These data are then used to show that the elongation rate of actin polymerization exhibits a linear dependence on the amount of actin dimer in the G-actin solution (Figs. 3A and 3B).

4-Bromomethyl-3-nitrobenzoic acid cross-linked actin containing several actin monomers has been shown to cap the ends of actin filaments, thereby preventing their depolymerization in low salt buffer.[17] Irradiation of actin filaments stabilized with these actin oligomers cleaves the cross link and generates ends with normal actin monomers, which then dissociate from the filament. We envision that photoactivation of these photocleavable cross-linked actin oligomers, microinjected into motile cells, will prove useful in understanding the role of actin filament capping proteins in the regulation of actin filament dynamics during cell motility.

A New Class of Photocleavable Cross-Linking Reagent

A new class of photocleavable cross-linking reagent has been synthesized to prepare protected amino compounds with improved spectroscopic and photocleavage properties for applications in cell biology (Scheme 1).[14] In this class of reagent, the two reactive groups are introduced off the benzylic carbon atom. These reagents are designed with the following operational considerations in mind: (1) the caged cross-linking reagents should react with the functional group of interest in an aqueous solution and at a slightly alkaline pH; (2) the photoactivation reaction should exhibit a good quantum yield and an action spectrum in the near-ultraviolet wavelength region (340–400 nm) to avoid interference with other biomolecules; (3) for photoactivation reactions performed on optically thin samples, the chromophore of the caged group should exhibit a high molar absorptivity; and (4) the photoproducts of the photoactivation reaction should not be toxic or reactive with other functional groups in the preparation. The new reagents presented in Scheme 1 fulfill most of these requirements.

---

[18] P. Knight and G. Offer, *Biochem. J.* **175,** 1023 (1978).
[19] S. C. Mokrin and E. D. Korn, *J. Biol. Chem.* **256,** 8228 (1981).

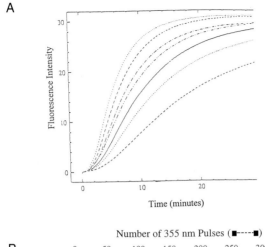

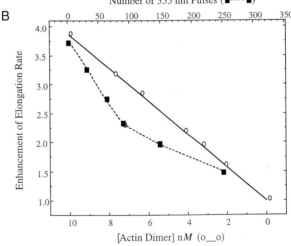

FIG. 3. Effect of BNBA cross-linked actin dimer on the polymerization kinetics of prodan-labeled G-actin. Of a 3.4 μM solution of prodan–G-actin in G-buffer, 195 μl was mixed with 5 μl of a 0.85 μM solution of BNBA cross-linked actin dimer that had been subjected to irradiation with 355-nm light using the following number of pulses (from top to bottom trace): 0, 30, 60, 90, 150, and 250. The lowest trace (250 pulses) represents the kinetics of the polymerization of 3.4 μM prodan–G-actin without the addition of the cross-linked actin dimer. (B) Dependence of the rate of elongation in the polymerization of prodan–G-actin calculated using data presented in A on the number of light pulses delivered to the cross-linked actin dimer sample (filled squares) and on the concentration of actin dimer (open circles) remaining in samples irradiated with 355-nm light. The actin dimer concentration was calculated from a densitometric analysis of the acrylamide gel, shown in Fig. 2A.

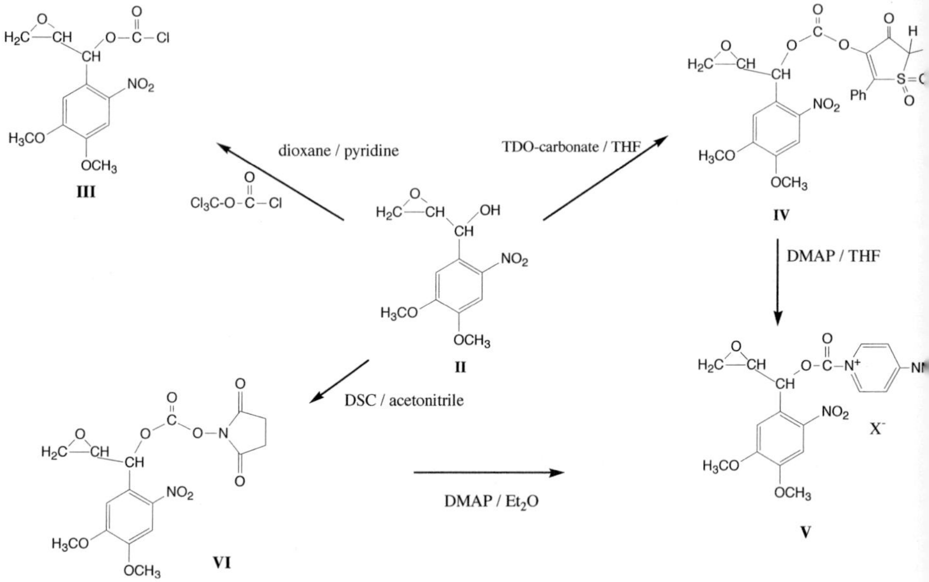

SCHEME 1

Syntheses

Preparation of 1-Hydroxy-1-(3,4-dimethoxy-6-nitrophenyl)-2-propene: compound I. 3,4-Dimethoxy-6-nitrobenzaldehyde, 1.88 g (8.95 mmol), freshly recrystallized out of toluene, is dissolved in 50 ml dry tetrahydrofuran (THF) under an argon atmosphere. The yellow solution is cooled to −70°, and 10.7 ml of a 1 M solution of vinylmagnesium bromide in THF is added dropwise. The deep red solution is stirred for 3 hr, slowly warmed to 20°, and then 40 ml of a saturated NH$_4$Cl solution added dropwise. The red solution is extracted three times with ethyl acetate, and the organic phase is washed five times with saturated NaCl, dried over MgSO$_4$, and the solvent removed. The red, oily residue (2.44 g) is taken up in a little ethyl acetate and eluted through a silica gel column developed in hexane:ethyl acetate (3:1, v/v). The solvent is removed from the product fractions, producing an orange–red powder. Yield: 1. 53 g (6.33 mmol; 72%). Molecular weight = 239.23, C$_{11}$H$_{13}$NO$_5$; mp 101–102° ^{1}H NMR (CDCl$_3$ in ppm): 2.63 (s, broad, 1H); 3.94 (s, 3H); 3.96 (s, 3H); 5.24 (dd, J = 11 Hz, 2 Hz, 1H); 5.42 (dd, J = 18 Hz, 2 Hz, 1H); 5.92 (d, J = 6 Hz, 1H); 6.07 (ddd, J = 18 Hz, 11 Hz, 6 Hz, 1H); 7.20 (s, 1H); 7.57 (s, 1H). MS (70 eV), m/z (%): 239 (42) M$^+$, 162 (100) C$_9$H$_8$NO$_2^+$.

Preparation of 1-(3,4-Dimethoxy-6-nitrophenyl)-2,3-epoxypropyl Hydroxide: compound II. meta-Chloroperoxybenzoic acid (MCPBA, technical grade, about 80% pure) is washed three times with phosphate buffer, pH 7.5, and dried. Purified MCPBA, 1.311 g (7.60 mmol), is dissolved in 20 ml CH_2Cl_2 and slowly dropped into a precooled solution of 1.80 g (7.52 mmol) of **I** in 15 ml CH_2Cl_2. The bright yellow solution is stirred for 48 hr at room temperature. The white precipitate is filtered and washed with CH_2Cl_2, and then 70 ml of a saturated solution of $NaHCO_3$ is added to the solution, which is stirred for 30 min. The organic phase is washed seven times with a saturated solution of $NaHCO_3$ and then three times with a saturated solution of NaCl. The organic phase is dried over $MgSO_4$ and the solvent removed. The yellow, oily residue (1.90 g) is recrystallized out of ethyl acetate/hexane. Yield: 1.55 g (6.10 mmol, 81%, orange needles). Molecular weight: 255.23, $C_{11}H_{13}NO_6$; mp, 118–121·. ^1H NMR ($CDCl_3$ in ppm): isomer a: 2.62 (s, 1H); 2.71 (dd, J = 5 Hz, 2 Hz, 1H); 2.90 (dd, J = 5 Hz, 2 Hz, 1H); 3.32 (ddd, J = 5 Hz, 5 Hz, 2 Hz, 1H); 3.96 (s, 3H); 3.98 (s, 3H); 5.57 (dd, J = 5 Hz, 2 Hz, 1H); 7.23 (s, 1H); 7.66 (s, 1H); isomer b: 2.63 (s, 1H); 2.82 (dd, J = 5 Hz, 2 Hz, 1H); 3.02 (dd, J = 5 Hz, 2 Hz, 1H); 3.62 (ddd, J = 5 Hz, 5 Hz, 2 Hz, 1H); 3.96 (s, 3H); 4.02 (s, 3H); 5.76 (dd, J = 5 Hz, 2 Hz, 1H); 7.28 (s, 1H); 7.67 (s, 1H). (70 eV), m/z (%): 255 (35) M^+, 164 (100) $C_9H_{10}NO_2^+$. FAB-MS m/z 256.1 MH^+MS.

Preparation of 1-(3,4-Dimethoxy-6-nitrophenyl)-2,3-epoxypropyl Chloroformate: compound III. In a well-ventilated hood, 566 mg of **II** (2.22 mmol) dissolved in 10 ml of water-free dioxane in a dried 50-ml flask is treated with 0.18 ml of water-free pyridine (2.25 mmol) together with 0.26 ml diphosgene (2.20 mmol), which is slowly dropped into the yellow solution. The reaction mixture is stirred for 2 hr at room temperature, the precipitate filtered off, the solvent removed, and the flask left for 1 hr under a high vacuum to remove excess diphosgene. The oily, orange–brown residue is taken up in a little ethyl acetate and eluted through a silica gel column developed in hexane : ethyl acetate (3 : 1). The solvent is removed from the product fractions, yielding a yellow powder. Yield: 620 mg (1.95 mmol, 88%). Molecular weight: 317.68, $C_{12}H_{12}NO_7Cl$; mp, 178–181°. ^1H NMR ($CDCl_3$, in ppm): isomer a: 3.57 (dd, J = 12 Hz, 4 Hz, 1H); 3.76 (dd, J = 12 Hz, 4 Hz, 1H); 3.94 (s, 3H); 3.97 (s, 3H); 5.52–5.56 (m, 1H); 6.32 (d, J = 8 Hz, 1H); 7.20 (s, 1H); 7.71 (s, 1H); isomer b: 3.92 (s, 3H); 3.96 (s, 3H); 4.00 (dd, J = 12 Hz, 2 Hz, 1 H); 4.19 (dd, J = 12 Hz, 3 Hz, 1H); 4.66–4.68 (m, 1H); 6.13 (d, 3 Hz, 1 H); 7.00 (s, 1H); 7.73 (s, 1H). MS (70 eV), m/z (%): 317 (82) M^+, 136 (100) $C_8H_8O_2^+$.

Preparation of 4-[1-(3,4-Dimethoxy-6-nitrophenyl)-2,3-epoxypropyl-1-oxycarbonyloxy] - 3 - oxo - 2,5 - diphenyl - 2,3 - dihydrothiophene 1,1-dioxide: compound IV. Five hundred milligrams of **II** (1.959 mmol) dissolved in 4 ml water-free THF is treated with 672 mg of freshly recrystallized, 4,6-diphenylthieno[3,4-d][1,3]dioxol-2-one 5,5-dioxide (TDO, 2.062 mmol). The orange solution is refluxed for 4 hr in the absence of base, and the solvent is evaporated to give an orange residue that recrystallizes out of dry toluene. Yield: 186.7 mg (0.321 mol, 82%). Molecular weight: 581.56, $C_{28}H_{23}NO_{11}$; mp, 176–180°. ^{1}H NMR (in CDCl$_3$, in ppm); isomer a: 2.35 (dd, J = 4 Hz, 3 Hz, 1H); 3.06 (dd, J = 4 Hz, 3 Hz, 1H); 3.53 (ddd, J = 4 Hz, 4 Hz, 2 Hz, 1H); 3.90 (s, 3H); 3.95 (s, 3H); 5.14 (s, 1H); 6.62 (d, J = 4 Hz, 1H); 7.01 (s, 1H); 7.27–7.32 (m, 4H); 7.43–7.48 (m, 4H); 7.60 (s, 1H); 7.92–7.97 (m, 2H); isomer b: 2.39 (dd, J = 4 Hz, 3 Hz, 1H); 3.08 (dd, J = 4 Hz, 3 Hz, 1H); 3.64 (ddd, J = 4 Hz, 4 Hz, 2 Hz, 3H); 3.92 (s, 3H); 3.97 (s, 3H); 5.15 (s, 1H); 6.67 (d, J = 4 Hz, 1H); 7.07 (s, 1H); 7.27–7.32 (m, 4H); 7.43–7.48 (m, 4H); 7.63 (s, 1H); 7.92–7.97 (m, 2H); isomer c: 2.83 (ddd, J = 4 Hz, 4 Hz, 2 Hz, 1H); 3.48 (ddd, J = 4 Hz, 4 Hz, 2 Hz, 1H); 3.64 (ddd, J = 4 Hz, 4 Hz, 2 Hz, 3H); 3.96 (s, 3H); 3.98 (s, 3H); 5.16 (s, 1H); 6.78 (d, J = 3 Hz, 1 H); 7.08 (s, 1H); 7.43–7.48 (m, 4H); 7.52–7.57 (m, 8H); 7.65 (s, 1H); 7.92–7.97 (m, 2H); isomer d: 2.92 (ddd, J = 4 Hz, 4 Hz, 2 Hz, 1H); 3.48 (ddd, J = 4 Hz, 4 Hz, 2 Hz, 1H); 3.64 (ddd, J = 4 Hz, 4 Hz, 2 Hz, 3H); 3.96 (s, 3H); 3.99 (s, 3H); 5.18 (s, 1H); 6.81 (d, J = 3 Hz 1H); 7.14 (s, 1H); 7.43–7.48 (m, 4H); 7.52–7.57 (m, 8H); 7.67 (s, 1H); 7.92–7.97 (m, 2H). FAB-MS m/z: 582.4 MH$^+$.

Preparation of N-1-([3,4-Dimethoxy-6-nitrophenyl)-2,3-epoxypropyl)oxy]carbonyl)4-(N,N-dimethylamino)pyridinium 3-oxo-2,5-diphenyl-2,3-dihydrothiophene 1,1-dioxide 4-hydroxide: compound V. One hundred fifty milligrams of **IV** (0.258 mmol) dissolved in 0.4 ml of dry THF is treated with a solution of 35 mg DMAP (0.285 mmol) in 0.1 ml of dry THF. After a 2-min sonification treatment, an orange precipitate forms that is left to stand for 1 hr at room temperature, centrifuged, and the orange residue washed with THF. Yield: 166 mg (0.237 mmol, 92%). Molecular weight: 704.7 (cation, 404.4) (cation, $C_{19}H_{22}N_3O_7$); mp 85° (decomposed). ^{1}H NMR (CDCl$_3$ in ppm): isomer a: 2.84 (dd, J = 5 Hz, 2 Hz, 1H); 3.02 (dd, J = 5 Hz, 2 Hz, 1H); 3.25 (s, 6H); 3.58 (ddd, J = 5 Hz, 5 Hz, 2 Hz, 1H); 3.93 (s, 3H); 4.04 (s, 3H); 6.63 (d, J = 4 Hz, 1H); 6.70 (d, J = 7 Hz, 2H); 7.05 (s, 1 H); 7.64 (s, 1H); 8.12 (d, J = 7 Hz, 2H); isomer b: 2.89 (dd, J = 5 Hz, 2 Hz, 1H); 3.11 (dd, J = 5 Hz, 2 Hz, 1H); 3.25 (s, 6H); 3.63 (ddd, J = 5 Hz, 5 Hz, 2 Hz, 1H); 3.96 (s, 3H); 4.06 (s, 3H); 6.72 (d, J = 7 Hz, 2H); 6.75 (d, J = 3 Hz, 1H); 7.09 (s, 1H); 7.66 (s, 1H); 8.14 (d, J = 7 Hz, 2H); counterion: 4.91 (s, 1H); 7.10–7.40 (m, 8H); 7.95–8.30 (m, 2H). FAB-MS m/z: 404.5 M$^+$ (cation).

Preparation of 1-(3,4-Dimethoxy-6-nitrophenyl)-2,3-epoxypropylsuccinimidyl Carbonate: compound VI. One hundred milligrams (0.392 mmol) of **II** dissolved in 2 ml of acetonitrile is treated with 110 mg (0.431 mmol) of di-(N-succinimidyl) carbonate. N-Ethyldiisopropylamine, 70 µl (0.431 mmol), is added, and the reaction mixture is stirred for 5 hr at room temperature. After evaporation of the solvent, the residue is dissolved in ethyl acetate and washed three times with 20% citric acid, sodium bicarbonate solution, and saturated sodium chloride. The organic phase is dried over magnesium sulfate, the solvent evaporated, and the yellow residue recrystallized out of chloroform/hexane, producing yellow–orange crystals. Yield 138 mg 89% (0.348 mml). Molecular weight: 396.32. $C_{16}H_{16}N_2O_{10}$. ^{1}H NMR (CDCl$_3$ in ppm): isomer a: 2.79 (s, 4H); 2.82 (dd, J = 5 Hz, 2 Hz, 1H); 3.02 (dd, J = 5 Hz, 2 Hz, 1H); 3.56 (ddd, J = 5 Hz, 5 Hz, 2 Hz, 1H); 3.95 (s, 3H); 4.04 (s, 3H); 6.66 (d, J = 4 Hz, 1H); 7.05 (s, 1H); 7.64 (s, 1H); isomer b: 2.79 (s, 4H); 2.90 (dd, J = 5 Hz, 2 Hz, 1H); 3.10 (dd, J = 5 Hz, 2 Hz, 1H); 3.63 (ddd, J = 5 Hz, 5 Hz, 2 Hz, 1H); 3.96 (s, 3H); 4.06 (s, 3H); 6.77 (d, J = 3 Hz, 1H); 7.09 (s, 1H); 7.67 (s, 1H). MS (70 eV), m/z (%) = 396 (51) M$^+$, 238 (100) $C_{11}H_{12}NO_5^+$. FAB-MS m/z 397.4 MH$^+$.

Spectroscopic and Photochemical Properties

The Grignard reaction of vinylmagnesium bromide on 3,4-dimethoxy-6-nitrobenzaldehyde introduces a hydroxyl and vinyl group off the chiral benzylic carbon and provides the starting point for the synthesis of the two reactive groups of the cross-linking reagent (Scheme 1). The first of these groups, an activated carbamate or carbonate, is used to react with and cage an amino group of the compound of interest to its corresponding photolabile carbamate. The second group, an oxirane, is used to alkylate thiol groups and serves to cross link the caged amino compound to a second thiol-containing biomolecule. This synthetic approach also allows the physical and chemical separation of the two nucleophilic reactive groups from the photoisomerizations functionality. The reagents are designed for applications in living cells, and this is facilitated by the use of the 3,4-dimethoxy-6-nitrophenyl chromophore, which has a high molar absorptivity and a red-shifted action spectrum around 340–420 nm that helped reduce irradiation-based cell damage.

The absorption spectrum of a defined concentration of an ethanolic solution of compound **II** shows the lowest energy transition in the near-ultraviolet region and has an extinction coefficient at 350 nm of 5000 M^{-1} cm^{-1}. Excitation of this ethanolic solution with near-ultraviolet light (340–420 nm) leads to an irradiation dose-dependent loss of the 350-nm band, whereas two new absorption transitions evolve with maxima at 268 and 378 nm. These absorption bands most likely belong to the photoproduct, a derivative of 3,4-dimethoxy-6-nitrosoacetophenone. Evidence that the reaction occurs via photoisomerization of the nitrophenyl group is provided by an analysis of the infrared spectra of a film of **II** recorded before and after near-ultraviolet irradiation, which shows the loss of a hydroxyl absorption band at 3450 cm^{-1} and a gain of a ketone absorption band at 1705 cm^{-1}.

Reactivity of Oxirane Group with Thiols

To show that the oxirane group of compound **II** can react selectively with thiol groups, a qualitative analysis of the reaction of compound **II** with 2-mercaptoethanol is performed under various conditions of pH and solvent composition. To summarize these data, alkylation of mercaptoethanol with compound **II** occurs in aqueous solutions at a pH value of 8.0–9.5 or in organic solvents in the presence of Al_2O_3 or tetrabutylammonium fluoride. Compound **II** does not react with primary amines under this condition. Investigations are also made on the reaction of the oxirane group of compound **II** with the single reduced thiol group of BSA and a 2-iminothiolane conjugate of BSA with several thiol groups. Dialysis of these reaction mixtures removes excess compound **II** and allows a calculation, based on absorption spectrophotometric analysis, of the labeling ratio of **II**/BSA for both conjugates. Native BSA is found to harbor a single molecule of **II** per BSA molecule, whereas the iminothiolane–BSA is labeled with six molecules of **II** per BSA. An independent determination of the number of reactive thiol groups in the iminothiolane–BSA conjugate using the thiol-labeling reagent acrylodan[16] also shows it contains six free thiol groups (data not shown). Evidently, at this slightly basic pH the oxirane group is able to react selectively with reduced thiol groups of proteins.

Reactivity and Properties of the New Class of Reagents

Compound **II** proves to be a key intermediate in the synthesis of four different cross-linking reagents (Scheme 1). The first of these, compound **III**, reacts poorly with the amino groups of polylysine or G-actin in bicarbonate buffer, pH 8.5. We suspect the substituent on the enzylic carbon atom of compound **III** is responsible for the deactivation of the chloroformate group because the unsubstituted chloroformate (NVOC-Cl) reacts efficiently with amino groups in both organic and aqueous solvents.[2,12] To improve on the poor reactivity of compound **II**, we use Steglich's reagent (TDO)[20] in order to make a more reactive carbonate. This reaction proceeds smoothly in THF under reflux and produces compound **IV** in high yield. The two isomers of compound **IV** cannot be resolved using the purification scheme described in the Methods section. However, the photoisomerization

---

[20] R. Kirstgen, A. Olbrich, H. Rehwinkel, and W. Steglich, *Liebigs Ann. Chem.* 437 (1988).

rates of the isomers of compound **IV** should be similar by analogy to the similar rates found for the isomers of compound **III**.

To improve the water solubility of heterobifunctional photocleavable cross-linking reagents, and to increase their reactivity toward the ε-amino group of lysine residues of proteins, compound **V**, the DMAP salt of compound **IV**, is prepared via two synthetic routes, outlined in Scheme 1. In the first approach, the succinimide carbonate of **II** is prepared in a reaction of **II** with di-N-succinimidyl carbonate in acetonitrile, which produces the crystalline product **VI** in high yield. Treatment of compound **VI** in THF with DMAP leads to the precipitation of the DMAP salt of compound **V**. In the second synthetic approach, compound **V** is prepared in a single-step reaction of compound **IV** with DMAP, as described in the Methods section. Compound **V**, a highly activated, water-soluble carbamate, reacts rapidly with the amino groups of proteins at a pH value between 8 and 9. Compound **V** is found to hydrolyze in water, and therefore high concentrations of both protein and compound **V** are recommended in labeling reactions. Compound **V** can be added to protein labeling reactions as a solid. It should be noted, however, that in organic solvents and in the presence of nucleophiles, DMAP-activated salts such as **V** undergo esterification or even reaction to isocyanates,[21] and therefore stock solutions of **V** in acetonitrile should be prepared just before the protein-labeling reaction.

Selected Applications

Members of the new class of cross-linking reagents are used to prepare a caged fluorophore and a caged G-actin.

Thiol-Reactive Caged Rhodamine 110

Preparation of Di-[1-(3,4-dimethoxy-6-nitrophenyl)-2,3-epoxypropyl]-N,N'-rhodamine Carbonate: compound VII. Compound **IV**, 100 mg (0.180 mmol), dissolved in 0.5 ml of water-free THF is treated with 40 μl of N-ethyldiisopropylamine (0.350 mmol) and 32 mg of rhodamine 110 hydrochloride (0.086 mmol). An orange precipitate develops over a 2-min sonification treatment. The reaction is left to stand for 4 hr at room temperature, centrifuged, the solvent evaporated, and the orange residue taken up in 5 ml of ethyl acetate. Successive washing of this organic solution with saturated $NaHCO_3$, 20% citric acid and with saturated NaCl is used to remove the sulfonate. The organic phase is then dried over $MgSO_4$, the solvent removed, and the product recrystallized from the residue out of toluene.

[21] H. J. Knölker, T. Braxmeir, and G. Schlechtingen, *Angew. Chem.* **107**, 2746 (1995).

Yield: 58 mg (0.065 mmol, 75%). Molecular weight: 892.5, $C_{44}H_{36}N_4O_{17}$; mp 188–193°. (CDCl$_3$ in ppm): isomer a: 2.35 (dd, J = 4 Hz, 3 Hz, 1H); 3.06 (dd, J = 4 Hz, 3 Hz, 1H); 3.53 (ddd, J = 4 Hz, 4 Hz, 2 Hz, 1H); 3.90 (s, 3H); 3.95 (s, 3H); 6.62 (d, J = 4 Hz, 1H); 7.01 (s, 1H); 6.68 (d, 8 Hz, 1H); 7.10 (s, 1H); 7.15 (d, 8Hz, 1H); 7.17 (d, 7 Hz, 1H); 7.19 (dd, 7 Hz, 7 Hz, 1H); 7.30 (dd, 7 Hz, 7 Hz, 1 H); 7.60 (s, 1H); 8.10 (d, 7 Hz, 1 H); isomer b: 2.39 (dd, J = 4 Hz, 3 Hz, 1H); 3.08 (dd, J = 4 Hz, 3 Hz, 1H); 3.64 (ddd, J = 4 Hz, 4 Hz, 2 Hz, 3H); 3.92 (s, 3H); 3.97 (s, 3H); 6.65 (d, J = 4 Hz, 1H); 6.68 (d, 8 Hz, 1H); 7.07 (s, 1H); 7.10 (s, 1H); 7.15 (d, 8 Hz, 1H); 7.17 (d, 7 Hz, 1H); 7.19 (dd, 7 Hz, 7 Hz, 1H); 7.30 (dd, 7 Hz, 7 Hz, 1H); 7.63 (s, 1H); 8.10 (d, 7 Hz, 1H); isomer c: 2.83 (ddd, J = 4 Hz, 4 Hz, 2 Hz, 1H); 3.48 (ddd, J = 4 Hz, 4 Hz, 4 Hz, 1H); 3.64 (ddd, J = 4 Hz, 4 Hz, 2 Hz, 3H); 3.96 (s, 3H); 3.98 (s, 3H); 6.68 (d, 8 Hz, 1H); 6.78 (d, J = 3 Hz, 1H); 7.08 (s, 1H); 7.10 (s, 1H); 7.15 (d, 8 Hz, 1H); 7.17 (d, 7 Hz, 1H); 7.19 (dd, 7 Hz, 7 Hz, 1H); 7.30 (dd, 7 Hz, 7 Hz, 1H); 7.65 (s, 1H); 8.10 (d, 7 Hz, 1H); isomer d: 2.92 (ddd, J = 4 Hz, 4 Hz, 2 Hz, 1H); 3.48 (ddd, J = 4 Hz, 4 Hz, 2 Hz, 1H); 3.64 (ddd, J = 4 Hz, 4 Hz, 2 Hz, 3H); 3.96 (s, 3H); 3.99 (s, 3H); 6.18 (d, J = 3Hz, 1H); 6.68 (d, 8 Hz, 1H); 7.10 (s, 1H); 7.14 (s, 1H); 7.15 (d, 8 Hz, 1H); 7.17 (d, 7 Hz, 1H); 7.19 (dd, 7 Hz, 7 Hz, 1H); 7.30 (dd, 7 Hz, 7 Hz, 1H); 7.67 (s, 1H); 8.10 (d, 7 Hz, 1H). FAB-MS m/z: 893.3 MH$^+$.

Amidation of the two aromatic amino groups of the highly fluorescent rhodamine 110 fluorophore forms the nonfluorescent, lactone form of the dye, which lacks a visible absorption band and fluorescence emission.[22] Similarly, compound **IV** reacts with rhodamine 110 and produces the nonfluorescent, bis-substituted carbamate (**VII;** Scheme 2). Irradiation of the nonfluorescent (caged) rhodamine 110 with near-ultraviolet light cleaves the two carbamate bonds and liberates free rhodamine 110 with a concomitant recovery of its visible absorption band and intense fluorescence emission centered at 523 nm (Fig. 4). This reaction is quite efficient because the photoactivation of a 500-μl sample of 50 μM solution of caged rhodamine 110 in the presence of 5 mM DTT using the near-ultraviolet output of a 75-W xenon arc lamp is complete within 240 sec (Fig. 4).

Labeling of G-Actin with Thiol-Reactive Caged Rhodamine 110 Reagent

The following method is used to couple one of the oxirane groups of compound **VII** to the thiol group on a cysteine residue of G-actin: 2 ml of 25 μM F-actin is depolymerized by dialysis over 2 days at 4° against nitrogen-purged G-buffer in the absence of DTT. The G-actin solution is clarified by centrifugation for 1 hr at 100,000 g at 4°. A 20 mM stock solution of

---

[22] S. P. Leytus, L. L. Melhado, and W. F. Mangel, *Biochem. J.* **209**, 299 (1983).

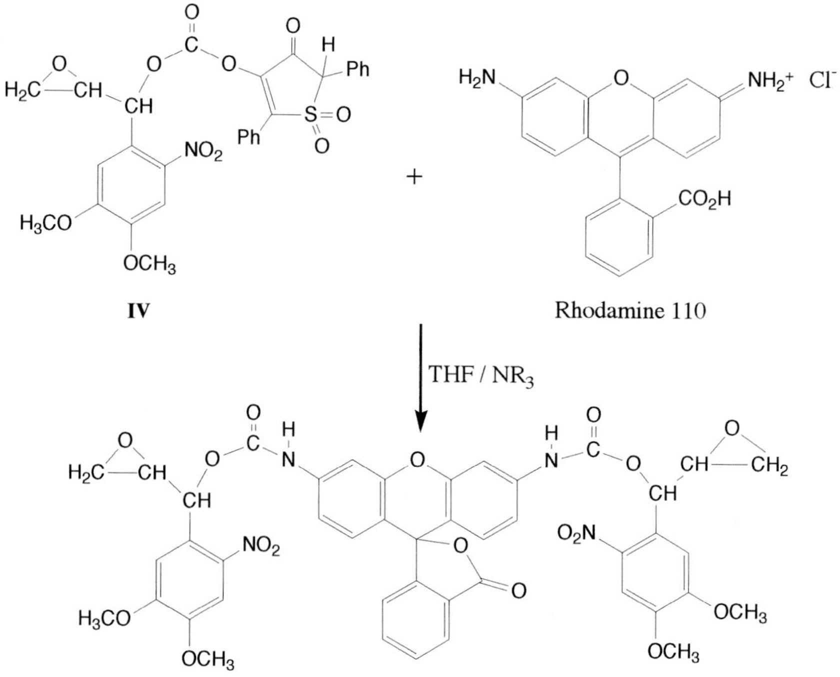

Scheme 2

IV + **Rhodamine 110** → (THF / NR₃) → **VII**

VII, 20 μl, is added to 500 μl of the G-actin solution after adjusting its pH to a value of 9.0. After a 1-hr reaction at room temperature the actin conjugate is clarified by centrifugation for 30 min at 10,000 rpm and dialyzed for 2 days against G-buffer, pH 8.0, containing 1 mM DTT. The sample is centrifuged for 1 hr at 100,000 g at 4°, and the labeling ratio of **VII** to G-actin is determined by absorption spectroscopy using an extinction coefficient for **VII** of 5000 M^{-1} cm^{-1} at 350 nm and 3400 M^{-1} cm^{-1} at 290 nm.[12]

The labeling ratio of 1 : 1 determined for compound **VII**/G-actin suggests that the caged fluorophore is attached to the highly reactive cysteine-374 residue.[16] The caged fluorescent G-actin conjugate polymerizes in physiological salt, as seen in a comparison of the absorption spectra of the supernatant and pellet fractions of a high-speed centrifugation run, because the majority of the photoactivated actin is found in the pellet fraction (Fig. 5). Irradiation of the nonfluorescent G-actin conjugate with near-ultraviolet light cleaves the carbamate bond that links the caged groups to rhodamine 110 and liberates the highly fluorescent rhodamine 110 and a polymeriza-

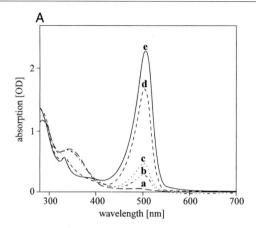

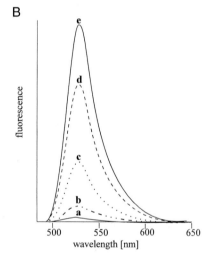

Fig. 4. (A) Absorption spectra of approximately 500 μl of a 50 μM solution of caged rhodamine 110 in ethanol/H$_2$O (10:1) containing 2 mM DTT. (a) Before and after near-ultraviolet irradiation (340–400 nm) for the following times: (b) 5 sec, (c) 10 sec, (d) 180 sec, and (e) 240 sec. (B) Fluorescence emission spectra of caged rhodamine 10 with excitation at 488 nm of approximately 50 μM of a 500-μl solution of caged rhodamine 110 in ethanol/H$_2$O (10:1). (a) Before, and after near-ultraviolet irradiation (340–400 nm) for the following times: (b) 5 sec; (c) after 20 sec; (d) 60 sec; (e) 240 sec.

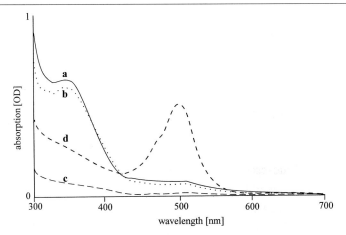

FIG. 5. (a) Absorption spectrum of G-actin-labeled caged rhodamine 110 before irradiation with near-ultraviolet light in G-buffer containing 2 mM DTT. The G-actin was treated with MgCl$_2$ and KCl to 2 and 100 mM, respectively, and, after 2 hr to allow for polymerization, the protein was centrifuged at 100,000 g for 60 min. (b) The absorption spectrum of the pellet fraction of this centrifugation run after resuspension in an equal volume of F-buffer. (c) The absorption spectrum of the supernatant fraction of this centrifugation run. (d) The absorption spectrum of the resuspended pellet fraction [from (b)] after irradiation with near-ultraviolet light for 10 min in the presence of 5 mM DTT.

tion-competent G-actin that harbors a 3,4-dimethoxy-6-nitrosophenyl ketone-based photoproduct (Fig. 5).

Preparation of Caged G-Actin Complex

Compound **V** is used as part of a new labeling procedure to prepare caged G-actin, i.e., a G-actin conjugate that does not polymerize in physiological salt but can polymerize normally after irradiation with near-ultraviolet light.[12]

A 30 μM solution, 800 μl, of IATMR-labeled G-actin in G-buffer without DTT is mixed with 20 μl of a freshly prepared 20 mM stock solution of **V** in acetonitrile. After a 45-min reaction in the dark at room temperature, the protein is centrifuged for 20 min at 14,000 rpm at 4° and dialyzed for 16 hr at 4° against nitrogen-purged G-buffer without DDT with two buffer changes. The protein is clarified by ultracentrifugation at 100,000 g for 1 hr and the absorption spectrum recorded. The protein-labeling ratio is calculated using an extinction coefficient for **V** of 5000 M^{-1} cm^{-1} at 350 nm and 3400 M^{-1} cm^{-1} at 290 nm.[12] Aminodextran labeled with 2-iminothiolane, as described earlier, is added to a 20 μM solution of IATMR-labeled G-actin in a borate-based G-buffer, pH 9.5. After a 1-hr reaction the complex is centrifuged for 20 min at 14,000 rpm at 4° and dialyzed overnight against two changes of G-buffer containing 1 mM DTT. The fluorescent actin–dextran complex is clarified by centrifugation for 20 min at 14,000 rpm at 4°.

Tetramethyl rhodamine iodoacetamide-labeled G-actin is labeled at an average of 3.5 lysine residues with compound **V**. One or more of the oxirane groups of the fluorescent G-actin conjugate of compound **V** is then cross linked to a thiolated dextran. Sixty minutes after the addition of physiological salt to this caged G-actin cross-linked complex, the sample is subjected to high-speed centrifugation and the absorption spectra recorded for the supernatant and resuspended pellet fractions (Fig. 6). These data show that the majority of the fluorescently labeled G-actin conjugate is contained in the supernatant fraction and is, therefore, polymerization incompetent (Fig. 6). Exposure of a 500-μl sample of this caged G-actin complex in F-buffer to near-ultraviolet light for 10 min in the presence of 5 mM DTT cleaves the cross link and released both G-actin and the dextran. After a period of 60 min at room temperature, to allow for polymerization of the photoactivated G-actin, the sample is subjected to high-speed centrifugation (100,000 g for 60 min at 4°), and absorption spectroscopy of the supernatant and pellet fractions now reveal that the majority of the fluorescent actin is present in the resuspended pellet fraction (Fig. 6). The inability of the G-actin cross-linked dextran complex to polymerize in physiological salt is most likely due to the physical masking of the actin-binding site on caged G-actin by the dextran.

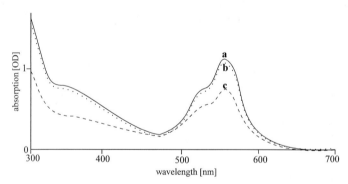

FIG. 6. (a) Absorption spectra of IATMR-labeled G-actin dextran complex in G-buffer containing 2 mM DTT. After addition of MgCl$_2$ and KCl to 2 and 100 mmol, respectively, and 2 hr at room temperature the sample was centrifuged at 100,000 g at 4° for 60 min. (b) The absorption spectrum of the supernatant fraction. No pellet fraction was found after the centrifugation, i.e., the actin was caged. The supernatant fraction was then irradiated with near-ultraviolet light for 10 min, and after 2 hr at room temperature the sample was centrifuged at 100,000 g for 60 min. (c) The absorption spectrum of the pellet fraction resuspended in an equal volume of F-buffer.

This mode of inhibition, however, is relieved after light-directed dissociation of the carbohydrate polymer from G-actin. Importantly, the photocleavage reaction releases native, functional G-actin, whereas the photoproduct remains covalently attached to the thiolated dextran. This new approach to prepare caged G-actin requires the modification of fewer lysine residues compared to the NVOC-Cl method.[12]

Summary

This study designed, synthesized, and characterized a number of new heterobifunctional photocleavable cross-linking reagents that may be used to photomodulate the activity of proteins or to prepare caged fluorescent dyes. Biomolecules or fluorophores caged via a thiol group with the BNBA-SE reagent can be covalently linked to a second protein, ligand, or derivatized surface through the activated carboxyl group. Members of the new class of photocleavable cross-linking reagent can be used to cage amino groups in the molecule of interest, which can then be covalently linked to a second molecule through the thiol-reactive oxirane group. These cross-linking reagents may be used for the following applications: (1) to cage the activity of a protein by masking its active site with a second macromolecule, e.g., aminodextran; (2) to prepare a protein conjugate exhibiting an enhanced or new activity that is lost on irradiation with near-ultraviolet light, e.g., cross-linked actin dimer; (3) to target the caged compound to a specific site by cross linking to a specific antibody;[13] (4) to attach the caged compound to a thiol or amino derivatized surface; and (5) to render the caged compound fluorescent in order to image or to quantify the yield of the photoactivation reaction.

[10] The Use of Lasers for One- and Two-Photon Photolysis of Caged Compounds

By JAMES A. MCCRAY

Introduction

The first use of a laser for single-photon photolysis of a caged compound, P^3-1-(2 nitro) phenylethyl-adenosine triphosphate (caged ATP), was in the late 1970s by McCray, Herbette, Kihara, and Trentham.[1] The work utilized

[1] J. A. McCray, L. Herbette, T. Kihara, and D. R. Trentham, *Proc. Natl. Acad. Sci. U.S.A.* **77,** 7237 (1980).

a passive dye Q-switched, frequency-doubled ruby laser and was published in 1980. This research demonstrated that biochemical kinetics inside cells could be followed on the millisecond time scale, because the cell membrane diffusion problem was bypassed. The motivation for this work was the 1978 paper by Kaplan, Forbush, and Hoffman,[2] where they had synthesized caged ATP and used it in the study of Na^+,K^+-ATPase in red cell ghosts. With the shuttered Hg arc light source that they used, they were able to study only slower processes, on 1-sec and 100-msec time scales. Further studies involving lasers to photolyze caged compounds have been guided in the choice of the pulsed laser system to be used by the action spectrum for photolysis of caged ATP, which was published in the Ph.D thesis of L. Herbette[3] in 1980 (Fig. 1). This action spectrum should be typical for other caged compounds in which the nitrobenzyl blocking group is used. The main work horse for one-photon photolysis of caged compounds has been the frequency-doubled ruby laser at 347 nm,[4] but other laser systems have also been used, such as the frequency-doubled liquid-dye laser at 320 nm,[5] the XeF excimer laser at 351 nm,[6] and the nitrogen laser at 337 nm.[7] The XeCl excimer laser at 308 nm has also been used to photolyze caged ATP in muscle studies, but sample destruction problems occurred,[8] presumably because of absorption of light by aromatic amino acids in some of the proteins. Caution should, therefore, be exercised with this laser system.

If it is desired to focus a laser beam through a microscope objective to a very small volume, on the order of a few cubic microns, for one-photon photolysis of caged compounds in or near biological cells, a serious problem arises because photodamage of the biological material occurs. A possible means of overcoming this problem was demonstrated in 1990 by Denk, Strickler, and Webb.[9] They suggested that it would be better to use lower energy (red) photons with wavelengths in the range of 640 to 700 nm and the quantum mechancial two-photon absorption process to reach an energy state different than that involved in one-photon absorption, which would, however, still lead by nonradiative decay to the uncaging excited state. They demonstrated that two-photon photolysis of caged ATP was possible.

---

[2] J. H. Kaplan, B. Forbush, and J. F. Hoffman, *Biochemistry* **17,** 1929 (1978).
[3] L. G. Herbette, "Structure-Function Correlation Studies of Isolated and Reconstituted Sarcoplasmic Membranes." Ph.D Thesis, University of Pennsylvania, Philadelphia, PA, 1980.
[4] Y. E. Goldman, M. G. Hibberd, J. A. McCray, and D. R. Trentham, *Nature* **300,** 701 (1982).
[5] J. W. Walker, G. P. Reid, J. A. McCray, and D. R. Trentham, *JACS* **110,** 7170 (1988).
[6] J. W. Karper, A. L. Zimmerman, L. Stryer, and D. A. Baylor, *Proc. Natl. Acad. Sci. U.S.A.* **85,** 1287 (1988).
[7] F. Engert, G. G. Paulus, and T. Bonheffer, *J.Neurosci. Methods* **66,** 47 (1996).
[8] R. Goody, private communication (1994).
[9] W. Denk, J. H. Strickler, and W. W. Webb, *Science* **248,** 73 (1990).

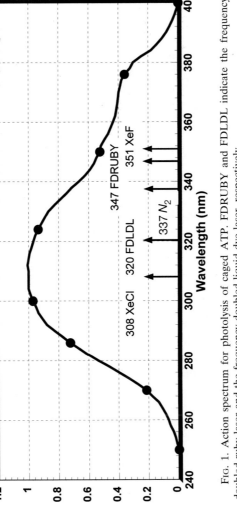

FIG. 1. Action spectrum for photolysis of caged ATP. FDRUBY and FDLDL indicate the frequency-doubled ruby laser and the frequency-doubled liquid dye laser, respectively.

The advantages of this method are that, in contrast to one-photon photolysis where UV photons not involved in uncaging can photo-oxidize proteins, leading to damage, the excess red photons not involved in uncaging can be absorbed only into vibrational states of proteins, leading to at most a temperature increase of the sample. This transient temperature rise problem, then, must be considered in the analysis of any experiment. The second advantage of two-photon photolysis is the highly localized photolysis volume because the two-photon absorption probably is highest in the region of highest peak power density. To have reasonable two-photon absorption, it is necessary not only to focus the laser beam to a small volume in space but also to have very narrow pulses in time so that the peak power density is high while the average power density is low, in order to minimize damage. This means that one should work with femtosecond lasers. The first femtosecond laser system used was a colliding-pulse, mode-locked dye laser[10,11] that produced 100-fsec-duration pulses at a repetition rate of 80 MHz and an average power of several milliwatts at 620 nm. This laser system, which was constructed at Cornell University, was rather difficult to keep running, so the next femtosecond laser system to be utilized for two-photon photolysis was a dye laser oscillator with gain jet and saturable absorber jet, compensated for group velocity dispersion and synchronously pumped by a mode-locked, frequency-doubled neodymium–yttrium–aluminum–garnet laser (neodymium-YAG). This commercial Antares/Satori laser system from Coherent, Inc.,[12] produced, for example, a train of 200-fsec pulses at 76 MHz with an average power of about 200 mW and a wavelength of 640 nm. The two-photon photolysis then occurs at an energy twice that of the original photons and hence at half the wavelength, i.e., at 320 nm. The corresponding region of the action spectrum (see Fig. 1) can then be covered using different dyes. This dye laser system does require some effort to operate properly and is not "turn-key," so for general usage it was thought that the argon ion-pumped, titanium-sapphire solid-state laser[12,13] might be easier for nonlaser specialists. This laser system is easier to use but can give femtosecond laser outputs only at wavelengths above 700 nm, which limits two-photon photolysis to the region above 350 nm. Preliminary work at Cornell University[14] with caged γ-aminobutyric acid (GABA), which released the neurotransmitter GABA, indicates that the two-photon absorption cross section is low at 700 nm but rises substantially as the wave-

[10] J. A. Valdemanis and R. L. Fork, *IEEE J. Quantum Electron.* **QE-22,** 112 (1986).
[11] F. W. Wise, I. A. Walmsley, and C. I. Tang, *Opt. Lett.* **13,** 129 (1988).
[12] Coherent, Inc., Laser Group, 5100 Patrick Henry Dr., Santa Clara, CA 95054.
[13] W. Koechner, "Solid-State Laser Engineering," 2nd ed., Springer-Verlag, Berlin, Germany, 1988.
[14] W. W. Webb, private communication (1997).

length is lowered. The Ti-sapphire laser is thus not the best choice for two-photon photolysis of caged compounds. A new laser system, the chromium-forstarite solid-state laser[15,16] with outputs in the 1230- to 1330-nm range, which frequency doubles to the 615–665-nm range, now being tested for two-photon photolysis of caged compounds (at 308–333 nm excitation), may turn out to be a good choice for the nonlaser specialist to use for two-photon photolysis. One of the most difficult problems with two-photon laser systems is, for many researchers, that of expense. A good discussion of two-photon laser excitation of molecular systems may be found in a 1995 article by Denk, Piston, and Webb.[17]

This article gives biologists, biochemists, physiologists, and other medical scientists the necessary basic knowledge they should have so that they can choose laser systems wisely and use them reliably and safely in experiments involving photolysis of caged compounds. A reasonable understanding of laser systems can be obtained by using rate equations, algebra, and trigonometry. Some general sources on optics and lasers can be found in Refs. 13 and 18–27.

Stimulated Emission

The basic concept involved in lasers is that of stimulated emission; in fact, the acronym laser means "light amplification by stimulated emission of radiation." It was Einstein[28,29] in 1917 who first showed that an excited state of a molecule can not only decay by spontaneous emission (fluorescence) but also must be able to decay by stimulated emission. He based his analysis on measurements and Max Planck's theory of black-body radia-

[15] V. Petricdvic, S. K. Gayen, and R. R. Alfano, *Appl. Phys. Lett.* **52,** 1040 (1988).
[16] V. P. Yanovsky and F. W. Wise, *Opt. Lett.* **19,** 1952 (1994).
[17] W. Denk, D. W. Piston, and W. W. Webb, in "Handbook of Biological Confocal Microscopy" (J. B. Pawley, ed.), p. 445. Plenum, New York, 1995.
[18] R. S. Longhurst, "Geometrical and Physical Optics," 2nd ed., Wiley, New York, 1967.
[19] M. V. Klein and T. E. Furtak, "Optics," 2nd ed., Wiley, New York, 1986.
[20] M. Born and E. Wolf, "Principles of Optics," 3rd ed., Pergamon, Oxford, England, 1965.
[21] A. Gerrard and J. M. Burch, "Introduction to Matrix Methods in Optics," Dover, New York, 1994.
[22] O. Svelto, "Principles of Lasers," Plenum, New York, 1976.
[23] A. E. Siegmen, "Lasers," University Science, Mill Valley, CA, 1986.
[24] F. P. Schäfer, ed., "Dye Lasers," 3rd ed., Springer-Verlag, Berlin, Germany, 1990.
[25] Ch. K. Rhodes, ed., "Excimer Lasers," 2nd ed., Springer-Verlag, Berlin, Germany, 1984.
[26] C. O. Weiss and R. Vilaseca, "Dynamics of Lasers," VCH, New York, 1991.
[27] J. R. Reitz, F. J. Milford, and R. W. Christy, "Foundations of Electromagnetic Theory," 4th ed., Addison-Wesley, New York, 1993.
[28] A. Einstein, *Physik. Zeitschr.* **XVIII,** 121 (1917).
[29] R. C. Tolman, *Phys. Rev.* **23,** 693 (1924).

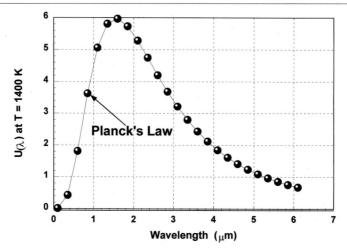

FIG. 2. Energy density of black-body radiation as a function of wavelength for the absolute temperature of 1400 Kelvin.

tion.[30] If the energy density, u (i.e., energy per unit volume), of a radiation field in equilibrium with the molecules on the inside of an enclosed cavity is measured as a function of wavelength of the radiation, a curve like that shown in Fig. 2 is obtained. This is the same curve that is measured for background cosmic radiation left over from the big bang by the COsmic Background Explorer (COBE) satellite, although in this case the temperature is very low, about 2.7 Kelvin.[31]

Einstein was trying to obtain this curve for equilibrium conditions by first writing the rate equations appropriate for absorption and emission of radiation by molecules. He considered molecules with two energy levels, E_1 and E_2, where $E_2 > E_1$, such that there are N_1 molecules in the ground state, E_1, and N_2 molecules in the excited state, E_2 (see Fig. 3). The total number of molecules is $N = N_1 + N_2$. In order to be consistent with the curve of black-body radiation, Planck found that the factor

$$\frac{1}{e^{(hc)/(\lambda kT)} - 1} \tag{1}$$

was essential in the energy density. In this expression, h is Planck's constant, c is the velocity of light in a vacuum, λ is the wavelength of the light, k is Boltzmann's constant, and T is the absolute temperature. Einstein, there-

---

[30] P. A. Tipler, "Modern Physics," Worth, New York, 1978.
[31] E. Chaisson and S. McMillan, "Astronomy Today," 2nd ed., Prentice-Hall, Upper Saddle River, NJ, 1997.

fore, also had to obtain this factor from the analysis. When he considered the rate equation for the upper state, he found that it was necessary to add an additional rate term that depended on energy density:

$$\frac{dN_2}{dt} = -A_{21}N_2 + B_{12}uN_1 - B_{21}uN_2 \qquad (2)$$

The first term on the right represents spontaneous emission (fluorescence), the second term represents absorption, and the third new term represents stimulated emission.

Setting dN_2/dt equal to zero for the case of equilibrium and solving for u gives

$$u = \frac{A_{21}}{B_{12}(N_1/N_2) - B_{21}} \qquad (3)$$

The ratio N_1/N_2 is given by the reciprocal Boltzmann factor $e^{(E_2-E_1)/kT}$. Use was also made of Bohr's energy–frequency relation,[30]

$$E_2 - E_1 = hf = \frac{hc}{\lambda} \qquad (4)$$

where hf is the energy per photon. Thus

$$u = \frac{A_{21}}{B_{12}e^{(hc)/(\lambda kT)} - B_{21}} \qquad (5)$$

Now, as the temperature becomes very large, approaching infinity, the energy density also approaches infinity. This requires that $B_{12} = B_{21}$, yielding the equation

$$u = \frac{A_{21}}{B_{21}} \left(\frac{1}{e^{(hc)/(\lambda kT)} - 1} \right) \qquad (6)$$

Einstein thus found that he could obtain the Planck factor only if he assumed that an excited state of a molecule must also be able to decay by a term that involves the number of photons already present. This is the motivation for calling this term stimulated emission. An immediate consequence of this conclusion is that photons can be added to those already in the field—an amplifier of electromagnetic radiation should be possible. However, in 1917, technology was not adequate to produce a light amplifier. In 1954, Gordon, Zeiger, and Townes[32] produced the first stimulated-emission device, the maser (microwave amplification by stimulated emission of radiation). By

---

[32] J. P. Gordon, H. Z. Zeiger, and C. H. Townes, *Phys. Rev.* **95**, 282 (1954).

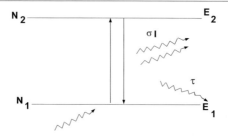

FIG. 3. Energy level diagram for a two-level system illustrating absorption, fluorescence, and stimulated emission.

1958 it was clear to Schawlow and Townes[33] that a light amplifier was technologically possible, and in 1960 Maiman[34] produced the first successful optical device, the ruby laser.

Dynamic Bleaching

When using lasers it is very important to realize that one is dealing with high-intensity light so that normal light filters used in the biochemical laboratory should not be used to protect the eyes from laser light. The point here is that stimulated emission must also be taken into account when high-intensity (laser) light is being used. For low-intensity light, only absorption and fluorescence need to be taken into account. Consider again the two-level molecular system with energy levels, E_1 and E_2, as shown in Fig.3.[35]

At any given time the population of the ground state is N_1 and that of the excited state is N_2. The time-dependent rate equation for the upper level in terms of the light intensity, I (number of photons per square centimeter per second) that impinges on the molecules is

$$\frac{dN_2}{dt} = \sigma I N_1 - \frac{N_2}{\tau} - \sigma I N_2 \tag{7}$$

where σ is the absorption and stimulated emission cross section (with units of area) and τ is the mean fluorescence lifetime. The total number of molecules is $N = N_1 + N_2$. At steady state the above derivative is zero, yielding the dependence of N_2 on the light intensity, I:

---

[33] A. L. Schawlow and C. H. Townes, *Phys. Rev.* **112**, 1940 (1958).
[34] T. H. Maiman, *Nature* **187**, 493 (1960).
[35] R. W. Keyes, *IBM J.* **October**, 334 (1963).

$$N_2 = \left(\frac{\sigma I}{2\sigma I + \frac{1}{\tau}}\right) N \tag{8}$$

A plot of Eq. (8) is given in Fig. 4.

At high light intensity Eq. (8), which is based on the steady-state assumption at the peak of a pulse, shows that N_2 is just one-half of the total number of molecules, N, which means also that one-half of the molecules are in the ground state. Just as many molecules emit light by stimulated emission as absorb light. The laser light then just passes through the material and provides very little filtering. This feature is called dynamic bleaching. Only filters labeled "laser filters" should be used to attenuate laser light for eye protection! It is also very important to make absolutely sure that laser goggles selected for eye protection are definitely for the particular wavelength of the laser being used. Particular care should be taken when frequency-doubled laser systems are being used because two different laser beams of widely different wavelengths are present. Dynamic bleaching is also used in passive Q-switching to hold off lasing while an inversion population is built up in Q-switched lasers and in saturable absorbers to help shape femtosecond laser pulses.

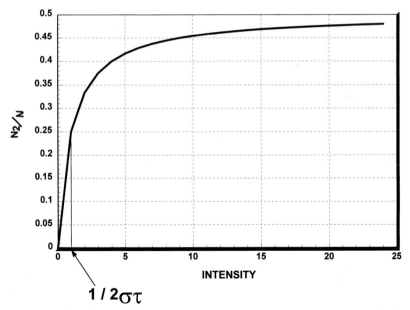

FIG. 4. Fractional occupation of the upper state with energy E_2 as a function of incident light intensity illustrating dynamic bleaching.

An estimate of the critical intensity, $I_c = 1/(2\sigma\tau)$, can be made if the size or cross section of an absorbing group on a molecule is taken as $\sigma \approx 2.5 \times 10^{-15}$ cm^2 and a fluorescent lifetime is taken as $\tau \approx 10^{-8}$ sec. Then $I_c \approx 2 \times 10^{22}$ photons per cm^2 per second. This corresponds to an energy flux or intensity, for $\lambda = 320$ nm, of about $I_c hc/\lambda \approx 12$ KW/cm^2. It should be noted that this is a peak power density, not average power. Most pulsed lasers used for photolysis have peak power densities above this value; for example, the frequency-doubled, Q-switched ruby laser generally used for one-photon photolysis of caged compounds has a pulse energy of 100 mJ in 30 nsec and a cross-sectional area of about 12 mm^2. These numbers result in a peak power density of $I \approx 30$ MW/cm^2, which is 2.5 thousand times the critical intensity. The peak power density directly out of two-photon laser systems used is also of the order of a few megawatts per square centimeter. After the microscope objective, the peak power density can reach gigawatts per square centimeter. Of course, to be useful as a photolysis source, there must also be nonradiative decay to a state in the absorbing molecule that leads to the uncaging of the caged compound. The combined spontaneous decay rates lead to a combined mean lifetime, τ_c, found from the equation

$$\frac{1}{\tau_c} = \frac{1}{\tau_f} + \frac{1}{\tau_n} \qquad (9)$$

where τ_f is the fluorescent lifetime (called τ above) and τ_n is the nonradiative lifetime. Because τ_c, which now takes the place of τ above, is less than the fluorescent lifetime, τ_f, the critical intensity is higher.

Dynamic bleaching can also been seen if the effect of stimulated emission on Beer's law of absorption is determined. The intensity loss, dI, in a passive material in a distance, dx, including stimulated emission is

$$dI = -\sigma I N_1 dx + \sigma I N_2 \, dx \qquad (10)$$

Using the constraint equation, $N_1 + N_2 = N$, and Eq. (8) for N_2 gives a differential equation for I and x that can be integrated to give a modified Beer's law.

$$\ln \frac{I(x)}{I_0} + A \left[\frac{I(x)}{I_0} - 1 \right] = -\sigma N x \qquad (11)$$

where $A = 2\sigma\tau I_0$. The intensity as a function of x for different values of A, plotted in Fig. 5, shows that for $A = 0$, Beer's law is obtained, but for larger values of A the intensity is much larger than one would expect from Beer's law. The situation for $A = 1$ is that for the critical intensity, I_c.

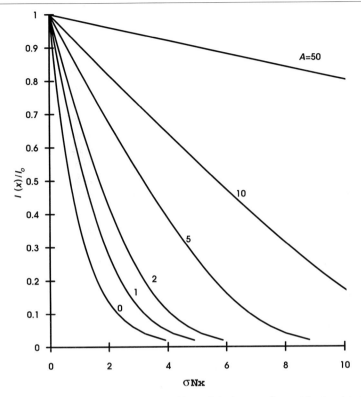

FIG. 5. Modified Beer's law absorption of laser light in a medium with stimulated emission included.

The Basic Laser

From the discussion on dynamic bleaching it can be seen that a passive material cannot be used as a light amplifier. It is necessary to have an inversion population where $N_2 > N_1$; then stimulated emission can add net photons to the original light intensity. Laser material hosts doped with metastable lasing ions or molecules are used as light amplifiers, but what we usually mean by a laser is really a laser oscillator. Figure 6 shows the layout of a basic laser oscillator. To achieve an inversion population, some means of pumping the ions or molecules to a higher, third state with energy, E_3, must be used, such as electronic discharge for a gas laser or optical pumping from a flash lamp, diode laser, or another laser for solid-state lasers or liquid-dye lasers. Nonradiative decay then leads to the long-lived second upper state of energy, E_2, making an inversion population possible. The stimulated emission may be increased by placing the lasing material

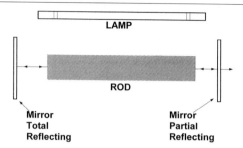

FIG. 6. The essential components of a basic laser.

in an optical cavity so that feedback is possible. This is done by placing a partially reflecting mirror and a totally reflecting mirror in front and behind the lasing rod. When the mirrors are aligned parallel to each other the optical cavity is "tuned" and feedback of the light can take place. Part of the laser light is extracted from the cavity through the partially reflecting mirror. The light makes two passes through the lasing rod for each round trip, where it is enhanced by the gain of the lasing rod. There are, however, losses of light during this round trip by absorption, scattering, diffraction, and extraction through the output mirror. Thus the round-trip gain must exceed the round-trip loss before laser light can appear at the output mirror. The result is that a threshold of some kind must be exceeded in order to get the laser to lase. For the flash-lamp pumped laser shown in Fig. 6, this is evident in the voltage across the flash lamp, which leads to more pump light. As will be discussed later in this article, it is necessary to operate sufficiently above threshold so that the transverse beam structure can be Gaussian. Some important safety points should be noted here. The voltages used in pulsed laser systems come from high-voltage power supplies. Users are well protected from these voltages, but if one takes off the protective covers, then extreme care must be exercised in order to avoid electrocution. Also, all laser beams should be covered, and any unused reflections or final main beams should be dumped in recessed blackened beam stops. Special care must be taken and fail-safe protection devised if laser beams are to be used in microscopes.

Gaussian Beam Optics

It is very important to realize that the transverse profile of a laser beam is not that of a plane wave but instead is usually Guassian, as given by Eq. (12):

$$f(x,y,z) = \frac{w_o}{w(z)} e^{-(x^2+y^2)/w^2} \qquad (12)$$

where w_o is the waist radius and

$$w(z) = w_o \sqrt{1 + \left(\frac{\lambda z}{\pi w_o^2}\right)^2} \qquad (13)$$

is the beam radius at a distance z from the waist. Equations (12) and (13) are the result of an analysis called paraxial or first-order beam optics. There are times when aberrations require the consideration of third-order beam optics, although for most work first-order beam optics suffices if the angles of the rays and beams from the optical axis are small.

Higher-order transverse modes are possible with a Gaussian profile but modified by either Hermit or Laguerre polynomials. If care is not taken to produce the fundamental Gaussian mode, e.g., with a small aperture, a multimode output is obtained. This is particularly likely to happen if the laser is operated near threshold.

As can be seen in Fig. 7, a Gaussian beam cannot be focused to a geometrical point but only to a small disk, called a waist. Close to the beam axis the wavefront is spherical. The tightness of the focusing is indicated by the Rayleigh range, which is the distance over which the beam radius increases to $\sqrt{2}$ times that at the waist and is given from Eq. (13) by the formula:

$$Z_R = \pi \frac{w_o^2}{\lambda} \qquad (14)$$

where w_o is the original waist radius and λ is the wavelength of the laser beam.

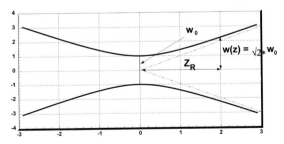

FIG. 7. Inherent diffraction of a Gaussian laser beam illustrating the concepts of the waist radius and the Rayleigh range.

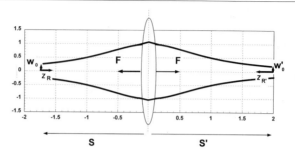

Fig. 8. Focusing of a Gaussian laser beam by a thin lens.

The propagation of a Gaussian laser beam through an optical system has been studied by Kogelnik and Li[36]; they used a complex beam parameter, matrices, and Kogelnik's "ABCD" law. Self,[37] however, has put the basic results in a simple form so that it is possible to calculate the positions and sizes of the various waists in a manner similar to that of ordinary ray tracing. In his paper Self treated only thin lenses with the same medium on both sides of the lens, but it can readily be shown from Kogelnik's results that Self's forms of the equations also hold for thick lenses, provided that waist object and image distances and focal lengths are measured from the principal planes. Self's equations follow directly from Kogelnik's and are, therefore, just as rigorous. The procedure is as follows.

The Gaussian beam is followed through each optical lens system, one lens at a time. The description is in terms of a thin lens, but the same procedure holds for a thick lens if the waist object and image distances and focal lengths are measured from the principal planes of the thick lens. A typical setup is shown in Fig. 8.

The position of the new waist is given by

$$\frac{1}{s + Z_R^2/(s-f)} + \frac{1}{s'} = \frac{1}{f} \qquad (15)$$

where s is the distance from the thin lens to the original waist, positive to the left and negative to the right; s' is the distance from the thin lens to the new waist, positive to the right and negative to the left; f is the focal length of the thin lens, positive for a converging lens and negative for a diverging lens; and Z_R is the Rayleigh range.

---

[36] H. Kogelnik and T. Li, *Appl. Opt.* **5**, 1550 (1966).
[37] S. A. Self, *Appl. Opt.* **22**, 658 (1983).

The waist radius magnification is found from

$$m = \frac{w_o'}{w_o} = \frac{1}{\sqrt{(1 - s/f)^2 + (Z_R/f)^2}} \quad (16)$$

The new Rayleigh range is determined from the equation

$$Z_R' = m^2 Z_R \quad (17)$$

This procedure is then repeated for the second optical lens system, carefully using the correct magnitude and sign of the distance from the new waist to the second optical lens system. The final waist radius from the first calculation is used as the new waist radius for the second calculation.

First-order ray (plane wave) optics equations are found from Self's equations by letting Z_R be equal to zero.

It should be stressed that the results obtained by ray optics can be grossly different from the correct Gaussian beam optics calculation, as can be seen from Fig. 9. For example, if the original waist is placed at the left focal point of a thin lens, the new waist is not at infinity but is at the right focal point.

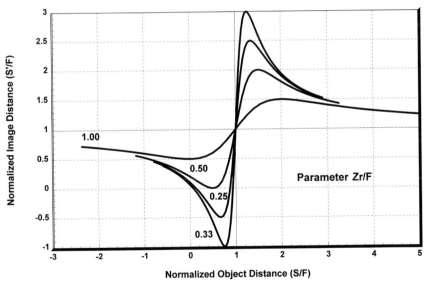

FIG. 9. Self's equation: Normalized image distance as a function of normalized object distance for an optical system with the normalized Rayleigh range as a parameter. All distances are normalized to the focal length of the optical system, where the medium on both sides of the optical system is the same.

A typical problem is to know the size and position of an original laser waist and to desire to find the focal length of a thin lens and where to place it in order to reduce the size of the waist by a given amount at a given distance from the first waist. Self's equations can be reduced to solving a quadratic equation whose solution is

$$f = -\frac{B}{2A} + \frac{1}{2A}\sqrt{B^2 - 4AC} \qquad (18)$$

where

$$A = 2m^2 - m^4 - 1; \quad B = -4m^2 l; \quad C = m^2[l^2 + (m^2 + 1)^2 Z_R^2] \qquad (19)$$

The distance of the new waist from the original waist is l.

The position of the lens from the position of the original waist, s, is found from

$$s = f\left[1 + \sqrt{\frac{1}{m^2} - \frac{Z_R^2}{f^2}}\right] \qquad (20)$$

The position of the new waist from the lens is given by $s' = l - s$. For example, suppose it is desired to reduce a HeNe (helium–neon) laser's (633 nm) waist radius of 0.5 mm by a factor of 10 at a distance of 1 m. According to Self's equations, a positive lens of focal length 14.3 cm should be placed 14.5 cm to the left of the new waist position at 1 m. A ray-tracing analysis predicts that a positive lens of focal length 8.26 cm should be placed 9.09 cm from the new waist position at 1 m. As can be seen, the ray tracing calculation gives answers that are quite wrong for a Gaussian beam.

Several important concepts must be considered in order for one to have a reasonable understanding of the laser systems used for photolysis of caged compounds. These concepts are considered one at a time and then put together for a description of the laser systems.

Wave Polarization[20]

Maxwell's equations predict that sinusoidal plane waves can be propagated in free space such that the electric, magnetic, and propagation vectors are all at right angles to each other. It is desired to see what curve the tip of the electric vector traces out in a plane perpendicular to the direction of propagation. The electric vector can be decomposed into its components along two perpendicular directions in that plane, and both components will oscillate sinusoidally. Maxwell's equations, however, do not predict the relative phase of these two components. The relative phase is set by initial conditions and can change after passing through various optical components. For example, if the two components have relative phase either 0° or 180°, then the wave is linearly polarized and the tip of the electric vector oscillates along a straight line. If the relative phase between the two components is 90° and the relative magnitudes of the two components are the same, then the wave is circularly polarized either right-handed or left-handed, depending on the sign of the relative phase. The general case is right- or left-handed elliptical polarization. This can be seen from the following equations for the oscillating electric field components of a traveling plane wave:

$$E_x = A_x \cos(\tau + \varphi_x)$$
$$E_y = A_y \cos(\tau + \varphi_y) \qquad (21)$$

where $\tau = kz - \omega t$; k is the propagation constant, equal to $2\pi/\lambda$, where λ is the wavelength; and ω is the angular frequency, equal to $2\pi f$, where f is the frequency. A_x and A_y are the amplitudes of the two components, and φ_x and φ_y are the phases of the two components. That these equations represent traveling waves can be seen from the following discussion: A point on the sinusoidal wave is to be followed in space as a function of time. Now, setting the absolute phase φ_x equal to zero for simplicity, the x component of the electric field has a space and time dependence, $E_x(z,t) = A_x \cos(kz - \omega t)$. If a point (phase) of the wave is followed, then at a time $t + \Delta t$ the wave point is at $z + \Delta z$. This requires that $\cos[k(z + \Delta z) - \omega(t + \Delta t)] = \cos(kz - \omega t)$. If the cosines are equal, then the angles must be equal, leading to $\Delta z = (\omega/k)\Delta t$. However, $\omega/k = (2\pi f)/(2\pi/\lambda) = \lambda f = v$, the speed of the wave in the material. Hence $\Delta z = v\Delta t$ and the particular point being followed on the wave has moved a distance in the positive direction $v\Delta t$. This type of wave is called a forward wave.

Recalling the trigonometric relations

$$\sin^2 A + \cos^2 A = 1$$
$$\cos(A + B) = \cos A \cos B - \sin A \sin B \qquad (22)$$
$$\sin(A - B) = \sin A \cos B - \cos A \sin B$$

and eliminating τ between the two equations (21) leads to Eq. (23) for the tip of the electric vector:

$$\left(\frac{E_x}{A_x}\right)^2 + \left(\frac{E_y}{A_y}\right)^2 - 2\frac{E_x E_y}{A_x A_y}\cos\varphi = \sin^2\varphi \qquad (23)$$

where $\varphi = \varphi_y - \varphi_x$ is the relative phase difference between the two components. These results can also be used for Gaussian laser beams.

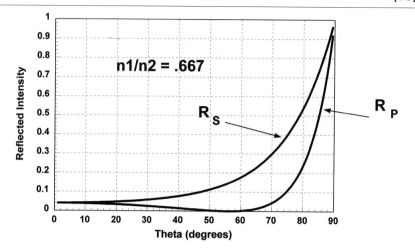

FIG. 10. Reflectivity of an air–glass interface for the component in the plane of incidence, R_p, and the component perpendicular to the plane of incidence, R_s.

Reflection and Refraction

The laws of reflection and refraction for plane waves at an interface between two dielectrics are well known and can still be used for Gaussian laser beams because the additional effects of the finite cross section of the laser beam are small and need to be considered only if there are a large number of multiple reflections.[38] Thus, the law of reflection (the angle of reflection is equal to the angle of incidence; $\vartheta_1 = \vartheta_0$) and the law of refraction (the ratio of the sines of the angle of refraction and the angle of incident is constant and equal to the ratio of index of refraction of the first medium to that of the second medium: Snell's law, $\sin \vartheta_2 / \sin \vartheta_0 = n_1/n_2$) can be used. The reflectivity of the component of the electric field perpendicular to the plane of incidence, R_S, and the reflectivity of the component of the electric field in the plane of incidence, R_p, can be found from Maxwell's equations and the appropriate boundary conditions for this problem, yielding the Fresnel relations. Figures 10 and 11 show the situation for the air–glass case and the glass–air case, respectively.

A striking result is that for the component of the electric field in the plane of incidence, there is no reflectivity at a certain angle, the Brewster angle, $\tan \vartheta_0 = n_2/n_1$ This means that unpolarized light reflected at the Brewster angle will be polarized perpendicular to the plane of incidence.

---

[38] Y. M. Antar and W. M. Boerner, *Can. J. Phys.* **52**, 962 (1974).

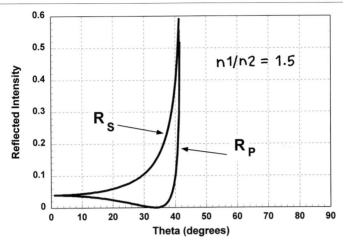

FIG. 11. Reflectivity of an glass–air interface for the component in the plane of incidence, R_p, and the component perpendicular to the plane of incidence, R_s.

There is also a Brewster angle for the case of going from glass to air at the complement of the former Brewster angle. An additional phenomenon appears in going from glass to air, as shown in Fig. 11, that of total reflection giving rise to a critical angle, $\sin \vartheta_0 = n_2/n_1$. Due to refraction, a beam of light passing through a plate with parallel sides is displaced laterally from its original direction. In order to achieve high linear polarization using the Brewster effect, several plates must be used, leading to an unacceptable lateral displacement. One can, however, arrange the plates so that half are pointing in one direction and the other half in the other so that the lateral displacement is in the opposite direction from the first, and the beam is returned to its original direction. Such a polarizing device is called a compensating Brewster stack.

If the incident angle is greater than the critical angle, there is total reflection, but this is for the situation when the intensities are averaged over one cycle of the wave. Maxwell's equations show that there actually is a surface wave that has an exponentially varying amplitude into the air, which can be detected and utilized. This is called the evanescent wave, and the process is termed frustrated total reflection. This effect is used, for example, to extract energy from a ring prism laser, as a light modulator, and has been able to excite molecular fluorescence. A necessary condition for the first two applications is that a second transparent plate, which is flat to within a fraction of a wavelength, must be brought to a distance of about a wavelength parallel to the original glass surface, which is also flat to a fraction of a wavelength.

If a wave arrives at the air–glass surface of a glass plate in air, the reflected wave, according to the Fresnel relations, will have a phase shift of 180°, whereas the transmitted wave, which reflects from the glass–air surface, has no phase shift. This second wave is transmitted back through the upper surface, and the two waves can then interfere coherently. What actually happens for a wave of wavelength λ in the glass depends on the path length in the glass because an additional phase shift occurs for the second wave because of the $(2\pi/\lambda)z$ term in the phase $(kz - \omega t)$ of the traveling wave. For a proper thickness there can be constructive interference with greater reflection or destructive interference leading to little reflection. If multiple, thin film layers are used with materials of the proper indices of refraction and proper thicknesses, then almost totally reflecting mirrors (multiple dielectric mirrors) can be constructed at a given wavelength, or antireflecting (AR) coatings can be deposited on various optical surfaces to achieve very little reflection at a given wavelength.

Birefringence[20]

If Maxwell's equations are solved for possible electromagnetic waves in an anisotropic medium, such as a crystal, it is found that in any given direction there are two traveling waves possible. These waves are linearly polarized and their electric displacement vectors are at right angles to each other, but they differ in a much more profound way. The phase speed of one of the waves is the same in all directions, whereas the phase speed of the other plane wave is dependent on the direction. The first wave is called an ordinary wave because it acts just like a wave in an isotropic medium, and the second wave is called the extraordinary wave because it is different. In the general case, one finds that there are two specific directions in a crystal in which the speeds of the waves are equal. The crystal is then called a biaxial crystal; however, some crystals have only one direction in space where the speeds are the same. In this case they are called uniaxial crystals, with the special direction termed the optical axis of the crystal. Uniaxial crystals are the crystals used most in optics. If light impinges on a uniaxial crystal surface whose optical axis is not perpendicular to the surface, then in a plane perpendicular to the surface the expanding Huygens wavefronts are circular for the ordinary wave but elliptical for the extraordinary wave, leading to two physically separated waves in the crystal. On leaving the crystal, the two waves carry two images of the same original object. This, then, is called double refraction. A typical demonstration of double refraction is to place a calcite crystal on top of a newspaper so that two distinct images of the newsprint are seen.

If the two perpendicular traveling waves in the crystal have different coefficients of absorption, then unpolarized light, when decomposed into the two components, becomes partially or wholly polarized, depending on the length of the crystal. This effect is called dichroism. Specially treated materials have been made that are also dichroic, such as polaroid sheets as polarizers, or dichroic films on substrates as dichroic mirrors.

Q-switching[13]

For resonant systems, a variable is used to express the quality of the resonant system, called the quality factor, Q, which is defined as the ratio of the average energy stored to the energy lost in one cycle. This definition applies to lumped parameter LCR circuits, microwave resonant cavities, and laser optical cavities. The Q of the circuit or cavity is frequency dependent, and it is usually desired to have a high Q at a given frequency. For the laser optical cavity, it is necessary to align the cavity optically so that efficient feedback can occur, leading to high stimulated emission and laser action. The losses in the cavity must be minimized so that the cavity has high Q. This can be done in different ways; for example, the cavity can be misaligned while the inversion population is maximized and then realigned in a short time. This is called Q-switching and can be produced by rotating either a mirror or a totally reflecting prism and synchronizing the pumping process.

Q-switching can also be achieved by using a compensating Brewster stack and a voltage-controlled birefringent device called a Pockels cell, which can rotate the line of linear polariation of the beam after passing through the cell. The basic laser of Fig. 6 is modified by placing the Pockels cell at the totally reflecting mirror, followed by the Brewster stack (to linearly polarize the laser beam). The Pockels cell can be operated so that initially a high voltage causes the crystal (KD*P, potassium dideuterium phosphate) to be birefringent such that the line of polarization is rotated 90° after two passes through the cell. The line of polarization is then at right angles to the original line of linear polarization, and the wave energy is dissipated in the Brewster stack by subsequent reflections. This condition is held while the flash lamp pumps the rod to achieve a high inversion population; at the appropriate time the high voltage on the Pockels cell is cut off, making the crystal transparent, and the laser light comes out in several round trips of the cavity, yielding a giant Q-switched pulse of the order of several Joules in tens of nanoseconds, depending on the length of the optical cavity. The peak power density of this pulse is high enough to achieve reasonable frequency doubling.

Frequency Doubling[39]

When an electric field is applied to a crystal, the center of the negative charge of the electrons, of otherwise neutral atoms, is displaced from the center of positive charge in the nucleus. This creates what is known as an electric dipole, which has a butterfly type of electric field pattern. The vector quantity consisting of the product of the magnitude of the charge of the nucleus and the vector displacement of the two charge centers is called the electric dipole moment of the atom. The net electric vector dipole moment per unit volume is called the electric polarization, $\vec{P}$, of the crystal (not to be confused with wave polarization). This variable obviously depends on the applied electric field, $\vec{E}$, but in an anisotropic crystal $\vec{P}$ is not in the same direction as $\vec{E}$. Because the electrical interaction is nonlinear in an atom, the components of $\vec{P}$ can be expressed as power series in the components of $\vec{E}$. The electric field in a laser beam is strong enough so that the nonlinear terms in the electric polarization can be significant. In fact, the first nonlinear terms, the terms quadratic in the electric field components, are used in frequency doubling. Maxwell's theory predicts that an accelerating charge will radiate electromagnetic energy, and therefore an oscillating electric dipole will also radiate electromagnetic energy. If the original electric wave is oscillating at a frequency ω, the electric polarization also oscillates, not only at the frequency ω but also at higher harmonics. The way this takes place can be seen from the following simple discussion. The original electric wave is assumed to be oscillating at frequency ω, so that at a given point in the crystal the time dependence of the electric field is $E(t) = E \cos \omega t$. For simplicity the relationship is given for an isotropic crystal. Thus the first nonlinear term in the electric polarization is given by

$$P_2 = \beta E^2 \cos^2 \omega t \tag{24}$$

where β is the second-order polarizability coefficient.

Recalling the trigonometric relation $\cos^2 A = (1/2)(1 + \cos 2A)$ leads to

$$P_2 = \frac{\beta E^2}{2} + \frac{\beta E^2}{2} \cos 2\omega t \tag{25}$$

There is thus a part of the electric polarization oscillating at the frequency 2ω, which leads to an induced electric wave at twice the frequency of the original frequency. This is called frequency doubling. Because of the relationship between frequency and wavelength, $\lambda f = v$ and $\omega = 2\pi f$, the new electromagnetic wave is at half the wavelength of the original wave. For example, it is said that the output of a ruby laser at 694 nm is frequency

---

[39] A. Yariv, "Quantum Electronics," Wiley, New York, 1967.

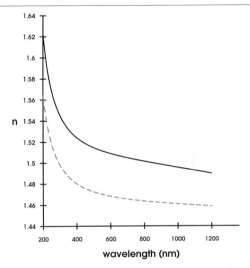

FIG. 12. Dependence of the index of refraction on wavelength for the ordinary wave (solid line) and the extraordinary wave (dashed line; at 90° from the optical axis) for a KDP crystal at a temperature of 24.8°.

doubled to 347 nm. An anisotropic crystal, however, must be used to generate the second harmonic. The original wave (fundamental) is introduced into the crystal as an ordinary wave, and the generated second harmonic wave is an extraordinary wave at right angle to the fundamental. The speed of the ordinary wave is the same in all directions, but the speed of the extraordinary wave varies with direction. For efficient transfer of energy between the fundamental and the second harmonic to take place, the speeds of the two waves must be the same. This means that the indices of refraction must also be the same because $n = c/v$, where c is the speed of light in a vacuum and v is the speed of light in the medium for a given wave. The indices of refraction are functions of frequency, and it turns out that for certain uniaxial crystals a direction in space can be found such that the index of refraction for the ordinary wave at frequency ω is the same as the index of refraction of the extraordinary wave at frequency 2ω. For a given uniaxial crystal this condition occurs for a direction in space at a definite angle with respect to the direction of the optical axis. In Fig. 12 it is seen that for a potassium dihydrogen phosphate (KDP) crystal, the index of refraction decreases with wavelength for both the ordinary wave and the extraordinary wave.[40] Because the index of refraction of the extraordinary wave is less than that of the ordinary wave, the crystal is called a negative uniaxial crystal. For data taken in Fig. 12, the crystal

[40] F. Zernike, *JOSA* **54,** 1215 (1964).

was at 24.8° and the light was propagating through the crystal at an angle of 90° with respect to the optical axis. It can be seen that for this case both the ordinary wave at 500 nm and the extraordinary wave at 250 nm have the same index of refraction, 1.515. If the crystal is transparent at both wavelengths, then frequency doubling can occur. This process is called index matching, and the crystal is said to be "angle tuned." For different angles of the beam with respect to the optical axis, the index of refraction of the extraordinary wave changes and the curve moves up in the figure until at 0° the two curves become one. Laser light of wavelengths above 500 nm can be frequency doubled by KDP for some angle between 0° and 90° as long as the crystal is transparent at both wavelengths. The angle must be set to within about a tenth of a degree from the correct angle if reasonable second harmonic output is to be obtained. Because the indices of refraction are temperature dependent, the crystals need to be temperature stabilized. Some of the crystals used, such as KDP, are hydroscopic, so they must also be kept in a sealed, dry nitrogen housing. Because matching the indices of refraction depends on both angle and temperature, the angle can be left fixed and the temperature changed to obtain index matching. This technique is called temperature tuning. The index matching angle may be calculated for an initial laser beam from data like those given in Fig. 12 and Eq. (26),[39]

$$\sin^2 \vartheta_m = \left[\frac{(n_0^\omega)^{-2} - (n_0^{2\omega})^{-2}}{(n_e^{2\omega})^{-2} - (n_0^{2\omega})^{-2}} \right] \quad (26)$$

For example, the index matching angle for ruby light (694 nm) and KDP is 50.4°. The description given earlier is called first index matching. Other conditions give rise to second index matching and vector index matching.

Mode Locking[13]

For an aligned optical cavity, the boundary conditions on the electromagnetic field determine that only standing waves of certain wavelengths can exist in the cavity (this is similar to the situation for sound standing waves in an organ pipe). For example, a half-wave can be fitted in as the largest wavelength possible and hence the lowest frequency or fundamental, ω_f. A full wave corresponding to twice the frequency of the fundamental can also fit, as well as higher harmonics. The difference in frequency between modes is $\Delta\omega = \omega_f$. These modes of oscillation are called longitudinal modes because the fitting is along the optical axis, in contrast to the transverse modes described earlier. The lasing ion or molecule of a laser does not lase at a precise frequency but has a certain spread of frequencies over which gain can be achieved. This spread in frequency of the gain curve covers many of the longitudinal mode frequencies, and in general these longitudinal modes are excited with random phases. If some means could be found so that the longitudinal modes could be made to oscillate with definite phase relations to each other, then the electric fields would be coherent, leading to constructive interference and very sharp pulses in time. How this might work can be seen from the following simple discussion.

If the laser is oscillating at frequency ω_0 and some means can be found to modulate this signal sinusoidally at a frequency ω_m, then the output will have a term of the form

$$E_m(t) = mE \cos \omega_0 t \cos \omega_m t \qquad (27)$$

where m is the fractional modulation coefficient. Recalling the trigonometric relation $\cos A \cos B = (1/2)[\cos(A + B) + \cos(A - B)]$ leads to

$$E_m = \frac{mE}{2} \cos[(\omega_0 + \omega_m)t] + \frac{mE}{2} \cos[(\omega_0 - \omega_m)t] \qquad (28)$$

Thus there are fields oscillating at the sum and different frequencies. A similar situation occurs in radio, where these are called side bands. If the modulation frequncy, ω_m, is set equal to the different frequency between longitudinal modes, ω_f, then the sum and difference signals excite their adjacent longitudinal modes, which now have a definite phase relation with the original signal of frequency ω_0. These signals are now modulated so that they excite adjacent modes in phase, and so on. Thus all of the longitudinal modes are oscillating in phase and the fields of the modes are coherent. The modes are then said to be locked. The coherence of the outputs of the modes leads to constructive interference and a train of narrow pulses in time. For example, the Antares neodymium–YAG laser is mode locked to give 70-psec wide pulses every 13 nsec.

Group Velocity Dispersion

As discussed in the section on frequency doubling, the index of refraction of a material is a function of frequency, which means it is a function of the wavelength of the light passing through the material. This dependence is normally a monotonically decreasing one, as shown in Fig. 12. This effect, called dispersion, is clearly seen when white light is passed through a glass prism. Blue light is bent more than red light. This is explained using Snell's law in the form

$$\sin \vartheta_2 = \frac{n_0}{n_2} \sin \vartheta_0 \qquad (29)$$

Because $n_b > n_r$, the angle $\vartheta_{2v} < \vartheta_{2r}$.

Two velocities (speeds) need to be considered in a homogeneous, isotropic, linear medium. The first is the phase velocity, described earlier, which gives the speed of a point on a sinusoidal wave; and the second is the speed of propagation of any modulation of the wave, the group velocity, which leads to a wave packet containing various frequency plane waves as a superposition. This can be seen by combining two traveling waves of the same amplitude but slightly different frequencies and propagation numbers:

$$E(z,t) = E[\cos(k_2 z - \omega_2 t) + \cos(k_1 z - \omega_1 t)] \quad (30)$$

Recalling the trigonometric relation $\cos A + \cos B = 2\cos[(A+B)/2]\cos[(A-B)/2]$ leads to

$$E(z,t) = 2E\cos\left(k_{ave}\left(z - \frac{\omega_{ave}}{k_{ave}}t\right)\right)\cos\frac{\Delta k}{2}\left(z - \frac{\Delta \omega}{\Delta k}t\right) \quad (31)$$

where

$$k_{ave} = \frac{k_2 + k_1}{2}; \quad \omega_{ave} = \frac{\omega_2 + \omega_1}{2}; \quad \Delta k = k_2 - k_1; \quad \text{and} \quad \Delta \omega = \omega_2 - \omega_1$$

The signal can be interpreted as a modulated carrier traveling wave, with the carrier having a phase velocity, $v_{ave} = \omega_{ave}/k_{ave}$, and the modulation having a group velocity, $v_g = \Delta\omega/\Delta k$. For a wave packet with many frequencies and propagation numbers, the group velocity is given by $v_g = d\omega/dk$. The group velocity can be expressed in terms of the wavelength and the index of refraction (see Fig. 12) using the relations $\omega = kv$ (from $v = \lambda f$); $k = 2\pi/\lambda$ and $n = c/v$. Then $v_g = d(kv)/dk = v + k\,dv/dk$, leading to $v_g = v(1 - (\lambda/n)|dn/d\lambda|)$. Because the absolute value of the derivative, $|dn/d\lambda|$, is positive, the slopes of the curves of n versus λ are negative, and it is seen that the group velocity is less than the phase velocity. Also if one compares the group velocity for part of the wave packet in the red region with that in the blue region using Fig. 12, the red group velocity is greater than the blue group velocity. Group velocity dispersion depends on the second derivative of the index of refraction with respect to wavelength. This means that the part of a narrow pulse corresponding to frequencies in the red moves faster than the part of the narrow pulse corresponding to frequencies in the blue—the pulse spreads, lowering its peak power density. This effect is called chirping the pulse. This is a disadvantage for femtosecond pulses and two-photon photolysis because high peak power density is desired. It can be overcome by letting the faster, red part of the pulse travel farther in space than the blue part, thereby bringing the parts together in space. This is called group velocity dispersion compensation and is used in femtosecond lasers.

Laser Systems

For one-photon photolysis, a Q-switched, frequency-doubled ruby laser is the principal laser system used. This system is composed of the basic laser with Pockels cell Q-switching and frequency-doubling in a KDP crystal, as discussed earlier.

For two-photon photolysis, a synchronously pumped, group velocity compensated, liquid-dye laser system with saturable dye is discussed. This laser is usually pumped by a mode-locked, frequency-doubled neodymium–YAG solid-state laser. The basic laser configuration of the pump laser shown in Fig. 6 is changed to a double elliptical cavity with two of the foci overlapping at the laser rod position and flash lamps at the other foci. A Gaussian beam (lowest transverse electric and magnetic mode (TEM_{00})) is achieved by placing a small aperture in the beam, and linear wave polarization is imposed with a Brewster plate. The laser is mode locked by having the laser beam modulated by passing it through a crystal that has a high-Q acoustical resonant mode driven by an electronic oscillator with a piezoelectric transducer to produce acoustic standing waves. The appropriate frequency-matching conditions for mode locking are obtained by first adjusting the temperature of the mode-locking crystal so that a resonant acoustical mode matches the fixed frequency oscillator and then adjusting the cavity length so that the longitudinal modes match the modulation frequency. The output of the mode-locked laser consists of a train of short pulses with a repetition rate equal to the cavity round-trip transit time.

The output of this solid-state laser is then frequency doubled, for example, by a potassium titanyl phosphate (KTP) crystal, by focusing the beam so that the waist is inside the doubling crystal, and the wave polarization is oriented with respect to the crystal so that it becomes an ordinary wave in the crystal. The beam is normal to the face of the crystal, which has been cut for the proper index matching angle. The crystal angle must be changed slightly for the corresponding index matching angle. The generated second harmonic is then an extraordinary wave that is linearly polarized at right angles to the linearly polarized original primary wave. The primary wave is then separated from the second harmonic wave using dichroic mirrors, leaving the second harmonic wave as a mode-locked train of pulses with the same pulse widths and repetition rate as the primary wave. This second harmonic wave is then amplitude and position stabilized so that it can be used as an optical pump source for the dye laser system.

The liquid-dye laser system consists of a laser cavity with a totally reflecting mirror and a partially reflecting mirror. The lasing medium is a continuously flowing, flat dye jet ribbon, oriented at the Brewster angle for the dye cavity. The length of the dye cavity is adjusted so that the produced dye laser pulses arrive at the gain jet at the same time as the pump pulses. This is called synchronous pumping. Two methods are used to reduce the pulse widths from the mode-locked, approximately 100-psec width to the desired pulse width of about 200 fsec. The first is to add a second flat, continuously flowing, saturable absorber dye jet in the dye cavity oriented also at the Brewster angle and to use dynamic bleaching so that the front part of the pulse brings the dye to dynamic saturation, thereby allowing the trailing part of the pulse to pass without absorption, thus shortening the pulse. The second method of pulse shortening involves group velocity dispersion. There are several elements in the dye cavity that the laser beam must pass through: the gain and saturable absorber jets, a birefringent tuning plate (set near the proper Brewster angle), and several group velocity dispersion compensation prisms, as well as numerous mirrors; all of which result in group velocity dispersion (chirping) and consequent spreading of the laser pulse in time. The prisms are adjusted for the proper angles and lengths so that the group velocity dispersion is compensated, leading to a dye laser output that has about 200-fsec-wide pulses with the same repetition rate as the pump beam.

Acknowledgments

The author thanks R. Chillingworth, J. Bunkenburg, and Drs. D. Trentham, T. Bliss, D. Ogden, N. Kiskin, M. Ferenzi, Y. Goldman, W. Webb, F. Wise, T. Asakura, T. Kihara, and B. Chance for helpful conversations and support. He also thanks Professor L. Narducci for critically reading the manuscript. Help with the figures by Anthony De Simone and Paul Heipp was much appreciated. This work was supported by Drexel University and MRC Grant E30/533-0300-574.

[11] Flash Lamp-Based Irradiation of Caged Compounds

By GERT RAPP

Introduction

The first part of this article contains a brief review of the various light sources that have been used for the photolysis of caged compounds, together with a discussion of their advantages and disadvantages. Flash lamp-based systems are then discussed in detail, with a particular emphasis on the individual components of these systems, including flashbulbs, the electronic driving circuit, and optical components. The final section contains a sum-

TABLE I
SUITABLE LIGHT SOURCES FOR PHOTOLYSIS OF CAGED COMPOUNDS[a]

Laser type	Wavelength (nm)	Approx. pulse width (nsec)	Typical pulse energy (mJ)
A. Lasers			
Ruby	347	25	100–300
Nd–YAG	355	10	100–200
Ti–Sapphire	350–370	10	90
Dye, laser pumped	From 300	10	50
Dye, flash-lamp pumped	From 300	1000	100
N_2	337	4	0.1–1
XeCl excimer	308	20	>200
XeF excimer	351	20	>200
B. Arc lamps			
Mercury, cw[b]			
Xenon, cw[b]			
Hg–Xe, cw[b]			
Xe flash lamps	From 300	1 msec	100

[a] Compiled from the "Laser Focus Buyers Guide," where the interested reader may find more detailed information.
[b] cw, continuously operated lamps.

mary of studies in which flash lamps were successfully applied in the areas of muscle contraction, protein crystallography, and electrophysiology.

Light Sources For Flash Photolysis of Caged Compounds

Photolysis of most caged compounds requires irradiation of the compound with light in the near-ultraviolet (UV) range, from about 300–370 nm. A compilation of suitable light sources is given in Table I. Important aspects to be considered are (1) the desired time resolution of the photoactivation reaction, (2) the required energy of the pulse, (3) the sample size and volume, (4) the experimental setup, and (5) the available budget.

Experiments that require a time resolution better than approximately 100 μsec must in general employ pulsed lasers; these devices are described in more detail in Refs. 1–3. Besides the short pulses, lasers have the advantage that their collimated beam can be guided over long distances with negligible loss of intensity and focused to a very small spot. Disadvantages

[1] J. A. McCray and D. R. Trentham, *Annu. Rev. Biophys. Biophys. Chem.* **18**, 239 (1989).
[2] J. A. McCray, *Methods Enzymol.* **291** [10] (1998) (this volume).
[3] E. Brown and W. Webb, *Methods Enzymol.* **291** [20] (1998) (this volume).

are that lasers are generally bulky, expensive, and require special safety precautions. In addition, the following points should be considered when deciding whether to use a laser or a flash lamp:

1. The efficiency of the photolysis of some caged compounds is lower for lasers than for flash lamps.[1]

2. Short pulses may lead to two-photon absorption,[2,3] which causes ionization of the sample if the incident wavelength is in the near-UV range.

3. If the wavelength is too close to the protein absorption band, the sample may be destroyed. Experiments on skeletal muscle using a XeCl excimer laser at 308 nm to photolyze caged ATP showed that the irradiation caused a physical disruption of muscle fiber bundle.[4] However, no signs of sample deterioration were observed when the same experiment was performed using the 351-nm-wavelength line of the XeF excimer laser at the same energy of 120 mJ. On the other hand, in a study of the kinetics of the phosphorylation of Na,K-ATPase,[5] an excimer laser running at 308 nm was reported to have successfully released caged phosphate without damaging the sample.

4. The damage threshold in flash photolysis experiments, in particular for time-resolved measurements in protein crystals, depends strongly on the power of the light source, i.e., the time interval within which a sample is irradiated with a given energy.

Flash Lamp Systems

The majority of kinetic studies employing caged compounds have been conducted with a time resolution of milliseconds to seconds. In this time regime, pulses delivered from flash lamps are generally satisfactory from the standpoints of both energy and time resolution. Given the importance of flash lamps in these studies, we describe the essential details of the components of these devices and provide an account of their performance. The principle layout of a flash lamp system is shown in Fig. 1. It basically consists of a capacitor, to store electrical energy, which is discharged over the flash lamp and optical components to collect the irradiated light.

The Lamp

The bewildering number of definitions and units used to define the emission properties of lamps complicates any assessment of the performance of light sources. Photometry is concerned with the response of the

[4] K. J. V. Poole, G. Rapp, and R. S. Goody, unpublished.
[5] H. J. Apell, M. Roudna, J. E. T. Corrie, and D. Trentham, *Biochemistry* **35,** 10922 (1996).

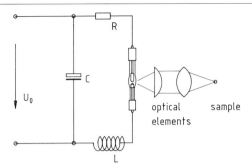

Fig. 1. Schematic layout of a flash lamp system.

human eye to a light source, and most lamps are specified by photometric quantities. However, for applications in photolysis experiments, radiometric quantities are more relevant. A compilation of the most important terms is provided in Table II. Detailed technical specifications are supplied only for continuously operating (cw) arc lamps. In combination with the use of a mechanical shutter, cw-operated xenon and mercury lamps may be a suitable alternative for some applications of caged compounds, provided a high time resolution is not required. Emission spectra of continuously operated xenon and mercury arc lamps are shown in Fig. 2. A 150-W xenon lamp has typical specifications of 3000 lm, 300 cd, and 20,000 cd/cm^2. Unfortunately, no useful quantities for the emission in the near-ultraviolet range can be deduced from these specifications, although the radiant flux in the wavelength region from 300 to 380 nm may be estimated from these spectra. Continuously operating arc lamps usually have an efficiency of 60%, and xenon lamps radiate about 6% of their electrical input power in the UV region below 380 nm. Special care has to be taken to eliminate the strong infrared irradiation of these lamps. Mercury lamps, which have strong emission lines between 320 and 370 nm, may be more suitable for photolysis experiments, although any emission below 300 nm should be filtered out using suitable bandpass filters. Using a 100-W mercury lamp about 25% caged Ca^{2+} has been photolyzed in approximately 2 sec.[6] Compared to other noble gases, xenon converts electrical energy most efficiently into light due to its lower ionization energy. The current density during the discharge is determined by the parameters of the driving circuit of the lamp. These factors can significantly influence the spectral distribution of the emission, as discussed in the next section. Different materials for flashbulbs are available. Pure silica or quartz is transparent down to the

---

[6] R. Zucker, in "Methods in Cell Biology," Vol. 40, p. 31. Academic Press, New York, 1994.

TABLE II
DEFINITIONS USED IN PHOTOMETRY AND RADIOMETRY[a]

Photometric quantities				Radiometric quantities			
Quantity	Symbol	Unit		Quantity	Symbol	Unit	Definition
Luminous energy	Q	lm sec		Radiant energy	Q_e	Joule, J	Quantity of light
Luminous flux	$\Phi = dQ/dt$	lumen		Radiant flux	Φ_e	Watt, W	
Luminous intensity	$I = d\Phi/d\omega$	lm = cd·sr candela, cd		Radiant intensity	$I_e = d\Phi_e/d\omega$	W/sr	Flux per solid angle, brightness
Luminance	$L = dI/dA \cos\theta$	stilb, sb = cd/cm^2		Radiance	$L_e = dI_e/dA \cos\theta$	W/sr/m^2	Flux per solid angle per irradiating area, brilliance
Illuminance		lux, lx = lm/m^2		Irradiance	$E = d\Phi_e/dA$	W/m^2	

[a] Photometric quantities are valid only for visible light and are related to the sensitivity of the human eye. Radiometric quantities, which are valid for all wavelengths, are often labeled with the index e (for energy).

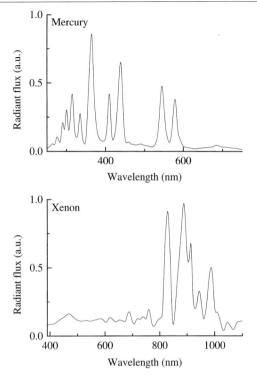

FIG. 2. Emission spectra of mercury and xenon arc lamps at continuous operation.

deep UV. In so-called ozone-free types, the quartz envelope is doped with titanium, which absorbs below 250 nm, as shown in Fig. 3. However, one should keep in mind that, depending on the driving circuit, a substantial portion of photons harmful to proteins might be irradiated.

The Driving Circuit

Basically the electrical circuit for flash lamps consists of a capacitor to store energy, which is discharged through the lamp as shown in Fig. 1. In the nonionized state the flash lamp has impedance of tens of megaohms. By applying a high voltage of 10–20 kV, xenon atoms ionize and cause the impedance to drop, thus allowing the capacitor to discharge. It is convenient to define the pulse length as the width at one-third of its maximum amplitude. Without considering the impedance of the circuit, the duration of the current pulse is given approximately by

$$t_{1/3} = \pi(LC)^{1/2}$$

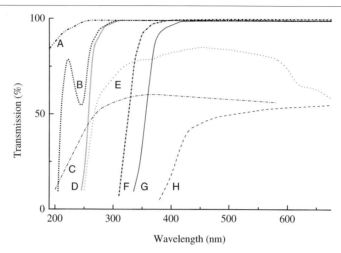

FIG. 3. Spectral transmission of selected materials important for caged compound applications. (A) Silica glass; (B) fused quartz; (C) quartz light guide, 1 m long, NA = 0.22; (D) titanium-doped quartz used for "ozone-free" flash lamp envelopes; (E) liquid light guide for UV applications, NA = 0.45; (F) BK7 optical glass, 25 mm thick; (G) cerium-doped quartz, used for flash lamp envelopes; (H) standard glass fiber. *Note*: Reflections at surfaces are not taken into account.

with an inductance of L and a capacitance of C. Using values of 20 μH for the inductance and 3500 μF for the capacitance, the pulse length is $t_{1/3} = 830$ μsec.

The stored energy E is given by

$$E = 1/2\ CU^2$$

with C the capacitance and U the charging voltage. For a given energy, reducing the capacitance and simultaneously increasing voltage shortens the pulse length. However, the peak current increases, which has the following consequences:

1. The lifetime of the flashbulb decreases due to a more rapid evaporation of electrode material.
2. At very short pulses (< 50 μsec at 200 J electrical) the shock wave associated with the discharge may lead to an explosion of the bulb.
3. Circuit losses increase and the conversion efficiency of electrical energy into light is reduced due to the creation of the shock wave.
4. The emission spectrum is shifted toward a shorter wavelength.

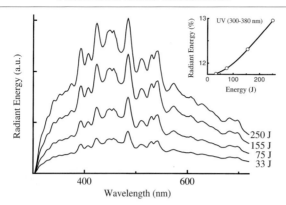

FIG. 4. Spectral distribution of a xenon flash lamp. Stored electrical energy from 33 to 240 J. There is a shift toward shorter wavelengths on increasing the energy and, hence, current density. The percentage of radiant flux is shown in the inset for wavelengths 300–380 nm.

The effect of different current densities on the efficiency and spectral distribution of flash lamps was the subject of an earlier investigation.[7] An electrical energy of 200 J was discharged in two electrical configurations. In the first, a bank of capacitors of 500 μF charged to 894 V was used, which resulted in a pulse of about 130 μsec full width at half maximum (FWHM). In the second configuration, a capacitor bank of 4500 μF charged to 298 V resulted in a pulse with a FWHM of about 750 μsec. The spectral distribution is shifted toward the ultraviolet at these high driving voltages, i.e., high current density, although the radiant energy over all emission wavelengths decreased due to the effects discussed earlier. Emission spectra obtained by operation of the xenon flash lamp at different electrical energies, and hence at different current densities, are shown in Fig. 4. At an input of 33 J the spectrum is almost flat, and at higher energies the center of gravity of the spectra is shifted toward lower wavelengths with a concomitant increase in the percentage of radiant energy in the UV range.

The Optical System

In many illumination systems ellipsoidal mirrors are used to collect the maximum amount (up to 80%) of the radiant flux (Fig. 5). In practice, a point-source lamp is placed at the first focus, and the sample, or one end of a light guide, is located at the second focal point. A variety of reflector geometries are available for this purpose. A ray-tracing analysis for a typical ellipsoidal mirror with a major axis of 135 mm and minor axis of 75 mm

[7] G. Rapp and K. Güth, *Pflügers Arch.* **411**, 200 (1988).

FIG. 5. Ellipsoidal mirror with flashbulb mounted.

is shown in Fig. 6. For an ideal point source, all reflected rays can be collected at the second focus. However, for a real source with finite dimensions, the image of the lamp is enlarged by about a factor of between 5 and 10, depending on the geometry of the mirror, and consequently the irradiance is reduced by the square of the magnification factor. The magnification properties of ellipsoidal mirror-imaging systems are illustrated in Fig. 7. It shows the radius of the cross section containing 50, 70, and 100%, respectively, of the rays reflected from the mirror as a function of distance around the focus, assuming the source to be a cube of 1 mm^3.

Illumination systems that use refractive optics consist of a condenser, an objective, and a back reflector. Both the condenser and the objective are usually assembled from an array of lenses. The ultraviolet transmission properties of standard glass optics, shown in Fig. 3, are not suitable for applications with caged compounds, and therefore quartz optics should be used. With an optical array of five lenses, reflection losses at the air–quartz interface accumulate to about 34% (4% reflection, 10 interfaces, transmission $T = 0.96^{10} = 66\%$). With appropriate ultraviolet antireflection coating,

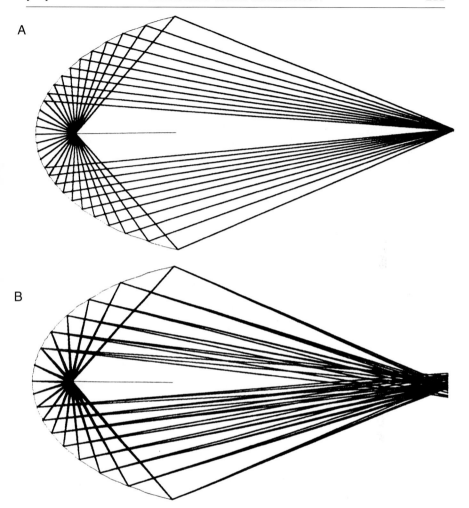

Fig. 6. Ray tracing with (A) a point source and (B) a real source of $1 \times 1 \times 1$ mm^3 in one focal point of the ellipsoid.

these losses can be reduced to less than 5% within the excitation bandwidth required for activation of caged compounds. The main distortions in the illumination system arise from spherical aberration. Due to chromatic aberration of the lens system, it is important to focus the beam using only UV-filtered light. Illumination of the sample using white light (1) results in heating of the preparation, which is certainly harmful for protein crystals and for photoactive samples, and (2) may cause mechanical artifacts due

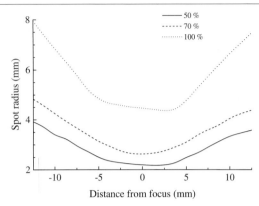

FIG. 7. Caustic of an ellipsoidal mirror using a real source of 1 mm³. The diagram shows the spot size containing 50, 70, and 100% of the reflected rays around the secondary focus, respectively, e.g., with the source of 1 mm³ 100% of the reflected rays are within a spot of about 9 mm in diameter at the back focal plane.

to momentum transfer of light. The second effect is a well-known problem in patch-clamping experiments.[6] A UG11 is a suitable UV filter, and its residual transmission in the near infrared (IR) can be blocked with a special coating. When necessary, harmful UV light at shorter wavelengths can be eliminated using an additional long-pass filter.

It should be kept in mind that the following relationship is an invariant of any optical system (Helmholtz–Lagrange invariant):

$$A_s \Omega_s n_e = A_i \Omega_i n_i$$

where A is the area of the source and its image, Ω is the solid angle determined by entrance and exit aperture, respectively, and n is the index of refraction; usually $n_e = n_i$. In other words, the radiance L is determined by the radiating source and the first optical element. It cannot be increased and, in practice, it is normally reduced. Optical elements can increase only the irradiance, I. One might think that the ideal system to collect more photons consists of a tapered fiber optic with a large aperture A_i at the input and a small aperture A_e at the exit.

Because $A_i/A_e = NA_e/NA_i =$ constant, light is emitted from the smaller area with a larger divergence, and so, in effect, the task of collecting more light is shifted to the next optical element (NA, numerical aperture $= n \sin \theta$, with θ the half-angle at the input and exit, respectively).

In an earlier investigation,[7] a comparison was made of the energy density in the focus using refractive optics versus ellipsoidal mirrors. As expected from the ray-tracing calculations, the irradiance is higher if refractive optics of short focal length are used compared to the ellipsoidal mirror system. In some applications though, the shorter working distance associated with a short focal length imposes technical problems. In these circumstances, or if samples under investigation are larger than about 3–4 mm (e.g., muscle fibers), an ellipsoidal mirror should be considered as an optical component.

Several reasons favor the use of light guides to deliver the light pulse to the sample: (1) Discharge lamps often create electrical artifacts in sensitive electronic equipment (e.g., amplifiers used in electrophysiology) if they are located close to the flash lamp; (2) the sample stage may be very sensitive to mechanical vibrations and must be isolated from the lamp housing; and (3) the space around the sample is limited. For applications using caged compounds, only quartz fibers or liquid light guides are suitable (see Fig. 3). Single quartz fibers are available up to about 1.5 mm in diameter with a NA up to 0.25. Standard liquid light guides have diameters of 2, 3, and 5 mm, with a NA of 0.45 or larger. Here, we are again faced with the problem of conserving brightness. The light-coupling efficiency of the light guide is much higher at the input side for liquid light guides compared to quartz fibers. It is, however, a challenge to collect and collimate or refocus light emitted from the liquid light guide, especially when large diameters are used. The optimal choice depends on the experimental conditions, and the best solution has to be determined for each case.

Light guides are frequently used to deliver light into microscopes for the reasons discussed earlier. For inverted microscopes, light is often coupled via the epifluorescence port. Quartz objectives are transparent over the entire wavelength range important for photolysis of most caged compounds, i.e., 300–370 nm, but they are expensive. Many experiments have been reported (see later) in which UV-transparent objectives have been used successfully. The transmission curves for these objectives are not available from any supplier, although suppliers claim their transmittance declines only around 340 nm. Note, however, that below 340 nm absorption, and hence photolysis yield, increases for most caged compounds.

Methods to Measure the Energy of a Flash Pulse

Calibrated joulemeters are easy to use, commercially available, and have an accuracy of about 5%. They have a large dynamic range (from mJ to several joules) and can either be directly attached to a storage oscilloscope or may have special readout electronics. The detector head consists of a coating, which has a uniform absorption from the UV to the infrared. A voltage proportional to the absorbed energy is generated at the detector, which can easily be recorded on a storage oscilloscope. Provided the pulse length of the flash is shorter than the response time of the joulemeter, energy can be calculated from the peak voltage and the calibration factor supplied from the manufacturer. An inexpensive detector may be built using a UV-sensitive silicon photodiode. However, in order to obtain absolute values of radiant energy, one must take into account the nonlinear radiant sensitivity of the diode and the time structure of the flash.

Applications

The following section summarizes applications wherein flash or continuously operated arc lamps were used for photolysis of caged compounds. This review is biased toward references that contain a more detailed description of technical aspects of the experiments. Reviews on the application of flash lamps for photolysis of caged compounds, include physiological and pharmacological experiments[8,9] and studies on smooth muscle[10] and striated muscle.[11]

Structural Studies on Noncrystalline Systems

X-ray diffraction is an ideal tool for determining the internal structure of macromolecular assemblies such as muscle fibers or microtubules. Whereas laboratory X-ray generators require exposure times on the order of minutes to hours, these times have been shifted to the millisecond region or even less using synchrotron radiation. In the first experiments of this type with caged compounds,[12] small strips of muscle fibers were incubated in a rigor solution (i.e., without ATP, see also Dantzig *et al.*[13]) and then relaxed by photolytic release of ATP from caged ATP. Further investigations along this line were described,[14,15] wherein a flash lamp was employed to induce photolysis of caged ATP, caged Ca^{2+}, or caged nucleotide analogs involved in muscle contraction. Experiments on muscle mechanics have shown that the relaxation rate from rigor induced by photolysis of caged ATP depends markedly on the ionic strength.[16] Diazo-2, a photolabile derivative of

---

[8] H. A. Lester and J. M. Nerbonne, *Annu. Rev. Biophys. Biophys. Eng.* **11,** 151 (1982).
[9] A. M. Gurney and H. A. Lester, *Physiol. Rev.* **67,** 583 (1987).
[10] A. P. Somly and A. V. Somlyo, *Annu. Rev. Physiol.* **52,** 857 (1990).
[11] E. Homsher and N. C. Millar, *Annu. Rev. Physiol.* **52,** 875 (1990).
[12] G. Rapp, K. J. V. Poole, Y. Maeda, K. Güth, J. Hendrix, and R. S. Goody, *Biophys. J.* **50,** 993 (1986).
[13] J. Dantzig, H. Higuchi, and Y. E. Goldman, *Methods Enzymol.* **291** [18] (1998) (this volume).
[14] K. J. V. Poole, G. Rapp, Y. Maeda, and R. S. Goody, in "Molecular Mechanism of Muscle Contraction" (H. Sugi and G. H. Pollack, eds.), p. 391, Plenum, New York, 1988.
[15] K. Horiuti, K. Kagawa, and K. Yamada, *Biophys. J.* **67,** 1925 (1994).
[16] C. Veigel, M. R. D. Maydell, R. Wiegand-Steubing, R. S. Goody, and R. H. A. Fink, *Pflügers Arch.* **430,** 994 (1995).

BAPTA that converts from a low to a high affinity chelator of calcium ions after photolysis, was used to study the calcium-dependent relaxation rate in intact striated muscle fibers[17] and in cardiac muscle.[18] Caged nucleotides were also used to measure the nucleotide-exchange kinetics of isolated synthetic myosin filaments using fluorescence microscopy and flash photolysis.[19]

Flash lamp-induced photolysis of caged ATP and caged AMPPNP was used with time resolved cryoelectron microscopy in order to study the dissociation of actomyosin.[20] Grids containing a suspension of sample were irradiated at given times (10–100 msec) prior to a rapid freezing. To avoid excess heating of the grid, it was important to limit the bandwidth of the flashlight to a narrow region between 320 and 370 nm.

Microtubules (MTs), elements of the eukaryotic cytoskeleton, consist of polymers of tubulin dimers. Microtubules are highly dynamic structures that undergo catastrophic depolymerization when the fast-growing end of the filament contains a certain number of GDP-bound tubulin molecules; the recovery of MT growth requires polymerization of dimers in the GTP-bound form. Oscillations in the growth and collapse of MTs have been induced by flash photolysis of caged GTP and investigated by X-ray scattering.[21] The authors of this particular study also showed that for each absorbed X-ray photon (8 keV) about 20 caged GTP molecules were activated, of which 7 were legitimate GTP molecules while the rest were nonfunctional fragments of caged GTP or GTP.

Protein Crystallography

For a detailed review of the application of caged compounds in time-resolved crystallography, see Scheidig *et al.*[22] The structural dynamics of a GTPase in crystals was studied using flash photolysis of caged GTP.[23] In X-ray diffraction studies, the measured diffraction intensities arise from spatially and temporarily averaged electron densities in the crystal. Therefore, to study the kinetics of a reaction in the crystal, the light pulse should photolyze 100% of the caged nucleotide. This efficiency is difficult to achieve, and even if a light source of sufficient energy were available to

---

[17] J. Lännergren and A. Arner, *J. Muscle Res. Cell. Motil.* **13**, 630 (1992).
[18] R. Zang, J. Zhao, A. Mandveno, and J. D. Potter, *Circ. Res.* **76**, 1027 (1995).
[19] P. B. Conibear and C. R. Bagshaw, *FEBS Lett.* **380**, 13–16 (1996).
[20] J. F. Menetret, W. Hoffmann, R. R. Schröder, G. Rapp, and R. S. Goody, *J. Mol. Biol.* **219**, 139 (1991).
[21] A. Marx, A. Jagla, and E. Mandelkow, *Eur. Biophys. J.* **19**, 1 (1990).
[22] A. Scheidig C. Burmester, and R. S. Goody, *Methods Enzymol.* **291** [14] (1998) (this volume).
[23] I. Schlichting, G. Rapp, J. John, A. Wittinghofer, E. Pai, and R. S. Goody, *Proc. Natl. Acad. Sci. U.S.A.* **86**, 7687 (1989).

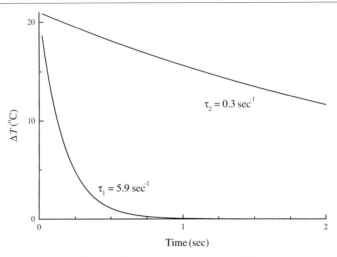

FIG. 8. Time course of heat dissipation in a protein crystal in an aqueous environment ($\tau = 5.9$ sec^{-1}) and in air ($\tau = 0.3$ sec^{-1}).

achieve this photolysis level, the thermal load sets the limitations. Indeed, heating of the crystal is the limiting factor in protein crystallography. The following example derives estimates of the temperature increase and the subsequent temperature reequilibration times in a protein crystal due to photolysis of a caged compound.

Consider a crystal of the protein Ras p21[24] of molecular mass 21 kDa, with a dimension of $300 \times 300 \times 500$ μm^3 and a solvent content of 40%. At a protein density of 1.23 g/cm^3 there are 9.2×10^{14} protein molecules in the crystal. To release the same number of molecules of GTP from caged GTP in this crystal would require 4.6×10^{15} photons, assuming a quantum efficiency of 20%, which is equivalent to the energy of 2.6 mJ for a monochromatic source at 347 nm ($E = nhc/\lambda$). At 347 nm, the wavelength of the ruby laser, the extinction coefficient of caged ATP $\varepsilon_{\text{caged ATP}} = 500$ M^{-1} cm^{-1} (at 320 nm $\varepsilon_{\text{caged ATP}} = 1000$ M^{-1} cm^{-1}). That is, to photolyze 100% of the caged GTP in our example, 4 mJ or 27 mJ/mm^2 are required. In practice, however, even more energy is required for photolysis than that calculated here.

Taking 1.5 J/g °C for the heat capacity of a protein,[25] the Ras p21 crystal (33 μg protein + 18 μg water) is heated by 21°. The amount of energy

[24] I. Schlichting, S. C. Almo, G. Rapp, K. Wilson, K. Petratos, A. Lentfer, A. Wittinghofer, W. Kabsch, E. F. Pai, G. A. Petsko, and R. S. Goody, *Nature* **345**, 309 (1990).
[25] G. I. Makhatadze and P. L. Privalov, *J. Mol. Biol.* **213**, 375 (1990).

TABLE III
PHOTOLYSIS EFFICIENCY OF CAGED ATP IN CRYSTALS OF GLUTATHIONE SYNTHETASE VS
ENERGY DENSITY USING A Nd–YAG LASER AT 355 nm[a]

Energy density (mJ/mm^2)	Number of flashes	Photolysis (%)	Comment
63	1	17	Crystal was burned
31	1	12	OK
31	10	33	Crystal and capillary were cracked
20	30	90	Crystal was cracked
10	60	50	OK

[a] Data from H. Kato, Institute for Chemical Research, Kyoto University, Japan, personal communication (1997).

dissipated as heat is calculated following an approach described by Holman.[26] For simplification we assume the crystal is instantaneously heated by 21° (which is certainly valid for nanosecond laser pulses) and remains suspended in an environment at the preflash temperature. The kinetics of cooling of the crystal in its mother liquor and air are shown in Fig. 8. These data are calculated assuming heat transfer coefficients of 1000 and 50 W/m^2/°C for water and air, respectively. From this example it is clear that photolysis of 100% of the caged compound with one intense flash leads to significant heating in the crystal. This heating effect is of even more concern in crystals that have a low solvent content, and it takes seconds for crystals suspended in air to dissipate the absorbed heat. This heating effect must be considered in time-resolved studies of enzyme reactions in crystals where the rate of the reaction is temperature dependent.

In practice, photolysis of caged compounds in crystals of p21[23,24] and phosphorylase b[27] require several flashes to achieve sufficient activation of the substrate. For glutathione synthetase,[28] the effect of the energy density on the efficiency of photolysis and survival of the crystals was investigated using a Nd–YAG laser at 355 nm and a xenon flash lamp.[29] The major differences between these photolysis sources are (1) 15 nsec versus 1 msec pulse length and (2) monochromatic versus broad wavelength band.

Data shown in Tables III and IV were kindly provided by Dr. Hiroaki Kato.[29] In these experiments the efficiency of photolysis was determined

[26] J. P. Holman, "Heat Transfer," McGraw-Hill, New York, 1986.
[27] E. M. H. Duke, S. Wakatsuki, A. Hadfield, and L. N. Johnson, *Protein Sci.* **3,** 1178 (1994).
[28] H. Yamaguchi, H. Kato, Y. Hata, T. Nishioka, A. Kimura, J. Oda, and Y. Katsube, *J. Mol. Biol.* **229,** 1083 (1993).
[29] H. Kato, Institute for Chemical Research, Kyoto University, Japan, personal communication (1997).

TABLE IV
PHOTOLYSIS OF CAGED ATP IN CRYSTALS OF GLUTATHIONE
SYNTHETASE USING A FLASH LAMP WITH A UV FILTER
DEPENDING ON (A) ENERGY DENSITY AND
(B) NUMBER OF FLASHES AT 18 mJ/mm^2 [a]

A		B	
Energy density (mJ/mm^2)	Photolysis (%)	Number of flashes	Photolysis (%) (SD)[b]
4	25	1	68 (5.7)
9	30	2	95 (4.8)
12	37	3	94 (3.9)
18	68	5	100 (nd)

[a] Data from H. Kato, Institute for Chemical Research, Kyoto University, Japan, personal communication (1997).
[b] SD, standard deviation; nd, not determined.

by high-performance liquid chromatography (HPLC) analysis of the crystal after its dissolution in buffer. At a very high energy density (63 mJ/mm^2) one shot destroys the crystal, and even with this high energy density only 17% of the caged ATP is photolyzed. A similar investigation using a flash lamp equipped with a UG11 filter is summarized in Table IV. Here much lower energy densities are required to produce the activated substrate: at 18 mJ/mm^2, 68% of caged ATP was photolyzed in one flash, whereas four to five flashes at 18 mJ/mm^2 are found to release almost 100% of the substrate. These two experiments demonstrate the dependence of the efficiency of photolysis on the pulse length and bandwidth, a property already addressed by McCray and Trentham.[1] A second potential problem using lasers with high energy density is the generation of a shock wave created by the rapid heating, which might be particularly harmful for protein crystals but does not appear to cause obvious damage to other samples, such as cells.

Binding of the inhibitor pyrone to chymotrypsin is initiated by flash photolysis of a protecting group covalently bound to the enzyme (caged enzyme).[30] Mosaicity in the crystals increases after photolysis, as indicated by streaks in the diffraction pattern. Laue crystallography, i.e., using white light, is more susceptible to this effect than standard Bragg diffraction. Increased mosaicity may arise from the temperature jump due to the flash or from major reorganizations in the crystal associated with the start of the reaction.

[30] B. L. Stoddard, P. Koenigs, N. Porter, K. Petratos, G. Petsko, and D. Ringe, *Proc. Natl. Acad. Sci. U.S.A.* **88**, 5503 (1991).

In all experiments it is important to know the concentration of the molecules of interest after photolysis. To achieve this goal the amount of photolytically released phosphate from caged phosphate is directly measured in a crystal of phosphorylase b using a diode-array spectrophotometer.[27] In these studies, 7 mM phosphate was released from 21 mM caged phosphate after five flashes.

Further Applications

The final part of this contribution summarizes a variety of further experiments in which flash lamps or continuous arc lamps equipped with a shutter were used. From the sound body of publications, only those articles were selected in which details of the experimental setup are described.

1. In studies on chemotaxis, the swimming behavior of bacteria was investigated by video microscopy following photolytic release of serine and protons from their caged precursors. High time resolution was not required, and light from a mercury lamp switched on for 30 msec was sufficient.[31]

2. Confocal microfluorimetry was used to measure intracellular-free [Ca^{2+}] induced in *Xenopus* oocytes by photoreleased inositol trisphosphate. A 100-W mercury lamp with a shutter attached to the epifluorescence port of an inverted microscope was used for photolysis. A small response was observed with 15 msec opening; with 60-msec flashes calcium liberation in oocytes persisted for several seconds.[32] Local increases in the calcium ion concentration around the nucleus have been found to trigger entry into mitosis in experiments combining flash photolysis of the calcium chelator nitrophenyl-EGTA and confocal microscopy.[33]

3. Regulation of sarcoplasmatic Ca^{2+} release was investigated following photolysis of caged cyclic ADP-ribose in intact cardiac myocytes.[34] In an inverted microscope with quartz optics, light for photolysis from a flash lamp was merged to the excitation light path of the microfluorimeter with a beam splitter.

4. In whole-cell patch-clamp experiments the effect of photolytically released ATP or InsP$_3$ on Ca^{2+} release in Purkinje neurons[35] and in cultured endothelial cells[36] was investigated. In both experiments, Ca^{2+} was mea-

[31] S. Khan, J. L. Spudich, J. A. McCray and D. R. Trentham, *Proc. Natl. Acad. Sci. U.S.A.* **92**, 9757 (1995).
[32] I. Parker, Y. Yao, and V. Ilyin, *Biophys. J.* **70**, 222 (1996).
[33] M. Wilding, E. M. Wright, R. Patel, G. Ellis-Davies, and M. Whitaker, *J. Cell Biol.* **135**, 191 (1996).
[34] X. Q. Guo, M. A. Laflamme, and P. L. Becker, *Circ. Res.* **79**, 147 (1996).
[35] K. Khodakhah and D. Ogden, *J. Physiol.* (*London*) **487**, 343 (1995).
[36] T. D. Carter and D. Ogden, *Proc. R. Soc. Lond. B Biol. Sci.* **250**, 235 (1992).

sured with a microspectrofluorimeter on an inverted microscope. Samples were exposed directly from the flash lamp housing. With 120 mJ of UG11 filtered light, about 50% photolysis of caged ATP was achieved.

5. Photolysis of caged glutamate was used to investigate whether L-glutamate can produce postsynaptic activation in the giant synapse of the squid.[37] Samples were flashed either through a silica condensor, giving a spot about 500 μm in diameter, or directly from the lamp housing, giving a spot about 4 mm in diameter. About 9 mM L-glutamate was released in the synapse.

6. The distribution of secretory granules with B cells was used in combination with flash photolysis of caged Ca (DM-nitrophen) to study exocytosis in mouse pancreatic B cells.[38] These studies used a combination of techniques, such as patch clamp to record whole cell or single-channel Ca^{2+} channel currents, capacitance measurements of exocytosis, and digital imaging to quantify cytoplasmic Ca^{2+} gradients.

7. The activation kinetics of L-type calcium current in frog cardiac myocytes was studied using flash photolysis of caged isoproterenol, a synthetic β-agonist.[39] In a different study, caged isoproterenol and caged cAMP were used to investigate sympathetic stimulation of rabbit cardiac cells. The flash was applied through the rear aperture of an inverted microscope to a selected cell using a Fluor 40× objective.[40]

8. The kinetics of potassium binding and conformational transitions in the enzyme were investigated using photolysis of caged K^+ together with microspectrofluorimetry on fluorescently labeled Na,K-ATPase.[41]

9. Whole-cell patch-clamp recordings were taken to investigate the effect of Ca^{2+} on the photoreceptors of *Drosophila*. Light for photolysis of DM-nitrophen or nitr-5 and for excitation of the calcium-sensitive dyes was directed onto the sample using a bifurcated light guide and an inverted microscope.[42] A mechanical shutter protected the photomultiplier during the light-directed activation of calcium from caged calcium

10. In studies on flash photolysis of caged cAMP and caged GTP-γ-S on Ca^{2+} currents in cardiac myocytes, light was coupled to the rear port of an inverted microscope with the use of a dichroic mirror and directed to the cell through a Fluor 40× objective.[43]

---

[37] J. E. T. Corrie, A. Desantis, Y. Katayama, K. Khodakhah, J. Messenger, D. Ogden, and D. R. Trentham, *J. Physiol.* **465**, 1 (1993).
[38] K. Bokvist, L. Eliasson, C. Ämmälä, E. Renström, and P. Rorsman, *EMBO J.* **14**, 50 (1995).
[39] A. M. Frace, P. F. Mery, R. Fischmeister, and H. C. Hartzell, *J. Gen. Physiol.* **101**, 337 (1993).
[40] H. Tanaka, R. B. Clark, and W. R. Giles, *Proc. R. Soc. Lond. B Biol. Sci.* **263**, 241 (1996).
[41] E. Grell, R. Warmuth, E. Lewitzki, and H. Ruf, *Acta Physiol. Scand.* **146**, 213 (1992).
[42] R. C. Hardie, *J. Neurosci.* **15**, 889 (1995).
[43] R. Z. Kozlowski, V. W. Twist, A. M. Brown, and T. Powell, *Am. J. Physiol.* **261**, 1665 (1991).

11. The kinetics of secretory response in bovine chromaffin cells following flash photolysis of caged Ca^{2+} (DM-nitrophen) was investigated via capacitance measurements with millisecond time resolution. Flashlight was directed through the epifluorescence port of an inverted microscope together with the excitation light for Ca^{2+} measurements by means of a sapphire window.[44] Using the formalisms of Zucker,[45] the photolysis efficiency was calculated for Ca^{2+}-free and Ca^{2+}-bound DM-nitrophen. At a maximum flash energy of 240 J, about 60% of Ca^{2+}-bound and 24% of Ca^{2+}-free DM-nitrophen was photolyzed. For neurophysiological studies with millisecond time resolution, Ca^{2+} chelators have been photolyzed using flash lamps,[46] whereas in other experiments in which time resolution was not critical, continuous lamps with a shutter were used to photolyze about 12% of nitr-5 in 1 sec.[47]

12. In the experimental setup used by Almers and co-workers[48] to study Ca^{2+}-triggered exocytosis, light from a flash lamp was coupled to the epifluorescence port of a microscope via a liquid light guide. Additionally, a shutter was used to protect the photomultiplier for Ca^{2+} measurements. At the maximum flash energy, 3.87 mM DM-nitrophen was photolyzed from a 10 mM solution. In a related study, a comparison was made of the effectiveness of photolabile calcium chelators nitrophenyl-EGTA (NPE) and DM-nitrophen in rising cytosolic calcium to trigger exocytosis.[49] Although NPE was found to be less efficient in rising Ca^{2+}, it was suggested to be a useful tool for time-resolved studies in the presence of physiological concentrations of magnesium.

13. Derivatives of inositol trisphosphate (InsP$_3$) and their photolabile precursors have been characterized and applied in a study on the kinetics of cytosolic calcium release.[50] Whole-cell recordings and fluorescence measurements were combined. Photolysis light from a flash lamp equipped with quartz condensor and objective was directed directly onto the sample without passing through the microscope. The objective had a working distance of about 4 cm and produced a spot of 4–5 mm in diameter. Photolysis of caged ATP was adjusted via the electrically stored energy and was reproducible within about 10%.

14. Caged derivatives of adrenergic receptor agonists, such as epinephrine, phenylephrine, and isoproterenol, have been described and photo-

---

[44] C. Heinemann, R. H. Chow, E. Neher, and R. S. Zucker, *Biophys. J.* **67**, 2546 (1994).
[45] R. S. Zucker, *Cell Calcium* **14**, 87 (1993).
[46] L. Lando and R. S. Zucker, *J. Gen. Physiol.* **23**, 1017 (1989).
[47] D. Neveu and R. S. Zucker, *Neuron* **16**, 619 (1996).
[48] P. Thomas, J. G. Wong, A. K. Lee, and W. Almers, *Neuron* **11**, 93 (1993).
[49] T. D. Parsons, G. C. R. Ellis-Davies, and W. Almers, *Cell Calcium* **19**, 185 (1996).
[50] J. F. Wootton, J. E. T. Corrie, T. Capiod, J. Feeney, D. R. Trentham, and D. C. Ogden, *Biophys. J.* **68**, 2601 (1995).

chemically characterized using repetitive pulse of arc lamps.[51] In studies on calcium antagonists and agonists, light-directed activation of caged nefedipine, a calcium ion antagonist, was achieved using light from a 100-W mercury arc lamp delivered through the epifluorescence port of an inverted microscope in cultured atrial myocytes.[52] Similar studies were performed using photolysis light from a flash lamp that had been directed onto frog skeletal muscle.[53]

15. The effect of GTP-γ-S and GDP-β-S on Ca^{2+} channel currents in neurons was investigated applying the whole-cell patch-clamp technique.[54] The flash lamp was mounted inside the Faraday cage. Photolysis efficiency for various caged GTP derivates at 3 mJ/mm^2 of bandpass-filtered UV light (300–380 nm) was around 10%. Using a 1-msec flash lamp, photolysis yields at various energy densities were determined for Ca^{2+}, ATP, GTP, and cAMP from their respective photolabile precursors in a study on Ca^{2+}-dependent Cl^- currents in cultured rat neurons following photolysis of DM-nitrophen.[55]

Conclusion

Flash lamp systems have proved useful to photolyze a large variety of caged compounds in diverse biological and biochemical systems. In some applications, especially those requiring very short pulses or sophisticated optics, flash lamps are not suitable and one must resort to the use of short, high-energy laser pulses, although these intense light pulses can damage the sample under study. For a given energy in the UV range the number of photoreleased molecules is higher for flash lamps compared to lasers. The typical pulse length of a flash lamp is around 0.5–1 msec, which is sufficiently short for many experiments in physiology. If caged compounds become available with their action spectrum shifted from the UV toward longer wavelengths, the efficiency of flash lamp-based photolysis will be enhanced even further due to increased photon flux of xenon flash lamps at longer wavelengths.

---

[51] S. Muralidharan and J. M. Nerbonne, *J. Photochem. Photobiol.* **27,** 123 (1995).
[52] M. Bechem and H. Hoffmann, *Pflügers Arch.* **424,** 343 (1993).
[53] D. Feldmeyer, W. Melzer, B. Pohl, and P. Zöllner, *J. Physiol.* **457,** 639 (1992).
[54] R. H. Scott, J. F. Wootton, and A. C. Dolphin, *Neuroscience* **38,** 285 (1990).
[55] K. P. Currie, J. F. Wootton, and R. H. Scott, *J. Physiol.* **482,** 291 (1995).

[12] Fourier Transform Infrared Photolysis Studies of Caged Compounds

By VALENTIN CEPUS, CAROLA ULBRICH, CHRISTOPH ALLIN, AGNES TROULLIER, and KLAUS GERWERT

Introduction

Time-resolved Fourier transform infrared (FTIR) difference spectroscopy has been established as a new tool to study molecular reaction mechanisms of proteins at the atomic level with nanosecond time resolution.[1,2] In photobiological proteins the reaction can be started directly by a short laser flash using the intrinsic chromophores. This approach has been successfully applied to the membrane proteins bacteriorhodopsin and the photosynthetic reaction center.[3] A much broader applicability can be achieved by the use of caged compounds. In this case biologically active molecules are released from inactive photolabile precursors. The use of caged compounds has become widespread in the last 10 years.[4-6] They allow the initiation of a protein reaction with a nanosecond UV laser flash.

This article presents FTIR photolysis studies of caged phosphate, caged GTP, caged ATP, and caged calcium. This should provide a good basis for further FTIR studies on molecular reaction mechanisms of proteins using caged compounds. Finally, as an example, FTIR studies on the GTP-binding protein H-*ras* p21 are presented.

Experimental Setup

Fourier transform infrared measurements are performed on an IFS 66v spectrometer (Bruker, Karlsruhe, Germany). The experimental setup is shown in Fig. 1. The MIR beam source is a water-cooled globar. The beam splitter consists of a KBr crystal. A HgCdTe (MCT) detector that is sensitive in the spectral region between 900 and 5000 cm^{-1} is used. Single-beam

---

[1] K. Gerwert, *Curr. Opin. Struct. Biol.* **3**, 769 (1993).
[2] R. Rammelsberg, B. Hessling, H. Chorongiewski, and K. Gerwert, *Appl. Spectrosc.* **51**, 558 (1997).
[3] K. Gerwert, *Biochim. Biophys. Acta* **1101**, 147 (1992).
[4] J. A. McCray and D. R. Trentham, *Annu. Rev. Biophys. Biophys. Chem.* **18**, 239 (1989).
[5] J. E. T. Corrie and D. R. Trentham, *in* "Bioorganic Photochemistry: Biological Application of Photochemical Switches" (H. Morrison, ed.), Wiley, New York, 1993.
[6] S. R. Adams and R. Y. Tsien, *Annu. Rev. Physiol.* **55**, 755 (1993).

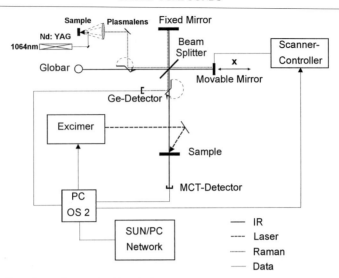

Fig. 1. Experimental setup for photolysis, FTIR, and Raman measurements of caged compounds.

spectra are measured. The spectrometer is controlled by a personal computer (PC) equipped with the program Opus (Bruker).

Photolysis of the caged compounds is performed by ultraviolet (UV) flashes at the wavelength λ of 308 nm [excimer laser, LPX 240 (Lambda Physics, Göttingen, Germany)]. The laser flashes are applied on the sample at an angle of 45° to the IR beam. The duration of a UV laser pulse is about 20 nsec, with the energy reaching up to 200 mJ per flash. Twenty to 60 flashes are used to achieve the complete photolysis of caged compounds, except in the case of DM-nitrophen, for which approximately 100 flashes are necessary. The home-built sample chamber of the FTIR spectrometer IFS 66v is purged with dried air. The metal sample holder is connected to a thermostat. The sample solution is prepared between two CaF_2 windows of 20 mm diameter. A spacer ring made of Mylar 2.5 μm thick is used in order to keep a defined distance between the windows. In the absence of other specification, spectral resolution of 4 cm^{-1} is used.

Fourier transform infrared measurements of H-*ras* p21 are performed with the truncated form of human H-*ras* p21c (1–166). For the nucleotide exchange of the protein-bound GDP for caged GTP, an excess of the desired caged GTP isotopomer in the presence of alkaline phosphatase is used.[7] During the GTPase reaction of H-*ras* p21-GTP, spectra are collected up to 200 min after photolysis.

Raman spectra are recorded on a IFS 88 (Bruker) spectrometer with an FRA106 Raman module equipped with a germanium detector. For excitation, a Nd–YAG (neodymium/yttrium–aluminum–garnet) laser (1064 nm) is used. The sample, in a quartz capillary 0.8 mm in diameter, is placed in the thermostatted sample holder.

Synthesis

The synthesis of 1-(2-nitrophenyl)ethyl phosphate (caged phosphate) is performed, as described by Dantzig *et al*.,[8] via the reaction of 1-(2-nitrophenyl)diazoethane and orthophosphate.[9] $^{18}O_4$-Labeled orthophosphate is synthesized as described by Hackney *et al*.[10]

P^3-[1-(2-Nitrophenyl)ethyl]guanosine 5′-triphosphate (caged GTP, see Scheme 3) and P^3-[1-(2-nitrophenyl)ethyl]guanosine 5′-O-(γ-thio)triphosphate (caged GTPγS) are synthesized following the procedure of Walker *et al*.[9] by the esterification of guanosine 5′-O-triphosphate (GTP) and the γS analog guanosine 5′-O-(γ-thio)triphosphate (GTPγS, Merck, Darmstadt, Germany), respectively, with 1-(2-nitrophenyl)diazoethane. The ^{18}O-labeled isotopomers are prepared starting with the sulfur analogs guanosine 5′-O-(α-thio)triphosphate (GTPαS), guanosine 5′-O-(β-thio)triphosphate (GTPβS), and P^3-[1-(2-nitrophenyl)ethyl]guanosine 5′-O-(γ-thio)triphosphate (caged GTPγS) by oxidation with N-chlorosuccinimide and hydrolysis with $H_2^{18}O$.[11,12] Guanosine 5′-O-[α-^{18}O]triphosphate (GTPα^{18}O, Scheme 3), guanosine 5′-O-[β-^{18}O]triphosphate (GTPβ^{18}O, Scheme 3), and guanosine 5′-O-[γ-^{18}O]triphosphate (GTPγ^{18}O, Scheme 3) are then esterified with 1-(2-nitrophenyl)diazoethane as described earlier.

Caged ATP (P^3-[1-(2-nitrophenyl)ethyl]adenosine 5′-triphosphate) and DM-nitrophen [1-(4,5-dimethoxy-2-nitrophenyl)-1,2-diaminoethane-N,N,N′,N′-tetraacetic acid] are available commercially (Calbiochem, La Jolla, CA).

[7] J. John, I. Schlichting, E. Schiltz, P. Rosch, and A. Wittinghofer, *J. Biol. Chem.* **264**, 13086 (1989).
[8] J. A. Dantzig, Y. E. Goldman, N. C. Millar, J. Lacktis, and E. Homsher, *J. Physiol.* **451**, 247 (1992).
[9] J. W. Walker, G. P. Reid, J. A. McCray, and D. R. Trentham, *J. Am. Chem. Soc.* **110**, 7170 (1988).
[10] D. D. Hackney, K. E. Stempel, and P. D. Boyer, *Methods Enzymol.* **64**, 60 (1980).
[11] B. A. Connolly, F. Eckstein, and H. H. Fuldner, *J. Biol. Chem.* **257**, 3382 (1982).
[12] J. Feuerstein, Dissertation, Universität Hannover, Germany (1987).

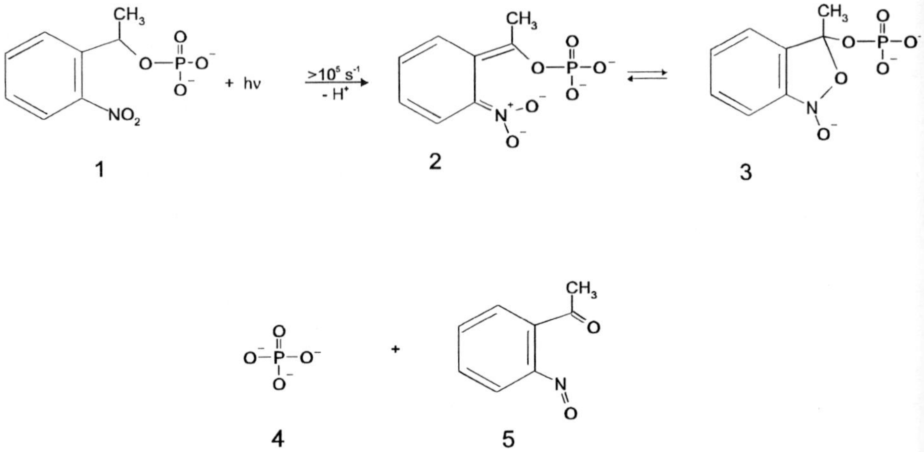

SCHEME 1. Reaction scheme for the photolysis of caged phosphate.[9]

Caged Phosphate

The 1-(2-nitrophenyl)ethyl moiety is used to protect phosphate, nucleotides, and nucleotide analogs.[9,13] The application of UV flashes leads to the release of the desired phosphate compound. The mechanism of photolysis of compounds containing the 2-nitrobenzyl group was the topic of several investigations.[9,14,15]

Scheme 1 shows the generally accepted reaction pathway developed by Walker et al.[9] for the photorearrangement of caged phosphate compounds. After photolysis, a rapid formation of intermediate **2**, called the *aci*-nitro anion, occurs. It decays subsequently in a dark reaction to orthophosphate **4** and the byproduct 2-nitrosoacetophenone **5**. The intermediate may comprise a number of rapidly interconvertible forms such as **3**, but UV–VIS studies[9] and single-wavelength time-resolved IR measurements[15] are able to resolve only a single intermediate for caged ATP.

A typical FTIR difference spectrum of caged phosphate photolysis is shown in Fig. 2. A spectrum is measured before the photolysis and is taken as a reference. From this spectrum and the spectrum taken after the photolysis of the caged phosphate, the difference spectrum is calculated. The flat baseline shows the difference between two spectra measured before the flash as a control. Only those vibrational modes cause bands in the

---

[13] J. W. Walker, G. P. Reid, and D. R. Trentham, *Methods Enzymol.* **172**, 288 (1989).
[14] S. Schneider, *J. Photochem. Photobiol. A* **55**, 329 (1991).
[15] A. Barth, J. E. T. Corrie, M. J. Gradwell, Y. Maeda, W. Mantele, T. Meier, and D. R. Trentham, *J. Am. Chem. Soc.* **119**, 4149 (1997).

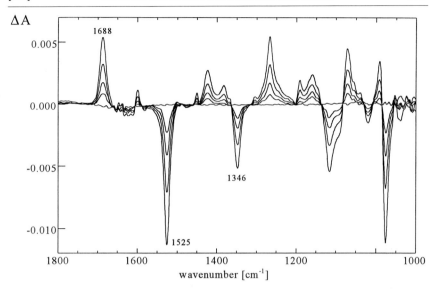

FIG. 2. Fourier transform infrared difference photolysis spectra of caged phosphate; 50 mM caged phosphate, 0.1 M MOPS, pH 7.0, baseline and a spectrum after the first, second, fourth, and ninth flash.

difference spectrum that undergo reaction-induced absorbance changes. Negative bands in the difference spectrum are due to the caged phosphate **1** in Scheme 1, whereas positive bands are due to the photolysis products, orthophosphate **4** and 2-nitrosoacetophenone **5**. Characteristic bands of 1-(2-nitrophenyl)ethyl derivatives are the disappearing asymmetric and the symmetric (NO$_2$) stretching vibrations [ν_{as}(NO$_2$) = 1525 cm^{-1}; ν_{sy}(NO$_2$) = 1346 cm^{-1}]; and the positive band is due to the carbonyl group of 2-nitrosoacetophenone **5** [ν(C=O) = 1688 cm^{-1}].[5,16–18]

Band Assignment by Isotopic Labeling

Unequivocal assignment of IR bands is made possible by isotopic labeling. For identification of phosphate bands we used ^{18}O$_4$-labeled caged phos-

[16] N. B. Colthup J. W. Daly, and S. E. Wiberly, "Introduction to Infrared Raman Spectroscopy," Academic Press, London, 1990.
[17] D. Lien-Vien, N. B. Colthup, and S. E. Wiberley, "The Handbook of Infrared and Raman Characteristic Frequencies of Organic Molecules," Academic Press, San Diego, 1991.
[18] H. Georg, A. Barth, W. Kreutz, F. Siebert, and W. Mantele, *Biochim. Biophys. Acta* **1188,** 139 (1994).

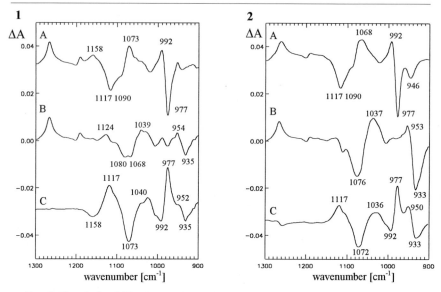

Fig. 3. Photolysis of 50 mM caged phosphate; **(1)** pH 7.0, 0.2 M MOPS; **(2)** pH 8.0, 0.2 M HEPES; (A) caged phosphate [$^{16}O_4$]; (B) caged phosphate [$^{18}O_4$]; (C) difference B − A.

phate. The higher mass of the oxygen isotope leads to a shift of all phosphate bands to lower wavenumbers.

Figure 3 **(1)** shows photolysis difference spectra of the unlabeled caged phosphate (A) and the ^{18}O-labeled caged phosphate (B) at pH 7.0. The double difference B minus A (C) represents the spectral differences between the unlabeled and the ^{18}O-labeled phosphate moiety of the caged compound. Figure 3 **(2)** shows the equivalent spectra measured at pH 8.0. The isotopical shifts are obvious. At pH 7.0, two species of orthophosphate, mono- and dihydrogen phosphate ($pK_{a1} = 2$, $pK_{a2} = 6.72$, $pK_{a3} = 11.96$, $I = 0.1\ M$, $25°$),[19] are present [Fig. 3 **(1)**]. To correctly assign the bands present at pH 7.0 to one of these two species, one has to measure the difference photolysis spectra at different pH values. Phosphate bands appear below 1300 cm^{-1}, and therefore only this spectral region is shown in Fig. 3.

The positive bands in spectra A and B correspond to the orthophosphate that is liberated by the photolysis of caged phosphate. The band at 1158 cm^{-1} [Fig. 3 **(1)**] is shifted to 1124 cm^{-1} and can be assigned to the (PO_2^-)

---

[19] "Critical Stability Constants. Vol. 4. Inorganic Complexes" (R. M. Smith and A. E. Marell, eds.), Plenum Press, New York, 1976.

TABLE I
ASSIGNMENT OF PHOSPHATE BANDS DUE TO ISOTOPIC LABELING

Caged phosphate	Wavenumber (cm^{-1})	Phosphate	Wavenumber (cm^{-1})
$(PO_3^{2-})_{deg}$	1117	$H_2PO_4^-$ $(PO_2^-)_{as}$	1158
$(PO_3^{2-})_{deg}$	1090	$H_2PO_4^-$ $(PO_2^-)_{sy}$	1073
$(PO_3^{2-})_{sy}$	977	HPO_4^{2-} $(PO_3^{2-})_{deg}$	1068
		HPO_4^{2-} $(PO_3^{2-})_{sy}$	992

asymmetric stretching vibration of $H_2PO_4^-$.[20] Indeed, this vibration is nearly completely absent at pH 8.0 [Fig. 3 **(2)**] where HPO_4^{2-} is the prevalent species. The broad band at 1073 cm^{-1}, which is shifted to 1039 cm^{-1}, is caused by the (PO_2^-) symmetric stretching vibration of $H_2PO_4^-$ and by (PO_3^{2-}) degenerate stretching vibration of HPO_4^{2-} [20,21] that appears at 1068 cm^{-1} at pH 8.0. The monohydrogen phosphate has an additional characteristic vibration at 992 cm^{-1} caused by the symmetric stretching vibration of the (PO_3^{2-}) moiety[20,21] that is shifted to 954 and 953 cm^{-1}, respectively.

The negative bands at 1117, 1090, and 977 cm^{-1} (Table I) belong to the unlabeled caged phosphate and are shifted to 1080, 1068, and 935 cm^{-1}, respectively, by isotopic labeling. The deprotonated monomethyl phosphate has two degenerate (PO_3^{2-}) stretching vibrations at 1115 and 1090 cm^{-1} and a (PO_3^{2-}) symmetric stretching vibration at 983 cm^{-1}.[22] Monoesters of phosphoric acid are reported to have two pK_a values, $pK_1 \approx 1.5$ and $pK_2 \approx 6.3$[23]; thus the FTIR spectrum of caged phosphate at pH 8.0 should be dominated by the completely deprotonated caged phosphate dianion. Because a band at 1117 cm^{-1} with a shoulder at 1090 cm^{-1} and a band at 977 cm^{-1} are observed in this spectrum, the same assignment could be given for caged phosphate.

Influence of Magnesium on Caged Phosphate

It is important to know the influence of magnesium on the FTIR spectrum of the phosphate moiety for the band assignments in protein measurements because magnesium is an essential cofactor for several phosphate and nucleotide binding enzymes, e.g., ATPases, kinases, and ion pumps.

---

[20] A. C. Chapman and L. E. Thirlwell, *Spectrochim. Acta* **20**, 937 (1964).
[21] E. Steger and K. Herzog, *Z. Anorg. Allg. Chem.* **331**, 169 (1964).
[22] T. Shimanouchi, M. Tsuboi, and Y. Kyogoku, *Adv. Chem. Phys.* **7**, 435 (1964).
[23] W. D. Kumler and J. J. Eiler, *J. Am. Chem. Soc.* **65**, 2355 (1943).

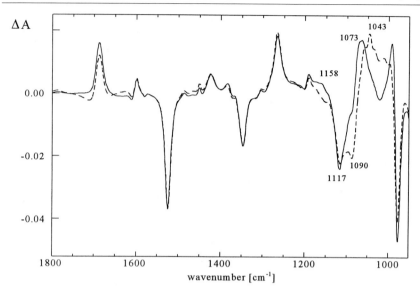

FIG. 4. Photolysis spectra of caged phosphate with Mg^{2+} (dashed line) and without Mg^{2+} (solid line); 50 mM caged phosphate, (100 mM $MgCl_2$), 200 mM HEPES, pH 7.0.

Figure 4 shows the photolysis spectrum of caged phosphate at pH 8.0 in the presence and absence of magnesium.

The positive band at 1073 cm^{-1}, which belongs to the orthophosphate, is shifted to 1043 cm^{-1}. This shift to lower wavenumbers is probably caused by the decrease of the partial double bond character of the P–O bonds involved in magnesium coordination. The band broadening in the orthophosphate region could also be due to a mixture of different magnesium orthophosphate complexes, e.g., $Mg_3(PO_4)_2$ and $MgHPO_4$.[24]

The small positive band at 1158 cm^{-1} missing in the presence of magnesium is due to a slightly higher concentration of $H_2PO_4^-$ in the absence of magnesium, which shifts the equilibrium of the orthophosphate species toward HPO_4^{2-}.

In the absence of magnesium the negative band belonging to the caged phosphate moiety at 1117 cm^{-1} has a shoulder at 1090 cm^{-1}. In the presence of magnesium a clearly resolved negative band is observed at 1090 cm^{-1}. The appearance of this band is probably caused by the shift of the positive band from 1073 to 1043 cm^{-1}. Therefore, it seems reasonable to assign this band to the degenerate stretching vibration of caged phosphate, which is masked by the positive band in the absence of magnesium. No further

[24] A. W. Taylor, A. W. Frazier, and E. L. Gurney, *Trans. Faraday Soc.* **59**, 1585 (1963).

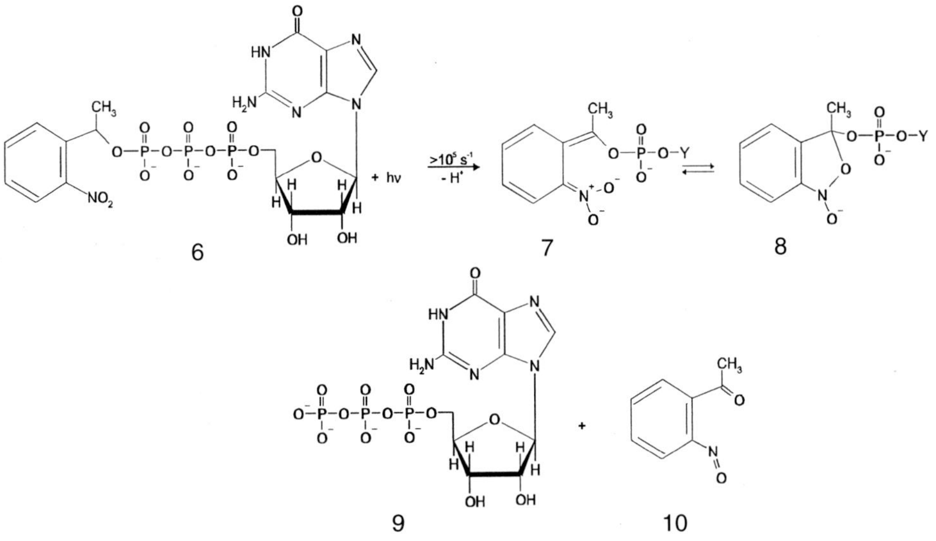

SCHEME 2. Reaction scheme for the photolysis of caged GTP.[9]

influence of magnesium on the bands corresponding to caged phosphate is detected.

Caged GTP

Further important compounds for the investigation of biological systems are caged nucleotides such as caged GTP and ATP.[4,5,9,13] The photolysis reaction of caged GTP is shown in Scheme 2 (Structures **6–10**).

The photolysis spectrum of caged GTP (Fig. 5, solid line) shows bands due to the 1-(2-nitrophenyl)ethyl moiety (1688, 1525, and 1346 cm^{-1}, as already discussed) and phosphate bands below 1300 cm^{-1}. Bond cleavage between the caged moiety and the leaving group GTP gives rise to the negative band at 1252 cm^{-1}. It is caused by the asymmetric stretching vibration of the γ (PO$_2^-$) group connected to the caging group. During photolysis this group is converted to the terminal γ (PO$_3^{2-}$) group of the released GTP. The degenerate stretching vibration of the γ (PO$_3^{2-}$) group leads to the positive band at 1118 cm^{-1}. Generally, the spectral region between 1050 and 1200 cm^{-1} is determined by a superposition of the degenerate γ (PO$_3^{2-}$) and the symmetric (PO$_2^-$) stretching modes.[25–27]

[25] H. Takeuchi, H. Murata, and I. Harada, *J. Am. Chem. Soc.* **110**, 392 (1988).
[26] A. Barth, W. Kreutz, and W. Mantele, *Biochim. Biophys. Acta* **1194**, 75 (1994).
[27] A. Barth, W. Mantele, and W. Kreutz, *Biochim. Biophys. Acta* **1057**, 115 (1991).

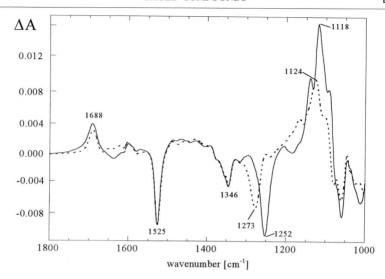

FIG. 5. Photolysis spectra of caged GTP with Mg^{2+} (dashed line) and without Mg^{2+} (solid line); 20 mM caged GTP (50 mM MgCl$_2$), 100 mM HEPES, pH 7.0.

Influence of Magnesium on Caged GTP

The formation of complexes among caged GTP, GTP, or other nucleotides with divalent cations, such as Ca^{2+}, Mg^{2+}, and Mn^{2+}, has a strong effect on FTIR spectra.[25,28]

The comparison of caged GTP spectra with and without Mg^{2+} (Fig. 5) shows no significant effect above 1300 cm^{-1}. The negative band at 1252 cm^{-1} in the absence of Mg^{2+} is less intensive and is shifted upward to 1273 cm^{-1} in the presence of the cation. This can be explained by an increase of the partial double bond character of the P–O bonds not directly involved in magnesium coordination. The positive band around 1118 cm^{-1} is shifted to 1124 cm^{-1} in the Mg^{2+}-bound case.

The similarity of caged GTP and caged ATP photolysis (Fig. 6) shows, as expected, that FTIR differences spectra are dominated by changes derived from the 1-(2-nitrophenyl)ethyl group and the phosphate chain.

Time-Resolved Measurements

Various FTIR techniques have been developed to collect time-resolved absorbance data. They are the rapid scan,[29] the stroboscopic,[30] and the

[28] H. Brintzinger, *Biochim. Biophys. Acta* **77**, 343 (1963).
[29] K. Gerwert, G. Souvignier, and B. Hess, *Proc. Natl. Acad. Sci. U.S.A.* **87**, 9774 (1990).
[30] G. Souvignier and K. Gerwert, *Biophys. J.* **63**, 1393 (1992).

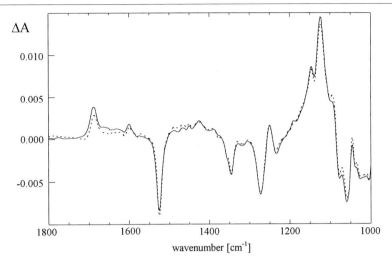

FIG. 6. Caged GTP (solid line), caged ATP (dashed line), 40 mM caged GTP/ATP, 200 mM MgCl$_2$, 500 mM HEPES/NaOH, pH 7.5.

step-scan techniques.[2,31,32] (and citations therein). Here we use the rapid scan technique with a 12-msec time resolution to record the IR difference spectra of the intermediate (Scheme 2, **7, 8**) and the photolysis products of caged GTP. Conditions close to physiological values, pH 7.5 and the presence of MgCl$_2$, are used for the measurements. The presence of dithiothreitol (DTT) is essential for the biological application of several caged compounds because the photolysis of the 1-(2-nitrophenyl)ethyl group yields 2-nitrosoacetophenone (Scheme 2, **10**). This compound can react with the cysteines in proteins and therefore inactivate or modify the biological system.[4] These complications can be overcome by the addition of thiols, such as DTT, which scavenge this nitrosoketone.[33]

Figure 7 compares photolysis spectra of caged GTP with and without DTT. The main bands are the same in the two spectra, which demonstrates the relatively small influence of DTT on the spectrum. The largest difference is the missing band at 1688 cm^{-1} in the measurement with DTT. This shows that the carbonyl group of the 2-nitrosoketone is absent in the final products, which can be explained by the reaction of 2-nitrosoketone with DTT. The lower part of the spectra shows small differences in the intensities at 1250,

---

[31] W. Uhmann, A. Becker, C. Taran, and F. Siebert, *Appl. Spectrosc.* **45**, 390 (1991).
[32] R. A. Palmer, J. L. Chao, R. M. Ditmar, V. G. Gregoriou, and S. E. Plunkett, *Appl. Spectrosc.* **47**, 1297 (1993).
[33] J. H. Kaplan, B. Forbush III, and J. F. Hoffman, *Biochemistry* **17**, 1929 (1978).

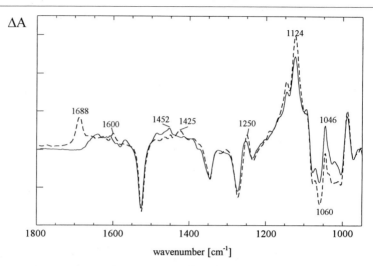

FIG. 7. Photolysis spectra of caged GTP with DTT (solid line, 50 mM caged GTP, 250 mM MgCl$_2$, 250 mM DTT, 500 mM HEPES/NaOH, pH 7.5, 10°, resolution 8 cm^{-1}) and without DTT (dashed line, 20 mM caged GTP, 50 mM MgCl$_2$, 100 mM HEPES/NaOH, pH 7.5, 20°, resolution 4 cm^{-1}).

1124, 1060, and 1046 cm^{-1} and a few smaller bands at 1600, 1452, and 1425 cm^{-1}.

In order to determine the IR difference spectra of the intermediate, spectra obtained between 1 and 26 msec, after a single laser flash are averaged (Fig. 8A). The region between 1600 and 1700 cm^{-1} shows a decreased signal-to-noise (S/N) ratio because of the large water background absorption. The product difference spectrum (Fig. 8B), obtained between 86 and 105 sec after the flash, shows the IR differences between caged GTP and the final products of the caged GTP photolysis, i.e., GTP and the reaction products of 2-nitrosoacetophenone with DTT.

As determined by a global fit analysis,[34] the intermediate decays exponentially with a decay constant of 8 sec^{-1} to the final products. Thus the average intermediate concentration in the spectrum (Fig. 8A) is 96%. Barth *et al.*[15] have reported the IR spectrum of an intermediate during the photolysis of caged ATP. In general, the difference spectra shown in A agree nicely with their published one, even though slightly different conditions (different buffer, pH 8.5, no magnesium) were used. Several bands, e.g., at 1379, 1328, 1244, and 1181 cm^{-1}, which they assign to the *aci*-nitro anion (Scheme 2, 7), can also be found in the spectrum (Fig. 8A). Barth *et al.*[15] could make

---

[34] B. Hessling, G. Souvignier, and K. Gerwert, *Biophys. J.* **65**, 1929 (1993).

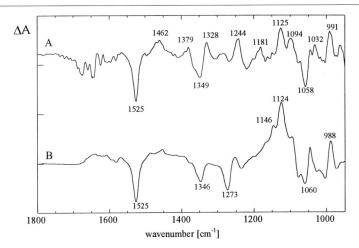

FIG. 8. Fourier transform infrared difference spectra of laser pulse photolysis of caged GTP, 50 mM caged GTP, 250 mM MgCl$_2$, 250 mM DTT, 500 mM HEPES/NaOH, pH 7.5, 10°, resolution 8 cm^{-1}. (A) Obtained 1–26 msec after a single laser flash; (B) obtained 86–105 sec after a single laser flash.

specific band assignments by the use of ^{13}C, ^{15}N, and ^{18}O isotopomers of caged ATP. Deviations to the difference spectra of Barth et al.[15] are observed for the disappearing asymmetric (NO$_2$) stretching vibration. They reported a shift of this educt band between the intermediate and the product difference spectrum. However, no difference at 1525 cm^{-1} is observed between spectra A and B. These differences could arise from the fact that Barth et al. used a different buffer that has a large background absorption in this spectral region.

Raman Spectroscopy

Raman spectroscopy, like infrared spectroscopy, probes the vibrational properties of a molecule. The selection rules between these two are different, however. This difference is a useful tool for the assignment of bands. Asymmetric vibrations give stronger bands in the IR spectrum, whereas bands due to symmetric vibrations are stronger in the Raman spectrum.

Figure 9 shows the solution Raman spectrum of caged GTP (A) and GTP (B). The Raman spectrum of the solvent was subtracted. The strongest band in the Raman spectrum of caged GTP is the symmetric (NO$_2$) stretching vibration at 1346 cm^{-1},[14] whereas the asymmetric (NO$_2$) stretching vibration gives rise only to a small band in the spectrum at 1525 cm^{-1}. By

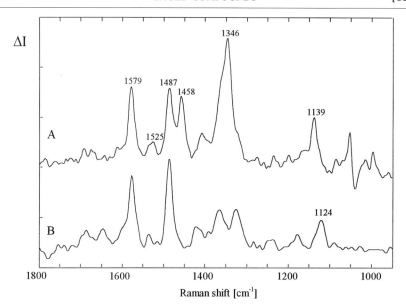

FIG. 9. Solution Raman spectrum. (A) Caged GTP; 100 mM caged GTP, 250 mM MgCl$_2$, 250 mM DTT, 500 mM HEPES/NaOH, pH 7.5, 4°. (B) GTP; 20 mM GTP, 60 mM MgCl$_2$, pH 7.5, 4°. The Raman spectrum of the solvent was subtracted.

comparison with the Raman spectrum of GDP of Weng et al.[35] the band at 1579 cm^{-1} can be assigned to the N3–C4 and C4–N9 stretching motions of the ring of the purine base. The band at 1487 cm^{-1} can be assigned to the N7=C8 and C4=C5 stretching vibration of the purine base. Another strong band is at 1458 cm^{-1}, which can tentatively be assigned to the phenyl ring of the 2-nitrophenyl group of caged GTP. The band at 1139 cm^{-1} can be caused by the symmetric (PO$_2^-$) vibration of the α-, β-, and γ-phosphate groups of caged GTP. The strong buffer scattering between 1000 and 1100 cm^{-1} causes a subtraction artifact at 1050 cm^{-1} and leads to a decreased S/N ratio in this spectral region. For Mg^{2+}-complexed GTP (Fig. 9B) the Raman band at 1124 cm^{-1} is assigned, as for magnesium complexed ATP,[25] to the in-phase symmetric (PO^{2-}) mode.

To demonstrate the different selection rules for IR and Raman spectroscopy, the Raman difference spectrum between GTP and caged GTP (Fig. 10A) and the IR difference spectrum of photolyzed caged GTP (Fig. 10B) are compared. In order to subtract the two Raman spectra (Figs. 9A and 9B), spectra are normalized using the band at 1579 cm^{-1}.

[35] G. Weng, C. X. Chen, V. Balogh-Nair, R. Callender, and D. Manor, *Protein Sci.* **3,** 22 (1994).

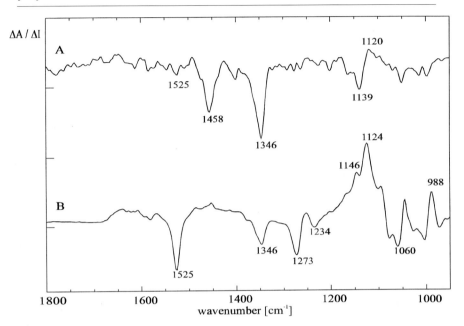

FIG. 10. (A) NIR-FT-Raman difference spectra of GTP (Fig. 9B) minus caged GTP (Fig. 9A); (B) FTIR difference spectra of laser pulse photolysis of caged GTP; 50 mM caged GTP, 250 mM MgCl$_2$, 250 mM DTT, 500 mM HEPES/NaOH, pH 7.5, 10°, obtained 86–105 sec after a single laser flash, resolution 8 cm^{-1}.

In contrast to FTIR difference spectra, Raman difference spectra between caged GTP and GTP show no large differences. Spectra show nicely that the symmetric vibration of the (NO$_2$) group at 1346 cm^{-1} [14] gives a much larger signal in the Raman spectrum, whereas the asymmetric (NO$_2$) stretching vibration at 1525 cm^{-1} is much stronger in the IR difference spectrum. The negative band at 1458 cm^{-1} is tentatively assigned to the phenyl ring of the 2-nitrophenyl group of caged GTP. The negative band at 1139 cm^{-1} is probably caused by the in-phase mode of the three (PO$_2^-$) groups of caged GTP. By comparison with the solution Raman spectrum of GTP (Fig. 9A), a small band at 1120 cm^{-1} can be visualized. This band is assigned to the in-phase mode of the (PO$_2^-$) groups of GTP.

Band Assignments of Caged GTP

For further band assignment, we synthesized caged GTP isotopomers that contain one ^{18}O isotope selectively in the α, β, and γ position, respec-

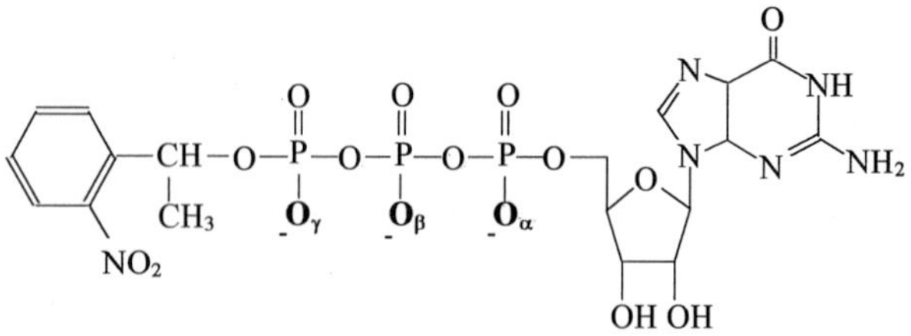

SCHEME 3. Position of the ^{18}O labels introduced. All oxygens the ^{16}O isotope (**1**); α: ^{18}O isotope, β and γ: ^{16}O isotope (**2**); β: ^{18}O isotope, α and γ: ^{16}O isotope (**3**); γ: ^{18}O isotope, α and β: ^{16}O isotope (**4**).

tively. Photolysis spectra of the four caged GTP isotopomers (Scheme 3, unlabeled and ^{18}O labeled) are shown in Fig. 11. To normalize spectra, the symmetric (NO_2) vibration at 1346 cm^{-1} is used (Fig. 11, dashed line).

The comparison of the four photolysis spectra shows downshifts for each of the ^{18}O-labeled compounds. Shifting of caged GTP bands is indicated by a light shading and shifting of GTP bands by a dark shading (Fig. 11, line 1). The black area symbolizes a shift that is observed only in the spectrum of the γ-^{18}O-labeled caged GTP. Spectra are termed by the numbers according to the respective caged GTP isotopomers in Scheme 3. The wave-

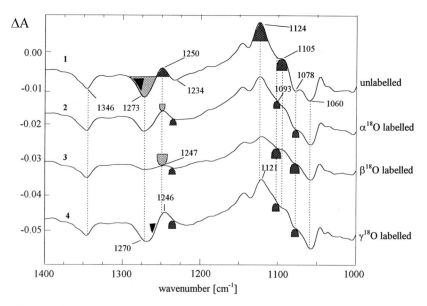

FIG. 11. Caged GTP photolysis spectra; 20 mM caged GTP, 50 mM MgCl$_2$, 100 mM HEPES, pH 7.5. (**1**) Unlabeled; (**2**) α-^{18}O-labeled; (**3**) β-^{18}O-labeled; and (**4**) γ-^{18}O-labeled.

numbers to which the bands are shifted in the spectra of the labeled compounds are indicated by the same shadings referring to line 1.

The negative band at 1273 cm^{-1}, representing the asymmetric modes of the three (PO$_2^-$) groups of caged GTP, is sensitive to all three labels. The intensity of this band is decreased by α- and even more by β-^{18}O-labeling, and the band is shifted to 1250 cm^{-1}. This is indicated by the light shading on lines 2 and 3. γ-^{18}O-Labeling causes a downshift of the band at 1273 to 1270 cm^{-1} but does not change its intensity (black area on line 4, Fig. 11). This interpretation is supported by the double differences (Fig. 12). Here the photolysis difference spectra of the labeled compounds are subtracted from that of the unlabeled one. For α- and β-labeled caged GTP, a shift can be detected from 1273 to 1253 and 1251 cm^{-1}, respectively. In contrast, caged GTPγ^{18}O shows a specific shift from 1277 to 1261 cm^{-1}.

The positive band at 1250 cm^{-1} represents the asymmetric stretching vibration of the α and β (PO$_2^-$) group of the free GTP (Fig. 11). Due to α, β, and γ labeling, the band at 1250 cm^{-1} is downshifted to around 1240 cm^{-1} (dark shading, Fig. 11). This becomes clearer in double difference spectra (Fig. 12). For the α, β, and γ label a shift is observed from 1253 to 1238, from 1251 to 1236, and from 1261 to 1240 cm^{-1}, respectively. For GTPβ^{18}O the shift is less obvious due to the masking shift of the band at 1273 cm^{-1}.

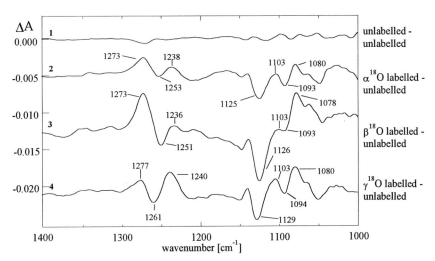

FIG. 12. Double differences of caged GTP photolysis spectra; 20 mM caged GTP, 50 mM MgCl$_2$, 100 mM HEPES, pH 7.5. **(1)** Unlabeled; **(2)** α-^{18}-O-labeled; **(3)** β-^{18}O-labeled; and **(4)** γ-^{18}-O-labeled.

TABLE II
Assignments of Changing Phosphate Vibrations During Caged GTP Photolysis Reaction[a]

Caged GTP	Wavenumber (cm^{-1})	Comment	GTP	Wavenumber (cm^{-1})	Comment
α-(PO$_2^-$)	1273	Coupled with β	α-(PO$_2^-$)	1253	α, β, and γ coupled
				1125	α, β, and γ coupled
				1093	α, β, and γ coupled
β-(PO$_2^-$)	1273	Coupled with α	β-(PO$_2^-$)	(1251)	α, β, and γ coupled
				1126	α, β, and γ coupled
				1093	α, β, and γ coupled
γ-(PO$_2^-$)	1277		γ-(PO$_3^{2-}$)	1261	
				1129	α, β, and γ coupled
				1094	α, β, and γ coupled

[a] Results are obtained by the use of the isotopomers 1 to 4 (see Scheme 3).

After photolysis the γ-phosphate is converted to a nonbridging (PO$_3^{2-}$) group. The degenerate stretching vibration of the γ (PO$_3^{2-}$) group should absorb below 1150 cm^{-1}. Furthermore, below 1150 cm^{-1} phosphate bands due to symmetric stretching vibrations of (PO$_2^-$) groups are superimposed.[15] The most intensive positive band is located at 1124 cm^{-1} (Fig. 11). It is downshifted by α-^{18}O-labeling. A better insight into the shifts is again provided by double difference (Fig. 12).[25] In principle, two different shift patterns are possible. One possibility is a downshift from 1125 to 1080 cm^{-1} (Fig. 12), but the maximal downshift due to ^{18}O-labeling in a diatomic molecule is 43 cm^{-1}. Therefore, it is more likely that the band is downshifted from 1125 to about 1103 cm^{-1} and a second shift is superimposed from 1093 to 1080 cm^{-1}. In this case, the shifted bands at 1103 and 1093 cm^{-1} cancel each other. Similar shifts are seen due to β-^{18}O-labeling from 1126 to 1103 and 1093 to 1078 cm^{-1} (Fig. 12). γ-^{18}O-Labeling seems to cause shifts from 1129 to 1103 and 1094 to 1080 cm^{-1}. The bands at 1124 and 1093 cm^{-1} seem to represent α and β (PO$_2^-$) that are coupled to the γ (PO$_3^{2-}$) vibrations. The observed sensitivity of the bands to the ^{18}O-labeling can be explained by a strong vibration coupling among all three phosphate groups (Table II).

DM-Nitrophen: A Photolabile Chelator of Divalent Cations

Another important group of caged compounds are photolabile derivatives of cation chelating reagents such as DM-nitrophen.[33,36–39] DM-Ni-

[36] J. H. Kaplan and G. C. R. Ellis-Davies, *Proc. Natl. Acad. Sci. U.S.A.* **85**, 6571 (1988).
[37] G. C. Ellis-Davies, J. H. Kaplan, and R. J. Barsotti, *Biophys. J.* **70**, 1006 (1996).

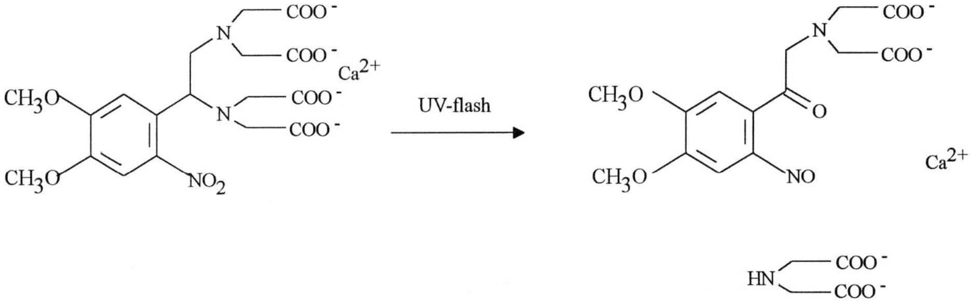

SCHEME 4. Photolysis of DM-nitrophen.

trophen contains an EDTA molecule and has a high affinity for divalent cations in the unphotolyzed state [K_D (Ca^{2+}) ≈ 5 × 10^{-9} M, K_D (Mg^{2+}) ≈ 2.5 × 10^{-6} M].[36] During the photolysis of DM-nitrophen, a C–N bond is broken and iminodiacetic acid is released (Scheme 4). Thus the K_D for the cations decreases to the millimolar range.[36,40]

This chelator has been used for investigating exocytosis,[41] muscle fibers,[42] and FTIR measurements on sarcoplasmic reticulum (SR) Ca^{2+}-ATPase.[18,40] Time-resolved FTIR difference spectroscopy has been applied to measure the kinetics of calcium binding to the SR Ca^{2+}-ATPase.[40]

A typical FTIR photolysis spectrum of DM-nitrophen is shown in Fig. 13. As is the case with the other caged compounds, the band at 1525 cm^{-1} is assigned to the asymmetric stretching vibration of (NO_2). This band is taken as reference to rescale spectra from different measurements when necessary. It could be expected that the bands due to the carboxylic groups are changed when different divalent cations are used.[43,44] As shown in Fig. 13, the wavenumber of the negative band at around 1585 cm^{-1} is different in the presence of Ca^{2+} (1585 cm^{-1}), Mg^{2+} (1588 cm^{-1}), and without divalent cation (1583 cm^{-1}). This band can be assigned to the asymmetric stretching vibration of the (COO^-) groups of the EDTA

---

[38] A. L. Escobar, F. Cifuentes, and J. L. Vergara, *FEBS Lett.* **364**, 335 (1995).
[39] J. A. McCray, N. Fidler-Lim, G. C. Ellis-Davies, and J. H. Kaplan, *Biochemistry* **31**, 8856 (1992).
[40] A. Troullier, K. Gerwert, and Y. Dupont, *Biophys. J.* **71**, 2970 (1996).
[41] T. D. Parsons, G. C. Ellis-Davies, and W. Almers, *Cell Calcium* **19**, 185 (1996).
[42] J. R. Patel, G. M. Diffee, and R. L. Moss, *Biophys. J.* **70**, 2333 (1996).
[43] M. Nara, M. Tasumi, M. Tanokura, T. Hiraoki, M. Yazawa, and A. Tsutsumi, *FEBS Lett.* **349**, 84 (1994).
[44] G. B. Deacon, R. J. Phillips, *Coord. Chem. Rev.* **33**, 227 (1980).

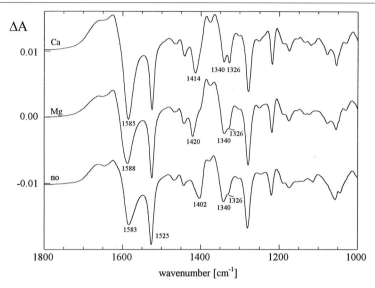

FIG. 13. 20 mM DM-nitrophen, 20 mM DTT, 10 mM Ca^{2+} (Mg^{2+} or without divalent cation), respectively, 0.3 M MOPS, pH 7.0, resolution 2 cm^{-1}.

moiety of DM-nitrophen.[16,17,45] Similar variation is observed for the band assigned to the symmetric carboxylate mode in the presence of Ca^{2+} (1414 cm^{-1}), Mg^{2+} (1420 cm^{-1}), and without divalent cation (1402 cm^{-1}). The symmetric vibration of arylic (NO$_2$) is usually found in the range of 1357 to 1318 cm^{-1}.[17] This vibration has been assigned to a broad negative band at 1332 cm^{-1} observed in the photolysis spectrum of DM-nitrophen.[18] We also observe a large band in this range that shows, in fact, two resolved peaks (Fig. 13): a band at 1340 cm^{-1} that is not modified in the absence or presence of divalent cations and a band at 1326 cm^{-1} that is sensitive to the presence of divalent cation. The last one is also present in EDTA absorbance spectra and is assigned to the (COO$^-$) moiety.[45] Therefore, only the band at 1340 cm^{-1} in the photolysis spectrum of DM-nitrophen should be assigned to the aromatic symmetric stretching vibration of the (NO$_2$) moiety.

Its ability to release a variety of cations by flash photolysis makes DM-nitrophen a useful compound for FTIR measurements on many biological systems.

[45] D. T. Sawyer and P. J. Paulsen, *J. Am. Chem. Soc.* **80,** 1597 (1958).

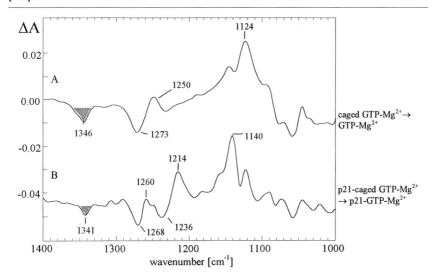

FIG. 14. (A) Photolysis spectrum of 20 mM caged GTP, 50 mM MgCl$_2$, 100 mM HEPES, pH 7.5 (B) Photolysis spectrum of 10 mM H-*ras* p21–caged GTP, 25 mM MgCl$_2$, 10 mM DTT, 50 mM HEPES, 1% glycerol, pH 7.5.

Biological Application: Fourier Transform Infrared Studies of GTPase Mechanism of H-*ras* p21

Caged GTP can be used as a substrate for H-*ras* p21. It is tightly bound similarly to GTP in close proximity to magnesium, but it is not hydrolyzed.[46] The photolysis spectrum of H-*ras* p21–caged GTP is shown in Fig. 14.

The shape of the H-*ras* p21–caged GTP photolysis spectrum (Fig. 14B) is remarkably different from the spectrum without protein (Fig. 14A). The difference bands of the free nucleotide (1273/1250 cm^{-1}) are split into two negative bands at 1268 and 1236 cm^{-1} and two positive bands at 1260 cm^{-1} and 1214 cm^{-1} in the enzyme-bound form. The broad positive band pattern originally around 1124 cm^{-1} is shifted to 1140 cm^{-1}. Also, below 1050 cm^{-1} small changes are observed. Compared to the photolysis without H-*ras* p21, the phosphate bands gain intensity and are shifted to higher wavenumbers. Because the degrees of freedom of the nucleotide are reduced by binding to the protein the bands appear sharper.[47,48]

---

[46] I. Schlichting, G. Rapp, J. John, A. Wittinghofer, E. F. Pai, and R. S. Goody, *Proc. Natl. Acad. Sci. U.S.A.* **86,** 7687 (1989).
[47] K. Gerwert, V. Cepus, A. J. Scheidig, and R. S. Goody, in "Proceedings of Time-Resolved Vibrational Spectroscopy VI" (A. Lau, F. Siebert, and W. Werncke, eds.), p. 185, Springer Verlag, Berlin, 1994.
[48] V. Cepus, R. S. Goody, and K. Gerwert, submitted (1998).

The GTPase reaction takes place after the photolytic liberation of GTP.[46,47] Fourier transform infrared spectra were then recorded from 30 sec to 2 hr 10 min. The difference calculation was referred to the last recorded spectrum (Fig. 15). Therefore, the positive bands represent the H-*ras* p21–GTP state and the negative ones the H-*ras* p21–GDP state and inorganic phosphate. At 1260 cm^{-1} a strong positive band is detected that is characteristic for the asymmetric stretching vibration of the GTP (PO_2^-) groups. The negative band at 1100 cm^{-1} represents the strongest GDP band. It is dominated by the degenerate stretching vibration of the terminal β (PO_3^{2-}) group.[25] Results show that kinetic analysis of the GTPase reaction of H-*ras* p21 can be performed with FTIR difference spectroscopy. Thus, a new experimental approach, time-resolved FTIR difference spectroscopy, is established that monitors p21 GTPase reaction with high structural and high time resolution.

Summary

Time-resolved FTIR difference spectroscopy is a powerful tool for investigating molecular reaction mechanisms of proteins. In order to detect, beyond the large background absorbance of the protein and the water, absorbance bands of protein groups that undergo reactions, difference spectra have to be performed between a ground state and an activated state of the sample. Because the absorbance changes are small, the reaction has to be started *in situ*, in the apparatus, and in thin protein films. The use of caged compounds offers an elegant approach to initiate protein reactions with a nanosecond UV laser flash. Here, time-resolved FTIR and

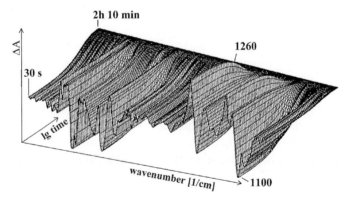

FIG. 15. Three-dimensional global fit representation of the GTPase reaction of H-*ras* p21; 10 m*M* H-*ras* p21–GTP, 25 m*M* MgCl$_2$, 10 m*M* DTT, 50 m*M* HEPES, pH 7.5, 1% glycerol.

FT–Raman photolysis studies of the commonly used caged compounds, caged P_i, caged ATP, caged GTP, and caged calcium are presented. The use of specific isotopic labels allows us to assign the IR bands to specific groups. Because metal ions play an important role in many biological systems, their influence on FTIR spectra of caged compounds is discussed. The results presented should provide a good basis for further FTIR studies on molecular reaction mechanisms of energy or signal transducing proteins. As an example of such investigations, the time-resolved FTIR studies on the GTPase reaction of H-*ras* p21 using caged GTP is presented.

Acknowledgments

This work was supported by the Deutsche Forschungsgemeinschaft, SFB 394, Teilprojekt B1. We thank Professor R. S. Goody for the gift of GTPβS and GTPαS and Dr. D. Kuschmitz for stimulating discussion.

[13] Use of Caged Compounds in Studies of Bioelectronic Imaging and Pattern Recognition

By C. W. WHARTON, R. A. MELDRUM, and R. S. CHITTOCK

Introduction

The main application of caged compounds has been to investigate the fast pre-steady-state kinetics of enzyme reactions. The photoactivatable nature of caged compounds opens up the possibility of manipulating the steady-state activity of enzymes using light, creating the potential for a new range of bioelectronics and molecular electronics applications. The bioluminescent enzyme firefly luciferase has been used to study possible methods of enzyme switching using caged compounds. The enzyme reaction can be summarized as[1]:

$$\text{ATP} + O_2 + \text{luciferin} \xrightarrow{\text{luciferase}} \text{AMP} + PP_i + \text{oxyluciferin} + CO_2 + \text{light}$$
$$(h\nu = 562 \text{ nm})$$

The heterocyclic substrate luciferin is converted to oxyluciferin in an excited state that emits the bioluminescent photon as it returns to the ground state.

Photoactivation of an enzyme is clearly possible using a caged substrate molecule, and photodeactivation is possible using a caged enzyme inhibitor.

[1] M. DeLuca and W. D. McElroy, *Methods Enzymol.* **57,** 3 (1987).

In the former case we describe the use of caged ATP to activate firefly luciferase, producing a light-in/light-out system. In the latter case, luciferase bioluminescence has been inhibited by the photolysis of caged acetic acid. Acetate is not itself an inhibitor of luciferase, but because the optimum pH of luciferase (7.8) is well above the pK_a of acetic acid, the net result of the uncaging process is the release of a proton (Fig. 1). If the reaction is carried out in a weakly buffered solution, the resultant change in pH will be sufficient to decrease enzymatic activity. Because the activity of all enzymes displays some pH dependence, this method will be generally applicable to the light-activated control of biological activity.

In order to move toward potential molecular electronics devices, a further development is required to allow spatially resolved switching of luciferase activity by immobilizing the luciferase molecules in an agarose gel containing luciferin and caged ATP. Illumination of the gel through a mask will activate the luciferase by the photolysis of caged ATP only in the illuminated areas to produce a bioluminescent image of the mask. Combination of this method with a lithographically produced active lucifer-

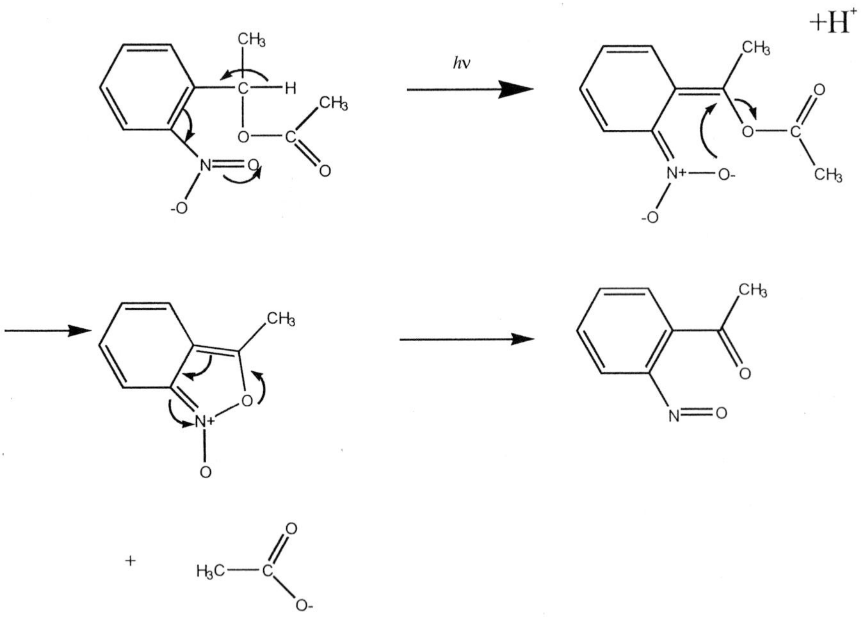

FIG. 1. Mechanism of caged acetic acid photolysis showing the light-induced release of a proton followed by a molecular rearrangement leading to the release of acetate. Physiological pH is much greater than the pK_a of acetic acid, so the overall reaction produces a proton plus acetate, allowing the modulation of enzyme activity by a light-induced pH change.

ase image within the gel has been used to demonstrate a rudimentary pattern correlator.[2] Inclusion of caged acetate in the agarose gel has been used to demonstrate two-wavelength switching by using chemical modification to shift the absorbance spectrum of one of the caging groups, allowing one wavelength to activate luciferase by photolysis of caged ATP, while a second wavelength can produce deactivation by uncaging of caged acetic acid.[3] The absorbance shift was achieved by synthesis of the 3,4-dimethoxy-substituted analog of caged ATP (dm-ATP), which has an absorbance maximum at 360 nm compared to 310 nm for the standard caging group. Luciferin has also been caged and provides a third potential method for light-mediated control of bioluminescence.[4]

This article describes the synthesis of caged acetate and the dimethoxy-substituted analog of caged ATP, methods by which luciferase may be immobilized in an agarose gel, the characterization of changes in pH induced by photolysis of caged acetate, and the switching of luciferase bioluminescence in an agarose gel.

Materials and Methods

Preparative Methods

Caging of molecules with the 1-(2-nitrophenyl)ethyl group (the "standard" caging group) uses nitroacetophenone (Aldrich, Milwaukee, WI) as the starting point for the synthesis. The corresponding dimethoxy compound required to produce the 1-(3,4-dimethoxy-6-nitro)ethyl caged compounds, 3,4-dimethoxy-6-nitroacetophenone, is not available commercially and so is synthesized by the nitration of dimethoxyacetophenone (Aldrich) by nitric acid.[5,6]

Both caging precursors are converted to the corresponding acetophenone hydrazone by the method of Walker et al.[7] as modified by Meldrum et al.[8] and subsequently to the diazoethane by manganese(IV) oxide reduction. The molecule to be caged is added immediately to produce the caged compound.

[2] D. G. Lidzey, R. S. Chittock, N. Berovic, C. W. Wharton, J. B. Jackson, and T. D. Beynon, *Adv. Mater. Opt. Elec.* **4,** 381 (1994).
[3] R. S. Chittock, D. G. Lidzey, N. Berovic, C. W. Wharton, J. B. Jackson, and T. D. Beynon, *Adv. Mater. Opt. Elec.* **4,** 349 (1994).
[4] J. Yang and D. B. Thomason, *Biotechniques* **15,** 848 (1993).
[5] C. A. Fetscher, *Organic Syntheses Collected Volumes* **4,** 735 (1963).
[6] J. H. Kaplan, B. Forbush, and J. F. Hoffman, *Biochemistry* **17,** 1929 (1978).
[7] J. W. Walker, G. P. Reid, J. A. McCray, and D. R. Trentham, *J. Am. Chem. Soc.* **110,** 7170 (1988).
[8] R. A. Meldrum, S. Shall, D. R. Trentham, and C. W. Wharton, *Biochem. J.* **266,** 885 (1990).

Caged ATP molecules are purified by high-performance liquid chromatography (HPLC) using a Technopak 10C18 reversed-phase column (HPLC Technology, Macclesfield, UK). A gradient of 0–40% methanol in 10 mM phosphate buffer gives a final product of greater than 98% purity. Unreacted ATP is eluted from the column after approximately 3 min; caged ATP is eluted after approximately 10 min, representing >99% of nucleotide present in the preparation as estimated from the relative areas of the HPLC peaks.

Caged acetic acid is purified using flash chromatography, using a silica gel column eluted with 80% hexane : 20% ethyl acetate. The product, recovered by rotary evaporation, is a yellow oil that runs as a single spot on thin-layer chromatography (TLC) with 80% hexane : 20% ethyl acetate. The product is dissolved in 1 ml dimethyl sulfoxide (DMSO) and stored in the dark at $-20°$ until required.

The luciferase used in these experiments is purchased from Sigma (St. Louis, MO). Stock enzyme solution is stored in 25 mM glycylglycine, 5 mM $MgCl_2$, 50 mM 2-mercaptoethanol, 0.2% Triton X-100, 50% glycerol, pH 7.8, at a concentration of 1 mg/ml. Under these conditions the enzyme is stable for up to 2 weeks.

Light-Induced pH Changes Using Caged Acetic Acid

In order to demonstrate the effect of uncaging caged acetic acid on solution pH, a 2.4 mM solution of caged acetic acid in 1 mM glycylglycine buffer is illuminated with a 100-W UV lamp, and the pH of the solution is recorded as a function of time using a pH meter (Fig. 2). Prior to UV illumination, the pH of the solution is stable at ambient light levels. On

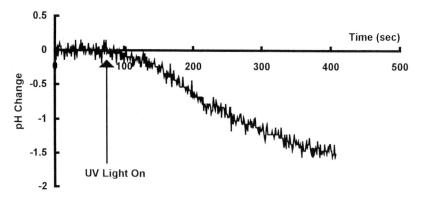

FIG. 2. Light-induced pH changes induced by the photolysis of caged acetic acid. The initial conditions were 2.4 mM caged acetic acid in 1 mM glyclyglycine, pH 7.8. The pH of the solution was measured using a standard pH meter.

UV illumination, a change of approximately 1.5 pH units is produced within 400 sec, sufficient to reduce luciferase activity to 9% of its initial optimum activity at pH 7.8. The slow initial rate of pH change corresponds to the peak buffering capacity of the glycylglycine buffer. At more acidic pH the buffering capacity of the solution becomes less, and so the rate of pH change at constant illumination becomes greater.

Immobilization of Luciferase and Caged Molecules in Agarose Gels

Immobilization of luciferase, a necessary prerequisite for the creation of bioluminescent patterns, is conveniently accomplished by encapsulation of the enzymes, substrates, and caged compounds in an agarose gel. To prevent heat denaturation of luciferase, it is necessary to use an agarose gel with a melting point below 30°, such as Agarose Prep (LKB, Bromma, Sweden). Gels are prepared by dissolving solid agarose in 25 mM glycylglycine, 5 mM magnesium chloride, 50 mM DTT to give a final volume of 1% (w/v) by heating. In preparations using caged acetic acid to produce light-induced pH changes, the concentration of glycylglycine is reduced to 1 mM. While still liquid the solution is transferred to a water bath at 28°, at which temperature the gel remains liquid but the luciferase is not thermally denatured. Luciferase, luciferin, dm-ATP, and caged acetic acid are added.

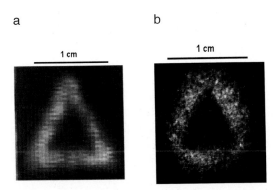

FIG. 3. (a) A triangular pattern written onto an agarose gel by photolysis of caged ATP. Conditions in the gel were 25 mM glycylglycine, 5 mM MgCl$_2$, 50 mM DTT, 1.6 μM luciferase, 40 μM luciferin, 80 μM caged ATP, 1% agarose, pH 7.8. Photolysis was achieved by illuminating the gel with a 100-W UV lamp for 30 sec. (b) A bioluminescent circular pattern written onto agarose gel by photolysis of dm-ATP by illumination through a mask using light of $\lambda > 350$ nm and a dark triangular pattern written onto the same gel by deactivation of luciferase by photolysis of caged acetic acid by illumination through a mask without a filter. The agarose gel consisted of 1 mM glycylglycine, 5 mM MgCl$_2$, 50 mM DTT, 90 nM luciferase, 100 μM luciferin, 2.8 mM dm-ATP, 5.7 mM caged acetic acid, 1% agarose prep, pH 7.8. Illumination time to achieve photolyse was 30 sec for dm-ATP and 120 sec for caged acetic acid.

For uncaging of caged ATP, final concentrations are 1.6 μM luciferase, 40 μM luciferin, and 80 μM caged ATP. For two wavelength experiments, final concentrations are 90 nM luciferase, 100 μM luciferin, 2.8 mM dm-ATP, and 5.7 mM caged acetic acid. This preparation is cast between glass plates separated by 0.75-mm spaces and allowed to set for 1 h at 4°. After removal of the top plate the solid gel is cut into 2- to 4-cm^2 samples for use. The gels may be stored in liquid nitrogen, losing approximately 50% of their bioluminescent activity after 250 hr.

Production of Bioluminescent Images

A collimated 100-W UV source fitted with appropriate filters for wavelength discrimination is used to illuminate the gel-encapsulated luciferase. Typical uncaging illumination times are 30 sec for dm-ATP and 120 sec for caged acetic acid. Images of the bioluminescence are recorded within 2 min of illumination of the gel using an intensified CCD camera (Photek, St. Leonards-on-sea, Sussex, UK) attached to a DL2867-LC frame grabber (Data Translation, Marlboro, MA).

Figure 3a shows a bioluminescent pattern of a triangular mask produced by a 30-sec illumination of a 0.75-mm-thick agarose gel containing 1.6 μM luciferase, 40 μM luciferin, and 80 μM caged ATP. Bioluminescence is observed only in those areas of the gel illuminated through the mask. The pattern decays with a half-life of 10 min at 18°. In Fig. 3a, the side of the triangular pattern is approximately 10 mm in length. The best edge resolution possible under these conditions is estimated to be approximately 0.7 mm using a variable-spaced grating.[2]

Figure 3b shows a bioluminescent pattern produced by illumination of a gel containing both dm-ATP and caged acetic acid. First, a bioluminescent pattern is produced by illumination of the gel through a circular mask with a 350-nm cutoff filter in front of the UV lamp. Under these conditions, dm-ATP is photolyzed, activating luciferase in the illuminated areas but not affecting caged acetic acid. Second, the gel is illuminated through a triangular mask with the cutoff filter removed. Under these conditions, caged acetic acid is photolyzed, lowering the pH of the gel in the illuminated areas and deactivating luciferase activity. The image of the triangular mask appears as a dark area within the bright area of bioluminescence. A control experiment demonstrates that the deactivation of luciferase by the second illumination is not the result of luciferase or luciferin degradation by UV light.[3]

These two experiments are a sample demonstration of the spatial resolved switching of luciferase activity using caged compounds, including a means of wavelength-selective uncaging of different molecules in the same preparation by chemical modification of the caging group.

[14] Use of Caged Nucleotides to Characterize Unstable Intermediates by X-Ray Crystallography

By Axel J. Scheidig, Christoph Burmester, and Roger S. Goody

Introduction

Despite the enormous contributions of X-ray crystallography to our understanding of the structure of biological macromolecules, unraveling the mechanism of the interaction of such a molecule with a ligand or substrate normally involves the indirect process of combining information from static high-resolution structures with dynamic information derived using other methods, including enzyme kinetics and spectroscopy, as well as molecular biological techniques, such as site-directed mutagenesis. The dramatic progress made in increasing the speed of X-ray crystallographic data collection has led to efforts to exploit this speed to obtain structural information on previously inaccessible states that occur as intermediates, particularly in enzymatic reactions. The advance in X-ray diffraction methods has been so profound that the time needed for data collection is no longer the main limitation in such experiments. The biggest remaining hurdle is that of triggering a reaction in a crystal of a macromolecule, and due to the nature of the difficulties involved, it is unlikely that there will be a general solution to this problem. One method that may have a somewhat more general application is the use of caged compounds for generating a ligand such as a substrate or other essential molecule in crystals of the macromolecule in order to start the reaction.

Requirements for Caged Substances Used in Starting Reactions in Crystals

Many potential applications of caged compounds require the release of a substrate for an enzyme in the crystalline form in such a manner that the reaction catalyzed by the enzyme can be started rapidly and efficiently. In principle, two extreme situations may be encountered. In one of these, the caged substance does not interact with the active site of the enzyme and the active site will be occupied only after photolytic release of the substrate. In principle, this situation would also allow monitoring of events associated with the binding of the substrate to the active site. For instance, substrate-induced conformational changes have been identified in many mechanistic studies and are often postulated as being required for catalysis to occur, but they have not, in most cases, been characterized structurally in detail. The second type of situation that might be encountered is that in which the caged ligand binds tightly to the active site of the protein or is covalently bound to the protein or nucleic acid, and the reaction is started by stoichiometric conversion of this molecule or group to its active form. Both types of situation have been encountered in attempts to perform time-resolved protein crystallography, and the photochemical requirements for the caged group must be considered with respect to the properties of the caged molecule as a potential ligand.

Most of this article is devoted to an example of the second type, but some of the potential questions arising with the first situation are discussed briefly. In the extreme situation in which there is no interaction between a caged substrate and a protein in a crystal, one of the main problems is concerned with the nature of the crystal packing and, in particular, whether there are solvent channels that allow easy access to the active site. Even if this is the case, the question has to be considered as to whether a sufficiently high concentration of the caged substance can be reached to allow, potentially, an amount of substrate to be released by photolysis that is sufficient to saturate the enzyme active site, both in terms of absolute amounts and in terms of concentration relative to the dissociation constant for binding to the active site. To appreciate that this is not a trivial point, it must be realized that typical protein concentrations in crystals are in the range of about 5–50 mM. A very high concentration of the caged version of the specific ligand would be needed to ensure reaching this concentration in the crystal, even if the photolysis reaction is highly efficient and the available light intensity is enough to photolyze a significant fraction with, ideally, one short flash.

Situations of this type have been encountered with glycogen phosphorylase using 3,5-dinitrophenyl phosphate (DNPP) as a caged phosphate molecule[1,2] and with isocitrate dehydrogenase using caged isocitrate [1-(2-nitrophenyl)ethyl 1-hydroxy-1,2-dicarboxy-3-propane carboxylate] as a substrate precursor.[3] In the first example, DNPP can be converted to inorganic phosphate and 3,5-dinitrophenol at a rate of about 10^4 sec^{-1} (the rate-limiting step in the dark reaction) by irradiation at 300–360 nm with a quantum yield of 0.67.[4] A crystal of the phosphorylase is soaked in 45 mM DNPP, excess mother liquor is removed, photolysis is performed with a continuous nitrogen laser at 337 nm, and the photolytic reaction is moni-

[1] A. T. Hadfield and J. Hajdu, *J. Appl. Crystallogr.* **26**, 839–842 (1993).
[2] A. Hadfield and J. Hajdu, *J. Mol. Biol.* **236**, 995–1000 (1994).
[3] M. J. Brubaker, D. H. Dyer, B. Stoddard, and D. E. Koshland, Jr., *Biochemistry* **35**, 2854–2864 (1996).
[4] J. E. T. Corrie and D. R. Trentham, *in* "Bioorganic Photochemistry," vol. 2 (H. Morrison, ed.) pp. 203–205. Wiley, New York, 1993.

tored spectrophotometrically. In these crystals, the protein concentration, and therefore the concentration of binding sites for phosphate, is about 7 mM. In a large crystal, most of the photolysis occurs after 1 min in a surface layer due to attenuation of the photolysis beam by absorption. In this situation, the actual rate of release of phosphate is not rate controlling; the rate of diffusion from the front of the crystal into regions that are inadequately illuminated is rate controlling. For slow reactions in the crystal following photolysis, the problem of nonuniform illumination and release can be overcome, or at least reduced, by rotation of the crystal during illumination. In addition, crystals that are significantly smaller than those used in these studies can be used for collecting data, especially using a synchrotron source. Structures are frequently determined using crystals with dimensions of significantly less than 100 μm in the thinnest direction, which considerably reduces the problem of absorption. Moving away from the wavelength for maximum absorption and correspondingly increasing the illuminating light intensity to excite the same number of molecules results in more uniform photolysis.

It should be noted here that the absorption spectrum of the photolysis product arising from the cage group is, in general, different from that of the caging group before photolysis. Depending on the exact nature of the light source for photolysis, this can be of significance. Using a continuous source, there is a build up of the photolysis product during the illumination procedure. If the source has a higher absorption at the wavelength used, it exacerbates the problem of nonuniform photolysis throughout the crystal. With a pulsed source, this problem is less important, although this may not apply to sources with pulses that are long in comparison with the rate of release. There is the additional factor that generation of a significant fraction of the photochemically excited state may lead to similar problems, depending on its spectral properties.

A potential problem arises with systems in which the affinity of the caged compound for the active site is not low enough so that there is no binding at all to the protein being examined, but is not high enough to form a stoichiometric complex. An example of this type is found in the use of caged ATP for the initiation of relaxation or contraction of muscle fibers, followed by monitoring of rapid structural changes by low-angle X-ray diffraction.[5-7] In early experiments of this type, it was assumed that caged ATP does not bind to actomyosin. In a typical experiment, 10 mM caged

---

[5] G. Rapp, K. J. V. Poole, Y. Maeda, K. Güth, and R. S. Goody, *Biophys. J.* **50**, 993–997 (1986).

[6] K. J. V. Poole, G. Rapp, Y. Maeda, and R. S. Goody, *in* "Topics in Current Chemistry" (E. Mandelkow, ed.), pp. 1–29, Springer-Verlag, Berlin, Heidelberg, Germany, 147 (1988).

[7] G. Rapp, K. J. V. Poole, Y. Maeda, G. C. R. Ellis-Davies, J. H. Kaplan, J. McCray, and R. S. Goody, *Berichte der Bunsen Gesellschaft* **93**, 410–415 (1989).

ATP is diffused into muscle fibers and approximately 2 mM ATP is released by a pulse from a flash lamp. The results are interpreted in terms of association of ATP release at a known rate free in solution with the active site of myosin. Caged ATP binds to actomyosin with a K_d of about 1 mM.[8] Thus, before photolysis, about 90% of the ATP-binding sites are occupied by caged ATP at the concentration used.

Assuming that the quantum yield for photolysis of caged ATP at the myosin active site is the same as that for caged ATP in solution, ATP is generated directly in 18% of the sites, whereas the remaining 82% recruit ATP from solution. This second group of active sites is, however, not free for direct association of ATP because this group is largely saturated by caged ATP. Depending on the magnitude of the rate constant of dissociation of caged ATP (which is not known), it is rate limiting for the association of ATP, or, if it is very rapid, caged ATP effectively buffers the rate of ATP association. In both cases, this slows the rate of ATP binding, which presents a potential problem, depending on the aim of the experiment. It should be noted that these problems apply not only to time-resolved X-ray diffraction studies, but to other studies using caged nucleotides as the starting point for investigations on muscle fibers or other actomyosin-containing systems, including time-resolved electron microscopy, spectroscopy, and kinetic and mechanical measurements.

The situation becomes much clearer if the caged compound binds with high affinity to the protein in the crystal. However, the presence of a stoichiometric amount of the cage group with respect to the protein also introduces new problems. The general considerations concerning uniformity of release of the substance of interest throughout the crystal still apply, particularly with respect to attenuation of the illuminating light due to absorption by the cage group, its products, or photolysis intermediates. The main complication is that if the reaction of interest is fast enough to require that the released ligand, substrate, or group be generated with a single light pulse, theoretically a quantum yield of 1 and enough light to excite all molecules of the cage group present are required to obtain complete conversion to the first complex of interest. In practice, often only an approximation to this ideal situation is possible. For reactions slow enough to allow use of a flash lamp with a duration of about 1 msec, this might be achievable in some cases. This arises from the fact that with relatively long light pulses, there is the possibility of multiple excitations, depending on the exact kinetics and mechanism of photolysis. When the time requirements are not so severe, and a series of pulses or continuous radiation can

---

[8] T. Thirlwell, J. A. Sleep, and M. A. Ferenczi, *J. Muscle Res. Cell Motil.* **16,** 131–137 (1995).

be used for photolysis, these problems become less significant, and this applies to the case discussed in detail later.

A similar situation arises in the case of caged enzymes. For chymotrypsin, caging of the enzyme has been achieved by esterifying the serine group that is involved in the reaction mechanism.[9,10] The *trans*-p-diethylamino-*o*-hydroxymethyl cinnamate group was used for this purpose, because on light-induced isomerization of this derivative, deacylation occurs. This would then allow a reaction to start in the crystal, although there might still be the problem of slow diffusion out of and into the active site. In the work reported on this system, the cinnamate group was photolyzed in the presence of a suicide substrate, which then reacts over a period of several hours with the enzyme.[11] Because this reaction is also very slow in solution, the experiments do not allow conclusions regarding the applicability of the method to more rapid reactions.

Data Collection Strategies

The enormous increase in X-ray intensity available from synchrotron sources which has been achieved in the last two decades allows diffraction data to be collected orders of magnitude more rapidly than using conventional sources. However, if monochromatic X-rays are used, there is still the problem that a large number of data frames must be collected at orientations of the crystal differing with respect to the beam in order to sample a major part of the reciprocal lattice. Although individual frames can be collected in milliseconds or less, the technical aspects of changing the crystal orientation and data readout can decrease the achievable temporal resolution by orders of magnitude. One potential solution to this problem is the use of Laue diffraction conditions, i.e., the use of polychromatic X-rays to allow collection of a large fraction of the diffraction data in one or a few exposures. This approach has been used in several applications, and the advantages and disadvantages are summarized briefly here. The main advantage is, of course, speed of data collection, although depending on the space group of the crystal, there will generally be a need to obtain more than one exposure at varying orientations. This often leads to the situation in which structures are calculated with significantly less than an optimal fraction of the theoretical data. Added to this is the fact that in Laue

---

[9] B. L. Stoddard, J. Bruhnke, N. A. Porter, D. Ringe, and G. A. Petsko, *Biochemistry* **29**, 4871–4879 (1990).

[10] B. L. Stoddard, P. Koenigs, N. Porter, D. Ringe, and G. A. Petsko, *Biochemistry* **29**, 8042–8051 (1990).

[11] B. L. Stoddard, P. Koenigs, N. Porter, K. Petratos, G. A. Petsko, and D. Ringe, *Proc. Natl. Acad. Sci. U.S.A.* **88**, 5503–5507 (1991).

data sets, low to medium resolution data are inherently underrepresented. Crystal quality is also more critical than for monochromatic data collection. Thus, in one example, about 100 crystals that were suitable for monochromatic diffraction, only a few were usable for Laue diffraction.[12] Despite these limitations, the Laue method in combination with flash photolysis has been used to obtain millisecond[13] and even nanosecond[14] resolution with inherent (built-in) photosensitive groups.

Recent work, some of which is described below, suggests that for relatively slow reactions (which are, however, still too fast to allow triggering by more conventional methods) it may be possible to adopt a different approach, which is to photolyze and then freeze crystals rapidly to a temperature at which the reaction of interest is so slow that monochromatic diffraction data can be collected. A variation on this theme is to perform the photolysis at a temperature at which the light-induced reaction can occur, but subsequent reactions are slow. An example of this is the photodissociation of carbon monoxide from myoglobin at temperatures of 20[15] or 40 K.[16] Under these conditions, relatively gentle constant illumination generates the first dissociated state of carbon monoxide and allows structural characterization before rebinding or subsequent steps in dissociation can occur.

Monitoring of Reactions in Crystals

In many applications involving the use of caged compounds for the initiation of reactions in crystals, knowledge of the degree of photolysis and of the extent of the reaction of interest will be needed. In most cases, this has been achieved after completion of the experiment by analysis of the crystal, for example, by high-performance liquid chromatography (HPLC) to detect photolysis products and where appropriate the products of an enzymatic reaction. Spectroscopic methods, mainly absorption spectroscopy, have been applied to protein crystals and have been used to monitor reactions in crystals.[2,17] For many applications of caged compounds,

---

[12] I. Schlichting, S. C. Almo, G. Rapp, K. Wilson, K. Petratos, A. Lentfer, A. Wittinghofer, W. Kabsch, E. F. Pai, G. A. Petsko, and R. S. Goody, *Nature* **345**, 309–315 (1990).

[13] U. K. Genick, G. E. O. Borgstahl, K. Ng, Z. Ren, C. Pradervand, P. M. Burke, V. Srajer, T. Y. Teng, W. Schildkamp, D. E. McRee, K. Moffat, and E. D. Getzoff, *Science* **275**, 1471–1475 (1997).

[14] V. Srajer, T. Y. Teng, T. Ursby, C. Pradervand, Z. Ren, S. Adachi, W. Schildkamp, D. Bourgeois, M. Wulff, and K. Moffat, *Science* **274**, 1726–1729 (1996).

[15] I. Schlichting, J. Berendzen, G. N. Phillips, Jr., and R. M. Sweet, *Nature* **371**, 808–812 (1994).

[16] T.-Y. Teng, V. Šrajer, and K. Moffat, *Nature Struct. Biol.* **1**, 702–705 (1994).

[17] Y. Chen, V. Srajer, K. Ng, A. LeGrand, and K. Moffat, *Rev. Sci. Instrum.* **65**, 1506–1511 (1994).

the ability to monitor photolysis and enzymatic reactions in real time during an experiment will allow better decisions to be made with respect to time windows for data collection and will also contribute to data analysis by removing uncertainty concerning the exact composition of the crystals at the time of diffraction data collection. Absorption (discussed briefly earlier for the case of glycogen phosphorylase) and fluorescence spectroscopy are two methods likely to be useful, but other approaches, such as Fourier transform infrared (FTIR) spectroscopy, could be of great value in certain systems.

Specific Example: Time-Resolved Crystallography on H-Ras GTPase

One of the first examples of the use of caged compounds for time-resolved crystallography invovled the small GTPase H-Ras, a protein that plays a key role in signal transduction in the cell. Because the most important property of this protein is a conformational change that occurs on GTP hydrolysis, it was of interest to characterize this change in crystals of the protein. There was also an expectation that the results could provide some evidence on the detailed chemical mechanism of the GTPase reaction. This is not only of importance at the fundamental level of understanding the hydrolysis mechanism, but also for understanding the mechanism of oncogenic transformation of this protooncogene because many transforming mutations result in the loss of ability to hydrolyze GTP.

Early experiments showed that the H-Ras protein, which has a very high affinity to GDP and GTP, also binds caged GTP [we refer to GTP bearing the 1-(2-nitrophenyl)ethyl group on the γ-phosphate as caged GTP] with high affinity (approximately $10^{10}\ M^{-1}$) so that it is relatively easy to prepare a 1:1 complex between the protein and the nucleotide.[18] As can be seen in Fig. 1, the presence of a chiral center in the cage group means that caged GTP exists as a mixture of two diastereomers, and the first experiments were performed with this mixture. It was shown by HPLC that the crystals contain an approximately equimolar mixture of the two diastereomers. Photolysis of the caged nucleotide could be achieved in the crystal using a xenon flash lamp.[18a] High-performance liquid chromatography analysis showed that, after photolysis, hydrolysis of GTP to GDP took place in the crystal at a rate constant that was similar to that seen in solution. As long as sufficient amounts of reducing agents [normally dithiothreitol (DTT)] were present to react with and inactivate the nitrosoacetophenone

---

[18] I. Schlichting, J. John, G. Rapp, A. Wittinghofer, E. F. Pai, and R. S. Goody, *Proc. Natl. Acad. Sci. U.S.A.* **86,** 7687–7690 (1989).

[18a] G. Rapp, *Methods Enzymol.* **291** [11] (1998) (this volume).

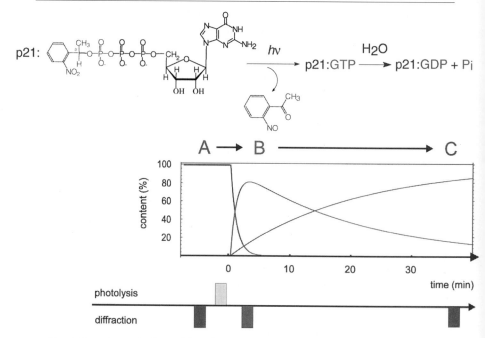

Fig. 1. Experimental scheme. Flow diagram for the generation of p21:GTP and p21:GDP complexes starting with p21:caged GTP. After photolysis the cage group 2-nitrosoacetophenone is released, and with a half-time of approximately 30 min (at room temperature) the generated GTP will be hydrolyzed to GDP and inorganic phosphate. Diffraction data were collected before photolysis, 2 min after photolysis, and after completion of the hydrolysis reaction.

photolysis product, the crystals survived both the photolysis and the hydrolysis reaction without macroscopic signs of damage. It could be confirmed by X-ray diffraction that the combined processes of photolysis and hydrolysis occurred without loss of crystal quality as long as several precautions were taken to prevent heating and damage due to light in the far-UV range. One of these was the use of a filter that allowed only light between 300 and 400 nm to illuminate the crystal. Another was to keep the crystal bathed in mother liquor to behave as a heat sink.[18a]

Our experiments with H-ras p21 showed that the photolysis yield in dilute solutions is much higher than in the crystallized system. Multiple factors contribute to this. Some of these are differences in surface reflection (flat cuvette versus curvature of capillary and shape of crystal); absorption effects of the capillary and crystallization mother liquor (the quartz material of UV/VIS cuvettes has a different absorption profile than quartz capillaries used for mounting crystals); and attenuation of intensity throughout the crystal arising from absorption due to the higher concentration of chromophore in the crystal. It is therefore not trivial to extrapolate from the solution results for complete photolysis to the situation in the crystal making use of the parameters of total effective energy per pulse, and the length and number of pulses. We have therefore optimized conditions for photolysis working only with crystals and analyzing the efficiency of photolysis, the activity of the crystallized protein, and the diffraction quality of the crystal. The actual procedure for photolysis in the earlier X-ray diffraction experiments was to use about 10 pulses of the xenon lamp separated by several seconds.

Because the half-life of the GTPase reaction of H-Ras is of the order of 30 min at room temperature, diffraction data must be collected within a few minutes of photolysis to characterize the GTP state. Clearly, this cannot be done by conventional methods, and data were collected at a synchrotron source using the white X-ray beam (actual range of wavelengths 0.6 to 2.0 Å). The first data set was collected at about 4 min after photolysis. This data set was used to calculate an electron density map at about 2.8 Å.[12] It showed that the overall structure of the protein was very similar to that which had previously been determined for the complex between H-Ras and GppNHp, a nonhydrolyzable GTP analog. The binding of GTP appeared to be identical at this resolution to that of GppNHp. In further experiments, crystals were aged after photolysis in order to allow hydrolysis to GDP to occur, and the structure of the resulting complex was determined by conventional monochromatic diffraction methods. This led to a detailed description of the structural changes accompanying GTP hydrolysis.[12]

Despite the success of these experiments, not all questions of interest could be answered. This arises mainly from the relatively low resolution of the structures determined and the less than optimal quality of the electron density maps. There are two main reasons for this. One is the limitation imposed by the Laue diffraction geometry and the general low completeness of the data sets obtained. The other is the intrinsically low resolution of the crystals used, which was not considerably better than 2.8 Å even under monochromatic conditions. It was suspected that the presence of the mixture of diastereomers of caged GTP in crystals used in the earlier work was a limiting factor in crystal quality, and this was strengthened by the results of structural determination of the H-Ras–caged GTP complex with the mixed isomers, which suggested an anomalous mode of binding.[12] In an attempt to overcome this problem, other cage groups were examined. The most easily available is the 2-nitrobenzyl group, which is not chiral. Its inferior photolysis properties (low quantum yield, relatively slow dark reaction) would probably be acceptable for slow GTPases if the crystallization properties of its complex with the protein were better than those of the classical cage group. Surprisingly, the best crystals that could be obtained diffracted only to 3.6 Å,[19] compared with 2.8 Å for the mixed isomers of caged GTP. A further group that was examined in an attempt to solve this problem is the bis(2-nitrophenyl)methyl group.[19] Like the 2-nitrobenzyl group, this group does not have a chiral center, but its photolysis properties are more similar to those of the classical cage group. Again, crystals were obtained, but the best resolution obtained was 2.6 Å, which was not a significant improvement over the crystals with mixed isomers of caged GTP. A solution was provided by the preparation of the pure diastereomers of caged GTP.[20] It was found that complexes of H-Ras with the separated diastereomers did indeed lead to better crystal quality than the mixed isomers.[21] The complex of H-Ras p21 with the *R*-isomer of caged GTP diffracted to 1.85 Å and that with the *S*-isomer to 2.2 Å.[21] An interesting but puzzling observation is that the *S*-isomer is degraded at a significant rate to GDP and caged phosphate in the crystals, whereas the *R*-isomer is much more stable. Determination of the structure at 1.85 Å showed that the caged nucleotide was bound normally at the active site of the protein, but that the cage group forces two loops to adopt new conformations.[21]

Crystals of the complex with the pure *R*- or *S*-isomers were used for time-resolved structural studies in a manner similar to that used earlier, with the modification that the crystal was rotated between flashes to counteract the problem of nonuniform release of GTP.[19] A further modification to the earlier experiments was that a mutant of H-Ras (G12P) was used. This mutant has similar GTPase properties to wild-type protein but a higher thermal stability. The wavelength range used in these experiments was 0.45–2.6 Å. Data sets were collected at 2–4, 11–13, 20–22, 30–32, and 90–92 min after photolysis. Despite the significant improvement in crystal quality under monochromatic conditions, under Laue conditions the resolution was limited to 2.8 Å. Despite this, the earlier interpretation could be extended. In particular, a conformation of the side chain of Gln-61 was identified that had not been previously detected. This residue is known to be directly involved in the mechanism of GTP hydrolysis. Importantly, it could be shown that the electron density representing the γ-phosphate group disappears smoothly during the reaction, confirming that a protein–product complex with a long-lived inorganic phosphate does not accumulate to a

[19] A. Scheidig, A. Sanchez-Llorente, A. Lautwein, E. F. Pai, J. E. T. Corrie, G. Reid, A. Wittinghofer, and R. S. Goody, *Acta Crystallographica D,* **D50,** 512–520 (1994).
[20] J. E. T. Corrie, G. P. Reid, D. R. Trentham, M. B. Hursthouse, and M. A. Mazid, *J. Chem. Soc. Perkin Trans.* **1,** 1015–1019 (1992).
[21] A. J. Scheidig, S. M. Franken, J. E. T. Corrie, G. P. Reid, A. Wittinghofer, E. F. Pai, and R. S. Goody, *J. Mol. Biol.* **253,** 132–150 (1995).

significant extent. Regions undergoing structural changes between the GTP and the GDP states displayed electron density that at this level of resolution and data quality could not be interpreted in terms of individual conformations present as a mixture. In addition, the relatively low resolution prevented an interpretation of the changes in positions and density of important water molecules during the GTPase reaction.

Because of the shortcomings of these experiments, other approaches have been sought to exploit the full potential of the basic principle. In most recent work, photolysis of caged GTP at the active site of H-Ras has been combined with rapid freezing of the crystals to boiling nitrogen temperature in order to stop the GTPase reaction at controlled time points. This approach is possible because of the relatively slow GTPase reaction, but it can be applied to considerably faster reactions, because the rate of cooling in a stream of nitrogen is likely to be high enough to effectively stop reactions that occur on the second time scale. In the case of H-Ras, as with many other protein crystals, the freezing procedure can be performed without the addition of cryoprotectants because the mother liquors contain about 30% polyethylene glycol (PEG) 400 and does not form ice crystals on freezing. Several different strategies for photolysis were examined before arriving at an optimal procedure. As a light source, both flash lamps and continuous irradiation from a mercury lamp were used in combination with filters, as previously described, or with a monochromator. The following procedure leads to the best results. Relatively small crystals (less than 150 μm in the largest direction, already mounted in a loop containing mother liquor) are illuminated continuously with light of wavelength 313 nm for 5 min. The procedure is performed at 2° to slow the rate of GTP hydrolysis and to reduce the negative effects of heating. The crystal is rotated continuously during illumination. After 5 min, the crystal is allowed to settle for 2 min and then is rapidly frozen in a stream of evaporating nitrogen and kept at this temperature (100 K) for the collection of diffraction data using monochromatic radiation from a rotating anode source using a multiwire detector.

Each of the factors just mentioned has been tested individually for its effect on the efficiency of GTP release and its effect on preservation of crsytal quality. The amount of GTP released was estimated after dissolving the crystal in water by HPLC.[18] It was found that although the total time required for photolysis was longer using monochromatic light, there was an advantage in terms of crystal quality. The low temperature during photolysis was also found to have a positive influence. Using the previously described protocol, frozen crystals of the H-Ras GTP state that diffracted to 1.6 Å could be obtained routinely. Surprisingly, in many cases there was an improvement in quality on photolysis of about 0.1–0.2 Å.

An important factor in the progress made is the small size of the crystals used. This is possible because data collection at liquid nitrogen temperature has the advantage that the crystals become very stable to radiation and data can therefore be collected over a long period. The small crystal size makes photolysis more efficient. We have also noted that small crystals survive the photolysis procedure more readily than large crystals. The combination of all the factors mentioned has led to an improvement in data quality from 2.8 Å resolution (<60% completeness) using the flash lamp and the Laue method at room temperature to 1.6 Å resolution (>90% completeness) using gentle monochromatic photolysis and monochromatic diffraction in combination with the cryotechnique. Data from several GTP experiments are presently being refined and evaluated. An example of the electron density is shown in Fig. 2.

In principle, the protocol described can be used to generate the GDP state from the GTP state. However, it is technically difficult to mount the crystal in a loop, photolyze, and then wait for a period of many hours for hydrolysis to take place, mainly because of the problem of drying. We have therefore photolyzed crystals in a batch process in a drop of mother liquor and allowed the crystals to age for about six half-lives of the GTPase reaction. This results in a loss of resolution to about 2 Å, with the best crystals so far diffracting to 1.8–1.9 Å. Further improvement of the protocol will probably be needed to obtain crystals of the quality of the GTP complex.

Conclusion

The few documented examples of the use of caged substances for initiating reactions in crystals allow tantalizing glimpses of the potential of the method but also demonstrate the substantial difficulties associated with application of the technique to answering meaningful biological questions. It is sobering to realize that the nearest approximation to this ideal has been achieved in a system that offers the advantage of a relatively long half-life of the enzyme–substrate complex (H-Ras–GTP) and that the logical extension of this work to the biologically more interesting complex between H-Ras–GTP and a GTPase activating protein molecule will require substantial technical advances.

The reasons for the relatively slow progress and limited number of applications should be clear from the discussion presented here. Comparing the caged substance approach to that of choosing systems that have built in triggers, it is clear that the technical problems involved are significantly more difficult to solve in the examples described here. In addition to the advantage that the stoichiometry of photoactivatable groups with respect to the protein is normally defined by the nature of the system when the

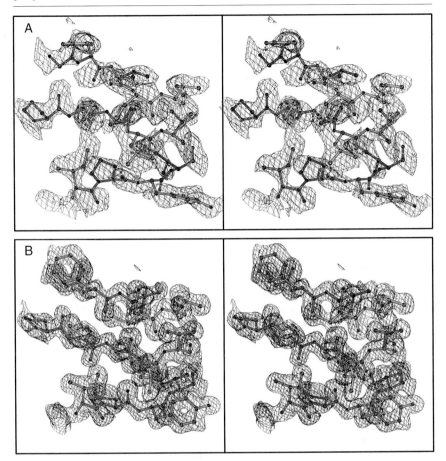

Fig. 2. Stereo view of a representative part of the electron density (shown at a 2σ cutoff level) of a β-sheet core region in H-ras p21:GTP (residues 77–81, 110–114, and 139–142). For this part of H-ras p21, no significant conformational changes on GTP hydrolysis are found. Therefore, the quality of the map can be related to the quality of the data set. (A) Data set collected using the Laue method; wavelength range: 0.45–2.6 Å; maximum resolution 2.8 Å, completeness: 60% (2.8–50 Å) and 20% (4.5–10 Å).[19] (B) Data set collected using monochromatic (λ = 1.54 Å) diffraction in combination with freeze trapping of the intermediate; maximum resolution 1.55 Å, completeness: 95% (1.55–50 Å) and 66% (1.55–1.7 Å).

trigger is an intrinsic component, there is also usually the advantage of reversibility. This offers the possibility of repetition of experiments in the same crystal or generation of pseudostable states, e.g., by continuous illumination. With caged substances, the photolysis reaction is generally irreversible, as is the subsequent reaction occurring in the protein or other macromolecule, thus excluding repetition. For the generation of steady states in a crystal, flow cells relying on diffusion will be simpler and preferable to caged compounds.

Despite these difficulties, it is important to continue to develop technology that will allow the caged approach to be applied to other systems because this will often be the only way of tackling the problem of structural characterization of unstable reaction intermediates. Because it will not always be possible to release enough of the active form of the substance uniformly throughout the crystal at a rate fast enough to allow structural data to be obtained in an appropriate time window, compromises will often have to be made. The combination of flash photolysis with cryocrystallographic methods is a promising development, and one example of this is described in this article. Further developments will include the use of low temperatures to slow the reaction of interest to the point where photolysis can be accomplished by less drastic procedures than single high-intensity pulses. The use of mutants with slower reaction kinetics can also bring the time scale into a more comfortable range, as can the use of modified substrates. However, in both cases evidence that the mechanism is not changed compared with the reaction of wild-type protein and the natural substrate or ligand will have to be obtained.

In summary, it seems likely that each specific application of caged substances to time-resolved crystallography will require extensive characterization of the system involved, its interaction with the caged substances used (unless the protein itself is caged), and the events occurring in the crystal on and after hydrolysis. It is clear that this is not achievable for all macromolecules of interest, but when the questions are of sufficient importance and are appropriately configured for such an approach, there is the potential reward of information at a mechanistic level that is not available from any other single method.

[15] Photoregulation of Cholinesterase Activities with Caged Cholinergic Ligands

By LING PENG *and* MAURICE GOELDNER

Introduction

Acetylcholinesterase (AChE) and butyrylcholinesterase (BuChE) both rapidly hydrolyze the neurotransmitter acetylcholine.[1] The principal role of AChE is to terminate the neurotransmission at the cholinergic synapses by rapid hydrolysis of acetylcholine, with a turnover number approaching 20,000 sec^{-1}.[2] Although BuChE plays an important role in breaking down the muscle relaxant succinylcholine,[3] its physiologic function remains unknown. This enzyme has been so named because it hydrolyzes butyrylcholine faster than actylcholine.[4] The description of the three-dimensional (3D) structure of AChE[5] as well as of several AChE–inhibitor complexes,[6–9] and of a model of BuChE,[10] based on the 3D structure of AChE, has permitted a better understanding of structure–function relationships in the cholinesterases. It has, however, also raised questions concerning the traffic of substrate and products to and from the active site in view of the high turnover rate. Time-resolved Laue crystallography[11] would present an ideal approach to investigate this issue by analyzing, at atomic level and in real time, the precise structural changes occurring at the enzyme active site during catalysis. Different from conventional X-ray diffraction techniques, Laue X-ray crystallography using high-intensity synchrotron polychromatic

---

[1] A. Chatonnet and O. Lockridge, *Biochem. J.* **260,** 625 (1989).
[2] D. M. Quinn, *Chem. Rev.* **87,** 955 (1987).
[3] F. Hobbiger and A. W. Peck, *Br. J. Pharmacol.* **37,** 258 (1969).
[4] K.-B Augustinsson, *Methods Biochem. Anal. Suppl.* 217 (1971).
[5] J. L. Sussman, M. Harel, F. Frolow, C. Oefner, A. Goldman, L. Toker, and I. Silman, *Science* **253,** 872 (1991).
[6] M. Harel, I. Schalk, L. Ehret-Sabatier, F. Bouet, M. Goeldner, C. Hirth, P. H. Axelsen, I. Silman, and J. L. Sussman, *Proc. Natl. Acad. Sci. U.S.A.* **90,** 9031 (1993).
[7] M. Harel, G. J. Kleywegt, R. B. G. Ravelli, I. Silman, and J. L. Sussman, *Structure* **3,** 1355 (1995).
[8] M. Harel, D. M. Quinn, H. K. Nair, I. Silman, and J. L. Sussman, *J. Am. Chem. Soc.* **118,** 2340 (1996).
[9] Y. Bourne, P. Taylor, and P. Marchot, *Cell* **83,** 503 (1995).
[10] M. Harel, J. L. Sussman, E. Krejci, S. Bon, P. Chanal, J. Massoulié, and I. Silman, *Proc. Natl. Acad. Sci. U.S.A.* **89,** 10827 (1992).
[11] D. W. J. Cruickshank, J. R. Helliwell, and L. N. Johnson, "Time-Resolved Macromolecular Crystallography," Oxford University Press, Oxford, England, 1992.

X-ray sources has reduced the exposure time to a microsecond or even nanosecond time scale.[12] Such studies, however, require a fast and efficient triggering of the dynamic reaction within the enzyme crystal, and, clearly, a temporally and spatially controlled photoregulation of enzyme activity represents the best adapted methodology.

Two photoregulation approaches have been developed.[13] The first method implies the "caging" of the protein and involves the blockade of the enzyme activity by modifying a crucial amino acid residue with a photolabile group allowing the regeneration of the enzyme activity after photolysis. The caging of the protein can be achieved either by chemical modification of the target amino acid residue in the enzyme or by protein biosynthesis using unnatural amino acids, namely the caged amino acid derivatives.[14] Although the synthetic pitfalls encountered during chemical synthesis may be overcome by the design of enzyme mutants possessing the appropriate chemical reactivity to permit an efficient and selective modification of the mutated amino acid residue (see H. Bayley, *et al.*, this volume[15]), the biosynthesis of protein might represent a very interesting but challenging alternative. The second photoregulation approach uses caged ligands that can act as reversible inhibitors of the enzyme and then photorelease the desired enzyme substrate or product at the active site, triggering the enzymatic reaction process under temporal and spatial control. This article deals with the second caging method and addresses the question of the photoregulation of ChE activities by means of caged cholinergic ligands and of their potential use in studies on ChE catalysis by time-resolved crystallography.

Caged Cholinergic Ligands

Scheme 1 shows the different types of caged cholinergic ligands that have been synthesized. These are 2-nitrobenzyl derivatives[16] of choline **1–3a**[17] (type A), carbamylcholine **1–3b**[18a,b] (type B), and noracetylcholine

---

[12] J. Hajdu and I. Andersson, *Annu. Rev. Biophys. Biomol. Struct.* **22**, 467 (1993).

[13] N. A. Porter, J. D. Bruhnke, and P. Koenigs, in "Bioorganic Photochemistry, vol. 2, Biological Applications of Photochemical Switches" (H. Morrison, ed.) p. 232, Wiley, New York, 1993.

[14] D. Mendel, J. A. Ellman, and P. G. Schultz, *J. Am. Chem. Soc.* **113**, 2758 (1991).

[15] H. Bayley, C.-Y. Chang, W. T. Miller, B. Niblack, and P. Pan, *Methods Enzymol.* **291** [7] (1998) (this volume).

[16] For photolabile protecting groups, see elsewhere in this volume.

[17] L. Peng and M. Goeldner, *J. Org. Chem.* **61**, 185 (1996).

[18a] J. W. Walker, J. A. McCray, and G. P. Hess, *Biochemistry* **25**, 1799 (1986).

[18b] T. Milburn, N. Matsubara, A. P. Billington, J. B. Udgaonkar, J. W. Walker, B. K. Carpenter, W. W. Webb, J. Marque, W. Denk, J. A. McCray, and G. P. Hess, *Biochemistry* **28**, 49 (1989)

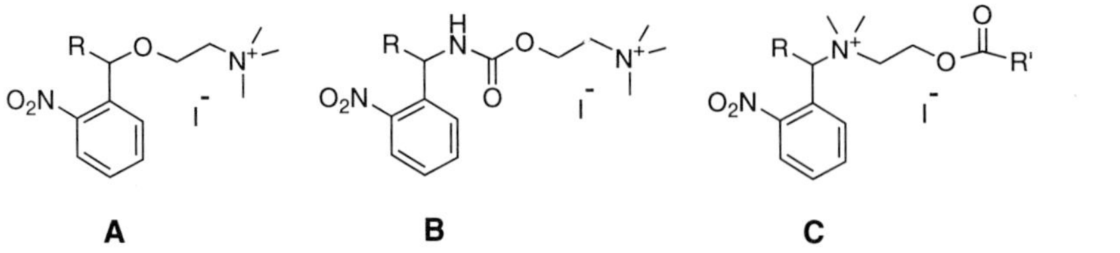

SCHEME 1. 2-Nitrobenzyl derivatives of choline, carbamylcholine, noracetylcholine, and norbutyrylcholine as caged cholinergic ligands.

A
1a: R = CH$_3$
2a: R = H
3a: R = CH$_2$NCS

B
1b: R = CH$_3$
2b: R = H
3b: R = COOH

C
1c: R = CH$_3$, R' = CH$_3$
2c: R = H, R' = CH$_3$
3c: R = CH$_3$, R' = CH$_2$CH$_2$CH$_3$
4c: R = H, R' = CH$_2$CH$_2$CH$_3$

1–2c[19] or norbutyrylcholine 3–4c[20] (type C). The different substitutions at the α-benzylic position were considered to improve the photochemical properties (i.e., rapid photofragmentation kinetics and high quantum yield) as well as to offer chemical variability (e.g., 3a was designed as a potential irreversible inhibitor by covalent interaction between its isothiocyanate group and the catalytical serine residue of ChE).

Hydrolysis of acetylcholine by ChEs (Scheme 2) involves at first a rapid acetylation of the catalytic serine with a concomitant release of choline, then an immediate deacetylation of the enzyme with an expulsion of the acetate.[2] As choline is the enzymatic product of ChEs, caged choline derivatives 1–3a were designed to be complexed at the ChE active site and to generate choline by subsequent photoactivation, allowing a study of the rapid clearance of choline from the active site (Scheme 3).

Carbamylcholine serves as a slow substrate for AChE: Rapid carbamylation with release of choline is followed by much slower decarbamylation (Scheme 2), which can be accelerated by dilution.[21,22] Thus, caged carbamylcholine 1–3b can offer the possibility to study the mechanism of carbamylation of AChE as well as the enzymatic release of choline (Scheme 3). However, the caged carbamylcholines 1–3b are not suitable for studies on BuChE because carbamylcholine has a very low affinity for BuChE, and carbamylation of BuChE is a very slow process that requires, in addition, a high concentration of carbamylcholine.[22]

Noracetylcholine is a close analog of acetylcholine, the endogenous substrate of AChE. Hydrolysis of noracetylcholine is chemically identical

[19] L. Peng, J. Wirz, and M. Goeldner, *Angew. Chem. Int. Ed. Engl.* **36**, 398 (1997).
[20] L. Peng, J. Wirz, and M. Goeldner, *Tetrahedron Lett.* **38**, 2961 (1997).
[21] I. B. Wilson, M. A. Hatch, and S. Ginsburg, *J. Biol. Chem.* **235**, 2312 (1960); I. B. Wilson, M. A. Harrison, and S. Ginsburg, *J. Biol. Chem.* **236**, 1498 (1961).
[22] L. Peng, I. Silman, J. Sussman, and M. Goeldner, *Biochemistry* **35**, 10854 (1996).

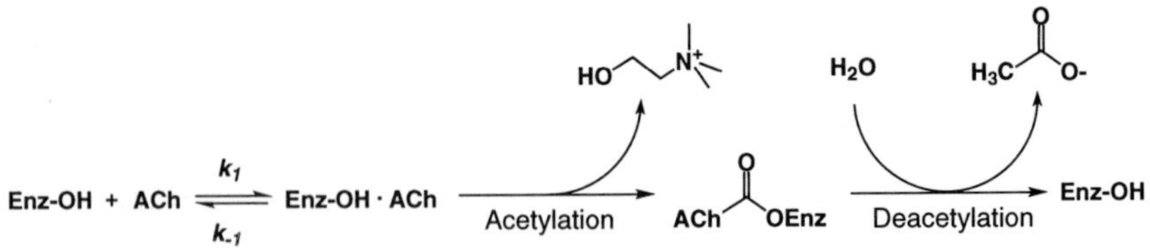

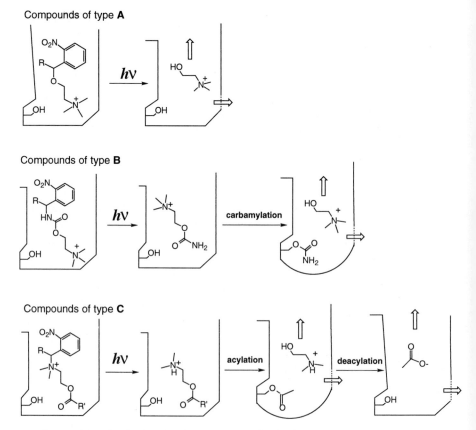

SCHEME 2. Mechanism of hydrolysis of acetylcholine (ACh) by cholinesterases.

SCHEME 3. Mode of action of different caged cholinergic ligands for controlling the catalytic reactions at different steps.

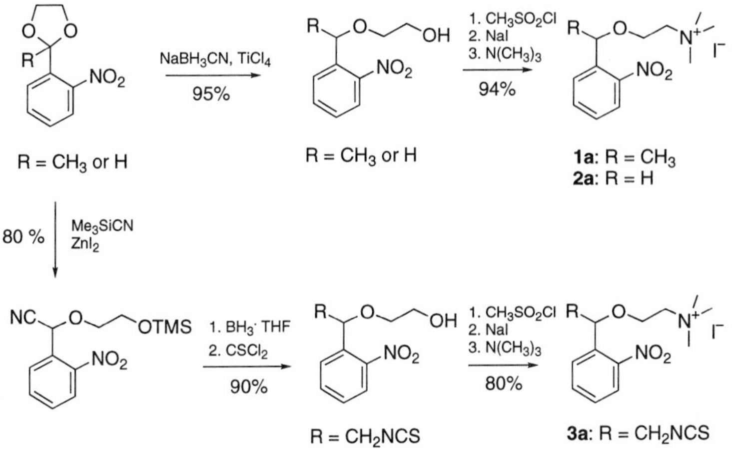

SCHEME 4. Synthesis of 2-nitrobenzyl derivatives of choline **1–3a**.

to that of acetylcholine.[23] After formation of the binary enzyme–substrate complex, successive acylation and deacylation steps follow, as illlustrated in Schemes 2 and 3. The advantage of caged noracetylcholine **1–2c** over the previously mentioned compounds **1–3a** and **1–3b** is that they offer the possibility of studying the mechanism of the entire catalytic process shown in Scheme 3. Following the same strategy, caged norbutyrylcholines **3–4c** were designed for the studies on BuChE.

Syntheses

The synthetic routes to compounds **1–3a** and **1–4c** are summarized in Schemes 4 and 5, respectively; the caged carbamylcholines **1–3b**[18a,b] were commercially available.

The key step in the synthesis of caged choline **1–3a** (Scheme 3) was a Lewis acid-catalyzed reductive opening of a cyclic acetal or ketal using sodium cyanoborohydride.[17] The syntheses of caged noracetylcholine **1–2c**[19] or caged norbutyrylcholine **3–4c**[20] (Scheme 4) started with a reductive amination reaction on 2-nitrobenzaldehyde or 2-nitroacetophenone, and the final products were obtained as described, in >80% overall yields.

Photochemical Properties

The application of caged compounds for the investigation of rapid kinetic processes depends critically on the ability of the photolysis reaction

---

[23] H. K. Nair, J. Seravalli, T. Arbuckle, and D. M. Quinn, *Biochemistry* **33**, 8566 (1994); I. B. Wilson and E. Cabib, *J. Am. Chem. Soc.* **78**, 202 (1956).

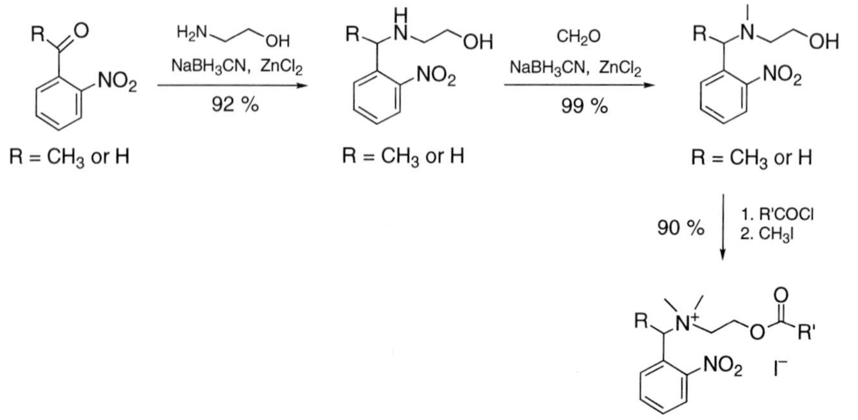

1c: R = CH_3, R' = CH_3
2c: R = H, R' = CH_3
3c: R = CH_3, R' = $CH_2CH_2CH_3$
4c: R = H, R' = $CH_2CH_2CH_3$

SCHEME 5. Synthesis of 2-nitrobenzyl derivatives of noracetylcholine **1–2c** and of norbutyrylcholine **3–4c.**

to give the desired products rapidly and with good yield. The kinetics and quantum yields of the photofragmentation of the caged cholinergic compounds are summarized in Table I.

The kinetic measurements took advantage of monitoring the decay of

TABLE I
PHOTOFRAGMENTATION PARAMETERS (HALF-TIME $t_{1/2}$ AND QUANTUM YIELD Φ) AND INHIBITION CONSTANTS OF CAGED CHOLINERGIC LIGANDS ON ChEs

Compound	$t_{1/2}$ (μsec)	Φ	K_I (M) AChE	K_I (M) BuChE	Ref.[a]
1a	10	0.27	1.30×10^{-5}	1.11×10^{-5}	17
2a	82,300	0.19	1.00×10^{-5}	1.94×10^{-5}	17
3a	560	0.26	3.73×10^{-6}	3.13×10^{-6}	17
1b	67	0.25	4.40×10^{-5}	7.84×10^{-5}	18, 22
2b	1,700	0.25	NT[b]	NT	18
3b	40	0.80	$>10^{-3}$	$>10^{-3}$	18, 22
1c	25	0.10	7.00×10^{-7}	NT	19
2c	24	0.01	4.52×10^{-6}	NT	19
3c	24	0.10	NT	3.44×10^{-5}	20
4c	23	0.01	NT	3.30×10^{-5}	20

[a] Numbers refer to text footnotes.
[b] Not tested.

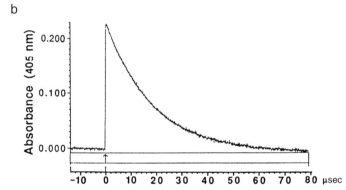

FIG. 1. Kinetic analysis of the laser flash photolysis of compound **1a**. A solution of 1 mM compound **1a** in 0.1 M phosphate buffer was exposed to a single 351-nm laser pulse at 20°. (a) The UV spectrum of a transient was observed by recording the spectral change before and after (0.2 μsec delay) laser flash photolysis of compound **1a**. (b) Kinetic record at 405 nm after a single laser flash photolysis of compound **1a** at pH 6.5. Arrow indicates the beginning of the laser flash. The transient was formed immediately and was followed by an exponential decay. [Reprinted with permission from L. Peng and M. Goeldner, *J. Org. Chem.* **61**, (1996). Copyright (1996) American Chemical Society.]

the presumed *aci*-nitro intemediate, which has a specific absorption around 400 nm. The decay of this *aci*-nitro intermediate was shown to be synchronous to the formation of the corresponding photoproducts.[24] Figure 1 illustrates the typical kinetic analysis for the photofragmentation reaction, show-

---

[24] J. W. Walker, G. P. Ried, J. A. McCray, and D. R. Trentham, *J. Am. Chem. Soc.* **110**, 7170 (1988).

ing the absorption spectrum of the transient and its corresponding decay.[17] Except for compounds **2a, 3a,** and **2b,** all the other molecules show excellent kinetic properties for photolysis (Table I), i.e., microsecond photofragmentation time range, which is compatible with the rate of the enzymatic reaction of cholinesterases ($\approx$50 μsec).[25]

The photoproducts of primary interest are choline, carbamylcholine, and noracetylcholine (or norbutyrylcholine), respectively. The release of carbamylcholine from probes **1–3b** was already described[18a,b]; the release of choline and noracetylcholine (norbutyrylcholine) from compounds of **1–3a** and **1–4c** was assessed and quantified by a specific enzymatic assay for choline and for noracetylcholine (norbutyrylcholine), respectively.[17,19,20] The amount of the formed photoproduct increased with increasing exposure to light and coincided with the disappearance of the parent compound, as quantified by a high-performance liquid chromatography (HPLC) analysis.[17,19,20] These results establish a stoichiometric conversion of cholinergic ligands from their precursors. Furthermore, the quantum yields for the photoconversion of caged choline and of caged noracetylcholine or norbutyrylcholine were determined by comparison with the photolysis of 1-(2-nitrophenyl)ethylcarbamylcholine (Φ = 0.25).[18] Although the low quantum yields of **2c** and **4c** may limit their further application in the studies of ChEs, the observed quantum yields for the other compounds (Table I) are sufficient to ensure an efficient photorelease of the desired ligands.

Biochemical Properties

All three types of compounds in Scheme 1 act as reversible inhibitors of both *Torpedo* AChE and human serum BuChE (Table I). Compound **3a** did not react covalently with the catalytic serine as anticipated, suggesting an unfavorable positioning of the isothiocyanato moiety within the active site. Except for **3b,** which is a very weak inhibitor of both enzymes (Table I), inhibition constants of all other compounds are in the micromolar range. The low affinity of **3b** for ChEs is presumably due to the negative charge of the α-carboxylic group at physiologic pH, which reduces the recognition of the positively charged quaternary ammonium ion by ChEs. In fact, this property has been taken advantage of by using **3b** as an ideal caged agonist for time-resolved studies of the nicotinic acetylcholine receptor,[18b,26] where no interaction between **3b** and the receptor was desired before photoactivation. In our case, the ligand must be positioned at the catalytic site of the

---

[25] M. Vigny, S. Bon, J. Massoulié, and F. Leterrier, *Eur. J. Biochem.* **85,** 317 (1978).
[26] G. P. Hess, *Biochemistry* **32,** 989 (1993).

target protein before light activation. Thus, the reduced affinity of **3b** might limit its further application for time-resolved crystallographic studies on the ChEs.

Whereas inhibition of AChE by the caged cholinergic ligands (except **3b**) is of the mixed type, that of BuChE is purely competitive. Both mixed-type and competitive inhibition imply that the ligands bind to the anionic substite of the active site. Acetylcholinesterase has two binding sites for quaternary ammonium ligands: the anionic subsite of the active site and the peripheral site.[2] The anionic subsite is adjacent to the catalytic triad, near the bottom of an aromatic gorge, which is revealed by the 3D structure of AChE.[5] The peripheral site is located at the entrance to this gorge.[6,7,9] The binding of compound **1a** at the active site of AChE was further assessed by ligand displacement experiments using site-specific fluorescent probes: N-methylacridinium, an active site-specific ligand,[27] and propidium, a peripheral site-specific ligand.[28] The fact that **1a** displaced N-methylacridinium, but not propidium, indicates that **1a** binds specifically at the anionic subsite of the active site.[22]

Although compounds of type C were designed as inhibitors of ChEs, both **1–2c** and **3–4c** can be hydrolyzed slowly by AChE and BuChE, respectively. This undesired property will limit their further utilization in the photoregulation of ChE activity.

Photoregulation of Cholinesterase Activities with Compounds **1a** and **1b**

The ideal caged compounds for photoregulation of the ChEs activities should meet the following criteria: (1) they should photofragment, rapidly and with high quantum yields, to the desired molecules; and (2) they should show sufficient affinity on the enzymes and at the same time should be chemically stable to the enzyme hydrolysis. Thus, among all the compounds presented in Scheme 1, **1a** and **1b** are the most promising candidates for photoregulating ChE activities. Both **1a** and **1b** display rapid kinetics of product release on pulsed illumination, with excellent photofragmentation half-times (10 μsec for **1a** and 24 μsec for **1b**, Table I), which are in the range required for studying both AChE and BuChE, whose turnover rates, depending on the species, vary in the range of 70–300 μsec.[25] The quantum yields of both **1a** and **1b** ensure satisfactory conversion to choline and carbamylcholine, respectively, and the by-product of photochemical fragmentation, 2-nitrosoacetophenone, has no effect on the enzymic activity

---

[27] G. Mooser, H. Schulman, and D. S. Sigman, *Biochemistry* **11**, 1595 (1972).
[28] P. Taylor, J. Lwebuga-Mukasa, S. Lappi, and J. Rademacher, *Mol. Pharmacol.* **10**, 703 (1974).

of either AChE or BuChE. Furthermore, both **1a** and **1b** show inhibitory effects on ChEs, with inhibition constants in the micromolar range, and are chemically stable to the enzyme hydrolysis.

We have established the photoregulation of ChE activities using either probe **1a,** by regenerating the enzymatic activity after photolysis of a ChE–**1a** complex (Fig. 2), or probe **1b,** by inactivating the AChE activity

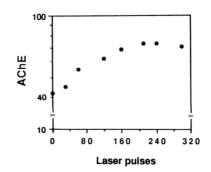

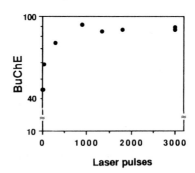

FIG. 2. Regeneration of enzymatic activity after laser flash photolysis of (a) AChE in the presence of **1a** and (b) BuChE in the presence of **1a.** Experimental conditions are given in the Appendix: Experimentation. [Reprinted with permission from L. Peng, I. Silman, J. Sussman, and M. Goeldner, *Biochemistry* **35,** (1996). Copyright (1996) American Chemical Society.]

through photoinduced carbamylation after photolysis of an AChE–**1b** complex (Fig. 3a).[22]

The experiments (see Appendix: Experimentation) in which we studied the recovery of enzymatic activity after flash laser photolysis of solutions of AChE or BuChE containing compound **1a** (Fig. 2) suggested that not

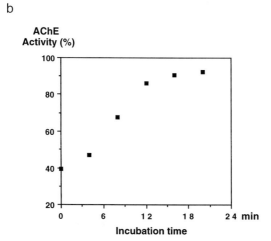

FIG. 3. (a) Time-dependent inactivation of AChE after laser flash photolysis of AChE in the presence of **1b**. ●, one pulse; ▲, 20 pulses. (b) Regeneration of AChE activity after subsequent dilution. Experimental conditions are given in the Appendix: Experimentation. [Reprinted with permission from L. Peng, I. Silman, J. Sussman, and M. Goeldner, *Biochemistry* **35**, (1996). Copyright (1996) American Chemical Society.]

only did the by-product 2-nitrosoacetophenone have no inhibitory effect on the two enzymes in the experimental conditions employed, but also both 2-nitrosoacetophenone and choline, which would be generated concomitantly within the active sites of the enzymes, were cleared from their active sites. Thus, the experimental system can indeed serve as a paradigm for studying the clearance of choline from the active site of AChE and of BuChE under the conditions of time-resolved crystallography.

In the experiment of photolysis of a solution of AChE containing **1b**, even a single pulse caused time-dependent loss of enzymatic activity to 60% of the control value, and 20 pulses decreased the enzyme activity by a further 15% (Fig. 3a). That this observed inactivation was indeed due to carbamylation by carbamylcholine photogenerated from **1b** was demonstrated by the fact that >90% of control activity could be regenerated in a progressive fashion on subsequent dilution of carbamylated enzyme (Fig. 3b), just as is observed for AChE carbamylated directly by carbamylcholine.[22]

Conclusion

The results we have obtained show that **1a** and **1b** possess photochemical and specificity characteristics that render them suitable to photoregulate ChE activities. They each release photochemically a different cholinergic modulator controlling the catalytic reactions at different steps. Compound **1a** photoreleases choline, the enzymatic product, whereas **1b** photogenerates carbamylcholine, an AChE substrate, allowing the carbamylation of the enzyme with a concomitant release of choline. Because these two compounds generate choline in two different ways, either by a direct photocleavage reaction **(1a)** or by enzymatic hydrolysis of a substrate generated by photocleavage **(1b),** they constitute complementary tools for the time-resolved crystallographic studies envisaged. Work is presently in progress to establish the dynamic structure of AChE during its catalysis by time-resolved crystallography using both **1a** and **1b** separately soaked into crystals of *Torpedo* AChE.

Perspective

The understanding of an enzymatic reaction generally involves kinetic studies in association with chemical methods, such as site-directed labeling and site-directed mutagenesis, designed to identify the important amino acid residues involved in the catalytic reaction. In addition, 3D structures of enzymes and different enzyme complexes have considerably improved the knowledge of such chemical reactions at the atomic level. However, important information is generally missing in such investigations, i.e., the description of the different interactions occurring along the reaction pathway, in other words, the dynamic picture of an enzymatic reaction. Time-resolved macromolecular crystallography offers the possibility of studying such issues at the atomic level provided that the time resolution is shorter than the lifetimes of transient key intermediates and that a fast and efficient triggering of the enzymatic activity in crystal is available.[11] Laue crystallography using a synchrotron X-ray source allows extremely rapid data collection, on the nanosecond time scale, and offers a direct experimental approach that may permit visualization of the putative conformational changes of the enzyme.[29] Photoregulation of enzyme activity with caged compounds can provide rapid photochemical release of enzyme product or enzyme substrate, and thus initiate the enzymatic reaction synchronously in enzyme crystals. Alternative methodologies, artificially prolonging the key intermediates' lifetimes, can be envisaged either by biological methods, i.e., production of slower mutants,[30] or by physical methods, i.e., a freeze-trapping approach.[31] The advantage of using caged compounds in time-resolved Laue crystallographic studies is that we could analyze directly the authentic, short-lived intermediates along an enzymatic reaction pathway, and thus we could observe the true dynamic picture of an enzyme while achieving its biological function.

Appendix: Experimentation

Materials

Acetylcholinesterase was purified from electric organ tissue of *Torpedo marmorata*[32] and BuChE from human plasma.[33] Compound **1a** was synthesized as described,[17] and *O*-[1-(2-nitrophenyl)ethyl]carbamylcholine iodide (**1b**) was purchased from Molecular Probes (Junction City, OR).

*Photoregulation of Cholinesterases Activities with Compounds **1a** and **1b***

Laser pulses of about 100 mJ and of about 20 nsec duration[17] are applied to a quartz cuvette (1 × 10 × 40 mm) containing 2 μg/ml AChE, together

---

[29] V. Srajer, T. Teng, T. Ursby, C. Pradervand, Z. Ren, S. Adachi, W. Schildkampf, D. Bourgeois, M. Wulff, and K. Moffat, *Science* **274,** 1726 (1996).
[30] J. M. Bolduc, D. H. Dyer, W. G. Scott, P. Singer, R. M. Sweet, D. E. Koshland, and B. L. Stoddard, *Science* **268,** 1312 (1995).
[31] K. Moffat and R. Henderson, *Curr. Biol.* **5,** 656 (1995).
[32] L. Ehret-Sabatier, I. Schalk, M. Goeldner, and C. Hirth, *Eur. J. Biochem.* **203,** 475 (1992).
[33] O. Lockridge and B. N. La Du, *J. Biol. Chem.* **253,** 361 (1978).

with 5.0×10^{-4} M probe **1a** in a total volume of 0.4 ml of 50 mM phosphate buffer, pH 6.5. Enzymatic activity is monitored spectrophotometrically, at 400 nm, using 1 mM p-nitrophenyl acetate as substrate, in 50 mM phosphate buffer, pH 7.2, 20°.[34] Butyrylcholinesterase (3 μg/ml) is incubated in a quartz cuvette ($5 \times 10 \times 40$ mm) with 5.4×10^{-5} M probe **1a** in 50 mM phosphate buffer, pH 7.2, in a total volume of 1.2 ml and similarly exposed to 351-nm laser flashes. Butyrylcholinesterase activity was monitored spectrophotometrically at 240 nm using 5.0×10^{-5} M benzoylcholine as substrate, in 50 mM phosphate buffer, pH 7.2, at 20°.[35]

Acetylcholinesterase (294 μg/ml) is incubated with 4.2×10^{-3} M probe **1b**, in 50 mM phosphate buffer, pH 6.5, total volume 0.4 ml, in a quartz cuvette ($1 \times 10 \times 40$ mm). After exposure to either 1 or 20 pulses of a 351-nm laser, aliquots are withdrawn for assay of enzymatic activity both by the Ellman method[36] and by using p-nitrophenyl acetate as substrate.[34] Regeneration of the enzymatic activity of carbamylated AChE was achieved by 500-fold dilution into 50 mM phosphate buffer, pH 7.2, at 20°.

Acknowledgments

This work was supported by the Association Française contre les Myopathies, the Centre National de la Recherche Scientifique, the Association Franco Israélienne pour la Recherche Scientifique et Technique, the Société de Secours des Amis des Sciences, and the European Community Biotechnology Programme under Contract 960081.

[34] R. K. Tripathi, J. N. Telford, and R. D. O'Brien, *Biochim. Biophys. Acta* **525**, 103 (1978).
[35] W. Kalow and H. A. Lindsay, *Can. J. Biochem. Physiol.* **33**, 568 (1953).
[36] G. L. Ellman, K. D. Courtney, V. Andres, and M. R. Featherstone, *Biochem. Pharmacol.* **7**, 88 (1961).

[16] Caged Substrates for Measuring Enzymatic Activity *in Vivo*: Photoactivated Caged Glucose 6-Phosphate

By ROBERT R. SWEZEY and DAVID EPEL

This article describes the use of caged substrate molecules to study the activity and regulation of enzymes *in vivo*. As stated by Van Noorden and Jonges,[1] "Estimations of metabolic rates in cells and tissues and their regulation on the basis of kinetic properties of enzymes in diluted solutions

[1] C. J. F. Van Noorden and G. N. Jonges, *Histochem. J.* **27**, 101 (1995).

may not be applicable to intact living cells or tissues." It is therefore preferable to study the activity of an enzyme while it resides in its physiological address, i.e., intracellularly.

Such endeavors to study enzyme activities *in vivo* have predominately used one of the following approaches: (a) Identification of enzymes at "crossover" points in a metabolic pathway (example in Ref. 2): Here cellular concentrations of metabolites are measured in an attempt to identify which substrate–product pairs in the pathway are not in thermodynamic equilibrium. This would indicate that the enzyme(s) involved in this catalysis is inhibited to some degree within the cell. (b) Quantitative histochemistry[1]: In this approach an exogenous substance that is chromogenic via an enzymatic activity (e.g., formazans with dehydrogenases) is administered to cells, and the rate of product formation in individual cells is monitored microscopically (e..g, with a scanning and integrating cytophotometer). (c) Radioactive precursors[3]: By administering a radiolabeled precursor for a metabolic pathway (e.g., $[^{32}P]P_i$ for glycolysis) and then measuring the changes with time in the specific radioactivities of all of the intermediates in a metabolic pathway, one can determine the fluxes of substrates through each enzyme of the pathway. This approach is precise, but it is particularly labor intensive.

In our strategy, an unreactive caged and radiolabeled substrate is first placed in a cell, and then initiation of the enzyme assay *in vivo* occurs by photolytic release of the substrate. A key advantage of this approach is that it permits a temporally defined radiolabeling of a particular intracellular substrate pool; the rate of conversion of this substrate to product within the cells is then followed by chemical, radiochemical, or spectroscopic means, and knowledge of the intracellular substrate concentration permits calculation of enzymatic activity *in vivo*. This approach has also been used by Meldrum *et al.*[4] to follow the rate of DNA repair in cells that have been exposed to ultraviolet (UV) irradiation damage.

This article describes the preparation and use of caged glucose 6-phosphate (G6P) to study changes in the glycolytic and pentose shunt pathways. We have used this to study the activity of glucose-6-phosphate dehydrogenase (G6PDH) in sea urchin eggs and the changes in activity of this enzyme that attend the fertilization of these eggs. We describe in detail the procedure for sea urchin eggs, realizing that although our protocols are highly specific for these eggs, the principles can be transferred to many other cell types.

---

[2] I. Yasumasu, K. Asami, R. L. Shoger, and A. Fujiwara, *Exp. Cell Res.* **80**, 361 (1973).
[3] J. G. Reich, U. Till, J. Gunther, D. Zahn, M. Tschisgale, and H. Frunder, *Eur. J. Biochem.* **6**, 384 (1968).
[4] R. A. Meldrum, S. Shall, and C. W. Wharton, *Biochem. J.* **266**, 891 (1990).

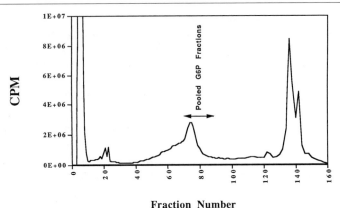

FIG. 1. High-performance liquid chromatography anion-exchange purification of radiolabeled glucose 6-phosphate. Fractions 70–90 were pooled, evaporated, and reacted with diazoalkane caging reagent (see text for details).

Materials and Methods

Preparation of Caged [^{3}H/^{14}C]Glucose 6-Phosphate

The chemical properties of the caging reagents and final caged products are covered more extensively in other articles in this volume. What follows are some practical pointers for those who do not have access to a synthetic organic chemistry laboratory.

Preparation of [5-^{3}H/1-^{14}C]Glucose 6-Phosphate. One millicurie of [5-^{3}H]glucose and 0.25 mCi of [1-^{14}C]glucose are combined and the solvents removed *in vacuo*. The residue is dissolved in 200 μl of a hexokinase reaction medium consisting of 50 mM HEPES buffer (pH 7.5), 20 mM adenosine triphosphate, 25 mM MgCl$_2$, and 20 units of commercial hexokinase from bakers yeast. This solution is incubated for 5 h at room temperature, and then the radioactive glucose 6-phosphate is purified by an anion-exchange high-performance liquid chromatography (HPLC) procedure that has been described elsewhere.[5] A representative chromatogram for one such synthesis is shown in Fig. 1. Fractions 70–90 were pooled for subsequent use. Using this procedure, approximately 15% of the radioactivity in the chromatogram is associated with G6P; the other major chromatographic peaks are most likely unreacted glucose (fraction 5) and the hexose diphosphates (fractions 140 and 144).

[5] R. R. Swezey, *J Chromatogr. B* **669,** 171 (1995).

Preparation of 2-Nitroacetophenone Hydrazone. In a round-bottom flask combine 2.065 g 2-nitroacetophenone (12.5 mmol), 25 ml 95% ethanol (v/v), 1.36 ml hydrazine hydrate, and 0.8 ml glacial acetic acid. Add boiling chips, attach a water-jacketed condenser to the flask, and reflux the mixture for 3 h using an electric heating mantle. Cool the solution to room temperature, and then remove the solvent under vacuum. Add 50 ml each of $CHCl_3$ and water to the residue, and then transfer to a separatory funnel. Mix immiscible phases well, allow to separate, and then retain the lower (organic) phase. Discard the aqueous phase, and wash the organic phase twice more with water in the separatory funnel. Evaporate the washed organic phase under vacuum to remove the $CHCl_3$.

The hydrazone is then purified chromatographically with standard silica gel, using ethyl acetate as the eluent as follows. Dissolve the $CHCl_3$ residue in a minimal volume of ethyl acetate (~5 ml) and chromatograph this on a column of silica gel (1.5 × 18 in.) equilibrated with and eluted by ethyl acetate. Collect the deep-yellow fraction that elutes first and remove the ethyl acetate under vacuum to obtain the final product, which is a viscous yellow oil. We typically recover 1.2 g of product (6.7 mmol), giving an overall yield of ~54%.

Conversion of Hydrazone to Diazo Caging Reagent and Reaction with [5-^{3}H/1-^{14}C]Glucose 6-Phosphate. Perform the following in subdued lighting (darkness or red lighting is not necessary). Transfer 65 µl of the hydrazone oil (0.4 mmol) to a glass test tube (16 × 100 mm), and add 5 ml diethyl ether and a 1-cm round, cross-headed stirring bar (e.g., Aldrich, Milwaukee, WI). With vigorous stirring carefully add 0.285 g MnO_2. Stopper the tube with a rubber septum impaled with a hypodermic needle to permit venting, cover the reaction tube with aluminum foil, and continue stirring 5 min. Filter the now deep orange–red ether solution through Whatman (Clifton, NJ) #1 filter paper under vacuum, releasing the vacuum as soon as filtration is complete to prevent drying down the solution.

We keep a 2-liter beaker containing approximately 800 ml of 20% (v/v) acetic acid in methanol nearby for immediate immersion of glassware, pipettes, filter paper, etc., that has come into contact with the ether solution of diazoalkane, and which is no longer needed in the synthesis. The acetic acid effectively quenches residual diazoalkane on these surfaces, thereby eliminating the formation of dry diazoalkane, which is potentially explosive.

Three milliliters of this diazoalkane ether solution are then transferred to a fresh test tube containing 1 ml of an aqueous solution of [5-^{3}H/1-^{14}C]glucose 6-phosphate at pH ~4–5 (pH paper is adequate for this adjustment) prepared as described earlier. This biphasic mixture is vigorously mixed for at least 3 hr at room temperature using a 1-cm round cross-headed stirring bar. The upper (ether) phase is then removed and replaced

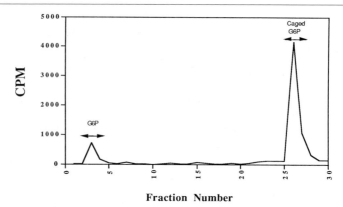

FIG. 2. Reversed-phase HPLC purification of caged glucose 6-phosphate. Fractions 25–28 were pooled, evaporated, and dissolved in the permeabilization medium for *in vivo* assays (see text for details).

with freshly prepared diazoalkane ether solution, and this is then stirred for at least 3 hr. After one more application of fresh diazoalkane ether solution and 3 hr stirring, the aqueous phase is removed and washed three times with 5 ml diethyl ether and then evaporated to dryness *in vacuo*. The residue is dissolved in a minimal volume of 50 mM ammonium bicarbonate (no pH adjustment), and the caged compound is purified by reversed-phase HPLC as has been described elsewhere.[6] As seen in Fig. 2, the unreacted G6P comes in the flowthrough fractions, whereas the more hydrophobic caged substrate elutes much later. Fractions 25–28 containing the caged G6P are pooled and taken to dryness *in vacuo*.

Testing Caged Substrates

After the caged substrate has been synthesized and purified it must be tested for inertness toward G6PDH and for conversion to the parent compound with resultant competence to be oxidized on photolysis. To do this a small portion of the caged compound is dissolved in a reaction medium for G6PDH and 6PGDH (6-phosphogluconic dehydrogenase), and the ability to form $^{14}CO_2$ before and after irradiation is tested. This medium contains 50 mM glycylglycine, pH 8.4, 5 mM MgCl$_2$, and 1 mM β-nicotinamide adenine dinucleotide phosphate (NADP). The reaction takes place in a small Warburg flask with 100% (w:v) trichloroacetic acid in the side arm and 0.5 M NaOH in the center well. The reaction is initiated by adding a G6PDH/6PGDH enzyme mixture that is prepared as follows: 100 μl of a commercially obtained ammonium sulfate suspension of 6-phosphogluco-

---

[6] R. R. Swezey and D. Epel, *Exp. Cell Res.* **201**, 366 (1992).

nate dehydrogenase is centrifuged for 2 min at 12,000 g (at room temperature), the supernatant is removed by aspiration, and the resulting pellet is dissolved in 100 μl of a 1-mg/ml solution of G6PDH. Ten microliters of the G6PDH/6PGDH enzyme mixture is then transferred to each Warburg flask; reaction with these two enzymes results in decarboxylation of the ^{14}C at the 1 carbon of G6P to form $^{14}CO_2$, which becomes trapped in the KOH in the center well.

We find that after 3 hr incubation at room temperature only 1.2% of the total ^{14}C label of the material not subjected to light flashes is present in the center cell, whereas 66.1% of the label is in the center well (i.e., is converted to $^{14}CO_2$) in the sample that had received two flashes of light from the xenon lamp (described in the Assay of Enzyme Activities *in Vivo* section). Therefore, this caged substrate demonstrates the desired properties of being inert prior to photolysis and enzymatically reactive afterward.

When we subjected the photolysis product to chromatographic analysis we made a surprising discovery. As seen in Fig. 3, the photolysis of caged G6P results in two products, G6P and a chemically unidentified side product labeled X. We found that X was capable of reacting with 6PGDH to form $^{14}CO_2$; however, the retention time of X in this HPLC procedure was different from that of 6-phosphogluconate.[5] The relevance of X and its reactivity with 6PGDH is discussed in the results section.

Placing Caged Substrate into the Cell: Electroporation

Once the caged substrate has been purified and demonstrated to be enzymatically inactive until photolysis, the next step is to place the caged

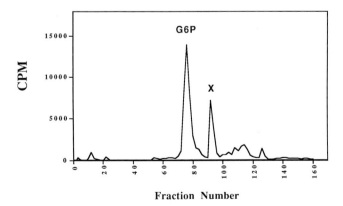

FIG. 3. High-performance liquid chromatography analysis of the photolysis product of caged glucose 6-phosphate. Chromatographic conditions identical to those in Fig. 1.

substrate into the cytoplasm. In our work with sea urchin eggs we have found that permeabilization of the cell membrane by mild electroporation is an effective means to introduce caged substrates (or other impermeant substrates) into the egg cytoplasm. The procedure for sea urchin eggs has been described in detail elsewhere,[7] but a brief description is given here. It should be noted that this procedure differs from most electroporation protocols in that the holes generated by the electroporation procedure remain open following its cessation, allowing permeation of substrates until the pores are resealed by addition of calcium ions to the media.

The medium surrounding the eggs during electroporation is crucial for successful electroporation of these cells. We have obtained good results utilizing a medium with the following composition: 300 mM glycine, 225 mM potassium gluconate, 185 mM mannitol, 20 mM NaCl, 10 mM spermidine, 5 mM MgCl$_2$, 4 mM ATP, 2 mM sodium bicarbonate, and 1.6 mM reduced glutathione; the pH of this medium is adjusted with HCl to 6.8 just prior to use in permeabilization of eggs. The composition of this medium, developed after much trial and error, reflects the natural osmolarity of the sea urchin egg, i.e., seawater, and other characteristics of this cell of a marine organism such as high K$^+$, low Na$^+$, low Cl$^-$, and a high concentration of an osmolyte such as glycine. The key features of this medium that may be transferable to other cell types include the high K$^+$, low Na$^+$ and Cl$^-$, and inclusion of MgATP at physiological levels. It is also crucial to keep the medium low in Ca^{2+}, in that Ca^{2+} ions impede electroporation, and as noted earlier are used later to reseal the egg cell membrane at the end of the electroporation loading process. (Inclusion of EGTA (ethylene glycol-bis(β amino ethyl ether)N,N,N',N'-tetraacetic acid) or other chelating agents is not recommended because chelators will enter the cell and could impede any Ca$^+$-dependent processes.)

A typical loading of sea urchin (species *Strongylocentrotus purpuratus*) eggs with caged (^{3}H/^{14}C) G6P proceeds as follows. To a tube containing the dry, purified caged G6P is added 400 μl permeabilization medium. Dissolution is accomplished by gentle repetitive pipetting, and the solution is then clarified by centrifugation for 3 min at approximately 12,000 g (at room temperature). Three hundred microliters of the supernatant are transferred to a test tube containing 200 μl of packed sea urchin eggs that have been equilibrated with permeabilization medium by cycles of suspension/centrifugation, first in calcium-free seawater and then in permeabilization media. This suspension is transferred to the electroporation

---

[7] R. R. Swezey and D. Epel, *Cell Regul.* **1**, 65 (1989).

chamber (described in Swezey and Epel[8]) where the eggs are exposed five times at 10-sec intervals to an electric field of 150 V/cm by capacitor discharge. Using a 0.5-μF capacitor, the resulting electric field decays by 1/e every 70 μsec in our system. The cells are then transferred to a test tube and incubated at 16° for 2 min with repeated gentle bubbling from a Pasteur pipette. The pores remain open, and this time period allows diffusion of the caged substrate into the permeabilized cell. After this incubation the eggs are placed in 10 ml of an artificial seawater (MBL formula) that has been modified to contain 1.1 mM Ca^{2+} ions (instead of the 11 mM normally present). The cells are allowed to settle in this low-calcium seawater, during which time the electrically generated pores reseal. The eggs are then transferred back to seawater for experimentation.

Assay of Enzyme Activities in Vivo

Once the cells are loaded with the caged substrate and placed back into an appropriate medium, the assay of enzyme activity within the cells can proceed. In our work, a 1.8-ml suspension of substrate-loaded sea urchin eggs is placed into the main compartment of a small Warburg flask containing 400 μl 0.5 M KOH in the center well and 200 μl 100% trichloroacetic acid (TCA) in the side arms. For photolysis, the Warburg flasks are held in the center of a torroidal xenon flash lamp (Speedotron Corp., Chicago, IL, 12 W-sec per flash, commercially available for photographic use). Aluminum foil placed over this unit maximizes light exposure to the cells and minimizes exposure to the experimenter.

Uncaging is accomplished by igniting the flashbulb twice, with 4 sec between flashes (the minimal recharge time). In our work with sea urchin eggs, this flashing is performed in three cases: (a) in unfertilized eggs, (b) in eggs that had been fertilized in the Warburg flask and allowed to develop 2 min prior to flashing, and (c) in eggs that had been fertilized in the Warburg flask and allowed to develop 20 min prior to flashing. In separate flasks, the reactions are stopped at various times postflashing by tipping the flasks to mix the TCA with the sea urchin egg suspensions (final concentration of TCA in the mixture is 10%). These flasks are then shaken on an orbital shaker for 30 min to allow quantitative accumulation of the $^{14}CO_2$ into the KOH solution in the center well; the amount of $^{14}CO_2$ formed is then quantitated by liquid scintillation counting of the contents of the center well.

[8] R. R. Swezey and D. Epel, in "Guide to Electroporation and Electrofusion" (D. C. Chang, B. M. Chassy, J. A. Saunders, and A. E. Sowers, eds.), p. 347. Academic Press, San Diego, 1992.

To measure the ^{3}H released from the [^{3}H]glucose 6-phosphate, the acidified egg suspensions are transferred from the flasks to centrifuge tubes, and the eggs are removed by centrifugation at approximately 500–1000 g for 2 min (at room temperature). The bulk of the TCA in the resulting supernatants (volume approximately 1.2 ml) is removed by extraction three times with 5 ml of diethyl ether, and small aliquots (10 to 25 μl) of a 1 M solution of ammonium bicarbonate are added until the pH is approximately neutral as evidenced with pH paper. All extracts are brought to the same volume (2 ml) by adding water, and then 0.6 ml are distilled until at least 0.2 ml is collected in the distillate (the distillations are never taken to dryness to avoid pyrolytic generation of ^{3}H$_2$O in the distillation process). Two hundred microliters of the distillates are then counted for ^{3}H, which is a measure the amount of ^{3}H$_2$O formed in the eggs from the [5-^{3}H]G6P. Therefore, in our studies the relevant rates in the cells being monitored are the rates of formation of ^{14}CO$_2$ and ^{3}H$_2$O.

Results

Our primary interest was to determine the activity of the enzyme G6PDH in eggs before and after fertilization. We already had evidence that the *in vivo* activity of the decarboxylating enzyme of the pentose shunt, 6PGDH, was so high as not to be rate limiting to the formation of ^{14}CO$_2$ from the uncaged G6P by the shunt.[6] However, because glycolysis and Krebs cycle activities also form ^{14}CO$_2$ from [1-^{14}C]G6P, we needed an independent assay of flux through these pathways. Loss of ^{3}H from the [5-^{3}H]G6P to form ^{3}H$_2$O at the enolase and/or triose-phosphate isomerase steps of glycolysis provides this information. Thus in these double-label experiments we could monitor flux rates through glycolysis/Krebs cycle enzymes and through the pentose shunt (=rate of G6PDH activity).

Figure 4 (data replotted from Swezey and Epel[9]) depicts the results for a typical experiment; here we have plotted the amounts of ^{14}CO$_2$ (Fig. 4A) and ^{3}H$_2$O (Fig. 4B), expressed as the percentage of the total ^{14}C or ^{3}H in each of the samples to normalize between flasks, formed as a function of time postuncaging for the three developmental states of the eggs. The relevant points obtained from experiments such as these are that 2 min after fertilization there is a large increase in the rate of ^{14}CO$_2$ generation, with very little change in the rate of ^{3}H$_2$O formation, and that by 20 min after fertilization this rate in ^{14}CO$_2$ formation subsides, but there is still little change in the rate of ^{3}H$_2$O formation. The changes in ^{14}CO$_2$ formation

---

[9] R. R. Swezey and D. Epel, *Dev. Biol.* **169,** 733 (1995).

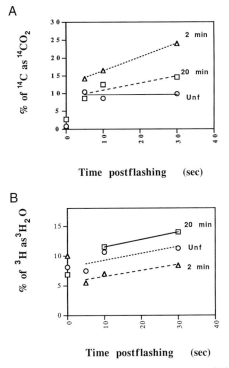

FIG. 4. Assay of G6P utilization *in vivo* in *S. purpuratus* eggs. (A) Formation of $^{14}CO_2$ from caged [1-^{14}C]glucose 6-phosphate after two light flashes. (B) Formation of ^3H_2O from caged [5-^{3}H]glucose 6-phosphate after two light flashes.

are therefore indicative of changes in G6PDH activity, and not in glycolytic/Krebs cycle activities (because ^3H_2O formation rates are relatively constant). These findings indicate that there is very little G6PDH activity operating in unfertilized eggs and that there is a pronounced activation of G6PDH shortly after fertilization. At a later developmental time (20 min) the activity of G6PDH is less.

Note that for $^{14}CO_2$ data that there is an apparent burst of $^{14}CO_2$ between time 0 (which represents an unflashed sample) and 5 sec after uncaging, followed by a slower rate of formation between 5 and 30 sec after flashing; as seen in Fig. 4B, there is no such burst in ^3H_2O formation. The burst-phase $^{14}CO_2$ formation has been attributed to the rapid oxidation of compound X by 6PGDH,[9] and therefore the rates of [1-^{14}C]G6P oxidation are taken from data subsequent to 5 sec postflashing. This shows that characterization of the photolysis products can be critical in the interpretation of data from *in vivo* experiments.

By knowing the concentration of intracellular G6P in eggs, 24 fmol/egg, a value that is fairly constant before and after fertilization of *S. purpuratus* eggs,[10] and knowing the efficiency of uncaging *in vivo* (34 ± 3.4%, Ref. 9), it is possible to calculate maximal estimates of G6PDH activity (see Refs. 9 and 11 for details). These maximum values of G6PDH (expressed as 10^{-8} units per eggs) are 0.0024 in unfertilized eggs, 1.10 at 2 min after fertilization, and 0.53 at 20 min after fertilization.[11] Relative to unfertilized eggs, G6PDH activity is 458 times greater than 2 min postfertilization and 221 times greater at 20 min.

In vitro estimates of maximal G6PDH activity in these eggs assayed at physiological pH and substrate concentrations indicated that there are 177×10^{-8} units per egg and that this *in vitro* value did not change on fertilization.[12] Therefore, even at the maximal rate observed *in vivo* (2 min) the enzyme exhibited only 0.62% of that seen *in vitro*. This fact underscores the quotation by Van Noorden and Jonges given at the beginning of this article.

Conclusions

Caged radiolabeled G6P placed into unfertilized sea urchin eggs and then uncaged by photolysis before or after fertilization showed that fertilization results in a pronounced increase in G6PDH activity, whereas glycolytic/Krebs cycle activity was constant during this same period.

The basic strategy described here could be adapted for use with other enzymatic substrates and with other cell types. One limitation is that this approach is best suited to cytoplasmic enzymes rather than those in other intracellular compartments, as the cytoplasmic pool is accessible to the electroporation protocol; analysis of compartmentalized enzymes could be complicated by the rate of the movement of the uncaged substrate from the cytoplasm to the relevant compartment.

This procedure could be useful for many other cell types, and we look forward to similar *in vivo* measurements of enzyme activity in other cells. It will be interesting to learn how well (or how poorly) the estimates from *in vitro* measurements compare to *in vivo* rates.

---

[10] D. Epel and R. M. Iverson *in* "Control of Energy Metabolism" (B. Chance, R. W. Estabrook, and J. R. Williamson, eds.), p. 267. Academic Press, New York, 1965.
[11] B. B. Rees, R. R. Swezey, H. Kibak, and D. Epel, *Invertebr. Reprod. Dev.* **30**, 123 (1996).
[12] R. R. Swezey and D. Epel, *Proc. Natl. Acad. Sci. U.S.A.* **85**, 812 (1988).

[17] Investigation of Charge Translocation by Ion Pumps and Carriers Using Caged Substrates

By K. FENDLER, K. HARTUNG, G. NAGEL, and E. BAMBERG

Introduction

Ion pumps create gradients of ions across cell or cell organelle membranes, while consuming light or chemical energy, e.g., by hydrolyzing ATP. Subsequently, the electrochemical gradients are used for passive transport via ion channels or carriers. The combined action of these membrane proteins serves to build up membrane potential, to regulate the intracellular pH, or to accumulate nutrients such as sugars or amino acids inside the cell.

Ion pumps and carriers pass through an enzymatic cycle during the transport process. The study of the charge translocation by ion translocating proteins is one of the most direct methods to study the mechanism of these enzymes. However, only in a few cases has it been possible to follow the ion translocation in the native membrane, as the turnover of pumps and carriers is low (between $1-1000$ sec^{-1}). Although electrical properties of single channels can be measured very precisely because of the large turnover of about 10^7 sec^{-1}, which corresponds to a current of 10^{-12} A per channel, the study of ion pumps requires the activation of about 10^4-10^7 molecules to produce a current equivalent to that of a single open channel.

Therefore, a high density of the protein within the membrane is a prerequisite to studying its ion transport. This condition can be met by using preparations that have a naturally high density of the transport system of interest or by purifying the protein. In addition, large membrane areas are required to produce a measurable pump current. This is possible by protein reconstitution on planar lipid bilayers or by the application of the so-called giant patch clamp technique.[1,2] Another technical difficulty is that many of the investigated enzymes occur in cell organelles or bacteria, where electrophysiological methods are not applicable because of the small size. In these cases ion pumps and carriers can be reconstituted on planar lipip membranes.[3]

Because of the low basic conductance of the planar lipid membranes, a high signal-to-noise (S/N) ratio can be expected for the registration of the charge translocation. The introduction of photolabile substrates (caged

---

[1] D. W. Hilgemann, *Pflügers Arch.* **415**, 247 (1989).
[2] T. Friedrich, E. Bamberg, and G. Nagel, *Biophys. J.* **71**, 2486 (1996).
[3] E. Bamberg, H.-J. Butt, A. Eisenrauch, and K. Fendler, *Quart. Rev. Biophys.* **26**, 1 (1993).

compounds) for ion carriers and ion-motive ATPases signifies an important advance in the study of transport properties of these enzymes.[4] Because the photoreaction does not produce mechanical disturbances of the system, the combination of caged compounds with electrical or electrophysiological methods on artificial or natural membranes is ideal for the studies of transport proteins with low turnover numbers. For kinetic investigations the application of caged compounds is advantageous because in absence of an active substrate, such as ATP, an ATPase is locked in a defined intermediate. This represents the ideal situation for the study of presteady-state transport.

Light-induced concentration jumps of ATP using caged ATP have been applied successfully on ion-motive ATPases to initiate the reaction and transport cycle for both cell membranes and planar lipid membranes. Substantial information about electrogenicity and kinetics has been obtained for the Na,K-ATPase, for the Ca-ATPase from sarcoplasmic reticulum, and for the H,K-ATPase from pig stomach (for references, see Table I). The latter is an electroneutrally operating 1:1 exchanging pump. Despite that, the application of caged ATP showed clearly that the H^+ translocating step is electrogenic,[5,6] as is the K^+ transporting step.[7] More recently, caged Ca^{2+} and caged ATP or ADP were applied to secondary active transporting systems such as the Na,Ca-exchanger of heart cells and the ADP/ATP-carrier from mitochondria. The following sections give some examples for the application of caged ATP, caged ADP, and caged Ca^{2+}. Also, problems associated with the inhibiting effect of the caged nucleotide on the Na,K-ATPase are demonstrated. This article presents studies on the application of caged compounds to planar lipid membranes as well as on giant patches of heart cell membranes.

Na,K-ATPase

Na,K-ATPase plays an important role in every animal cell. Using the free energy of ATP hydrolysis, it transports three Na^+ ions out of the cell in exchange for two K^+. The Na,K-ATPase thereby maintains the ion gradients across the cell membrane that are important for various functions of the cell. The generally accepted reaction mechanism assumes two conformations of the unphosphorylated (E_1 and E_2) and the phosphorylated

---

[4] J. H. Kaplan, B. I. I. I. Forbush, and J. F. Hoffman, *Biochemistry* **17,** 1929 (1978).
[5] H. T. W. M. v.d. Hijden, E. Grell, J. J. H. H. M. de Pont, and E. Bamberg, *J. Membr. Biol.* **114,** 245 (1990).
[6] M. Stengelin, K. Fendler, and E. Bamberg, *J. Membr. Biol.* **132,** 211 (1993).
[7] T. Lorentzon, G. Sachs, and B. Wallmark, *J. Biol. Chem.* **263,** 10705 (1988).

protein (E_1P and E_2P). Na^+ and K^+ translocation is thought to proceed via the $E_1P \rightarrow E_2P$ and $E_2 \rightarrow E_1$ conformational transitions, respectively.

Membrane fragments, liposomes, or whole cells containing Na,K-ATPase were used to investigate charge transport by this ion pump. Purified preparations,[8–12] as well as native plasma membranes[2] were studied successfully with different electrophysiological techniques using rapid activation via caged ATP.

Bilayer Experiments

Caged ATP [P^3-1-(2-nitro)phenylethyladenosine 5′-triphosphate] is prepared as described earlier.[4,8] To photolyse the caged ATP, light pulses of a high-pressure 100-W Hg arc lamp and a mechanical shutter (exposure time 125 msec) equipped with UV optics are used. Alternatively, an excimer laser flash with a duration of 10 nsec, a wavelength of 308 nm, and an energy density of 10–200 mJ/cm^2 is used. The fraction η of caged ATP converted to ATP from photo-release measurements is determined using the luciferin luciferase assay (Boehringer, Ingelheim) as described earlier.[10]

Membrane fragments or liposomes containing the Na,K-ATPase are adsorbed on a planar bilayer as shown in Fig. 1 (top). The front and rear compartments are separated by a lipid bilayer that is formed from a solution of diphytanoyl phosphatidylcholine (1.5%) and octadecylamine (0.025%) in decane. The currents are measured using Ag|AgCl electrodes. Details of the technique have been presented elsewhere.[8,13]

After release of ATP by irradiation with UV light a current could be measured that rises within ca. 100 msec and decays slowly. Under these conditions, the rise of the signal is limited by the release of ATP during the shutter opening time (125 msec) of the UV lamp. The polarity of the current corresponds to the transport of positive charge to the planar bilayer (Fig. 1). This is in agreement with the 3Na$^+$/2K$^+$ stoichiometry of cation transport by the Na,K-ATPase. The membrane fragments or liposomes are not integrated in the planar bilayer but rather adsorbed onto it [Fig. 1 (top) and Fig. 7 (left)]. This explains the slow decay of the current that is brought about by the charging of the membrane capacitance and can be rationalized by the equivalent circuit shown in Fig. 1 (top).

---

[8] K. Fendler, E. Grell, M. Haubs, and E. Bamberg, *EMBO J.* **4**, 3079 (1985).
[9] K. Fendler, E. Grell, and E. Bamberg, *FEBS Lett.* **224**, 83 (1987).
[10] G. Nagel, K. Fendler, E. Grell, and E. Bamberg, *Biochim. Biophys. Acta* **901**, 232 (1987).
[11] R. Borlinghaus, H.-J. Apell, and P. Läuger, *J. Membr. Biol.* **97**, 161 (1987).
[12] H.-J. Apell, R. Borlinghaus, and P. Läuger, *J. Membr. Biol.* **97**, 179 (1987).
[13] K. Fendler, S. Jaruschewski, A. Hobbs, W. Albers, and J. P. Froehlich, *J. Gen. Physiol.* **102**, 631 (1993).

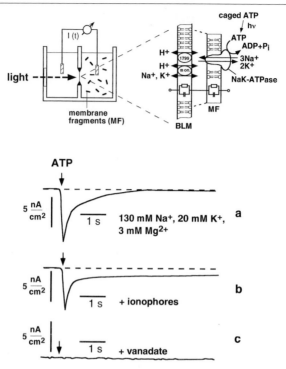

FIG. 1. (Top) Experimental setup for the bilayer experiments. Light enters the cuvette through a quartz window. ATPase-containing membrane fragments and caged ATP are only in the rear compartment. The membrane fragments adsorb to the planar bilayer as shown on the right-hand side. The capacitive coupling of the membrane fragments to the bilayer is described by the equivalent circuit. Stationary currents were measured in the presence of monensin and 1799. (Bottom) Pump currents of the Na,K-ATPase measured in a solution containing (a) 25 mM imidazole hydrochloride, pH 7.5, 130 mM NaCl, 20 mM KCl, 3 mM MgCl$_2$, 35 μM caged ATP; irradiation: 125-msec flash from a UV lamp; enzyme preparation: Na,K-ATPase from pig kidney. (b) 1 μM 1799 and 10 μM monensin were present. 1 mM vanadate inhibited the signal (c).

Addition of the H$^+$/Na,K-exchanging ionophore monensin and the protonophore 1799 renders the bilayer conductive for Na$^+$ and K$^+$ and allows the measurement of a stationary current (Fig. 1b), which demonstrates the continuous pumping activity of the enzyme. Inhibitors of Na,K-ATPase such as orthovanadate (Fig. 1c) or preincubation with ouabain[8] abolish the electrical signal.

The ion specificity of the transport activity of the Na,K-ATPase is shown in Fig. 2. In addition to Mg^{2+}, which is required for ATP binding, only Na$^+$ is necessary for the generation of an electrical signal. Based on these

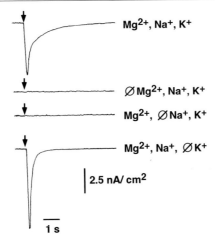

FIG. 2. Ion specificity of the signal. Conditions as in Fig. 1 except for the omission (∅) of certain cations as indicated.

results, an electrogenic step in the early Na^+-dependent steps of the reaction cycle is proposed.[8]

Caged ATP binds to the Na,K-ATPase and acts as a competitive inhibitor in the protein reaction.[10,14] The stability of the biological preparations and the power output of light sources limit the fraction of ATP that can be released from caged ATP to about 30%. Therefore, photoactivated ATP always has to compete with free caged ATP for the binding site on the protein. This is shown in Fig. 3 where a dissociation constant for caged ATP of ca. 30 μM was determined compared to a value of 1–2 μM for ATP.[10]

A rapid ATP concentration jump can be obtained by irradiation of caged ATP with a XeCl excimer laser flash (duration 10 nsec). However, the release of ATP from caged ATP is limited by the dark reaction of the decay of caged ATP, which depends on pH and on Mg^{2+} concentration.[15,16] At a temperature of 24° and a Mg^{2+} concentration of 3 mM, ATP is released with a time constant of 2.6 msec at pH 6.2 and 16 msec at pH 7.0 [for a consideration of the temperature correction see (17)]. Therefore, the best time-resolved measurements require an acidic pH.

[14] B. Forbush III, *Proc. Natl. Acad. Sci. U.S.A.* **84**, 5310 (1984).
[15] J. A. McCray, L. Herbette, T. Kihara, and D. R. Trentham, *Proc. Natl. Acad. Sci. U.S.A.* **77**, 7237 (1980).
[16] J. W. Walker, G. P. Reid, J. A. McCray, and D. R. Trentham, *JACS* **110**, 7170 (1988).
[17] K. Barabas and L. Keszthelyi, *Acta Biochim. Biophys. Acad. Sci. Hung.* **19**, 305 (1984).

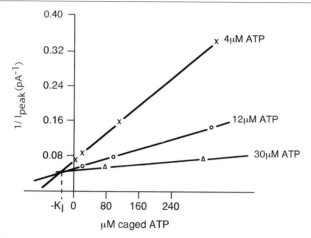

FIG. 3. Dixon plot of the peak current I_{peak} at different concentrations of caged ATP and released ATP after the light flash. The fraction of caged ATP released was varied by changing the light intensity of the UV lamp. Conditions as in Fig. 1.

Time-resolved measurements at pH 6.2 were performed with Na,K-ATPase from pig kidney[8] and eel electric organ.[13] In the absence of K⁺, a signal similar to the one shown in Fig. 2 was obtained. The reciprocal relaxation times of the rise and decay of the signal are shown in Fig. 4. In this experiment, the concentration of ATP was varied by adding different amounts of caged ATP to the electrolyte while maintaining the laser energy constant. This resulted in a constant fraction ($\eta = 0.23$) of released ATP. The influence of ATP (and caged ATP) on the reciprocal relaxation time of the decay of the signal (τ_2^{-1}) is immediately apparent whereas τ_1^{-1} remains unaffected. Therefore, the decaying phase of the signal (τ_2^{-1}) was assigned to the binding and exchange of ATP and caged ATP. This effect can be quantitatively accounted for by a simple kinetic model.[13] An independent demonstration of the competition of ATP and caged ATP comes from the slowing down of the apparent rate of phosphorylation with [γ-^{32}P]ATP in the presence of caged ATP.[13]

The rapidly rising phase, having a reciprocal relaxation time of $\tau_1^{-1} \approx$ 300 sec^{-1}, is independent of the concentration of caged ATP and is therefore assigned to the electrogenic step.[13] Different propositions have been put forward for a mechanistic explanation of an electrogenic reaction in the Na⁺-dependent steps of the reaction cycle: electrogenic Na⁺ translocation during the conformational transition $E_1P \rightarrow E_2P$ or electrogenic release of

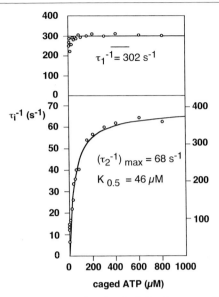

FIG. 4. Reciprocal time constants of the electrical signal generated by the release of ATP from different concentrations of caged ATP using a short (10 nsec) laser flash that released 23% ATP. Conditions: 25 mM imidazole hydrochloride, pH 6.2, 130 mM NaCl, 3 mM MgCl$_2$, 1 mM dithiothreitol; enzyme preparation: Na,K-ATPase from eel electric organ. The concentrations refer to the caged ATP concentration before the flash. The signals are similar to the ones shown in Figs. 1 and 2 except for a more rapid decay. τ_1 corresponds to the rapid rise, τ_2 to the slower decay of the signal.

Na$^+$ via a narrow extracellular access channel (for a review, see Rakowski et al.[18]).

Patch Clamp Experiments

Isolated heart cells from guinea pig (myocytes) were investigated using the giant patch technique.[1] Gigaohm seals are obtained on "blebs" from myocytes (Fig. 5).

After excision of the patch, the pipette is moved into the temperature-controlled perfusion and photolysis chamber (see Fig. 5). For more details, see Friedrich et al.[2] To photolyse the caged ATP, light pulses of an excimer laser are used.

Figures 6a and 6b show the response of the patch current to an ATP concentration jump using photolysis of caged ATP at pH 7.4 and pH 6.3,

[18] R. F. Rakowski, D. C. Gadsby, and P. de Weer, *J. Membr. Biol.* **155**, 105 (1997).

Fig. 5. Schematic view of the photolysis experiment on a giant excised plasma membrane patch from a ventricular heart cell. (Insert) Cardiac myocyte with a giant patch pipette (tip diameter = 20 μm) approaching a plasma membrane "bleb" (arrow).

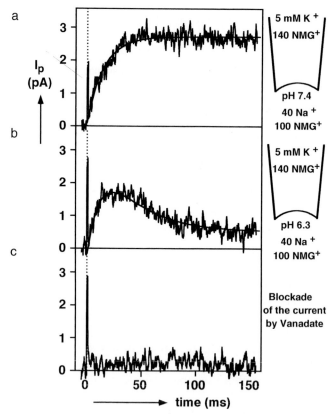

Fig. 6. Patch clamp current recordings. (a) Response to photolysis of 500 μM caged ATP at pH 7.4. (b) Response in a different patch at pH 6.3. Rat cell, 24°. Caged ATP concentration 250 μM. (c) Response to photolysis of 500 μM caged ATP on the same patch as in (a), after inhibition of the stationary pump current by 1 mM orthovanadate. Bath solution: B-P. Pipette solution: P-NMG-K. Guinea pig cell, 24°.

respectively. Figure 6c demonstrates the inhibition of this signal by pretreatment of the same patch with orthovanadate. In contrast to experiments at pH 7.4, photolytic release of ATP at pH 6.3 results in a fast transient outward current, followed by a stationary outward current, which best-fit using a sum of two exponential functions and a constant. The laser flash-induced signals are measured at different caged ATP concentrations, but at a constant fraction η of released ATP. The rise of the signal at pH 6.3 is characterized by an ATP-independent rate constant k_1 of ~200 sec^{-1} (at 24°) and the decay by an ATP-saturable rate constant k_2.

We found an exponentially increasing pump current on photolysis of caged ATP at pH 7.4 (Fig. 6a). The rate of this increase is 70 sec^{-1}, but this rate is dependent on the caged ATP concentration and light intensity. From this it has been concluded that the rise time of the signal at pH 7.4 is determined by the release of ATP from caged ATP. Therefore, all studies are performed at pH 6.3, where the release of ATP proceeds with a time constant of 3.2 msec.[16,17]

The current traces obtained with the patch clamp and the bilayer technique are similar in the fast time range. At $t > 100$ msec the patch clamp current has decayed to a stationary current whereas the bilayer signal decays further and becomes negative at $t > 200$ msec. The latter behavior is attributable to the capacitive coupling of the membrane fragments to the planar bilayer.[13] Capacitive coupling does not alter the rate constants but only adds the time constant of the system and distorts the amplitudes of the signal.[13] Therefore, an analysis of the rate constants is meaningful whether or not the ion pump is capacitively coupled to the measuring system.

The rate constants obtained by both methods at different concentrations of released ATP depend in the same way on the ATP concentration and have approximately the same values. Both techniques show a rapid rise of the electrical signal with a rate constant of ~200 sec^{-1}, which is independent of the ATP concentration at 24°. However, the decay of the signal is clearly ATP-dependent with an affinity of 10 to 20 μM. Based on experiments on planar lipid membranes, we have previously assigned the rising phase of the signal to a fast electrogenic step with a rate of ~200 sec^{-1}.[9,13] These conclusions are now confirmed by the results of our patch clamp study using cardiac myocytes. This rules out tissue-specific behavior as a reason for discrepancies found in the literature.

We conclude that phosphorylation of cardiac Na,K-ATPase is fast and the electrogenic E_1P-E_2P conformational change proceeds with a rate constant of ~200 sec^{-1} at 24°. This is not contradictory to an electroneutral E_1P-E_2P conformational change (with a rate constant of 200 sec^{-1}) followed by a fast electrogenic Na$^+$ release,[19] as put forward by access channel models for the Na,K-ATPase.[20] In our experiments this mechanism would yield the same results because it is kinetically equivalent to an electrogenic E_1P–E_2P conformational change.

ADP, ATP-Carrier

The ADP, ATP-carrier (AAC) is the most abundant protein in the inner mitochondrial membrane. Its function consists of the export of ATP from the mitochondria into the cytosol of the cell. The transport occurs in exchange with ADP, which is used for the production of ATP via oxidative phosphorylation. The function of the carrier has been studied in the past in mitochondria as well as in reconstituted proteoliposomes. It was shown that the AAC is potential sensitive and that the electrogenic event is related to the ATP translocation.[21,22] Roughly one net charge per ATP is transported.[23,24]

Because the ADP/ATP exchange was assumed to be electrogenic, the transport of the nucleotides has to be accompanied by electrical currents. The carrier is an ideal object for the application of caged ATP or caged ADP. To study its electrical transport, proteoliposomes reconstituted with AAC are adsorbed to a planar lipid bilayer as described earlier for Na,K-ATPase. The configuration of liposomes and planar bilayer is shown in Fig. 7. Caged ATP and caged ADP [P^3-1-2-nitro)phenylethyladenosine 5'-triphosphate or diphosphate] are prepared as described earlier.[4,8] To photolyse the caged ATP, 125-msec light pulses of a high-pressure 100-W Hg arc lamp are used.

Although ion pumps create ion gradients by the use of the chemical energy from ATP, secondary active transporters such as the AAC need gradients as a driving force for the transport. This implies that proteoliposomes preloaded with ATP or ADP in the case of the AAC have to be impermeable to nucleotides. Because of their charged nature the nucleotides can be retained in the liposomes over a period of days. Also, in the presence of external caged nucleotides, no exchange or loss of the internal nucleotides occurs. From this a second prerequisite for the successful appli-

---

[19] D. W. Hilgemann, *Science* **263**, 1429 (1994).
[20] D. C. Gadsby, R. F. Rakowski, and P. De Weer, *Science* **260**, 100 (1993).
[21] E. Pfaff and M. Klingenberg, *Eur. J. Biochem.* **6**, 66 (1968).
[22] M. Klingenberg and H. Rottenberg, *Eur. J. Biochem.* **73**, 125 (1977).
[23] K. LaNoue, S. M. Mizani, and M. Klingenberg, *J. Biol. Chem.* **253**, 191 (1978).
[24] R. Krämer and M. Klingenberg, *Biochemistry* **21**, 1082 (1982).

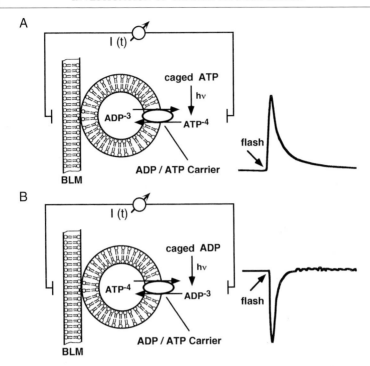

FIG. 7. Capacitive currents by the ADP,ATP carrier after a light-induced concentration jump of ATP or ADP. Proteoliposomes preloaded with ADP (A) and ATP (B). The sign of the current corresponds to a movement of negative charges from the outside into the lumen of the lipid vesicles. (A) The oppositely directed current in (B) corresponds to the reversed application of the nucleotides; 100 μM caged nucleotide was applied. One UV flash releases 30% of ATP. Flash duration was 125 msec with 2 W/cm^2 intensity. 0.1 M NaCl, pH 7.2.

cation of the caged nucleotides is fulfilled, namely that the caged nucleotides are not transported by the AAC.

In principle, six different transport modes can be studied. In the "normal" heteroexchange with either ADP or ATP inside the vesicles (caged ATP$_{ex}$/ADP$_{in}$ and caged ADP$_{ex}$/ATP$_{in}$), the current is initiated by photolysis either of caged ATP or caged ADP. In both cases, the obligatory counter exchange is observed. Figure 7 shows such an experiment. As expected, the transient capacitive currents are oppositely directed. From the sign of the current it was derived that negative charge during ATP/ADP exchange is translocated concomitant with ATP transport.[25] In addition to the hetero-

[25] N. Brustovetsky, A. Becker, M. Klingenberg, and E. Bamberg, *Proc. Natl. Acad. Sci. U.S.A.* **93**, 664 (1996).

exchange ATP/ADP, the homoexchange ADP_{ex}/ADP_{in} and ATP_{ex}/ATP_{in} are also of interest as well as the zero-trans conditions $ATP_{ex}/0_{in}$ and $ADP_{ex}/0_{in}$. From these results, information about the electrical nature of the transport is obtained. It was demonstrated that all transport modes are accompanied with an electrical current and therefore with charge movements. The current due to the ADP transport is connected with a movement of positive charge.

In all systems, whether in hetero- or homoexchange modes or in unloaded vesicles, the currents have an opposite sign depending on whether they started with caged ATP or with caged ADP. This agrees with the postulate that the charge difference between ATP^{4-} and ADP^{3-} is responsible for the currents. However, it contradicts the concept derived from other experimental evidence that ATP and not ADP carries the charge. Thus, it has been suggested that ATP^{4-} combines with three positive charges of the binding center forming $ATP^{4-}C^{3+}$ with one negative excess charge, whereas ADP^{3-} would neutralize the binding center forming $ADP^{3-}C^{3+}$.[26] The results obtained under homoexchange conditions lead to the conclusion that ATP^{4-} binds with a negative charge excess and ADP^{3-} binds with a positive charge excess to the carrier. With 2 ATP^{4-} one negative charge and with 2 ADP^{3-} one positive charge are moved across the membrane or, alternatively, one carrier site works alone, giving rise to the net transfer of a fractional charge equivalent to one-half of an elementary charge: $(ATP^{4-}C^{3.5+})^{0.5-}$ and $(ADP^{3-}C^{3.5+})^{0.5+}$.

The experiments just shown demonstrate the possibilities of measuring currents associated with AAC-catalyzed ADP/ATP transport or with interaction of AAC and ligands. Because of high time resolution, this will allow us to study the kinetics of the charge translocation in the AAC and thus give us more information about the single steps in the reaction cycle of the carrier similarly as obtained in the earlier work on ATP-driven ion pumps. The approach used in the present work may also be applied to study other carriers.

Caged Ca^{2+} and Na,Ca-Exchanger

Photolabile Ca^{2+} chelators (caged Ca^{2+}) have been used to study the Na^+,Ca^{2+} exchange current from mammalian heart[27-30] and lobster skeletal

---

[26] N. Brustovetsky and M. Klingenberg, *J. Biol. Chem.* **269**, 27329 (1994).
[27] T. A. Powell, A. Noma, T. Shioya, and R. Z. Kozlowski, *J. Physiol. (Lond.)* **472**, 45 (1993).

muscle.[31] In the experiments described in this article, the EDTA derivative DM-nitrophen[32] was used to study the presteady-state current of the Na,Ca-exchanger from rat and guinea pig ventricular cells by means of the giant patch clamp technique[1] (see also the section on the Na,K-ATPase) on the submillisecond time scale.[33] DM-nitrophen is suitable for these studies because it is photolyzed rapidly ($\tau \leq 30$ μsec[34]) and with a dramatic change of the affinity for Ca^{2+} from 0.5 nM to 3 mM. Using a laser flash (308 nm, 10 nsec, 480 mJ/cm^2), 20% of the chelator can be cleaved and the Ca^{2+} concentration jumps from 0.1 to 100 μM. Because the affinity of the unphotolyzed DM-nitrophen fits to the apparent affinity of the transport sites for Ca^{2+} on the cytoplasmic side of the Na,Ca-exchanger (5 μM), it allows for the adjustment of suitable free Ca^{2+} concentrations (≤ 0.5 μM) without an excess of chelator over total Ca^{2+} (see Fig. 8), which would lead to so-called "Ca^{2+} spikes"[35] after photolysis by a flash.

In addition to transport-binding sites, a regulatory Ca^{2+}-binding site has been identified on the cytoplasmic side of the exchanger[36,37] that inactivates the exchanger if no Ca^{2+} is bound. The regulatory action of this site can be eliminated by a mild treatment with chymotrypsin.[37a] Because the Ca^{2+} affinity of the regulatory site is higher than that of the transport site it has only been possible to study its action in the reverse mode of the Na,Ca-exchanger[38] or in mutants with drastically reduced Ca^{2+} affinity.[39] It could be shown, however, that photolytic Ca^{2+} concentration jumps allow the experimenter to distinguish between both Ca^{2+}-binding sites on the wild-type exchanger under conditions of forward Na^+, Ca^{2+}-exchange.[33]

Figure 8 shows the current signal obtained from an excised patch from a ventricular cell after the photolytic release of Ca^{2+} (0.5 → 100 μM) on

---

[28] E. Niggli and W. J. Lederer, *Nature* **349,** 621 (1991).
[29] E. Niggli and W. J. Lederer, *Biophys. J.* **65,** 882 (1993).
[30] E. Niggli and P. Lipp, *Biophys. J.* **67,** 1516 (1994).
[31] A. Eisenrauch, M. Juhaszova, G. C. R. Ellis-Davies, J. H. Kaplan, E. Bamberg, and M. P. Blaustein, *J. Membr. Biol.* **145,** 151 (1995).
[32] J. Kaplan and G. C. R. Ellis-Davies, *Proc. Natl. Acad. Sci. U.S.A.* **85,** 6571 (1988).
[33] M. Kappl and K. Hartung, *Biophys. J.* **71,** 2473 (1996).
[34] G. C. R. Ellis-Davies, J. H. Kaplan, and R. J. Barsotti, *Biophys. J.* **70,** 1006 (1996).
[35] R. S. Zucker, *Cell Calcium* **14,** 87 (1993).
[36] R. DiPolo and L. Beaugé, *Biochim. Biophys. Acta* **854,** 298 (1986).
[37] J. Kimura, A. Noma, and H. Irisawa, *Nature* **319,** 596 (1986).
[37a] W. Stürmer, H.-J. Apell, I. Wuddel, and P. Läuger, *J. Membr. Biol.* **110,** 67 (1989).
[38] D. W. Hilgemann, A. Collins, and S. Matsuoka, *J. Gen. Physiol.* **100,** 993 (1992).
[39] S. Matsuoka, D. A. Nicoll, L. V. Hryschko, D. O. Levitsky, J. N. Weiss, and K. D. Philipson, *J. Gen. Physiol.* **105,** 403 (1995).

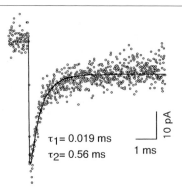

FIG. 8. Current recorded from an excised membrane patch (guinea pig ventricular cell) after a saturating Ca^{2+} concentration jump under conditions that promote inward Na–Ca exchange current (forward mode of Na–Ca exchange). The time course has been fit by using a sum of two exponential functions plus a constant. The rising phase is determined by the rise time of the recording system (15 μsec). The extracellular solution contained (in mM) 100 Na^+, 20 Cs^+, 10 TEA, 2 Mg^{2+}, 119 Cl^-, 10 EGTA, 10 HEPES, 0.02 Verapamil. The solution on the cytoplasmic side contained (in mM): 100 Li^+, 0.44 Ca^{2+}, 0.45 DM-nitrophen, 20 Cs^+, 20 TEA, 10 HEPES. The Ca^{2+} concentration before the flash was 0.5 μM. After the laser flash, Ca^{2+} increased to >100 μM, saturating the cytoplasmic transport sites. All solutions were adjusted to pH 7.1. The temperature was maintained at 21°. The cytoplasmic side of the membrane patch has been treated with α-chymotrypsin (1 mg/ml) for 1 min to eliminate the action of the regulatory-binding site.

the cytoplasmic side under conditions that promote forward Na^+,Ca^{2+}-exchange, i.e., Na_{ex} = 100 mM, Ca_{ex} = 0, Na_{in} = 0. For further experimental details, see the figure legend. Negative current flow indicates an inward current from the extracellular to the intracellular side. The current rises within 50 μsec to a peak and declines exponentially to a plateau with a time constant of about 0.6 msec. The rising phase cannot be resolved because it is disturbed by an initial artifact due to the dissipation of laser energy. This artifact, which varies in size from experiment to experiment, could be reduced by using petri dishes made from glass instead of from polystyrene. A further reduction could be achieved by short-circuiting the input of the Axopatch 200A amplifier (Axon Instruments, Foster City, CA) for 50 μsec beginning 20 μsec before the laser flash is triggered. Thus the rising phase of the current signal does not provide useful information on the kinetics of the Na,Ca-exchanger.

The direction of stationary current is compatible with the $3Na^+/1Ca^{2+}$ stoichiometry of the exchanger. With respect to the transient current we have suggested that the presteady-state current is mainly determined by the outward translocation of Ca^{2+}.[33] If this is true, a plausible assumption is that the exchanger moves negative net charge toward the extracellular

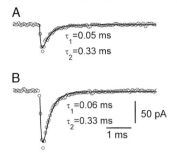

FIG. 9. Transient current signals recorded from an excised membrane patch under conditions promoting Ca–Ca exchange before (A) and after the application of chymotrypsin (B). The transient currents can be described by a sum of two exponential functions without a stationary component. The free Ca^{2+} concentration before the flash was adjusted to 0.1 μM; after the flash it was >100 μM. The pipette solution contained (in mM) 100 Li^+, 5 Ca^{2+}, 20 TEA, 20 Cs^+, 10 HEPES, 0.02 verapamil. The cytoplasmic solution contained 100 Li^+, 0.44 Ca^{2+}, 0.455 DM-nitrophen, 20 Cs^+, 20 TEA, 10 HEPES, 130 Cl^-. The bandwidth of the recording system was 5 kHz. Other conditions as described in Fig. 8.

phase during Ca^{2+} translocation. This idea is supported by Ca^{2+} concentration jumps under conditions that promote Ca–Ca exchange, which is electroneutral in the stationary state. Figure 9 shows a current signal generated by a Ca^{2+} concentration jump in the absence of extracellular Na^+ with 5 mM Ca^{2+} in the pipette. Under the conditions of Ca–Ca exchange, a transient inward current is observed that decays with a time constant of 0.3 msec. No stationary current is observed in the absence of extracellular Na^+ in agreement with the electroneutrality of Ca–Ca exchange. The observation of a transient inward current following a Ca^{2+} concentration jump in both transport modes, forward Na–Ca exchange and Ca–Ca exchange, is most easily explained by assuming that the initial inward current is related to the outward movement of Ca^{2+}. If net charge movement of the Ca-loaded enzyme occurs in the same direction as the Ca^{2+} translocation, then one has to assume that the enzyme carries negative charge in this state. A different situation has been observed in the Ca-ATPase from sarcoplasmic reticulum that carries positive charge if it is loaded with Ca^{2+}.[40]

The application of rapid Ca^{2+} concentration jumps offers a unique chance to study the action of the regulatory Ca^{2+}-binding site under conditions that allow for the binding of Ca^{2+} to both sites (i.e., forward Na–Ca exchange and Ca–Ca exchange). Partial inactivation of the Na,Ca-exchanger can be induced by lowering the preflash Ca^{2+} concentration to about 0.1 μM.[33] Under these conditions a Ca^{2+} concentration jump elicits

---

[40] K. Hartung, J. P. Froehlich, and K. Fendler, *Biophys. J.* **72**, 2503 (1997).

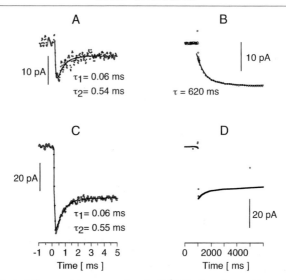

FIG. 10. Effect of the regulatory cytoplasmic Ca^{2+}-binding site on the stationary forward Na–Ca exchange current. Current signals following a Ca^{2+} concentration jump were recorded at two different time resolutions before (A,B) and after (C,D) treatment of the patch with α-chymotrypsin (1 mg/ml) for 40 sec. The preflash Ca^{2+} concentration was less than 0.1 μM. Other conditions as in Fig. 9. At high temporal resolution (A,C), the time course of the current signals is identical, but the amplitudes of transient and stationary current are considerably increased after the treatment with chymotrypsin. At a lower resolution another effect of the treatment with chymotrypsin is visible. In the untreated membrane, an initial rise, corresponding to the stationary amplitude shown in (A), is followed by a slowly rising phase that can be fit using an exponential function with a time constant τ = 90.6 sec. Because this secondary rise is abolished by the treatment with chymotrypsin (C), it has been attributed to the binding of Ca^{2+} to the regulatory binding site.

a signal as shown in Fig. 10A which, on the time scale of microseconds, is comparable to the signal shown in Fig. 8. However, on a slower time scale (Fig. 10B) it can be seen that there is a secondary increase of the stationary current with a time constant of about 0.6 sec. This secondary rise of outward current is attributed to the action of the regulatory Ca^{2+}-binding site because it is eliminated if either the preflash Ca^{2+} concentration is >0.5 μM or if the membrane has been treated with chymotrypsin (Fig. 10D). Both means are known to release the exchanger from the inactivated state.[41] These experiments show that Ca^{2+} binding to the regulatory binding sites is much slower than Ca^{2+} binding to the transport sites.

Treatment with chymotrypsin increases the amplitude of stationary and peak inward current in both transport modes without modification of the

[41] D. W. Hilgemann, *Nature* **344**, 242 (1990).

TABLE I
CHARGE TRANSPORT OF ION-TRANSLOCATING PROTEINS USING CAGED SUBSTRATES

Pump	Technique[a]	Caged compound	Ref.
Na,K-ATPase	MF	Caged ATP	8–13, 37a, 47–49
	INT		44
	PC		2, 50
H,K-ATPase	MF	Caged ATP	5, 6
Ca-ATPase	MF	Caged ATP	40, 45
	INT		52
F_0F_1-ATPase	LIP	Caged ATP	51
Kdp-ATPase	LIP	Caged ATP	46
Na,Ca-exchanger	PC	Caged Ca^{2+}	33
	MF		31
ADP, ATP-exchanger	LIP	Caged ATP	25
		Caged ADP	

[a] MF, Purified or native membrane fragments or vesicles adsorbed to a planar bilayer; LIP, purified ion pumps reconstituted into liposomes adsorbed to a planar bilayer; INT, ion pumps integrated in a planar bilayer; PC, giant patch technique.

kinetics of the transient current, indicating that a fraction of transporter molecules is fully inactivated at low cytosolic Ca^{2+} concentrations.[42]

The observation of a transient inward current after a saturating Ca^{2+} concentration jump in the Ca–Ca and the Na–Ca exchange mode, which is followed by a stationary inward current in the Na–Ca exchange mode, provides some insight into the reaction cycle of the Na,Ca exchanger (for details, see Kappl and Hartung[33]). First, it shows that Ca^{2+} binding to the transport sites on the cytosolic side is fairly rapid ($k_{on} > 10^8\ M^{-1}\ sec^{-1}$) because the time to peak is $<50\ \mu sec$. Second, it shows that Ca^{2+} translocation is electrogenic and, together with previous findings that Na^+ translocation is a major electrogenic transition, it indicates that the Na,Ca-exchanger contains at least two electrogenic steps. Third, it demonstrates that a transition following the first electrogenic step (Ca^{2+} translocation) is rate limiting, e.g., Na^+ translocation.

Conclusions

We have given a few examples of transport ATPases and carriers that have been investigated using caged substrates. The advantage of this approach is the high time resolution, which allows the investigation of pre-

[42] D. W. Hilgemann, *Biophys. J.* **71**, 759 (1996).

steady-state kinetics. In addition to the systems described here, a number of other ion-translocating ATPases were investigated using caged ATP: the Ca-ATPase from sarcoplasmatic reticulum, the H,K-ATPase from stomach, the F_0F_1-ATPase from chloroplasts, and the Kdp-ATPase from the plasma membrane of *Escherichia coli*. These systems are summarized together with the respective electrophysiological technique in Table I.

In the case of caged ATP the time resolution is limited by the time constant of ATP release from the caged compound used in those studies [P^3-1-(2-nitro) phenylethyladenosine 5'-triphosphate]. By working at a somewhat acidic pH (6.2), ATP is released rapidly enough (within ca. 3 msec) to study most of the enzymatic reactions of interest. Nevertheless, compounds with an improved time resolution at neutral pH are desirable and will be available in the near future.[43] In the case of caged Ca^{2+}, the release is extremely rapid so that the limit there is still set by the time resolution of the electrophysiological techniques.

Acknowledgments

The results presented in this article are based on the work of numerous colleagues and collaborators. Special thanks go to M. Klingenberg and N. Brustovetsky, University of Munich, and to M. Kappl and T. Friedrich, Max-Planck Institut für Biophysik, Frankfurt. The work was supported by the DFG (SFB 169 and 472) and the Max-Planck Society.

[43] R. S. Givens, J. F. W. Weber, A. H. Jung, and C.-H. Park, *Methods Enzymol.* **291** [29] (1998) (this volume).
[44] A. Eisenrauch, E. Grell, and E. Bamberg, "The Sodium Pump: Structure, Mechanism, and Regulation." Rockefeller Univ. Press, New York, 1991.
[45] K. Hartung, E. Grell, W. Hasselbach, and E. Bamberg, *Biochim. Biophys. Acta* **900,** 209 (1987).
[46] K. Fendler, S. Dröse, K. Altendorf, and E. Bamberg, *Biochemistry* **35,** 24, 8009 (1996).
[47] S. Heyse, I. Wuddel, H.-J. Apell, and W. Stürmer, *J. Gen. Physiol.* **104,** 197 (1994).
[48] I. Wuddel and H.-J. Apell, *Biophys. J.* **69,** 909 (1995).
[49] R. Borlinghaus and H.-J. Apell, *Biochim. Biophys. Acta* **939,** 197 (1988).
[50] T. Friedrich and G. Nagel, *Biophys. J.* **73,** 186 (1997).
[51] B. Christensen, M. Gutweiler, E. Grell, N. Wagner, R. Pabst, K. Dose, and E. Bamberg, *J. Membr. Biol.* **104,** 179 (1988).
[52] A Eisenrauch and E. Bamberg, *FEBS Lett.* **268,** 152 (1990).

[18] Studies of Molecular Motors Using Caged Compounds

By JODY A. DANTZIG, HIDEO HIGUCHI, and YALE E. GOLDMAN

Introduction

The dynamic nature of cell motility and the macromolecular organization of contractile structures in cells make the study of muscle contraction and nonmuscle cell motility natural targets for the application of photolabile compounds and flash photolysis. After the synthesis of caged ATP and its application to Na^+-K^+-ATPase pumps were reported by Kaplan *et al.*,[1] the study of muscle contraction was one of the first fields to benefit from this new technology.[2]

Cell motility includes contraction of skeletal, cardiac, and smooth muscles, locomotion of nonmuscle cells, dynamic alteration of cell shape by rearrangements of the cytoskeleton, intracellular motions and targeting of organelles, karyokinesis (congression and segregation of the chromosomes to the daughter cells in cell division), cytokinesis (splitting of the progenitor cell into two daughter cells), and other motions. The motor proteins that carry out these essential functions are members of three superfamilies: the myosins, kinesins, and dyneins.

Myosin, in muscle or in the cytoplasm of nonmuscle cells, slides along actin filaments (F-actin) toward their "barbed" (plus) ends using the energy liberated from the hydrolysis of cellular ATP to ADP and orthophosphate (P_i). More than 13 subgroups of myosins have been identified with homologous N-terminal globular regions, termed the heads or motor domains, but with highly variable molecular mass (<100–500 kDa), assembly, enzymatic activity, and function.[3] In muscle, two-headed myosin molecules polymerize into bipolar filaments that interdigitate with F-actin to form almost crystalline cylindrical organelles (~1 μm in diameter) termed myofibrils.

Kinesins are smaller motor proteins (~120 kDa) that translocate along microtubules (stiff cytoskeletal structures polymerized from α- and β-tubulins) toward the plus end (the fast growing end, generally peripheral in cells). Other members of the kinesin superfamily translocate toward the minus (slower growing, central) end of microtubules. Dozens of kinesin-like proteins have been identified with roles in mitosis and targeted intracellular

---

[1] J. H. Kaplan, B. Forbush III, and J. F. Hoffman, *Biochemistry* **17,** 1929 (1978).
[2] Y. E. Goldman, M. G. Hibberd, J. A. McCray, and D. R. Trentham, *Nature* **300,** 701 (1982).
[3] M. S. Mooseker and R. E. Cheney, *Annu. Rev. Cell Dev. Biol.* **11,** 633 (1995).

transport.[4] Dynein is a giant (1000–2000 kDa) motor protein that slides toward the minus end of microtubules.[5,6] Flagellar dyneins drive the bending of eukaryotic axonemes and cytoplasmic dyneins perform targeted transport of intracellular vesicles and nuclear migration. Kinesins and dyneins also transduce the energy liberated by the hydrolysis of ATP to perform mechanical work.

The cytoskeletal structures, filamentous actin (F-actin) and microtubules, are tracks that guide and direct the translocation motions. A diverse group of proteins that bind to actin or tubulin control the remarkably dynamic formation and remodeling of these cytoskeletal structures and regulate the activity of molecular motors.[7,8]

Solution of the crystal structures of actin[9] and the motor domains of myosin[10] and kinesin[11,12] has extended research on the functional mechanisms of these proteins to the amino acid and even the atomic level. The catalytic ATP-binding core of kinesin and myosin have surprisingly similar secondary structure even though there is little sequence homology. The atomic structure of this catalytic core also maps closely onto the nucleotide-binding regions of many otherwise unrelated enzymes, including creatine kinase,[13] the mitochondrial ATP synthase,[14] and heterotrimeric and small GTP-binding proteins, such as α-transducin,[15] EF-Tu,[16] and the H-*ras* p21 oncogene.[17] Thus understanding the operation of motor proteins may reveal general principles in enzyme biophysics and help to unravel the molecular mechanisms of these other signaling and energy transducing enzymes.

Some features of the chemomechanical events in the energy converting ATPase cycle are similar among the various motor proteins (Fig. 1). Each protein associates and dissociates with its cytoskeletal track during the

[4] L. S. B. Goldstein, *Annu. Rev. Genet.* **27**, 319 (1993).
[5] E. L. F. Holzbaur and R. B. Vallee, *Annu. Rev. Cell Biol.* **10**, 339 (1994).
[6] I. R. Gibbons, *Cell Struct. Funct.* **21**, 331 (1996).
[7] T. D. Pollard, S. Almo, S. Quirk, V. Vinson, and E. E. Lattman, *Annu. Rev. Cell Biol.* **10**, 207 (1994).
[8] R. B. Vallee, *Cell Motil. Cytoskel.* **15**, 204 (1990).
[9] W. Kabsch, H. G. Mannherz, D. Suck, E. F. Pai, and K. C. Holmes, *Nature* **347**, 37 (1990).
[10] I. Rayment, W. R. Rypniewski, K. Schmidt-Bäse, R. Smith, D. R. Tomchick, M. M. Benning, D. A. Winkelmann, G. Wesenberg, and H. M. Holden, *Science* **261**, 50 (1993).
[11] F. J. Kull, E. P. Sablin, R. Lau, R. J. Fletterick, and R. D. Vale, *Nature* **380**, 550 (1996).
[12] E. P. Sablin, F. J. Kull, R. Cooke, R. D. Vale, and R. J. Fletterick, *Nature* **380**, 555 (1996).
[13] K. Fritz-Wolf, T. Schnyder, T. Wallimann, and W. Kabsch, *Nature* **381**, 341 (1996).
[14] J. P. Abrahams, A. G. W. Leslie, R. Lutter, and J. E. Walker, *Nature* **370**, 621 (1994).
[15] J. P. Noel, H. E. Hamm, and P. B. Sigler, *Nature* **366**, 654 (1993).
[16] H. Berchtold, L. Reshetnikova, C. O. A. Reiser, N. K. Schirmer, M. Sprinzl, and R. Hilgenfeld, *Nature* **365**, 126 (1993).
[17] E. F. Pai, U. Krengel, G A. Petsko, R. S. Goody, W. Kabsch, and A. Wittinghofer, *EMBO J.* **9**, 2351 (1990).

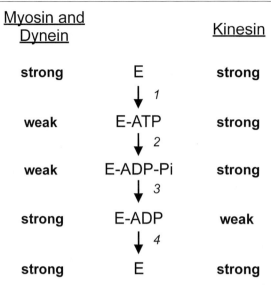

FIG. 1. Biochemical scheme for the ATPase activity of motor proteins (E). "Strong" and "weak" refer to the affinity of the motor domain of the proteins for their respective cytoskeletal tracks, actin filaments in the case of myosin and microtubules for dynein and kinesin. In strongly bound states, the motor proteins are mostly bound to the track and can exert sliding force to translocate their cargo.

enzymatic cycle, and the biochemical transitions are probably associated with structural changes that lead to the generation of force or filament sliding. Myosin typically is tuned for rapid filament sliding, which requires the cooperative action of many molecules. Myosin heads are strongly bound to actin in the nucleotide-free and ADP-bound states and associate weakly with actin in the ATP- and ADP–P_i- bound states.[18,19] The transition to force generation seems to be associated with the release of P_i (Fig. 1, step 3) from actomyosin–ADP–P_i.[20–22]

Kinesin translocates more slowly than myosin, but a single kinesin molecule, with two motor domains, can slide along a microtubule without diffusing away, presumably by alternate power strokes of the two heads.[23] The

[18] R. W. Lymn and E. W. Taylor, *Biochemistry* **10,** 4617 (1971).
[19] E. Eisenberg and L. E. Greene, *Annu. Rev. Physiol.* **42,** 293 (1980).
[20] M. G. Hibberd, J. A. Dantzig, D. R. Trentham, and Y. E. Goldman, *Science* **228,** 1317 (1985).
[21] J. A. Dantzig, Y. E. Goldman, N. C. Millar, J. Lacktis, and E. Homsher, *J. Physiol.* **451,** 247 (1992).
[22] R. Cooke and E. Pate, *Biophys. J.* **48,** 789 (1985).
[23] D. D. Hackney, *Nature* **377,** 448 (1995).

kinesin head is strongly associated with a microtubule in the nucleotide-free, ATP- and ADP–P_i-bound states and weakly associated in the ADP-bound state.[24,25] The pathway of the dynein ATPase is more similar to that of myosin than kinesin.[26]

In organized systems capable of transducing chemical to mechanical energy, the kinetics and pathway of the elementary reaction steps within the ATPase cycles can be probed by photorelease of specific ligands, such as ATP,[27] ADP,[28] P_i,[21,29] and Ca^{2+},[30,31] or by the photochelation of Ca^{2+}.[31] Sudden photoliberation of these biological activators and inhibitors initiates the motor protein reactions from well-defined starting configurations or perturbs the dynamics of their enzymatic cycles. Photolysis avoids temporal limitations due to diffusion of ligands into the macromolecular assemblies and mechanical disruption associated with exchange or mixing of solutions. The mechanical and structural responses initiated by photolysis indicate the course of events and their rates. Varying the mechanical conditions, such as external load, identifies which steps are dependent on mechanical strain and thereby likely to be associated with force generation and which reactions control the overall ATPase rate. Another potential application of photolysis in motility studies is to deliver substrate or signaling molecules at highly localized regions in functioning intact cell systems.

This article summarizes methods developed to study cell motility using photolysis of photolabile precursors of nucleotides, nucleotide analogs, and Ca^{2+}. Applications of caged compounds to signalling in muscle are described elsewhere in this volume. We describe preparations of motor proteins, apparatus, appropriate solutions, reagents, caged compounds, photolysis light sources, special precautions required for photolysis experiments on motor proteins, optimization of the caged molecule concentration and photolysis chamber dimensions, binding of Ca^{2+} and Mg^{2+} to caged ATP, and inhibition of sliding velocity of myosin and kinesin by caged ATP. Many of these considerations should be applicable to other macromolecular systems.

---

[24] L. Romberg and R. D. Vale, *Nature* **361**, 168 (1993).
[25] I. M.-T. C. Crevel, A. Lockhart, and R. A. Cross, *J. Mol. Biol.* **257**, 66 (1996).
[26] D. D. Hackney, *Annu. Rev. Physiol.* **58**, 731 (1996).
[27] Y. E. Goldman, *Annu. Rev. Physiol.* **49**, 637 (1987).
[28] Z. Lu, R. L. Moss, and J. W. Walker, *J. Gen. Physiol.* **101**, 867 (1993).
[29] N. C. Millar and E. Homsher, *J. Biol. Chem.* **265**, 20234 (1990).
[30] T. D. Lenart, T. St. C. Allen, R. J. Barsotti, G. C. R. Ellis-Davies, J. H. Kaplan, C. Franzini-Armstrong, and Y. E. Goldman, in "Mechanism of Myofilament Sliding in Muscle Contraction" (H. Sugi and G. H. Pollack, eds.), p. 475. Plenum Press, New York, 1993.
[31] C. C. Ashley, I. P. Mulligan, and T. J. Lea, *Quart. Rev. Biophys.* **24**, 1 (1991).

Actomyosin ATPase Activity in Muscle Fibers

Preparation of Muscle Fibers

The most common sample for photolysis experiments on muscle is a segment of a single fiber, 2–3 mm in length, dissected from a bundle of glycerol-extracted fibers of rabbit psoas muscle.[32] This preparation is convenient, retains good mechanical integrity, and can be compared to many experiments in the literature on isolated, purified proteins from rabbit muscle. Glycerol extraction effectively removes the diffusion barrier of the surface membrane (sarcolemma), allowing rapid (1–2 min) equilibration of bathing constituents, such as caged molecules, with the interior of the myofilament lattice.[32] The procedure for preparing muscle is to expose the psoas muscle anteriorly, tie small (<0.5–1 mm diameter) bundles of fibers to glass or wooden sticks and soak them in a membrane disruption solution containing a high concentration of ATP, low free [Ca^{2+}] and a cocktail of protease inhibitors. The solution is gently stirred at 4° overnight, taking care not to mechanically agitate the muscles. The bundles are then transferred to a solution with Mg-ATP and 50% glycerol to further extract the membranes and stored at $-20°$ for up to 6 weeks.[32] Single fiber segments are separated from bundles under silicone oil or in glycerol relaxing solution, with low free [Ca^{2+}] and high Mg-ATP concentration, maintained at 10–12° to minimize proteolysis and bacterial contamination. Glycerol-extracted psoas muscle fibers generate ca. 200 kN·m^{-2} of isometric (fixed length) force at 20° and shorten with maximum filament sliding velocities[33] of 2.1 μm·sec^{-1} at 15°. These values are similar to the performance of intact mammalian muscle fibers, indicating that the energy transduction mechanism is preserved through removal of the membrane and cold storage.

Single fibers from rabbit and frog muscle are also permeabilized by treatment with detergents[34] or by removal of the sarcolemma by mechanical dissection.[35] Insect fibrillar flight muscle fibers are prepared by glycerol extraction.[36] Bundles of cardiac[37] or smooth muscle[38] cells are also amenable

---

[32] Y. E. Goldman, M. G. Hibberd, and D. R. Trentham, *J. Physiol.* **354,** 577 (1984).
[33] R. L. Moss, *J. Muscle Res. Cell Motil.* **3,** 295 (1982).
[34] L. C. Yu and B. Brenner, *Biophys. J.* **55,** 441 (1989).
[35] Y. E. Goldman and R. M. Simmons, *J. Physiol.* **350,** 497 (1984).
[36] M. Yamakawa and Y. E. Goldman, *J. Gen. Physiol.* **98,** 657 (1991).
[37] H. Martin and R. J. Barsotti, *Biophys. J.* **66,** 1115 (1994).
[38] K. Horiuti, A. V. Somlyo, Y. E. Goldman, and A. P. Somlyo, *J. Gen. Physiol.* **94,** 769 (1989).

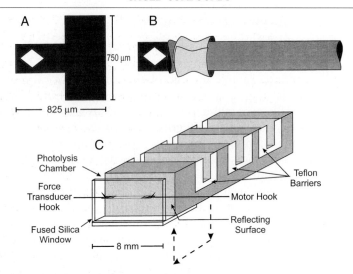

FIG. 2. Mechanical arrangements for mounting single muscle fibers in a photolysis setup and for exchanging solutions. (A) A T-shaped aluminum foil clip (~180 μg) used for attaching the fiber to the setup. Tapered hooks on the force transducer and length-step motor pass through the diamond-shaped hole in each T clip (C). (B) Single muscle fiber held in a T clip near its end. The T clip is crimped onto the fiber away from the foil edges, causing the edges to flare away from the fiber to avoid damage. (C) Temperature-controlled trough assembly used for solution exchange. Solution troughs (60 μl) are milled into a stainless-steel block. A chilled mixture of ethylene glycol and water flows through the hollow center of the block to adjust the temperature. Slotted Teflon barriers are glued, with epoxy, to the ends of each trough to retain the aqueous contents by surface tension. The "front trough" of the assembly is the photolysis chamber made from an L-shaped section of a quartz cuvette attached underneath the block (see also Fig. 3). Laser light traversing the photolysis trough is reflected back from the polished surface of the steel block to increase photolysis yield. Dashed lines indicate the path of the trough assembly for solution exchange.

to photolysis experiments after permeabilization by detergents, glycerol,[39] or bacterial α-toxin.[40]

Because the functional output of a muscle is mechanical force and shortening, attachment of its ends to the experimental setup must minimize physical damage and extraneous mechanical compliance. A common method is to fold T-shaped aluminum foil clips (Fig. 2A), made by photolithography, directly around the fiber (Fig. 2B).[35] A clip applied with the optimum pressure holds the muscle fiber reliably during full contraction while minimizing crushing of the muscle ends. A hole in the stem of the

[39] A. V. Somlyo, Y. E. Goldman, T. Fujimori, M. Bond, D. R. Trentham, and A. P. Somlyo, *J. Gen. Physiol.* **91,** 165 (1988).
[40] T. Kitazawa, S. Kobayashi, K. Horiuti, A. V. Somlyo, and A. P. Somlyo, *J. Biol. Chem.* **264,** 5339 (1989).

clips slips over the tapered tips of hooks attached to the mechanical transducers of the apparatus (Fig. 2C). Single fibers can also be mounted by gluing the ends of the fiber segment to hooks or pins on the apparatus[41] or by using specialized mounting clamps.[42,43]

Mechanical compliance at the fiber attachment points compromises quantitative measurements of fiber elasticity and kinetics of mechanical transients initiated by photolysis.[44] The series compliance can be reduced by stiffening the ends of the fiber by fixation of a localized region with glutaraldehyde and attaching the fixed regions to the setup.[45]

Because cardiac and smooth muscle fibers are much smaller than skeletal muscle fibers, photolysis experiments typically involve bundles of these cells. Such preparations are attached to the experimental apparatus by T clips, clamps, or glue. Substantial mechanical compliance between the cells of these preparations cannot be eliminated by treating the ends of the preparation. However, cardiac muscle is cross-striated, enabling monitoring of extent of sarcomere length changes during photolysis-initiated transients. The effects of series compliances can be estimated and taken into account in smooth muscle experiments as well, but this is more difficult because the direct measure of filament sliding velocity from the sarcomere length is not available.

Experimental Apparatus

Apparatus for photolysis experiments on muscle fibers must enable convenient, complete, and reliable exchange of media bathing the muscle, optically couple the photolysis chamber to the ultraviolet (UV) photolysis light source, and allow samples to be collected after photolysis for determination of the photochemical conversion of the caged compound. The volume of solution required to exchange or add the caged compound should be minimized because caged compounds are often costly or time-consuming to prepare. The ends of the muscle fiber must be available for attachment to fixed supports or mechanical transducers. The requirements of the other signals being recorded from the preparation often dictate some of the mechanical and optical arrangements of the apparatus.

Two types of muscle fiber setups for photolysis experiments are shown schematically in Figs. 2 and 3. The muscle fiber is typically connected to a tension transducer at one end and a fixed support or a length driver (motor)

---

[41] B. Brenner, *Biophys. J.* **41,** 99 (1983).
[42] Y. Zhao and M. Kawai, *Am. J. Physiol.* **266,** C437 (1994).
[43] R. L. Moss, *J. Physiol.* **292,** 177 (1979).
[44] Y. Luo, R. Cooke, and E. Pate, *Am. J. Physiol.* **265,** C279 (1993).
[45] P. B. Chase and M. J. Kushmerick, *Biophys. J.* **53,** 935 (1988).

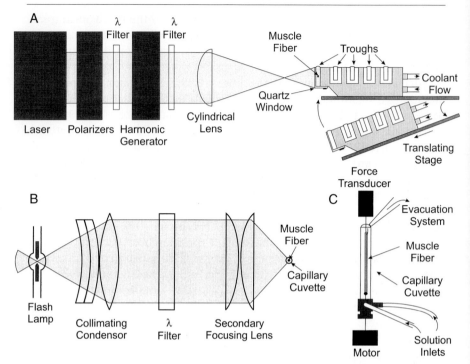

Fig. 3. Examples of experimental setups used in photolysis studies of single skeletal muscle fibers or cardiac or smooth muscle fiber bundles. (A) Schematic side view of a frequency-doubled ruby laser system, optics, and translating trough assembly. The ruby rod (1/2-inch diameter) is optically pumped with a helical xenon flash lamp (not shown). The laser cavity is formed by the partially reflecting front face of the ruby rod and a rear fully reflecting mirror. A triggered Pockel cell and Brewster stack polarizer assembly (not shown) Q-switch the laser resonator to output a 50-nsec "giant pulse" at 694 nm. Brewster stack (clean up) polarizers and a red Wratten (Kodak) filter remove scattered pump light and unpolarized laser light. A temperature-tuned rubidium dihydrogen arsenate crystal or an angle-tuned potassium dihydrogen* phosphate (KD*P) crystal serves as a harmonic generator to convert the 694-nm laser light to 347 nm radiation. A U350 glass filter (Hoya, Fremont, CA), a UG11 filter (Schott, Duryea, PA), or a 1 M copper sulfate liquid filter after the harmonic generator is used to block the 694-nm primary beam. A fused silica cylindrical lens focuses the 347-nm beam to the height of the trough (3 mm). The working distance between the lens and the trough is ~100 mm. Masks on the lens or near the trough reduce the width of the beam to the fiber length (2–3 mm). Four to six troughs are milled into the stainless-steel block (see also Fig. 2). The quartz trough nearest the UV light source is illuminated from the front by the photolysis pulse and from below by other light sources for structural signals such as fluorescence polarization or light scattering (Table I). A microcomputer-driven stepper motor lowers, translates, and raises the trough assembly to exchange solutions while the fiber remains in place. (B) Schematic side view of a flash lamp and capillary flow system. Light from a xenon flash lamp is shaped into a collimated 320- to 360-nm beam (1-inch diameter) using a spherical mirror, fused silica condenser, and bandpass filter such as a U350 with an interference

at the other end (Figs. 2C and 3C). Solutions are exchanged either by a flow arrangement (Fig. 3C) or by a mechanism to exchange solution troughs while keeping the muscle fiber in place (Figs. 2C and 3A). The photolysis trough is constructed of fused silica for transparency at the near-UV wavelength (320–360 nm) used for photolysis. Often another optical system, such as a fluorometer or a sarcomere length detector, uses the axis orthogonal to the fiber and the photolysis optics.

Several techniques, listed in Table I, have been used in conjunction with photolysis to determine the relationship among biochemical, mechanical, and structural events in the actomyosin ATPase cycle. Mechanical signals, such as force, stiffness, shortening velocity, and sarcomere length, give direct macroscopic indications of the functional output. Biochemical assays provide transient or steady-state information on progress of the ATPase. The other signals are structural probes that indicate protein motions. Thus methods are available using photolysis of caged molecules to elucidate the temporal and functional relationship among the enzymatic turnover, mechanical steps, and structural changes within the contractile proteins during energy transduction.

Photolysis Light Sources

Near-UV light sources used for photolysis with motor proteins are listed in Table II. The wavelength and pulse duration are important criteria for choosing a photolysis light source. The wavelength must be matched to the photolysis action spectrum of the caged compound, but as discussed in "Uniformity of Photolysis in the Sample Chamber," the peak of the absorption spectrum is not always the best choice because propagation of the light into the sample is limited due to absorption by the caged molecule. Wavelengths above 310 nm are less injurious to muscle fiber proteins than shorter wavelengths.

Pulse duration should be shorter than the intended time resolution

---

heat reflecting coating. The pulse is directed onto the muscle fiber cuvette by a fused silica astigmatic secondary focusing doublet consisting of a planoconvex spherical lens and a cylindrical lens. The working distance between the refocusing lens and the sample is 15–25 mm. (C) Schematic top view of the capillary cuvette and solution exchange system used in a flash lamp photolysis system available commercially from Scientific Instruments, Heidelberg, Germany. The muscle fiber is wrapped around a hook on a force transducer and both ends are glued to a pin on a length driver (motor) within a quartz capillary cuvette (1-mm inner diameter). The cuvette is perfused through a solution inlet at the motor end. A seal to the motor pin directs flow through the cuvette. Usually one of two inlets is dedicated to caged compounds. Outflow is removed through an evacuated pipette.

TABLE I
Biophysical Signals Recorded from Skinned Muscle Fibers with Laser Photolysis

Signal	Instrument or detector	Time resolution	Information obtained
Mechanical techniques			
Force[a]	Strain gauge at fiber end	20 μsec	Main functional output of muscle fibers held at fixed length. (Number of cross-bridge attachments) × (Average force per cross-bridge)
Striation spacing (sarcomere length)[b,c]	Optical diffractometer using laser light or white light	200 μsec	Filament sliding, degree of overlap between thick and thin filaments, uniformity of the cross-striations
Sarcomere length clamp[d]	Feedback from striation spacing detector to length driver servo controls sarcomere length	1 msec	Removes effects of mechanical compliance outside of sarcomeres on kinetics of force transients initiated by photolysis
Shortening velocity[e–g]	Optical diffractometer detects rate of sarcomere shortening during servo control of force at values below maximum isometric force	1 msec	Main functional output of muscle fibers. Filament sliding velocity, work production, length of interaction between actin and myosin, dependence of velocity on substrate, products and mechanical load
In-phase sinusoidal stiffness[a]	Strain gauge detects oscillations of force in phase with small imposed sinusoidal oscillations of fiber length	10 msec	Relative number of cross-bridges interacting with actin, filament compliance, end compliance
Quadrature stiffness[a]	Strain gauge detects force oscillations 90° out of phase with fiber length oscillations	10 msec	Viscoelasticity of active or weakly attached cross-bridges
Biochemical techniques			
Linked enzymatic assay[h]	NADH fluorescence decreases as ADP is rephosphorylated by pyruvate kinase and lactate dehydrogenase	40 msec	Steady-state ATP hydrolysis rate

Technique	Description	Time	Information obtained
Time to return to rigor[f,i,j]	Mechanical indication of rigor, such as decline of quadrature stiffness or shortening velocity, after known amount of ATP release from caged ATP	1 sec	Steady-state ATP hydrolysis rate
Freeze quenching[k]	High-performance liquid chromatography of radiolabeled nucleotides in fiber rapidly frozen at predetermined times after photolysis	25 msec	Transient and steady-state ATP hydrolysis rate, burst kinetics
Phosphate binding protein[h]	Fluorescent derivative of bacterial protein changes fluorescence intensity on binding product phosphate	1 msec	Product release, transient and steady-state ATP hydrolysis rate, burst kinetics
Structural techniques			
Electron paramagnetic resonance spectroscopy[l,m]	Microwave bridge detects resonance and hyperfine splitting of quantum states of unpaired electrons dependent on orientation with respect to applied magnetic field	10 msec	Orientation, rotation, and mobility of labeled protein domains and mobility on the picosecond to millisecond time scale. Order and disorder of spin label probes
X-ray diffraction[n]	Muscle is irradiated by an intense collimated X-ray beam. Electronic detectors or film record scattered X-rays	20 μsec	Periodicities of protein mass including axial, radial and helical arrangements, filament lattice spacing and radial mass distribution, changes in form factor (shape) of attached heads
Rapid freezing electron microscopy[o,p]	Electron micrographs of stained sections or etched and replicated fracture planes after rapid freeze-quenching of structure	1 msec	Images of myofibrillar protein disposition at predetermined times after photolysis. Computed power density spectra are comparable to X-ray diffraction patterns
Birefringence[q]	Optical transmission through crossed polarizers 45° to muscle fiber axis	20 μsec	Changes in cross-bridge orientation, filament spacing
Fluorescence polarization[r,s]	Polarized excitation and detection of fluorescent reporter groups on myosin or other myofibrillar proteins	50 μsec	Orientation, degree of order and mobility of labeled protein regions. Relative motions between domains

(*Continued*)

[a] Y. E. Goldman, M. G. Hibberd, and D. R. Trentham, *J. Physiol.* **354**, 577 (1984).
[b] Y. E. Goldman, *Biophys. J.* **52**, 57 (1987).
[c] J. A. Dantzig, Y. E. Goldman, N. C. Millar, J. Lacktis, and E. Homsher, *J. Physiol.* **451**, 247 (1992).
[d] T. D. Lenart, T. St. C. Allen, R. J. Barsotti, G. C. R. Ellis-Davies, J. H. Kaplan, C. Franzini-Armstrong, and Y. E. Goldman, *in* "Mechanism of Myofilament Sliding in Muscle Contraction" (H. Sugi and G. H. Pollack, eds.), pp. 475–487, Plenum Press, New York, 1993.
[e] H. Thirlwell, J. A. Sleep, and M. A. Ferenczi, *J. Muscle Res. Cell Motil.* **16**, 131 (1995).
[f] H. Higuchi and Y. E. Goldman, *Biophys. J.* **69**, 1491 (1995).
[g] E. Homsher, J. Lacktis, and M. Regnier, *Biophys. J.* **72**, 1780 (1997).
[h] Z.-H. He, R. K. Chillingworth, M. Brune, J. E. T. Corrie, D. R. Trentham, M. R. Webb, and M. A. Ferenczi, *J. Physiol.* **501**, 125 (1997).
[i] A. Ishijima, H. Kojima, H. Higuchi, Y. Harada, T. Funatsu, and T. Yanagida, *Biophys. J.* **70**, 383 (1996).
[j] C. S. Cook, H. Higuchi, and Y. E. Goldman, *Biophys. J.* **70**, A291 (1996).
[k] M. A. Ferenczi, H. Homsher, and D. R. Trentham, *J. Physiol.* **352**, 575 (1984).
[l] C. L. Berger, E. C. Svensson, and D. D. Thomas, *Proc. Natl. Acad. Sci. U.S.A.* **86**, 8753 (1989).
[m] E. M. Ostap, H. D. White, and D. D. Thomas, *Biochemistry* **32**, 6712 (1993).
[n] K. J. V. Poole, Y. Maeda, G. Rapp, and R. S. Goody, *Adv. Biophys.* **27**, 63 (1991).
[o] T. Funatsu, E. Kono, and S. Tsukita, *J. Cell Biol.* **121**, 1053 (1993).
[p] K. Hirose, T. D. Lenart, J. M. Murray, C. Franzini-Armstrong, and Y. E. Goldman, *Biophys. J.* **65**, 397 (1993).
[q] M. Peckham, M. A. Ferenczi, and M. Irving, *Biophys. J.* **67**, 1141 (1994).
[r] T. St. C. Allen, N. Ling, M. Irving, and Y. E. Goldman, *Biophys. J.* **70**, 1847 (1996).
[s] C. L. Berger, J. S. Craik, D. R. Trentham, J. E. T. Corrie, and Y. E. Goldman, *Biophys. J.* **68**, 78s (1995).

TABLE II
ILLUMINATION SOURCES AVAILABLE FOR PHOTOLYZING CAGED COMPOUNDS WITH MUSCLE FIBERS AND *in Vitro* MOTILITY ASSAYS[a]

Illumination source	Wavelength (nm)	Maximum output pulse energy (mJ)	Emitting medium	Pulse duration	Maximum repetition rate (Hz)
Lasers					
Fiber experiments					
Frequency-doubled ruby[b]	347	300	Ruby crystal	30–50 nsec	1
Frequency-doubled liquid dye laser[c]	320	100	Rhodamine 640G dye	500 nsec	1
Frequency-tripled neodymium: YAG	355	200	Nd:YAG crystal	10 nsec	10
Excimer[d]	308	400	Xenon chloride gas	10 nsec	100
Excimer	351	200	Xenon fluoride gas	10 nsec	100
In vitro motility assays					
Nitrogen	337	9	Nitrogen gas	3 nsec	50
Frequency-tripled neodymium: YAG	355	3	Nd:YAG crystal	4–6 nsec	10
Neodymium: YLF	349	0.1	Nd:YLF crystal	30 nsec	10
Helium: cadmium[e]	354	12 mW	He:Cd gas	—[f]	—
Other light sources					
Continuous arc lamp[b]	300–400[g]	200 mW	Mercury vapor	—[f]	—
Flash lamp[h,i]	300–400	100–200	Xenon gas	200 μsec–1 msec	0.2

[a] Manufacturers for lasers are listed in *Laser Focus Magazine* (Nashua, NH) and the *Photonics Corporate Guide to Profiles and Addresses* (Pittsfield, MA). ⟨http://www.laurin.com⟩.
[b] Y. E. Goldman, M. G. Hibberd, and D. R. Trentham, *J. Physiol.* **354**, 577 (1984).
[c] J. A. Dantzig, Y. E. Goldman, N. C. Millar, J. Lacktis, and E. Homsher, *J. Physiol.* **451**, 247 (1992).
[d] C. L. Berger, E. C. Svensson, and D. D. Thomas, *Proc. Natl. Acad. Sci. U.S.A.* **86**, 8753 (1989).
[e] H. Higuchi, E. Muto, Y. Inoue, and T. Yanagida, *Proc. Natl. Acad. Sci. U.S.A.* **94**, 4395 (1997).
[f] The pulse duration is dependent on the speed of the mechanical shutter.
[g] In the range of 300–400 nm, most of the radiation is emitted at 312, 335, and 365 nm, the intense lines of the mercury spectrum. The intensity of the light at these three wavelengths is similar and can be selected for with the appropriate filters.
[h] K. J. V. Poole, Y. Maeda, G. Rapp, and R. S. Goody, *Adv. Biophys.* **27**, 63 (1991).
[i] G. Rapp, *Methods Enzymol.* **291** [11] (1998).

of the experiment, which is often limited by the dark photochemical reaction rate (the rate of release of the active compound after the light pulse, see Table III). If the caged compound has a low product quantum yield (Q_p, the proportion of optically excited caged molecules that photolyze), a theoretical advantage of longer light pulses is the potential for repeated excitation of unphotolyzed molecules, thereby increasing the extent of photolysis.[46] If the photolysis pulse duration is comparable to the optically excited state, repeated excitation cannot occur. Lifetimes of optically excited states of 2-nitrobenzyl compounds are subnanosecond,[46,46a] so the light sources in Table II have sufficient pulse durations to allow multiple excitations.

Laser sources have the advantages of high power and simple optical interface to the sample due to the monochromatic, collimated nature of laser radiation. A simple fused silica lens shapes the beam and projects it onto the sample (Fig. 3A). Many of the muscle fiber experiments have used frequency-doubled ruby lasers[27,46] or frequency-doubled tunable dye lasers,[21,46] but these two sources are not readily available commercially. In the trough of our muscle fiber setup (Fig. 3A), with a 3 × 8-mm optical window and 1-mm path length, a 100-mJ pulse of 347 nm radiation photolyzes ~10% of 10 mM NPE-caged ATP to liberate ~1 mM ATP.

Shuttered continuous arc lamps or flash lamps (Fig. 3B) are also convenient light sources.[47] The choice between a laser and a flash lamp system weighs considerations of wavelength, working distance, pulse duration, energy, and cost. Flash lamp systems have the disadvantage that a relatively high aperture secondary focusing lens is required to project the photolysis light onto the fiber, reducing the working distance to typically 15–25 mm. This constraint may limit access to the preparation by other mechanical or optical components. However, the cost of a flash lamp system is much less than high-powered lasers. Flash lamp systems for photolysis experiments are described in more detail in another chapter in this volume.[48]

Light sources for *in vitro* experiments are not as demanding as with muscle fibers because the volume irradiated (30-μm diameter field × 10 μm–2 mm path length) is much smaller and focusing the photolysis light through a microscope objective increases the power density in the small photolysis volume. Low-power lasers are listed in Table II for *in vitro* experiments.

---

[46] J. A. McCray and D. R. Trentham, *Annu. Rev. Biophys. Biophys. Chem.* **18**, 239 (1989).
[46a] S. R. Adams, J. P. Y. Kao, G. Grynkiewicz, A. Minta, and R. Y. Tsien, *J. Am. Chem. Soc.* **110**, 3212 (1988).
[47] G. Rapp and K. Güth, *Pflügers Arch.* **411**, 200 (1988).
[48] G. Rapp, *Methods Enzymol.* **291** [11] (1998) (this volume).

Caged Compounds

Table III lists the principal caged components used to study cell motility and some of their properties. They are available commercially except where indicated. 1-(2-Nitrophenyl)ethyl (NPE) esters of nucleotides and phosphate were the first caged compounds used in muscle fiber studies. They have high product quantum yields (Q_p) but the leaving group, 2-nitrosoacetophenone, is reactive and forms covalent adducts with reactive sulfhydryls, such as protein cysteine residues. Thus, reduced glutathione (GSH), dithiothreitol (DTT), or 2-mercaptoethanol must be included in the photolysis media to protect the contractile apparatus, as discussed later.

Typical samples of NPE-caged nucleotides are mixtures of diastereoisomers due to a chiral center at the benzylic carbon of the 1-(2-nitrophenyl)ethyl group.[46,48a] The R- and S-diastereoisomers of NPE-caged nucleotides are separable using reversed-phase chromatography, but the separation does not confer any advantage because the diastereoisomers have the same properties in photolysis experiments on muscle fibers.[49]

Desyl (diphenyloxoethyl)[50] and dimethoxybenzoin[51] derivatives have faster dark reactions than the NPE compounds and the benzofuran photoproducts are less reactive. However, the quantum yield of some of the derivatives is lower and they are less stable in the dark.[50–52]

NP-EGTA and DM-nitrophen are Ca^{2+} chelators that photolyze to iminodiacetic acids that have Ca^{2+} affinities several orders of magnitude lower, thus releasing bound Ca^{2+}. DM-nitrophen is a derivative of EDTA and binds Mg^{2+} quite tightly (Table IV), enabling it to serve as a caged Mg^{2+} or else necessitating low free Mg^{2+} concentrations when used as a caged Ca^{2+}. NP-EGTA is selective for Ca^{2+} (Table IV). Nitr-5 and Nitr-7 are highly selective Ca^{2+} compounds that photolyze at longer wavelengths (>360 nm), but they have much lower quantum yields than NP-EGTA or DM-nitrophen. DMNPE-caged BAPTA and Diazo-2 are caged Ca^{2+} chelators used to abruptly reduce free $[Ca^{2+}]$.

Effects of impurities or contaminants should be considered in all samples of caged compounds. Muscle fibers are particularly sensitive to contaminant nucleotides because actomyosin has high affinity for ATP and ADP. For

---

[48a] J. E. T. Corrie, G. P. Reid, D. R. Trentham, M. B. Hursthouse, and M. A. Mazid, *J. Chem. Soc. Perkin Trans. I*, 1015 (1992).

[49] H. Thirlwell, J. A. Sleep, and M. A. Ferenczi, *J. Muscle Res. Cell Motil.* **16**, 131 (1995).

[50] R. S. Givens, P. S. Athey, B. Matuszewski, L. W. Kueper III, J. Xue, and T. Fister, *J. Am. Chem. Soc.* **115**, 6001 (1993).

[51] J. E. T. Corrie and D. R. Trentham, *J. Chem. Soc. Perkin Trans. I*, 2409 (1992).

[52] H. Thirlwell, J. E. T. Corrie, G. P. Reid, D. R. Trentham, and M. A. Ferenczi, *Biophys. J.* **67**, 2436 (1994).

TABLE III
CAGED COMPOUNDS USED TO STUDY MOTOR PROTEINS[a]

Caged compound	Absorptivity (M^{-1} cm^{-1})	Wavelength[b] (nm)	k_p (sec^{-1})	Q_p	Notes and caveats
Caged nucleotides					
NPE-caged ATP[c,d]	660	347	86	0.63	Weakly binds to motor proteins Nitrosoacetophenone by-product binds to -SH groups[e]
DMNPE-caged ATP[f]	5,000	347	18	0.07	Nitrosoacetophenone by-product binds to -SH groups
Desyl-caged ATP[g]	500	320	~10^8	~0.3	May spontaneously hydrolyze to form ADP and desyl-caged P$_i$ Photolysis by-product is biologically inert
DMB-caged ATP[h,i]	170	347	10^5	0.3	Not commercially available
NPE-caged ATPγS[c,j]	660	347	35	0.57	See NPE-caged ATP
NPE-S-caged ATPγS			35		1-(2-nitrophenylethyl) group on either the γ-sulfur (S-caged) or a nonbridging γ-oxygen (O-caged)
NPE-O-caged ATPγS			109		
NPE-caged ADP[k]	660	347	86	0.63	See NPE-caged ATP
Caged phosphates					
NPE-caged phosphate[c,l]	660	347	2.4 × 10^4	0.54	See NPE-caged ATP
Desyl-caged phosphate[g]	400	323	~10^8	~0.3	See desyl-caged ATP
DMB-caged phosphate[h]	170	347	>10^5	0.78	Not commercially available
Caged calcium compounds					
Nitr-5[m,n]	4,500	390	3000	0.035	Selectivity for Ca^{2+} ≫ Mg^{2+}
Nitr-7[m,n]	4,500	390	550	0.042	Selectivity for Ca^{2+} ≫ Mg^{2+}
DM-nitrophen[o,p,q]	4,200	295	3.8 × 10^4	0.18	Mg^{2+} chelator Requires precautions against contaminant Ca^{2+} and Mg^{2+} loading

Compound					Notes
NP-EGTA[o,q]	974	347	6.8×10^4	0.23[r]	Selectivity for $Ca^{2+} \gg Mg^{2+}$; Requires precautions against contaminant Ca^{2+}
Caged calcium chelators					
DMNPE-caged Bapta[s,t]	5,000	347		300	0.075 Not commercially available
Diazo-2[n,s]	18,000	369		2000	0.6 Low Ca^{2+} affinity before photolysis

[a] k_p, the dark reaction rate at which photolysis products appear; Q_p, the quantum yield. NPE, 1-(2-nitrophenyl)ethyl; DMNPE, 1-(4,5-dimethoxy-2-nitrophenyl)ethyl; desyl, desoxybenzoinyl; DMB, 3',5'-dimethoxybenzoin; NP, nitrophenyl; DM-nitrophen, dimethoxy-2-nitrophenyl EDTA. Except where indicated, caged compounds are available from Calbiochem-Novachem International, La Jolla, California; Molecular Probes, Inc., Eugene, Oregon; and Dojindo Molecular Technologies, Inc., Bethesda, Maryland.
[b] Wavelength at which absorptivity was measured.
[c] J. W. Walker, G. P. Reid, J. A. McCray, and D. R. Trentham, *J. Am. Chem. Soc.* **110**, 7170 (1988).
[d] Y. E. Goldman, M. G. Hibberd, and D. R. Trentham, *J. Physiol.* **354**, 577 (1984).
[e] For details see text.
[f] J. F. Wootton and D. R. Trentham, in "Photochemical Probes in Biochemistry" (P. E. Nielson, ed.), NATO ASI Ser. C, Vol. 272, pp. 277–296, Kluwer, Dordrecht, The Netherlands, 1989.
[g] R. S. Givens, P. S. Athey, B. Matuszewski, L. W. Kueper III, J. Xue, and T. Fister, *J. Am. Chem. Soc.* **115**, 6001 (1993).
[h] J. E. T. Corrie, Y. Katayama, G. P. Reid, M. Anson, and D. R. Trentham, *Philos. Trans. R. Soc. Ser. A* **340**, 233 (1992).
[i] H. Thirlwell, J. E. T. Corrie, G. P. Reid, D. R. Trentham, and M. A. Ferenczi, *Biophys. J.* **67**, 2436 (1994).
[j] J. A. Dantzig, J. W. Walker, D. R. Trentham, and Y. E. Goldman, *Proc. Natl. Acad. Sci. U.S.A.* **85**, 6716 (1988).
[k] J. A. Dantzig, M. G. Hibberd, D. R. Trentham, and Y. E. Goldman, *J. Physiol.* **432**, 639 (1991).
[l] J. A. Dantzig, Y. E. Goldman, N. C. Millar, J. Lacktis, and E. Homsher, *J. Physiol.* **451**, 247 (1992).
[m] S. R. Adams, J. P. Y. Kao, G. Grynkiewicz, A. Minta, and R. Y. Tsien, *J. Am. Chem. Soc.* **110**, 3212 (1988).
[n] C. C. Ashley, I. P. Mulligan, and T. J. Lea, *Quart. Rev. Biophys.* **24**, 1 (1991).
[o] J. H. Kaplan and G. C. R. Ellis-Davies, *Proc. Natl. Acad. Sci. U.S.A.* **85**, 6571 (1988).
[p] G. C. R. Ellis-Davies and J. H. Kaplan, *Proc. Natl. Acad. Sci. U.S.A.* **91**, 187 (1994).
[q] G. C. R. Ellis-Davies, J. H. Kaplan, and R. J. Barsotti, *Biophys. J.* **70**, 1006 (1996).
[r] Determined in the presence of Ca^{2+}.
[s] S. R. Adams, J. P. Y. Kao, and R. Y. Tsien, *J. Am. Chem. Soc.* **111**, 7957 (1989).
[t] M. A. Ferenczi, Y. E. Goldman, and D. R. Trentham, *J. Physiol.* **418**, 155P (1989).

TABLE IV
Apparent pK_a Values[a] for Binding of Ca^{2+} and Mg^{2+}

Caged compound	Before photolysis		After photolysis[b]
	Ca^{2+}	Mg^{2+}	Ca^{2+}
NPE-caged ATP[c]	2.52	2.45	3.65
DM-nitrophen[d,e]	8.30	5.65	0.21
NP-EGTA[e,f]	6.93	2.40	0.11
Nitr-5[g]	6.84	2.07	2.22
Nitr-7[g]	7.30	2.27	2.07
diazo-2[h]	2.66	—	7.14

[a] Apparent pK_a values for NPE-caged ATP and NP-EGTA were calculated for pH 7.1. All others were measured at pH 7.2.
[b] Most apparent pK_a values for Mg^{2+} binding after photolysis are not available.
[c] Apparent pK_a values for NPE-caged ATP were determined as described in the text. Values were adjusted to pH 7.1.
[d] J. H. Kaplan and G. C. R. Ellis-Davies, *Proc. Natl. Acad. Sci. U.S.A.* **85,** 6571 (1988).
[e] Apparent pK_a values for the photolysis by-product are for iminodiacetic acids.
[f] G. C. R. Ellis-Davies and J. H. Kaplan, *Proc. Natl. Acad. Sci. U.S.A.* **91,** 187 (1994).
[g] S. R. Adams, J. P. Y. Kao, G. Grynkiewicz, A. Minta, and R. Y. Tsien, *J. Am. Chem. Soc.* **110,** 3212 (1988).
[h] S. R. Adams, J. P. Y. Kao, and R. Y. Tsien, *J. Am. Chem. Soc.* **111,** 7957 (1989).

example, ADP runs very close to NPE-caged ATP in anion-exchange chromatography, so it is a common contaminant. With only 1% ADP contamination, a photolysis solution with 10 mM caged ATP contains 100 μM ADP, in the range of the 20–200 μM dissociation constant for ADP binding to actomyosin.[22,53–55] Contaminant ADP at this level would markedly inhibit the rate of binding of photoreleased ATP to actomyosin and the ATP-induced dissociation (step 1, Fig. 1).[52] Purity of caged nucleotides is evaluated using high-performance liquid chromatography (HPLC).[32,46,52] A colorimetric assay is used to determine phosphate (P_i) contamination in caged P_i.[21]

---

[53] S. Marston, *Biochim. Biophys. Acta* **305,** 397 (1973).
[54] J. A. Dantzig, M. G. Hibberd, D. R. Trentham, and Y. E. Goldman, *J. Physiol.* **432,** 639 (1991).
[55] M. Schoenberg and E. Eisenberg, *J. Gen. Physiol.* **89,** 905 (1987).

Due to the severe requirements for purity of caged ATP, we synthesize and purify NPE-caged ATP in the laboratory using 1-(2-nitrophenyl)-diazoethane to modify ATP according to a procedure modified from that of Walker et al.[56,56a] Purification by alternating ion exchange (Whatman DE-52, DEAE) and reversed-phase (Whatman C_{18}, μBondpack, bead size 50–100 μm) HPLC chromatography (totaling four column runs) reduces nucleotide contamination to <0.02%.[54] The purified caged compound is divided into aliquots for single experiments, rapidly frozen using liquid nitrogen or a mixture of 2-propanol and dry ice, and then stored at $-80°$.

Considerations of purity for caged Ca^{2+} compounds include contaminant Ca^{2+} and Mg^{2+}. The concentration of Ca^{2+} that brings a caged Ca (e.g., NP-EGTA) solution to the threshold for activation of muscle fiber force can vary considerably among batches of nominally the same chemical. This variability is presumably an indication of Ca^{2+} contamination or impurities in the caged chelator stock.

Experimental Solutions

Experimental solutions for muscle fiber experiments are designed to mimic the intracellular milieu in ionic strength, pH, free Ca^{2+}, free Mg^{2+}, and nucleotides. Buffering systems should minimize changes in these parameters that occur on photorelease of the active experimental compound. Because these ionic and biochemical factors all influence the performance of the fully constituted and regulated contractile system, the design of the solutions must take account of the multiple metal–ligand interactions among the constituents and regulate the total ionic strength, $\Gamma = (\Sigma_i\ C_iV_i^2)/2$, where C_i is the concentration and V_i is the net charge of each charged species in the solution. The interaction of myosin with actin is partly ionic, so the affinity is strongly affected by Γ.

Computer programs that analyze or design solutions to specified concentrations and ionic strength have been described.[57,58] We use a program originally written by J. Thorson, D. C. S. White, and Y. E. Goldman and then upgraded to a modern user interface by R. J. Barsotti and A. Fielding (personal communication). It uses stability constants for each of the metal–ligand complexes in the solution to solve the multiple binding equilibria.

[56] J. W. Walker, G. P. Reid, J. A. McCray, and D. R. Trentham, *J. Am. Chem. Soc.* **110**, 7170 (1988).
[56a] J. W. Walker, G. P. Reid, and D. R. Trentham, *Methods Enzymol.* **172**, 289 (1989).
[57] A. Fabiato and F. Fabiato, *J. Physiol. (Paris)* **75**, 463 (1979).
[58] Commercially available software for determining solution constituents: Max Chelator version 4.61 supplied by C. Patton, Hopkins Marine Station, Stanford University, Pacific Grove, California.

TABLE V
COMPOSITIONS OF SOLUTIONS USING CAGED ATP AND CAGED CALCIUM DURING PHOTOLYSIS TRIALS[a]

Solution	MgCl$_2$	ATP	Ca^{2+}	EGTA	HDTA	Na$_2$CP	Caged compound
High ATP relax	7.7	5.4	—	25	—	19.1	—
Prerigor relax	2.7	0.12	—	30	—	21.6	—
Rigor	3.1	—	—	53.4	—	—	—
Ca rigor	1.2	—	20	20	34.3	—	—
Rigor + NPE-caged ATP	4.6	—	—	35.9	—	—	10
Ca rigor + NPE-caged ATP	3.2	—	36.5	36.5	—	—	10
Pre-NP-EGTA	6.3	5.4	—	—	48.1	—	—
NP-EGTA	6.5	5.4	1.256	—	46.0	20	2.0

[a] ATP, adenosine 5'-triphosphate; EGTA, ethyleneglycol-bis(β-aminoethyl ether)-N,N,N',N'-tetraacetic acid; HDTA, hexamethylenedinitrilotetraacetic acid; TES, N-tris(hydroxymethyl)methyl-2-aminoethanesulfonic acid; CP, phosphocreatine; NP-EGTA, nitrophenylethyleneglycol-bis(β-aminoethyl ether)-N,N,N',N'-tetraacetic acid. Solution constituents were designed using computer software originally written by J. Thorson, D. C. S. White, and Y. E. Goldman and upgraded by R. J. Barsotti and A. Fielding. All values are total concentrations in mM. All solutions contain 100 mM TES and 10 mM reduced glutathione. Ionic strength of the solutions is 200 mM, pH 7.1, 20°. The K$^+$ concentration is 120–140 mM and the Na$^+$ concentration is 10–30 mM. For photolysis solutions, caged compounds, concentrated stocks, and H$_2$O are combined on the day of an experiment. Free [Mg^{2+}] is calculated to be 1 mM in all solutions. Free [Ca^{2+}] is calculated to be ~30 μM in the Ca rigor and Ca rigor + NPE-caged ATP solutions. The NP-EGTA photolysis solution has 1.256 mM Ca added and 250 nM calculated free Ca^{2+}. All other solutions contain <10^{-8} M free Ca^{2+}. Creatine phosphokinase (1 mg/ml, 100–150 units/ml) is added to all solutions with creatine phosphate.

Given the desired free metal (Ca^{2+}, Mg^{2+}) concentrations, free, bound, or total concentration of each anion, pH, temperature, and total ionic strength, the program calculates the total concentration of each constituent and a recipe for making the solution from given stock solutions (Table V). The metal–ligand stability constants used in this calculation are mostly obtained from published data, tabulated in Martell and Smith.[59]

Ca^{2+} and Mg^{2+} affinities for caged ATP and the caged calcium compounds are listed in Table IV. Measurements of Ca^{2+} and Mg^{2+} binding to NPE-caged ATP are described later in this article. Caged ATP has a valence of -3 at neutral pH and a pK_a of 4.23 for proton binding to form H–caged ATP^{2-} (J. W. Walker, personal communication). pK_a values for NP-EGTA are 9.95, 7.41, 2.64, and 2.0 (G. C. R. Ellis-Davies, personal communication). For the photoproducts of NP-EGTA, we use pK_a values for iminodiacetic acid, 9.45 and 2.4.[59] Thus the photoproducts have a valence of -1 at neutral pH.

[59] A. E. Martell and R. M. Smith, "Critical Stability Constants." Plenum Press, New York, 1974.

Examples of solutions used for single muscle fiber experiments with caged ATP and NP-EGTA-caged Ca^{2+} are listed in Table V. Design criteria for these solutions are 1 mM free Mg^{2+}, 200 mM total ionic strength, pH 7.1 at 20°, and full activation either by photolysis of caged ATP from rigor (no nucleotide) or by photolysis of NP-EGTA from full relaxation (low [Ca^{2+}], + MgATP). Creatine phosphate (CP) and creatine phosphokinase (CPK) buffer the concentration of ATP by rephosphorylating ADP produced by the contractile proteins. Ca^{2+} is buffered by EGTA. pH is buffered by TES buffer.

Caged compounds contribute significantly to ionic strength. For instance, the -3 valence of caged ATP at pH 7.1 implies that a 10 mM concentration with its 30 mM associated cation (TEAB$^+$; triethylamine bicarbonate salt) contributes 60 mM to Γ [(10 mM × 3^2 + 30 mM)/2]. Therefore, to keep Γ constant when the caged compound is included in a solution, other constituents must be reduced.

A fast pH buffer with a pK_a near the working pH limits changes of pH on photolysis of the caged compound. The pH buffering capacity is maximal when pH = pK_a. BES buffer[21] (2-[bis(2-hydroxyethyl)amino]ethanesulfonic acid with pK_a of 7.12 at 20° and MOPS buffer [3-(N-morpholino)propanesulfonic acid], pK_a of 7.2 at 20°, are potential choices. We prefer TES, pK_a of 7.5 at 20°, for the following reason: the caged molecules and other constituents required to buffer ATP and Ca^{2+} make considerable contributions to Γ. The amount of pH buffer that can be added to the solution is limited by the requirement to keep total Γ within 200 mM. Therefore, a figure of merit for the pH buffer is the buffering capacity *per unit contribution* to Γ. When pK_a is somewhat greater than pH, more than half of the buffer molecules are liganded with protons and do not contribute to Γ. At pH 7.1, only 25% of TES molecules are charged, leading to high buffering capacity per unit contribution to the ionic strength.

The species of anions present in the solution affect the longevity of the muscle fiber preparation. In solutions with high concentrations of chloride, the strength of contraction markedly diminishes after a few activations.[60] This decline is less prominent with propionate and acetate.[60] In the solutions shown in Table V, the main anions are EGTA, HDTA, CP, and caged ATP, which do not seem to be deleterious.[32,60,61]

GSH or DTT is added to the photolysis solutions to decrease damage to the contractile system from reaction with the photolysis leaving group, 2-nitrosoacetophenone. This by-product of NPE-caged compounds reacts

---

[60] M. A. W. Andrews, D. W. Maughan, T. M. Nosek, and R. E. Godt, *J. Gen. Physiol.* **98**, 1105 (1991).
[61] D. G. Moisescu, *Nature* **262**, 610 (1976).

with free sulfhydryls, such as cysteine residues in the proteins. In early caged ATP experiments using NPE-caged ATP in the absence of Ca^{2+}, photoreleased ATP failed to relax rigor fibers when GSH was not present. The concentration of ATP required to relax the fiber was irreversibly increased, indicating modification of the contractile apparatus by the photolysis by-products. When sulfhydryl reagents such as GSH or DTT are included in the photolysis medium in excess, these compounds react with the photoreleased by-product, thereby avoiding modification of the proteins.[61a] The second order rate constant for reaction of 2-nitrosoacetophenone with DTT is 3.5×10^3 M^{-1} sec^{-1} at pH 7.0, 21°, and 0.18 mM ionic strength.[56] At 10 mM DTT, the time constant for removal of the reactive by-product is thus $\tau = 1/(10$ m$M \times 3.5 \times 10^3$ M^{-1} $sec^{-1}) = 29$ msec.

The concentration of caged compound added to the solution is based on several factors. The desired amount of biologically active compound photoreleased depends on the number of photons absorbed during the flash and Q_p for photolysis, as discussed later. Other factors that bear on the amount of caged compound to be added to the medium are spatial uniformity of photorelease and effects of the caged compound on the biological preparation before photolysis. Caged compounds bind to the contractile proteins, generally with lower affinity than the intended photoproduct, and either activate or inhibit the reaction. Uniformity of photolysis and inhibition of motor proteins by NPE-caged ATP are discussed later.

In some experiments, it is necessary to reduce nucleotide concentrations before photolysis below the levels caused by nucleotide contamination of the caged compound. Glucose (200 mM) and hexokinase (5–10 U/ml),[54] diadenosine pentaphosphate (250 μM),[54] or apyrase (10–50 U/ml)[37,52,62] have been used for this purpose in muscle fiber experiments. Apyrase hydrolyzes adenine nucleotides to adenosine and P_i. Apyrase coupled to CNBr-activated Sepharose beads retains activity and can be removed from a treated solution by centrifugation.[52] Apyrase has also been useful in removing tightly bound nucleotides from the contractile proteins. It diffuses rapidly into muscle fibers, but washes out poorly.[37,62]

Given the starting concentrations of the solution constituents and the amount of photolysis, the final concentrations of ionic species and bound complexes can be calculated. Table VI shows the changes expected for typical NPE-caged ATP and caged Ca^{2+} experiments. For the experiment starting with 10 mM caged ATP, assuming 1 mM total ATP is photoreleased, 0.9 mM ATP complexes Mg^{2+} and is available for binding to the

---

[61a] A. Barth, J. E. T. Corrie, M. J. Gradwell, Y. Maeda, W. Mäntele, T. Meier, and D. R. Trentham, *J. Am. Chem. Soc.* **119**, 4149 (1997).

[62] H. Martin and R. J. Barsotti, *Biophys. J.* **67**, 1933 (1994).

TABLE VI
CHANGES IN SOLUTION CONSTITUENTS FOLLOWING THE PHOTOLYSIS OF CAGED ATP OR NP-EGTA

Constituent[a]	Before photolysis	After photolysis
Activation solution with NPE-caged ATP		
ATP (total)	0	1.0
Free Mg^{2+}	1.0	0.74
Free Ca^{2+}	0.053	0.056
Mg-ATP	0	0.9
Caged ATP (total)	10.0	9.0
Mg-caged ATP	2.17	1.54
pH	7.10	7.05
Ionic strength	200	199
Activation solution with NP-EGTA		
ATP (total)	5.39	5.39
Free Mg^{2+}	1.0	1.28
Free Ca^{2+}	250 nM	0.023
Mg-ATP	5.0	5.005
NP-EGTA (total)	2.0	1.2
Ca-NP-EGTA	1.26	1.19
Mg-NP-EGTA	0.15	0.001
IDA (total)	0	1.6
Ca-IDA[b]	0	47 nM
Mg-IDA	0	0.005
pH	7.1	7.09
Ionic strength	200	199

[a] All concentrations are in mM except where indicated. The total concentrations of the constituents before photolysis are given in Table V.
[b] IDA, iminodiacetic acid.

contractile proteins. The free [Mg^{2+}] declines from 1 mM starting concentration to 0.74 mM. Free Ca^{2+} changes only slightly; ionic strength decreases by 1 mM. Protons (1 mM) liberated on photolysis are mostly buffered and the pH changes from 7.1 to 7.05. Heating of the solution caused by absorption of the laser pulse is less than 2°.

A similar calculation for a typical NP-EGTA caged Ca^{2+} experiment gives a final free [Ca^{2+}] of 23 μM, corresponding to pCa of 4.6. Free [Mg^{2+}] increases from 1.0 to 1.28 mM due to release from the NP-EGTA. Ionic strength decreases by 1 mM, and pH decreases 0.01 unit. The changes in concentrations other than the intended increases of ATP and Ca^{2+} in these examples are minimal, suggesting that the transient phenomena induced by the photolysis pulse are due primarily to the released ATP or Ca^{2+}, not to incidental changes or by-products of the photolysis reaction. Control experiments liberating ineffective compounds, such as P_i in rigor, support this conclusion.[63]

Probing the Actomyosin ATPase in Fibers

ATP Binding, Cross-Bridge Detachment, and Relaxation of Tension

To determine the second-order rate constant for ATP binding and detachment of myosin from actin, the muscle fiber is started from rigor with all of the myosin heads in the AM nucleotide-free state. ATP released from caged ATP binds to the myosin heads, promoting detachment and then relaxation of tension and stiffness in the absence of Ca^{2+} or development of active tension in the presence of Ca^{2+}. The protocol to induce rigor for a caged ATP photolysis experiment with a skinned psoas fiber is to start with the fiber relaxed at high ATP (Table V) and incubate for >5 min in prerigor solution (0.1 mM ATP). The fiber is then transferred to the rigor solution for ~30 sec and then to a second rigor wash to reduce the ATP concentration around the fiber rapidly. Tension and stiffness increase in the rigor solution and stabilize after 2–3 min.

Inhibitors of force such as P_i[20,21] or butanedione monoxime (BDM)[64] improve the uniformity of striation spacing by reducing filament sliding as the fiber goes into rigor. The tension is very low when rigor is induced in the presence of P_i, BDM, low Mg^{2+} concentration, or low temperature. After the ATP is removed and the fiber becomes mechanically stable, it can be stretched slightly if necessary to increase tension. The solution is then exchanged for a rigor solution with 2–10 mM caged ATP and >2 min are allowed for the caged ATP to equilibrate in the fiber interior before the photolysis pulse is triggered.

Photoreleased ATP complexes Mg^{2+} quickly and then is available for binding to actomyosin. Following the release of ATP into the rigor fiber in the absence of Ca^{2+}, tension and stiffness decline biphasically with NPE-caged ATP[32,52] (Fig. 4A) and monotonically with DMB-caged ATP in the presence of apyrase.[52] The rate at which tension and stiffness fall depends on the concentration of ATP released, which is adjusted by varying the laser pulse energy or by altering the concentration of caged ATP. These data provide estimates of the second-order rate constant for ATP binding and ATP-induced dissociation of actomyosin.[32] Methods to analyze the tension transients in these experiments can be found in Goldman et al.[32] and Sakoda and Horiuti.[65]

---

[63] J. W. Tanner, D. D. Thomas, and Y. E. Goldman, J. Mol. Biol. **223,** 185 (1992).
[64] H. Higuchi, T. Yanagida, and Y. E. Goldman, Biophys. J. **69,** 1000 (1995).
[65] T. Sakoda and K. Horiuti, J. Muscle Res. Cell Motil. **13,** 464 (1992).

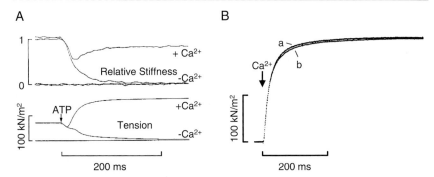

FIG. 4. Mechanical recordings from single glycerol-extracted fibers relaxed and activated by photolysis of caged ATP or caged calcium. (A) Transient changes in tension (lower traces) and 500-Hz in-phase stiffness (upper traces, stiffness relative to that in the rigor state where all of the myosin heads are attached to actin). Following liberation of ~1 mM ATP from NPE-caged ATP (arrow), the fiber relaxes ($-Ca^{2+}$) or isometrically contracts ($+Ca^{2+}$). The flat lines show the baseline tension and stiffness in the absence of Ca^{2+}. Solution conditions are given in Table V. (B) Transient changes in tension during the activation of a single fiber from the relaxed state by photolysis (arrow) of NP-EGTA [trace a in (B), see Table V for solution constituents] and DM-nitrophen [trace b]. (B) was modified from Fig. 6 in G. C. R. Ellis-Davies and J. H. Kaplan, *Proc. Natl. Acad. Sci. U.S.A.* **91,** 187 (1994). Copyright (1994) National Academy of Sciences, U.S.A. The total ATP (3 mM) and Mg (1.2 mM) concentrations for DM-nitrophen were lower than for NP-EGTA, but the time courses and plateau tensions were very similar.

Activation by Photolysis of Caged ATP and Caged Calcium

The fiber is put into rigor as described earlier, then rigor with ~30 μM free Ca^{2+}, and, finally, rigor solution with Ca^{2+} and caged ATP (Table V). Two washes of rigor solution with Ca^{2+} are necessary to ensure that the Ca^{2+} regulatory system is saturated with Ca^{2+}. Release of ATP causes a transient dip in tension followed by the development of full active isometric tension (Fig. 4A).[66] This tension increase reflects the attachment and force generation of partly synchronized cross-bridges, so it has been combined with many other signals that indicate biochemical or structural changes in the contractile proteins (Table I). After steady tension is reached, the fiber is transferred to relaxing solution and the procedure can be repeated 10–20 times with active tension declining typically to 80% of the original value.

Creatine phosphate and creatine phosphokinase can be added to the photolysis medium to buffer the ATP concentration and keep the ADP concentration low during the steady contraction initiated by photolysis. However, very pure caged ATP is required (≤0.02% contamination) to

---

[66] Y. E. Goldman, M. G. Hibberd, and D. R. Trentham, *J. Physiol.* **354,** 605 (1984).

obtain reliable results in the presence of CP because ADP present at submicromolar concentration is converted by CPK to ATP, which causes slow cross-bridge cycling.

In caged ATP experiments, the Ca^{2+} regulatory system is fully switched on and the kinetics are determined by the chemomechanical events of actomyosin. Kinetics of the Ca^{2+} regulatory system become relevant when caged Ca^{2+} is used. For activation of a fiber by photolysis of NP-EGTA, the fiber is initially in relaxing solution and is incubated in two exchanges of pre-NP-EGTA solution (Table V) containing low Ca^{2+} buffering capacity and then the photolysis solution containing 2 mM NP-EGTA and typically 1–1.8 mM Ca^{2+}. The amount of Ca^{2+} in the photolysis solution is adjusted to poise the fiber just below the threshold for tension development. Photolysis of the NP-EGTA or DM-nitrophen rapidly activates full tension development (Fig. 4B).

ATPase Rate

The overall turnover rate of the actomyosin ATPase in skinned muscle fibers has been measured in steady contractions by linked enzymatic assays[67,68] or by accumulation of the product ADP using HPLC.[69] Analysis of the rate of ATPase or product release shortly after initiation of contraction by photolysis of caged molecules identifies transient biochemical steps of the reaction pathway that are faster, or earlier, than the rate-determining step of the cycle. Ferenczi et al.[70] rapidly froze muscle fibers shortly after activation by photolysis of caged [^{3}H]ATP and determined the production of [^{3}H]ADP by HPLC fractionation followed by liquid scintillation counting. They identified a burst of ADP production followed by a steady-state rate of $\sim$2 sec^{-1} per myosin head.

A fluorescent-labeled P_i-binding protein incorporated within the muscle fiber has been used to monitor P_i release after photolysis of caged ATP and caged Ca^{2+} in fibers.[68] The rate of P_i release was markedly higher during the first several turnovers of the ATPase than in other studies. The basis of this discrepancy remains unclear since the signals in these studies were carefully calibrated.

A third method for transient ATPase measurements, described here, is much simpler than freeze quenching or fluorescent product-binding pro-

---

[67] E. J. Potma and G. J. M. Stienen, *J. Physiol.* **496,** 1 (1996).
[68] Z.-H. He, R. K. Chillingworth, M. Brune, J. E. T. Corrie, D. R. Trentham, M. R. Webb, and M. A. Ferenczi, *J. Physiol.* **501,** 125 (1997).
[69] M. G. Hibberd, M. R. Webb, Y. E. Goldman, and D. R. Trentham, *J. Biol. Chem.* **260,** 3496 (1985).
[70] M. A. Ferenczi, E. Homsher, and D. R. Trentham, *J. Physiol.* **352,** 575 (1984).

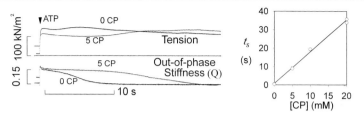

FIG. 5. Determination of the ATPase rate in a single fiber. Transient tension (upper traces) and quadrature (out-of-phase) stiffness (Q, lower traces) were recorded during activation from rigor by photolysis of caged ATP in the presence of 32 μM free Ca^{2+}, 5 mg/ml (1500 U/ml) creatine phosphokinase, and 0 or 5 mM creatine phosphate (CP), temperature 10°. At the arrow, 1.7 mM ATP was liberated from NPE-caged ATP. As the CP was depleted, ATP was no longer regenerated from ADP and the fiber returned to rigor (declining phase of Q). The duration of the steady-state activation (t_s) was determined from the intersection of lines fitted to the steady and declining phases of Q. The ATPase rate is given by the reciprocal of the slope of a line fitted to t_s data plotted against the initial CP concentration.

teins, but with a lower time resolution. The muscle fiber is loaded in rigor with 30 μM Ca^{2+}, 5 mM caged ATP, 1500 U/ml CPK, and various selected CP concentrations. The fiber is transferred to a photolysis trough filled with silicone oil so that ATP is photoreleased only within the fiber. A photolysis pulse initiates isometric contraction of the fiber and the consequent ATPase activity. The CPK reaction buffers the ATP concentration until the CP is depleted, at which point the fiber returns to rigor.[71,72] Figure 5 shows recordings of tension and quadrature stiffness (Q) 90° out-of-phase with an imposed 500-Hz length change on a skinned psoas fiber.[72] The quadrature stiffness is 0.1–0.2 of the in-phase stiffness during the initial contraction and then declines to zero when the CP runs out. Lines are fit to the steady-state and declining phases of Q to determine the duration of the steady-state contraction (t_s). On a plot of t_s vs [CP] (Fig. 5), the reciprocal of the slope of a line fitted to the data points gives the ATPase rate, which is constant from 0 to 20 mM CP. The mean ATPase rate in this assay was 0.355 ± 0.164 mM sec^{-1} (mean ± SD, n = 16,[71]) at 10°. Taking the concentration of myosin heads in the muscle fiber[70] to be 154 μM and the space available to CP within the fiber to be 0.84 of the fiber volume,[73] the ATPase rate is calculated to be 1.93 sec^{-1}, in agreement with Ferenczi et al.[70] and Hibberd et al.[69] No initial burst of ATPase activity is detected with this method, but that may be due to its limited sensitivity.

[71] A. Ishijima, H. Kojima, H. Higuchi, Y. Harada, T. Funatsu, and T. Yanagida, *Biophys. J.* **70**, 383 (1996).
[72] C. S. Cook, H. Higuchi, and Y. E. Goldman, *Biophys. J.* **70**, A291 (1996).
[73] I. Matsubara, Y. E. Goldman, and R. M. Simmons, *J. Mol. Biol.* **173**, 15 (1984).

Shortening Velocity and Interaction Distance

To initiate steady shortening by photolysis of caged ATP, the fiber is put into rigor, Ca^{2+} rigor, and then Ca^{2+} rigor with caged ATP as described earlier. Photolysis of caged ATP causes cross-bridge detachment, reattachment, and initial redevelopment of isometric tension. For a shortening velocity experiment, when tension reaches a preset threshold below the isometric force level, the feedback controlling the motor switches to clamp tension constant. The fiber shortens at the preset isotonic tension until an electronic or mechanical stop is reached that limits the amount of shortening. Tension redevelops and the fiber is then relaxed. Curves of velocity vs force or velocity vs [ATP] are assembled from shortening velocities in a series of such trials. For isotonic force levels below the steady rigor tension, the fiber is released slightly or the fiber is put into rigor in the presence of BDM to lower rigor tension below the preset isotonic level.

A variation of the steady velocity experiment is to transfer the fiber to a trough containing silicone or paraffin oil before photolysis.[74] In this case, the caged ATP is contained within the fiber and a limited, predetermined quantity of ATP is released by the photolysis pulse. The isotonic shortening continues only as long as the ATP remains, after which the fiber returns to rigor. From the amount of shortening (D_s), the ATP concentration utilized by actomyosin ($[ATP_u]$), and the average stiffness during the isotonic phase (S_a), the interaction distance (i.e., the isotonic sliding distance per ATP molecule hydrolyzed by myosin heads while interacting with actin filaments, D_i) can be calculated according to $D_i = D_s S_a [M_o]/[ATP_u]$, where $[M_o]$ is the concentration of myosin heads within the muscle fiber.[70,74] At high velocity (low load), D_i is surprisingly high (≥ 60 nm) relative to the length of the myosin head (16 nm), leading to the suggestion that myosin heads interact with actin multiple times for each ATP molecule hydrolyzed.[74]

Application of Caged ATP to in Vitro Motility Systems

Development of assays of motor protein activity reconstituted *in vitro* from purified proteins has contributed extensively to recent progress in understanding the function and biophysical characteristics of these important enzymes. For actomyosin-based motility and microtubule-based dynein and kinesin motility, assays of motor protein sliding velocity, the unit distance traversed per biochemical cycle (termed step size) and force production by one or a few molecules have been measured.[71,75-81] Photolysis of

---

[74] H. Higuchi and Y. E. Goldman, *Biophys. J.* **69**, 1491 (1995).
[75] S. J. Kron, Y. Y. Toyoshima, T. Q. P. Uyeda, and J. Spudich, *Methods Enzymol.* **196**, 399 (1991).

caged compounds in conjunction with these *in vitro* assays of motor protein function provides an opportunity to link the mechanical events of energy transduction with the elementary biochemical steps in the ATPase reaction of these systems (Fig. 1) and to determine the coupling between the ATPase reaction and motion. Only a few experiments of this kind have been published to date,[81–83] but this is a promising area for further progress. As an example we list methods used for studying force and sliding velocity of kinesin on microtubules. Myosin and actomyosin have also been studied using caged ATP *in vitro* with evanescent wave video microscopy[82] and in suspension with electron paramagnetic resonance spectroscopy.[83,84]

Preparation of Kinesin and Tubulin

Kinesin is purified from bovine brain by microtubule affinity followed by DEAE chromatography (Fractogel DEAE, Merck, Darmstadt, Germany) and sedimentation.[85] Tubulin is prepared from bovine brain by repeated temperature-dependent polymerization/depolymerization cycles and phosphocellulose chromatography, and then labeled with tetramethylrhodamine[86] (Molecular Probes, Eugene, OR). Purified kinesin (~0.2 mg/ml) and tubulin (~10 mg/ml) are rapidly frozen in liquid nitrogen and stored at −80°.

Microtubules are prepared by copolymerization of fluorescent and nonfluorescent tubulin in a molar ratio of 1:20, total protein concentration ~10 mg/ml, for 30 min at 35° in a solution containing 1 mM GTP, 100 mM MES buffer, pH 6.7, 5 mM MgCl$_2$, 1 mM EGTA, and 33% glycerol. After polymerization, microtubules are stabilized by adding 40 μM taxol,[81] pelleted at 250,000 g for 10 min at 25°, resuspended at 10 mg/ml in a solution containing 80 mM PIPES, pH 6.8, 2 mM MgCl$_2$, 40 μM taxol, and 1 mM EGTA, and stored at 4° for up to a few months.

---

[76] Y. Harada, K. Sakurada, T. Aoki, D. D. Thomas, and T. Yanagida, *J. Mol. Biol.* **216,** 49 (1990).
[77] A. Ishijima, T. Doi, K. Sakurada, and T. Yanagida, *Nature* **352,** 301 (1991).
[78] A. Ishijima, Y. Harada, H. Kojima, T. Funatsu, H. Higuchi, and T. Yanagida, *Biochem. Biophys. Res. Commun.* **199,** 1057 (1994).
[79] K. Svoboda, C. F. Schmidt, B. J. Schnapp, and S. M. Block, *Nature* **365,** 721 (1993).
[80] J. T. Finer, R. M. Simmons, and J. A. Spudich, *Nature* **368,** 113 (1994).
[81] H. Higuchi, E. Muto, Y. Inoue, and T. Yanagida, *Proc. Natl. Acad. Sci. U.S.A.* **94,** 4395 (1997).
[82] P. B. Conibear and C. R. Bagshaw, *FEBS Lett.* **380,** 13 (1996).
[83] C. L. Berger, E. C. Svensson, and D. D. Thomas, *Proc. Natl. Acad. Sci. U.S.A.* **86,** 8753 (1989).
[84] E. M. Ostap, H. D. White, and D. D. Thomas, *Biochemistry* **32,** 6712 (1993).
[85] B. J. Schnapp and T. S. Reese, *Proc. Natl. Acad. Sci. U.S.A.* **86,** 1548 (1989).
[86] A. Hyman, D. Drechsel, D. Kellogg, S. Salser, K. Sawin, P. Steffen, L. Wordeman, and T. Mitchison, *Methods Enzymol.* **196,** 478 (1991).

Apparatus for Combined Motility Assays and Photolysis of Caged ATP

To detect force or motion from kinesin interacting with microtubules, either fluorescent-labeled microtubules or kinesin-coated fluorescent beads are observed by fluorescence microscopy (Fig. 6). Fluorescence is excited by light from a mercury arc lamp through a commercial rhodamine epifluorescence filter block or an intense laser such as a frequency-doubled Nd:YAG [neodymium–(yttrium–aluminum–garnet)] green laser ($\lambda = 532$ nm). Fluorescence images are recorded by a high sensitivity camera such as a SIT camera.

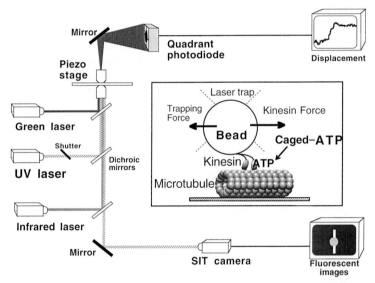

FIG. 6. Schematic diagram of an experimental arrangement for laser trapping nanometry of displacement and force generation by single kinesin molecules initiated by flash photolysis of caged ATP. The green laser (frequency-doubled Nd:YAG, 100 mW at $\lambda = 532$ nm; Model 140-0534-100, Lightwave Electronics, Mountain View, CA) for fluorescence excitation, the UV laser (He–Cd, 12 mW at $\lambda = 354$ nm; Model 7212I, mfg. Liconix, Santa Clara, CA), and the infrared laser (diode pumped, Nd:YAG, 300 mW, $\lambda = 1064$ nm; Model MYLS-300-S OEM, Santa Fe Laser Co., CA) are all focused onto the sample through an objective lens (Plan NCF Fluor 100×, Nikon). The UV laser is equipped with a rapid mechanical shutter (response time 0.2 msec, Cambridge Technology, Inc.). Fluorescence images are recorded by a SIT camera (C2741, Hamamatsu Photonics). The transmitted bright-field image of the sample, illuminated by a 50-W tungsten–halogen lamp or the green laser, is projected onto a quadrant photodiode (s944-13, Hamamatsu Photonics) by an objective lens (Plan 100×, Nikon). (Inset) A representation of a kinesin molecule producing active translocation force along a microtubule against the force applied to a latex bead by the optical trap. [Modified from H. Higuchi, E. Muto, Y. Inoue, and T. Yanagida, *Proc. Natl. Acad. Sci. U.S.A.* **94**, 4395 (1997). Copyright (1997) National Academy of Sciences, U.S.A.]

To measure the forces produced by single motor protein molecules bound to beads, an optical trapping system is combined with fluorescence microscopy[79,81] (Fig. 6). A diode-pumped, Nd:YAG infrared laser (λ = 1064 nm) is focused at the sample plane by an objective lens to form the optical trap. The transmitted bright-field image of the bead, illuminated by a 50-W tungsten–halogen lamp or the green laser, is projected onto a quadrant photodiode by an objective lens. Displacements of the bead in orthogonal directions are determined from the outputs of the quadrant photodiode boosted by differential amplifiers. Data are recorded using a computerized data acquisition system at a sampling rate of 200 Hz–20 kHz.

For photolysis of caged ATP, the sample is illuminated through the lower microscope objective in Fig. 6 with a pulse from a continuous wave, near-ultraviolet laser such as a He–Cd laser (λ = 354 nm) equipped with a rapid mechanical shutter or from a Nd:YAG or Nd:YLF pulsed laser (Table II). The laser beam passes through a 3-mm-diameter field diaphragm and is then focused to a diameter of ~30 μm at the sample plane by the objective lens. With the He–Cd laser propagating 3.8 mW of energy through the objective lens, the energy density in the 707-μm^2 laser spot at the sample is 5400 mW mm^{-2}.

In Vitro Assays of Motor Protein Activity

Microtubule Gliding Assay

A 3- to 10-μl flow chamber is made of large (24 × 32 mm^2) and small (18 × 18 mm^2) coverslips separated by 2-mm-wide slivers of 10- to 30-μm-thick polyester film. To reduce adhesion of proteins, the surfaces of the coverslips are coated with casein by incubating in a standard solution (80 mM PIPES buffer, pH 6.8, 2 mM MgCl$_2$, 1 mM EGTA, and 0.7 mg/ml casein) for 2 min. Kinesin is rapidly thawed, diluted to 10 μg/ml (~ 25 nM) in the standard solution, and introduced into the flow chamber for 2 min. After unbound kinesin is washed out with the standard solution, fluorescent microtubules (~1 μg/ml) are put into the flow chamber in a gliding assay solution containing an enzymatic oxygen scavenging system[76] [80 mM PIPES, pH 6.8, 2 mM MgCl$_2$, 1 mM EGTA, 0.5–2 mM ATP or caged ATP, 10 μM taxol, 10 mM glucose, 0.5% (v/v) 2-mercaptoethanol, 50 μg/ml (~5 U/ml) glucose oxidase, and 9 μg/ml (~100 U/ml) catalase (Sigma, St. Louis, MO), ionic strength ~ 50 mM and temperature, 25°–27°]. Finally, the four edges of the flow chamber are sealed with nail polish. The rate of microtubule sliding (after photolysis in caged ATP experiments) is determined by fluorescence video microscopy.

Bead Assay for Force Measurement

Kinesin-coated beads are prepared as follows: the surface of 0.5- to 1-μm-diameter fluorescent latex beads [blue fluorescent, Molecular Probes, final concentration of 5% (w/v) or 140 pM] is coated with casein by incubating on ice in the standard solution for >30 min. The beads (final concentration, 70 pM) are then rapidly mixed for about 30 sec with an equal volume of the standard solution and kinesin (final concentration, 70–1000 pM) and incubated on ice for >30 min.

A flow chamber is made with coverslips spaced ~10 μm apart. Fluorescent microtubules are attached to the coverslips by incubation for 2 min in the flow chamber at ~10 μg/ml in the standard solution with 10 μM taxol added. The remaining coverslip surfaces are then coated with casein by incubation for 2 min with the standard solution plus 10 μM taxol. With this treatment, the beads do not bind to the coverslips. Finally, the chamber is filled with a bead–assay solution containing the beads and a gliding assay solution [~0.1 pM kinesin-coated beads, 0.7 mg/ml casein, 1 mM caged ATP, 1–25 U/ml apyrase (Sigma)]. The apyrase is added to hydrolyze any ATP or ADP contamination in the solution. To measure force, kinesin-coated beads are trapped with the infrared laser and brought into contact with microtubules attached to the coverslip (inset in Fig. 6A). Then, active movement is initiated by photoliberating ATP from caged ATP with a 5-msec pulse of UV laser light. The concentration of released ATP is varied by changing the caged ATP concentration. After photolysis of caged ATP and a variable time lag, the beads actively translate in a stepwise manner and then stabilize at a high force. The rate and magnitude of these events are measured from the quadrant photodiode records.

Quantitating Photochemical Conversion

Photochemical Conversion in Muscle Fibers

Techniques to analyze the extent of photochemical conversion depend on the volume of sample that can be collected following the experiment. In a typical muscle fiber experiment, the fiber is bathed in a 10- to 50-μl trough that is illuminated fairly evenly by the photolysis light source. For caged ATP or caged ADP, this macroscopic sample is collected and analyzed for released nucleotide by HPLC. Two microliters of photolyzed solution is loaded onto an anion-exchange column (Radial Pak 8PSAX, bead size 10 μm, Waters Corp., Milford, MA) and eluted with an isocratic running buffer of 0.3 M (NH$_4$)H$_2$PO$_4$ adjusted to pH 4.3 and mixed with acetonitrile (97:3, v/v). Signal amplitude and chromatographic separation are sufficient to enable detection of 2–30% photolysis of 10 mM caged nucleotide.

For caged P_i, HPLC analysis or a colorimetric assay[21] is used to determine the amount of photolysis. For the colorimetric assay, P_i is first separated from phosphoesters, such as ATP and CP, that increase background P_i by hydrolysis. P_i in the photolysis solution is precipitated as calcium phosphate using cold $Ca(OH)_2$ (0.2 M), pelleted, and resuspended twice. The pellet is then dissolved in 1 N HCl and assayed for P_i using a molybdate-type phosphorus reagent (Sigma). Comparing the absorption measurements in this assay with those on standards made by serial dilution of a 10 mM stock of P_i, the method is linear up to 10 mM and can detect 0.05–0.1 mM photoreleased P_i.

For photolysis of caged ATP in fibers immersed in silicone or paraffin oil, the volume of photolysis solution within the fiber (~20 nl) is not sufficient for these chemical methods. In this case, the ATP concentration liberated by photolysis inside the fiber is determined using caged [^{3}H]ATP. The fiber is prepared for photolysis using the same solutions and procedures as described earlier except that the Ca rigor solution contains ~2 mM caged [^{3}H]ATP (loading solution). ATP is released by photolysis, and the fiber is allowed to contract isometrically in the oil for 1–2 min after photolysis. The fiber is then immersed for ~4 min in a washout solution, relaxing solution with 0.2 mM unlabeled caged ATP, 0.2 mM ADP, and 0.1 mM AMP added to reduce loss of radioactive counts. The fiber is immersed for 2 min each in two more aliquots of the washout solution. The three postphotolysis washes, the loading solution, paraffin oil after the photolysis, and the fiber after washing are saved for counting of radioactivity. Nucleotides and caged nucleotides in the solutions are separated using a SAX analytical HPLC column as described earlier. The elution is monitored at 254 nm. One-milliliter fractions are collected, mixed with scintillation fluid, and each sample is counted for 1 min in a liquid scintillation counter along with standards, the loading solution, and the oil collected from the photolysis trough. The concentration of photolyzed caged ATP is calculated from the initial amount of labeled caged ATP in the fiber, the proportion photolyzed, and the bulk concentration of caged ATP in the Ca rigor solution. It is important to measure the laser energy actually striking the fiber in this technique because inhomogeneity of the laser beam can otherwise markedly increase variability between experiments.

In experiments using DM-nitrophen or NP-EGTA to release Ca^{2+}, the muscle fiber itself provides a convenient Ca^{2+} sensor. The relationship between steady active tension and the free Ca^{2+} concentration is determined before or after the photolysis trials. Ionic conditions during this control experiment are chosen to be identical to those in the photolysis experiment, except a Ca^{2+} chelator with well-characterized affinity, such as EGTA, is used to buffer the free $[Ca^{2+}]$ at saturating and various intermediate values. The fiber is activated at saturating $[Ca^{2+}]$ at the beginning of the experiment and several times later to guard against, or correct for, decline of the fiber performance. Comparison of the fiber tension on photolysis with this standard curve provides an estimate of the [Ca^{2+}] released. Of course, this estimate depends on the similarity in ionic conditions between the control and test portions of this experiment and it assumes that photolysis does not alter the Ca^{2+} response of the muscle fiber. When DM-nitrophen or NP-EGTA is photolyzed, bound Ca is released rapidly, creating an initial spike of free Ca^{2+} in the range of 100–500 μM. Some of the released Ca^{2+} binds to unphotolyzed caged Ca, reducing the free [Ca^{2+}] to 10–50 μM within a few milliseconds.[87] This initial transient of high free Ca^{2+} does not affect the steady tension reached in a contraction, but it might alter the kinetics. The effects of this initial spike in caged Ca experiments are controversial in the excitation–contraction coupling field.[88,89]

Fraction of Caged ATP Photolyzed in Vitro

The concentration of photoreleased ATP for *in vitro* experiments is measured by illuminating a sample chamber through an objective lens at a known energy density. In Fig. 6, the photolysis pulse propagates through the lower objective lens and a restricting aperture is removed.[81] The chamber, made of fused silica (path length 2 mm), is filled with 90 μl of a standard motility assay solution (described earlier) containing caged ATP at 0.2 mM or less. The concentration of caged ATP should be kept low to minimize the light gradient due to absorption by the caged ATP (see further discussion of this point later). ε_{354} for caged ATP is ~600 M^{-1} cm^{-1}, giving an optical density of 0.012 (97% transmission) through the 2-mm path length at 0.1 mM caged ATP. The chamber is illuminated by the UV laser at known low energy density, e.g., 0.13 mW mm^{-2}, for 1, 2, and 3 sec at 25–27°. The photoreleased ATP is then hydrolyzed by adding 90 μl of a solution containing hexokinase (~10 U/ml, Sigma) or 0.2 mg ml^{-1} skeletal muscle myosin in 200 mM HEPES buffer, pH 7.8, 0.6 M KCl, 20 mM EDTA and incubating at 25–27° for 10–60 min. The amount of P$_i$ produced is measured by the malachite green method.[76] The percentage photolysis, R, was calculated to be 8.7% of the starting caged ATP per mJ mm^{-2} of illumination.[81] This value is consistent with that obtained by photolysis of caged ATP within an 80-μm-diameter muscle fiber (7.4% per mJ mm^{-2}) using a pulsed 347-nm laser.[74] At the energy density used in the motility experiments, the rate constant k_r for removal of caged ATP is calculated to be 470 sec^{-1}

[87] G. C. R. Ellis-Davies, J. H. Kaplan, and R. J. Barsotti, *Biophys. J.* **70,** 1006 (1996).
[88] S. Györke and M. Fill, *Science* **260,** 807 (1993).
[89] G. D. Lamb, M. W. Fryer, and D. G. Stephenson, *Science* **263,** 986 (1994).

(0.087 mm² mJ⁻¹ × 5400 mJ mm⁻² s⁻¹). The extent of photolysis is given by $(1 - e^{-k_r t})$, where t is the duration of illumination. For a 5-msec shutter time, the fraction of photolyzed caged ATP is calculated to be 0.90.

Uniformity of Photolysis in Sample Chamber

As the photolysis beam propagates through the sample chamber, its intensity decreases due to absorption by the caged compound or scattering by the biological sample. The amount of photolysis thus decays through the sample. The consequences of nonuniform release of the active molecule depend on the experimental preparation and the geometry. The concentration of photoreleased biological effector R_0 at the entrance face of the sample chamber is given by $R_0 = (I_0 Q_p \lambda C \varepsilon)/(abhcN_A)$, where I_0 is the photolysis pulse energy at wavelength λ, a and b are the height and width of the photolysis chamber, Q_p is the quantum yield for photolysis, C is the concentration of the caged molecule, ε is its molar absorption coefficient at λ, h is Planck's constant, c is the velocity of light, and N_A is Avogadro's number. In the case of a skinned muscle fiber, the decay of intensity is due largely to absorption by the caged compound in the bathing medium. The light intensity decreases according to $I = I_0 e^{-C \varepsilon x}$, where x is the distance into the sample chamber. Assuming photolysis is proportional to light intensity (valid except for very short, focused laser pulses or high percentage photolysis), $R(x) = (I_0 Q_p \lambda C \varepsilon)/(abhcN_A) e^{-C \varepsilon x}$. Figure 7A shows the profile of relative light intensity expected for the typical conditions of a muscle fiber experiment in which NPE-caged ATP is photolyzed using a frequency-doubled ruby laser, $\lambda = 347$ nm, $\varepsilon = 660\ M^{-1}$ cm⁻¹, $Q_p = 0.63$, and $d = 0.1$ cm. The average concentration of photoreleased ligand over a path ligand d is $R_{av} = I_0 Q_p \lambda (1 - e^{-C \varepsilon d})/(abhcN_A d)$, which is shown by a horizontal dashed line in Fig. 7A. The position, x_{av}, within the trough where the local photolysis is equal to the average value is the intersection of these two lines 470 μm from the front surface of the trough, very close to the center. The photolysis at the back of the 1-mm trough is 0.5 of that in the front, but this can be increased to 0.7 with a polished stainless-steel surface that reflects one-half of the light incident on it[32] (see Fig. 2C). The decay of intensity across a 100-μm muscle fiber (circle in Fig. 7A) is only 6% at 10 mM caged ATP.

The optimal concentration of caged ATP is dictated by the concentration of ATP to be released and by agonist or antagonist action of the caged compound before photolysis. Taking the position of the fiber to be fixed at a certain depth, d_f (typically 500 μm), the released concentration of ATP [$R(d_f)$ at the fiber] increases with caged ATP concentration up to a

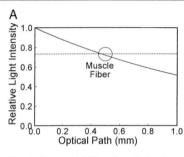
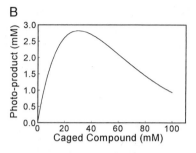

FIG. 7. Characteristics of spatial and concentration dependence of photorelease in a typical muscle fiber experiment. (A) Light intensity distribution along a 1-mm optical path in a muscle fiber photolysis trough. The solid line shows the exponential delay of light intensity and photorelease in a 1-mm optical path photolysis trough containing 10 mM NPE-caged ATP, ε_{347} = 660 M^{-1} cm^{-1}. The average concentration of photoreleased ATP is shown by the dashed line. A 100-μm-diameter muscle fiber (circle) in the center of the trough receives almost the same light intensity and photolysis as the average value. (B) Concentration dependence of photolysis. The concentration of ATP released at d_f = 500 μm depth in the trough (the position of the fiber) by photolysis of NPE-caged ATP at 347 nm is plotted vs the initial [caged ATP]. Photolysis quantum yield (Q_p) is 0.63. Assuming that photolysis is proportional to light absorption, the maximum photolysis occurs at caged ATP concentration, C_{max} = $1/(\varepsilon\, d_f)$ = 30 mM in this example, but at a lower concentration if ε or d_f is higher.

maximum (Fig. 7B) and then declines due to excessive absorption of the light in the first half of the sample chamber. The maximum release occurs at $C_{max} = 1/(\varepsilon\, d_f) = 30$ mM. In the fiber experiment, caged ATP concentrations are kept lower (<10 mM) because caged ATP competitively inhibits ATP binding (see later).

In general, the concentration of caged molecule is limited by the sample depth and the absorptivity at the photolysis wavelength because $R(d_f)$ peaks at $C_{max} = 1/(\varepsilon\, d_f)$. This phenomenon also bears on the choice of illumination wavelength. For example, NPE-caged ATP absorbs very strongly at 260 nm, suggesting that a laser emitting at wavelengths less than 300 nm would be more suited to the experiment. However, the absorptivity, $\varepsilon_{260} = 19,600$ M^{-1} cm^{-1}, is too high for efficient photolysis at $d_f = 500$ μm. The caged ATP concentration giving maximal photolysis [$1/(\varepsilon\, d_f)$] is only 1 mM in this case, making it difficult to approach 1 mM ATP release. At given light levels, increasing the caged ATP concentration above 1 mM would decrease rather than increase the amount of ATP released at the fiber. Higher photolysis wavelengths avoid this problem because ε is smaller and there is the further advantage that absorption by proteins and consequent optical damage is lower for wavelengths above 310 nm.

Prephotolysis Effects of Caged ATP on Motor Proteins

NPE-caged ATP binds weakly to the active site of myosin[52] and kinesin[81] and thereby inhibits ATP binding and sliding velocity. The extent of this inhibition measured by the "slack test"[90] in muscle fibers and in a kinesin microtubule gliding assay is presented below.[81]

Inhibition of Muscle Shortening Velocity

Glycerol-extracted single psoas muscle fibers are prepared from rabbit muscle and mounted in the photolysis setup as described earlier. Relaxed fibers are stretched by 10% of slack length to ~2.5 μm sarcomere length. They are incubated for ~2 min in a preactivating solution containing low Ca-buffering capacity and then activated at 30 μM free $[Ca^{2+}]$, $\Gamma = 200$ mM (Table V), $T = 10°$, and 0.3–5 mM [ATP]. After active tension reaches a steady level, a quick release, typically 5–20% of the fiber length, is applied and the fiber briefly goes slack until active shortening causes redevelopment of tension (Fig. 8A, inset). The fiber is then relaxed again. From pairs of slack releases of different amplitudes, the times elapsed until the initial redevelopment of tension are measured.[90] Shortening velocity is calculated from the difference in amplitudes of the releases and the differences in slack times.

Shortening velocity increases with MgATP concentration and saturates in the mM range.[91,92] 5 mM caged ATP suppresses shortening velocity (Fig. 8A). When plotted on an inverse plot (Lineweaver–Burk plot, not shown), data fall on two straight lines that intercept the ordinate at approximately the same maximum velocity, indicating that caged ATP is a competitive inhibitor of ATP. Equation (1) is fit to velocity data to give a half-saturation value for ATP concentration

$$V/V_{max} = [ATP]/\{[ATP] + K_{ATP}(1 + [cATP]/K_I)\} \quad (1)$$

where K_I is an inhibition constant for caged ATP (cATP), $K_{ATP} = 280 \pm 50$ μM ($n = 21$ data points from 10 fibers), and $K_I = 1.93 \pm 0.74$ mM ($n = 16$ points from 9 fibers). In a series of experiments using free $[Mg^{2+}]$ from 300 μM to 10 mM, the inhibitory effect of caged ATP did not depend appreciably on $[Mg^{2+}]$, suggesting that both cATP^{3-} and Mg-cATP$^-$ are equally effective inhibitors. The R-isomer of caged ATP is slightly (20%) less effective at inhibiting shortening velocity than the S-isomer.

[90] K. A. P. Edman, *J. Physiol.* **291**, 143 (1979).
[91] M. A. Ferenczi, Y. E. Goldman, and R. M. Simmons, *J. Physiol.* **350**, 519 (1984).
[92] E. Pate, M. Lin, K. Franks-Skiba, and R. Cooke, *Am. J. Physiol.* **262**, C1039 (1992).

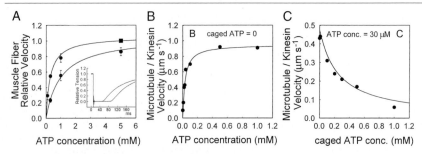

FIG. 8. Inhibition of motor protein activity by NPE-caged ATP. (A) Shortening velocity measured by the "slack test" (see inset) plotted against ATP concentration in the absence (upper curve) and presence of 5 mM caged ATP. All velocities are relative to that at 5 mM ATP, 0 caged ATP (square) at 10°, corresponding to 4.25 ± 0.38 μm sec^{-1} per half sarcomere (mean ± SEM, n = 19 fibers). The curves represent Eq. (1) of the text plotted with K_{ATP} = 280 μM and the inhibition constant for NPE-caged ATP, K_I = 1.93 mM. (Inset) Force recordings for a single fiber activated at ~32 μM free Ca^{2+}, 0.3 mM ATP, 0 caged ATP at 10°. Quick releases of 12 and 18% were applied at the vertical line, causing the fiber to go slack. The slack time is measured by recording the times elapsed between the release and until the initial redevelopment of tension. (B) Velocity of microtubule gliding driven by kinesin plotted against ATP concentration in the gliding assay solution (see text) with no added caged ATP at 25–27°. The curve represents Eq. (1) of the text with K_{ATP} = 30 μM and V_{max} = 0.95 μm sec^{-1}. (C) Velocity of microtubule gliding as in (B), but plotted against concentration of NPE-caged ATP (mixed R- and S-isomers) in the presence of 30 μM total ATP and 2 mM total Mg. The curve represents Eq. (1) with K_{ATP} and V_{max} as described earlier and K_I = 126 μM.

According to these values for inhibition of muscle fiber shortening velocity, unphotolyzed caged ATP remaining after a flash in laser photolysis experiments would inhibit shortening. For example, when 1 mM caged ATP is released from 5 mM starting caged ATP, the remaining 4 mM of caged ATP would suppress the shortening velocity to 70% of its value at 1 mM ATP in the absence of caged ATP.

Caged ATP binding to the active site also affects the time course of isometric force development following activation of a fiber by photolysis of caged ATP. The expected slowing of reaction kinetics by this effect is illustrated in Fig. 9, which shows simulations of the simple reaction scheme (Scheme 1).

The effective rate through the cross-bridge detachment step in Scheme I is given by Eq. (2):

$$[ATP][AM]k_T = [ATP][AM_t]K_I/([cATP] + K_I) \qquad (2)$$

where K_I is the inhibition constant of caged ATP for detachment and [AM$_t$] is the total actomyosin concentration (= [AM] + [AM·cATP]) at the

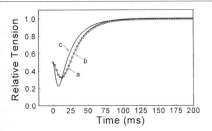

FIG. 9. Simulations of tension transients of a muscle fiber activated by photolysis of caged ATP. Differential equations corresponding to Scheme I of the text were integrated with a fourth-order Runge–Kutta algorithm. Photoreleased ATP concentration, 1 mM; total myosin head concentration, 200 μM; k_p, 100 sec^{-1}; k_R, 50 sec^{-1}; k_C, 1 × 10^6 M^{-1} sec^{-1}; k_{-C}, 2 × 10^3 sec^{-1}. For trace a, [cATP], 6 mM; k_T, 1 × 10^6 M^{-1} sec^{-1}. For trace b, [cATP], 0; k_T, 2.5 × 10^5 M^{-1} sec^{-1}. For trace c, [cATP], 0, k_T, 1 × 10^6 M^{-1} sec^{-1}.

beginning of the experimental trial. Including the dark photolysis reaction (k_p), the photolysis-induced detachment is a two-step series reaction whose rate (k_{obs}) is approximately the product divided by the sum of k_p and ([ATP]k_T[AM]/{[AM] + [AM·cATP]}).

$$k_{obs} = \frac{\left(\dfrac{K_I}{[cATP] + K_I}\right) k_T [ATP] k_p}{\left(\dfrac{K_I}{[cATP] + K_I}\right) k_T [ATP] + k_p} \qquad (3)$$

As an example of the expected effect of caged ATP binding on the detachment kinetics, let k_p = 100 sec^{-1}, k_T = 1 × 10^6 M^{-1} sec^{-1}, released [ATP] = 1 mM, K_I = 2 mM, and [cATP] = 6 mM. k_{obs} is then calculated from Eq. (3) to be 71.4 sec^{-1}, and a simulation of detachment and reattachment is shown in Fig. 9 (trace a). If caged ATP did not inhibit detachment (set [cATP] = 0 in Eq. (3)), then the same k_{obs} would be obtained with k_T = 2.5 × 10^5 M^{-1} sec^{-1}, and a kinetic simulation (Fig. 9, trace b) shows that the time course of cross-bridge detachment and reattachment would be the same. Trace c in Fig. 9 shows the expected faster time course at k_T = 1 × 10^6 M^{-1} sec^{-1} with no inhibition by caged ATP. k_{obs} in this case is 90.9 sec^{-1}. Thus caged ATP inhibits the rate of detachment in typical

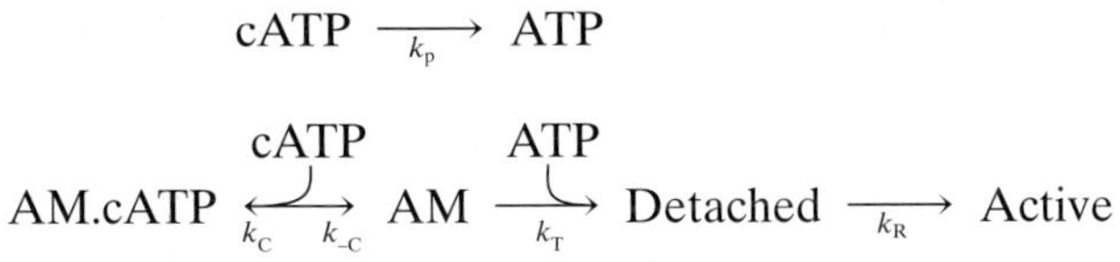

$$\text{cATP} \xrightarrow{k_p} \text{ATP}$$

$$\text{AM·cATP} \underset{k_{-C}}{\overset{\text{cATP}}{\underset{k_C}{\rightleftarrows}}} \text{AM} \xrightarrow[k_T]{\text{ATP}} \text{Detached} \xrightarrow{k_R} \text{Active}$$

SCHEME I

conditions of a photolysis experiment. This inhibition must be taken into account in estimating k_T, either by fitting a model, such as Scheme I, which includes the AM.cATP state, to the transient data or by adjusting k_T upward by the factor ([cATP] + K_I)/K_I) implied by Eq. (3). This adjustment increases k_T fourfold in the case considered here. Suppression of the detachment rate would be less at lower [cATP].

Inhibition of Kinesin Translocation Velocity by Caged ATP

To determine the inhibition constant of caged ATP on the velocity of microtubule sliding driven by kinesin, concentrations of ATP and caged ATP were varied in the incubation medium of the *in vitro* microtubule gliding assay. Microtubule sliding velocity increased with [ATP], giving a half-saturation value of $K_{ATP} = 30$ μM and $V_{max} = 0.95$ μm sec^{-1} (Fig. 8B, correlation coefficient $r = 1.0$). At constant total [ATP] = 30 μM and constant total [Mg] = 2 mM, increasing [cATP] up to 1 mM inhibited velocity (Fig. 8C). Assuming competitive inhibition, fitting Eq. (1) to these data gave a $K_I = 126$ μM ($r = 0.99$) (Y. Inoue and H. Higuchi, personal communication).

Binding of Mg^{2+} and Ca^{2+} to Caged ATP

In calculating the free [Mg^{2+}] and free [Ca^{2+}] in solutions containing caged ATP, the affinities of these ions to caged ATP must be taken into account. This section presents measurements of these affinities. The predominant species of NPE-caged ATP near neutral pH is cATP^{3-} and that of the calcium complex of caged ATP is Ca-cATP$^-$. To determine the affinity

$$K_{Ca} = [\text{Ca-cATP}^-]/([\text{Ca}^{2+}][\text{cATP}^{3-}]) \quad (4)$$

we made up test solutions containing known total concentrations of Ca, Mg, and cATP (Ca$_t$, Mg$_t$ and cATP$_t$, respectively) and measured free [Ca^{2+}] using a Ca-sensitive electrode (WPI Inc., New Haven, CT, type TIPCA). The caged ATP stock was a pure sample (<0.01% contamination) of a mixture of both enantiomers of NPE-caged ATP.

A solution containing 5 mM caged ATP, 146 mM KCl, and 100 mM TES buffer, pH 7.1, 20° was mixed with 10 mM CaCl$_2$, 5 mM caged ATP, 137 mM KCl, and 100 mM TES buffer in various ratios to result in test solutions with 0.63–10.0 mM Ca$_t$. cATP$_t$ was determined from the absorption at 260 nm, assuming $\varepsilon_{cATP} = 19.6$ M^{-1} cm^{-1}. The free [Ca^{2+}] is given by the quadratic equation

$$[\text{Ca}^{2+}]^2 K_{Ca} + [\text{Ca}^{2+}](1 + \text{cATP}_t K_{Ca} - \text{Ca}_t K_{Ca}) - \text{Ca}_t = 0 \quad (5)$$

and the expected output potential difference of the Ca-sensitive electrode from that in a solution with 1 mM free [Ca^{2+}] is given by

$$V - V_{mM} = S \log \{[\text{Ca}^{2+}]/(1 \text{ m}M)\} \tag{6}$$

Before and after each calcium titration, the response of the electrode was tested in calibrating solutions containing 0, 0.1, 1, and 10 mM free Ca^{2+} (WPI, CALBUF). The slope (S) of the electrode standard curve was 30.1 mV per decade change of [Ca^{2+}]. After adjusting K_{Ca} in Eqs. (4) and (5) for the best fit, Eqs. (5) and (6) predicted calcium titration data quite well (correlation coefficients $r = 0.99$).

The affinity of Mg^{2+} for caged ATP,

$$K_{Mg} = [\text{Mg-cATP}^-]/([\text{Mg}^{2+}][\text{cATP}^{3-}]) \tag{7}$$

was similarly measured by displacing Ca from caged ATP with Mg and measuring free [Ca^{2+}] with the calcium-sensitive electrode. Mg$_t$ was altered from 0 to 10 mM in solutions containing 5 mM caged ATP, 100 mM TES buffer, 137–146 mM KCl, and Ca$_t$ at 0.5, 1.0, or 2.0 mM. The free [cATP^{3-}] in solutions containing Mg is given by the cubic equation [Eq. (8)]:

$$[\text{cATP}^{3-}]^3 K_{Ca} K_{Mg} + [\text{cATP}^{3-}]^2 (K_{Ca} + K_{Mg} - \text{cATP}_t K_{Ca} K_{Mg} + \text{Ca}_t K_{Ca} K_{Mg} + \text{Mg}_t K_{Ca} K_{Mg}) + [\text{cATP}^{3-}](1 - \text{cATP}_t K_{Ca} - \text{cATP}_t K_{Mg} + \text{Ca}_t K_{Ca} + \text{Mg}_t K_{Mg}) - \text{cATP}_t = 0) \tag{8}$$

Free [Ca^{2+}] is then given by

$$[\text{Ca}^{2+}] = \text{Ca}_t/(1 + K_{Ca}[\text{cATP}^{3-}]) \tag{9}$$

After adjusting K_{Ca} and K_{Mg} in Eqs. (7) and (8) for the best fit, Eqs. (6) and (9) predicted magnesium titration quite well (correlation coefficients $r = 0.99$).

pK_{Ca} = $-\log K_{Ca}$ was found to be 2.52 ± 0.03 (mean ± SD, $n = 6$) at 20°, pH 7.1, corresponding to a dissociation constant of 3 mM for Ca-cATP$^-$. pK_{Mg} = $-\log K_{Mg}$ was found to be 2.45 ± 0.02 ($n = 4$), corresponding to a dissociation constant of 3.5 mM for Mg-cATP$^-$. The dissociation constant for Mg for the reaction intermediate populated upon UV irradiation of caged ATP (*aci*-nitro intermediate) is also 3 mM.[56] pK_{Ca} and pK_{Mg} for binding of Ca^{2+} and Mg^{2+} to ATP at pH 7.1 were found to be 3.55 ± 0.02 ($n = 6$) and 3.81 ± 0.02 ($n = 3$), respectively, in agreement with values obtained by Moisescu and Thieleczek[93] under very similar conditions, thus validating this potentiometric method.

Conclusions

Photolysis of caged molecules, particularly caged ATP and caged Ca^{2+}, provides a very flexible experimental method to activate muscle fibers and

---

[93] D. G. Moisescu and R. Thieleczek, *Biochim. Biophys. Acta* **546**, 64 (1979).

motor proteins *in vitro*. Hence many protocols for laser pulse photolysis with these proteins have been described. Combination of photolysis with other biophysical signals that probe molecular events is a fertile area for investigation. Many of the considerations in photolysis experiments on skeletal muscle and motor proteins are applicable to other types of muscle and other enzymes.

Acknowledgments

This work was supported by NIH Grant HL15835 to the Pennsylvania Muscle Institute and the Yanagida Biomotron Project ERATO JST. We thank Drs. J. W. Walker and G. C. R. Ellis-Davies for access to unpublished data, Drs. J. E. T. Corrie and D. R. Trentham for helpful comments on the manuscript, and Ms. Kimberly L. Dopke for help with the manuscript.

[19] Application of Caged Fluorescein-Labeled Tubulin to Studies of Microtubule Dynamics and Transport of Tubulin Molecules in Axons

By TAKESHI FUNAKOSHI and NOBUTAKA HIROKAWA

Introduction

Microinjection of caged fluorescein-labeled cytoskeleton associated proteins into cultured cells and subsequent photoactivation of the fluorescence in these conjugates has proved to be a very powerful tool in studying the dynamics of the microtubule and actin cytoskeleton in living cells.[1–4] Caged fluorescein-labeled tubulin has been successfully applied to study microtubule dynamics in mitotic spindles and neuronal axons. This approach has allowed us to investigate molecular mechanisms of mitosis and the origin of the slow transport of tubulin molecules in nerve axons. This article describes methods and techniques using caged fluorescein-labeled tubulin for investigations of the dynamics and transport of axonal tubulin molecules in neurons.

In these studies, caged fluorescein-labeled tubulin is introduced into cultured neurons by microinjection and, after a recovery period that also allows for diffusion of the caged fluorescein tubulin throughout the cell,

[1] T. J. Mitchison, *J. Cell Biol.* **109**, 637 (1989).
[2] J. A. Theriot and T. J. Mitchison, *Nature* **352**, 126 (1991).
[3] S. Okabe and N. Hirokawa, *J. Cell Biol.* **117**, 105 (1992).
[4] J. Sabry, T. P. O'Connor, and M. W. Kirschner, *Neuron* **14**, 1247 (1995).

small segments of axons are exposed to a brief pulse of near-ultraviolet light. The caged protection groups on the fluorescein absorb these photons, and in a dark reaction the ether bond that links the caged groups to the phenolic groups of fluorescein is cleaved, liberating highly fluorescent tubulin conjugates only in the irradiated segment. The fate of the photoactivated tubulin molecules was observed with low-light level fluorescence microscopy or with electron microscopy by staining with antibody directed against fluorescein.

Preparation of Caged Fluorescein-Labeled Tubulin

Caged fluorescein-labeled tubulin is prepared according to the method of Mitchison[1] with slight modifications. Briefly, phosphocellulose-purified hog brain tubulin is labeled with Bis caged-fluorescein Sulfo-OSu (Dojindo Laboratories, Japan) in solution at high pH (pH 8.5). This fluorescein derivative is not in itself fluorescent, but generates highly fluorescent carboxyfluorescein after irradiation with near-ultraviolet light (photoactivation). After the labeling reaction, the tubulin preparation is subjected to two cycles of polymerization and depolymerization to select for assembly competent conjugates. Labeled tubulin preparations are analyzed after photoactivation by sodium dodecyl sulfate gel electrophoresis to confirm that the protein is labeled with the caged fluorescein reagent and is free of unlabeled caged fluorescein. Free caged carboxyfluorescein runs at the front of the gel and can be easily detected by irradiating the gel under an ultraviolet light transilluminator.

Cell Culture

Cultured mouse dorsal root ganglion (DRG) neurons are used in these experiments. These neurons have large cell bodies (25–60 μm), and the growth rate of their axons plated on laminin-coated coverslips is very fast, with a protrusion rate of 10–80 μm/hr.[3]

1. About 20 dorsal root ganglions are isolated from adult mice[5,6] and kept in Hanks' balanced salt solution (HBSS, GIBCO, Grand Island, NY) at 4°. To minimize injury to these neurons, this step is completed within 45 min.

2. Dorsal root ganglions are washed twice with Ca^{2+}, Mg^{2+}-free HBSS (Gibco).

[5] S. S. S. Goldenberg and U. De Boni, *J. Neurobiol.* **14**, 195 (1983).
[6] S. Okabe and N. Hirokawa, *J. Neurosci.* **11**, 1918 (1991).

3. Ganglions are treated with 0.25% (w/v) collagenase in Ca^{2+}, Mg^{2+}-free HBSS for 1 hr at 37°.

4. Ganglions are further incubated with 0.25% (w/v) trypsin in Ca^{2+}, Mg^{2+}-free HBSS for 15 min.

5. After washing with culture medium (MEM, Nissui, Tokyo, Japan) supplemented with glutamine and 15 mM HEPES, 5% horse serum, and 5% new calf serum, the ganglions are homogenized in the culture medium by pipetting several times through a Pasteur pipette. The pipetting should be performed gently to avoid damaging cells. If the ganglions cannot be resuspended easily, harsher treatment with protease (with shaking) is recommended.

6. Suspended cells are plated on laminin-coated glass coverslips. Healthy cells should be smooth and round, whereas heavily damaged cells have a wrinkled appearance. The addition of Nerve Growth Factor (100 ng/ml) to the culture medium is optional.

Glass coverslips are secured with vacuum grease on the bottom of plastic dishes that have been drilled with a hole smaller than the diameter of the coverslip (Dow Corning, high vacuum silicone grease). Care should be taken as some kinds of adhesive agents are toxic to the cells. For electron microscopic analysis, we use CELLocate micro-grid coverslips (Eppendorf) in order to identify the axons of cells irradiated with near-ultraviolet light. The square meshes etched onto these coverslips are transferred to the resin block and allow us to easily and precisely identify the photoactivated axons after embedding.

Microinjection of Caged Fluorescein-Labeled Tubulin Into Neurons[3,7]

Neurons that do not show any axonal processes or those with minor sprouts are microinjected with bis-caged carboxyfluorescein-labeled tubulin 3–8 hr after plating and are incubated for a further 10–20 hr. This ensures that microinjected-caged fluorescein tubulin is equally distributed throughout the neuron. To identify microinjected neurons, cells are coinjected with caged fluorescein-labeled tubulin (50–100 μM) and rhodamine-bovine serum albumin (BSA) (0.5 mg/ml) in injection buffer (50 mM potassium-glutamate, 100 mM KCl, and 1 mM $MgCl_2$).

Photoactivation and Low Light Level Video Microscopy

Photoactivation is performed as described by Mitchison and colleagues (see also the chapter in this volume) with slight modifications.[1,3,8] Essen-

---

[7] T. Funakoshi, S. Takeda, and N. Hirokawa, *J. Cell Biol.* **133**, 1347 (1996).

[8] T. Umeyama, S. Okabe, Y. Kanai, and N. Hirokawa, *J. Cell Biol.* **120**, 451 (1993).

tially, the 365-nm line of a mercury lamp selected with a bandpass filter (390-nm long pass) is introduced into the epifluorescent light path of an inverted microscope (Axiovert; Carl Zeiss, Inc.) via a dichroic mirror positioned between the original mirror box and the field diaphragm (Fig. 1). A handmade slit is placed at the point where the slit makes an image at the field diaphragm. The beam of the mercury lamp is focused onto the

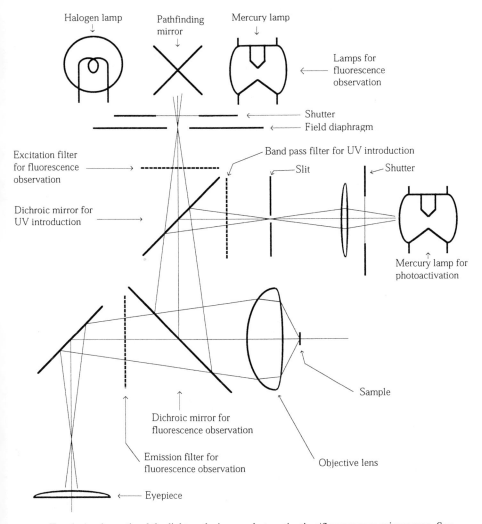

FIG. 1. A schematic of the light paths in our photoactivation/fluorescence microscope. See text and the chapter in this volume by Mitchison *et al.* for further details.

slit using a lens. We can easily illuminate a small region of an axon (3–5 μm wide) using this microscope with a 100× objective lens (Plan-Neofluar, Zeiss).

To observe the fluorescence of cells we use a 100-W halogen lamp, the excitation–emission filter set for fluorescein and/or rhodamine, and a cooled CCD camera (C3640; Hamamatsu Photonics) with an exposure time of 1 sec. The halogen lamp is turned on only to record the fluorescence region in order to limit photobleaching of the dye and any accompanying damage to the cells. Mercury lamp epi-illumination of the sample is used in combination with a low power objective lens to identify injected cells.

Photomicroscopic Observation Following Photoactivation[3,8]

Although fluorescence microscopy does not allow us to resolve a single microtubule in an axon, as microtubules are so densely packed, we can observe a single cell many times after photoactivation.

Approach

1. Illuminate a small segment of the axon of the microinjected cell with near-ultraviolet light. Fluorescein-labeled tubulin molecules are generated in the illuminated segment.
2. Observe the fluorescence images immediately after and at appropriate time intervals following photoactivation.

Narrow bars of fluorescent tubulin marked in the axon by photoactivation do not move from the irradiation site even if in extending axons, although the fluorescence intensity of the marked tubulin does decay gradually (Fig. 2). These observations strongly imply that most of the axonal microtubules are stationary, and we suggest that the intensity decay reflects the polymerization–depolymerization turnover of microtubules in the photoactivated segments. The limited spatial resolution and sensitivity of fluorescence light microscope make it difficult to detect any population of tubulin molecules that move out of the photoactivated segments.

Electron Microscopic Observation Following Photoactivation[7]

This method allows us to observe the fate of photoactivated molecules at very high resolution with high sensitivity. To detect the population of mobile tubulin molecules that might have been overlooked in observations with the fluorescence microscope, we determined the distribution of fluoresceinated tubulin molecules in the same photoactivated axons using elec-

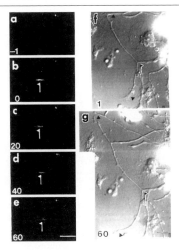

FIG. 2. Photoactivation of an extending axon of a mouse DRG neuron. (a–e) Fluorescence images showing the marks (arrows) made on the microtubules by the photoactivation technique remain stationary as the axon moves forward. (f and g) DIC images of the same axon. Arrows indicate the position of the photoactivated region. During this experimental run, there was a rapid extension of the axon. Elapsed time (minutes) after photoactivation is shown in the lower left-hand corner. Reproduced from S. Okabe and N. Hirokawa, *J. Cell Biol.* **117,** 105 (1992) by copyright permission of Rockefeller University Press. Bars: 10 μm.

tron microscope-based detection of an antiflorescein antibody. Although this method does not allow us to observe a single axon at many time points after the photoactivation, it provides both the resolution and the sensitivity to detect transported tubulin molecules in a single axon. Antiserum against fluorescein was produced using fluorescein-labeled keyhole limpet hemocyanin for the antigen. The antibody directed against the fluorescein hapten was affinity purified using a CNBr-Sepharose affinity column (Pharmacia, Piscataway, NJ) coupled with fluorescein-conjugated BSA.

We employ two different procedures to address the issue of whether tubulin molecules are transported in the axon. To unambiguously observe single microtubules, we permeabilize cells using taxol before fixation, as previously described[9] with some slight modifications. This procedure extracts soluble cytosolic protein from axons (Fig. 3). To preserve the soluble population of tubulin molecules (tubulin oligomers or heterodimers), cells are fixed without a permeabilizing agent. Under this condition, using high-resolution electron microscopy, a transported population of tubulin molecules can be seen (Fig. 4).

[9] P. W. Baas and M. M. Black, *J. Cell Biol.* **111,** 495 (1990).

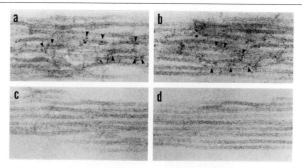

Fig. 3. Behavior of photoactivated microtubules in extending axons. One hour after the photoactivation of small regions of the axon, the cells were permeabilized, fixed, and stained by antifluorescein antibody. (a and b) Photoactivated region. Microtubules decorated with gold particles (arrowheads) were observed. (c and d) Examples of the areas outside the photoactivated region: (c) an area approximately 4 μm distal to the photoactivated region and (d) an area approximately 4 μm proximal to the photoactivated region. No microtubules with gold labels were observed outside of the photoactivated regions. Reproduced from T. Funakoshi, S. Takeda, and N. Hirokawa, *J. Cell Biol.* **133,** 1347 (1996) by copyright permission of Rockefeller University Press.

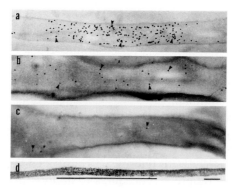

Fig. 4. Transported tubulin molecules in the axon. (a–c) One minute after the photoactivation of a small section of the axon; cells were fixed without the preceding permeabilization and visualized using an antifluorescein antibody. Many gold particles were observed outside of the photoactivated region. (a) Photoactivated region. Many gold particles can be seen. (b) An area about 20 μm distal and (c) an area about 20 μm proximal to the photoactivated region. (a–c) Arrowheads indicate some of the gold particles. (d) An axon photoactivated after fixation. The photoactivated region is underlined. (a–c) Many gold labels appear outside of the photoactivated regions; the density of gold label is higher in the distal region than in the proximal region. (a–c) Large spots are used to exaggerate the positions of the gold particles, as they are too small to be resolved at this magnification. Reproduced from T. Funakoshi, S. Takeda, and N. Hirokawa, *J. Cell Biol.* **133,** 1347 (1996) by copyright permission of Rockefeller University Press. Bars: 1 μm.

The procedure to observe single microtubules is as follows:

1. Illuminate a small segment of the axon with a beam of near-ultraviolet light.
2. After appropriate time intervals, permeabilize and fix the cells according to the method described previously.[9] Permeabilize the cells for 5 min with 1% Triton X-100 in PHEM buffer (60 mM PIPES, 25 mM HEPES, 10 mM EGTA, 2 mM MgCl$_2$, pH 6.9) containing 10 μM taxol and 0.2 M NaCl.
3. Fix the cells for 30 min by adding an equal volume of 1% glutaraldehyde in PHEM.
4. Wash with phosphate-buffered saline (pH 7.2) (PBS) and treat the cells with 50 mM glycine in PBS for 30 min.
5. Incubate the cells with blocking solution containing 5% skimmed milk in PBS.
6. Incubate the cells overnight at 4° with the antifluorescein antibody. Antiserum is used at a dilution of 1:500 in PBS with 0.5% skimmed milk.
7. Wash the cells for 1 hr with PBS.
8. Incubate the cells with a 5-nm gold-conjugated anti-rabbit antibody (Amersham Life Science) for 5 hr at a dilution of 1:10 in PBS at 37°.
9. Wash the cells completely with PBS and fix again with 1% glutaraldehyde in PBS and process for silver enhancement (Amersham).
10. Osmify (for 1 min), block stain, dehydrate, and embed the cells.
11. Serially section the cells and examine the block.

The procedure to preserve the soluble population of tubulin molecules is as follows:

1. Illuminate a small segment of the axon with a UV beam.
2. Fix the cells 1–20 min after photoactivation with 0.1% glutaraldehyde and 2% paraformaldehyde in PEM.
3. Treat the neurons with 1% Triton X-100 in PBS for 30 min.
4. Wash with PBS.
5. Incubate with 50 mM glycine in PBS for 30 min.
6. Incubate the cells with blocking solution containing 5% skimmed milk in PBS for 30 min.
7. Incubate the samples overnight with affinity-purified antifluorescein antibody in BSA–PBS (0.5% (w/v) BSA and 0.1% (w/v) gelatin in PBS) containing 0.5% (w/v) skimmed milk at 4°. Completely wash the cells with BSA–PBS containing 0.5% skimmed milk.
8. Incubate the cells overnight at 4° with a 1.4-nm gold-conjugated anti-rabbit Fab' fragment (Nanoprobes, Inc., Nanogold anti-rabbit) diluted 500 times in BSA–PBS containing 1% skimmed milk, 500 mM NaCl, and 0.05%

Tween 20 (the 5 n*M* gold-conjugated antibody usually employed for immunoelectron microscopy does not penetrate DRG neurons that have not been permeabilized before fixation).

9. Completely wash (about 1 hr) the cells with BSA–PBS containing skimmed milk, NaCl, and Tween 20 and then with PBS.

10. Fix the cells again with 1% glutaraldehyde in PBS.

11. Silver enhancement in developer solution (30% gum arabic, 0.85% hydroquinone, 0.11% silver lactate, citric acid monohydrate, 25.5 g/liter, sodium citrate dihydrate, 23.5 g/liter)[10] for about 1 hr.

12. Block stain, dehydrate, and embed cells without osmification for serial sectioning and observation.

To evaluate this procedure and to check whether labels found outside of the photoactivated regions really represent moving tubulin molecules in the axons, we illuminated small axonal regions of previously fixed cells with near-ultraviolet light, stained and observed in the electron microscope as described earlier. In these axons, no gold labels were found out of the photoactivated regions (Fig. 4d). No translocated microtubules were detected in cells that had been fixed after permeabilization. In photoactivated axons of cells fixed without permeabilization, where fluorescently labeled oligomers, heterodimers, and polymers are preserved, a significantly higher amount of gold label was found in regions distal to the photoactivated regions than in the proximal region. These data indicate that tubulin molecules are not transported as polymers, but as heterodimers or oligomers by an active mechanism rather than by simple diffusion.

[10] M. A. Hayat, "Principles and Techniques of Electron Microscopy." CRC Press, Boca Raton, FL, 1989.

[20] Two-Photon Activation of Caged Calcium with Submicron, Submillisecond Resolution

By EDWARD B. BROWN and WATT W. WEBB

Introduction

Photolabile calcium chelators, or "cages," are a popular tool to explore the calcium dynamics of a variety of biological systems. On absorption of a single UV or short-wavelength visible photon a calcium cage can undergo a conformational shift and release a calcium ion. This release can take place as rapidly as 10 μsec after photoexcitation,[1] allowing calcium levels to be locally altered with high temporal resolution. Photolysis of calcium cages has been used to study cytosolic calcium oscillations[2] and membrane ion channels.[3] The field of muscle contractility has also benefited from the use of calcium cages,[4–6] as has the study of exocytosis.[7,8]

Calcium cages have commonly been photolyzed using UV flash lamps or focused laser beams.[9] A focused laser beam generates a double cone of illumination with the narrowest point at the focal plane of the objective lens, so within the focal plane the spatial resolution of conventional calcium uncaging can be quite high. However, significant excitation of cage molecules can occur above and below the focal plane. If a series of slices are envisioned running through the double cone of focused illumination, parallel to the focal plane and at increasing distances from it, conservation of energy dictates that the same number of excitation photons must pass through each slice. This means that the same total amount of cage excitation will also be generated in each slice, with the result that conventional excitation of calcium cages (or any chromophore) has essentially no resolution along the axis of the focused excitation light.

Two-photon excitation (TPE) is a recently developed microscopy technique that can excite chromophores, including calcium cages, with intrinsic three-dimensional resolution.[10] Two-photon excitation of a UV-excitable cage can be accomplished using a single near-IR (~700 nm) focused laser beam with a wavelength chosen such that each individual photon generated by the laser has roughly half the energy required to excite the cage. Two such photons can be absorbed by the cage in a single quantum event,[11] provided that the photons interact with the cage at essentially the same time (within ~10^{-16} sec). The requirement for simultaneous absorption of two photons gives TPE a quadratic dependence on the incident intensity,

---

[1] G. Ellis-Davies, J. Kaplan, and R. Barsotti, *Biophys. J.* **70,** 1006 (1996).
[2] G. M. Calder, V. E. Frank-Tong, P. J. Shaw, and B. K. Drobak, *Biochem. Biophys. Res. Commun.* **234,** 690 (1997).
[3] P. Velez, S. Gyorke, A. Escobar, J. Vergara, and M. Fill, *Biophys. J.* **72,** 691 (1997).
[4] J. R. Patel, K. S. McDonald, M. R. Wolf, and R. L. Moss, *J. Biol. Chem.* **272,** 6018 (1997).
[5] B. K. Hoskins, P. J. Griffiths, C. C. Ashley, and R. Rapp, *Biophys. J.* **72,** A275 (1997).
[6] T. D. Lenart, J. M. Murray, C. Franzini-Armstrong, and Y. Goldman, *Biophys. J.* **71,** 2289 (1996).
[7] A. F. Oberhauser, I. M. Robinson, and J. M. Fernandez, *Biophys. J.* **71,** 1131 (1996).
[8] T. D. Parsons, G. C. R. Ellis-Davies, and W. Almers, *Cell Calcium* **19,** 185 (1996).
[9] J. P. Y. Kao and S. R. Adams, *in* "Optical Microscopy: Emerging Methods and Applications" (B. Herman and J. L. Masters, eds.), p. 27. Academic Press, London, 1993.
[10] W. Denk, J. Strickler, and W. Webb, *Science* **248,** 73 (1990).
[11] M. Göppert-Mayer, *Ann. Phys.* **9,** 273 (1931).

with the consequence that excitation is intrinsically confined to a small region in the immediate vicinity of the focal point of the focused laser beam. A brief flash of focused near-IR light can therefore generate a localized elevation of calcium over a volume of less than a femtoliter. Chromophores generally have extremely small multiphoton absorption cross sections, requiring pulsed ("mode-locked") lasers in order to generate photon densities high enough to accomplish significant TPE. The extremely localized nature of this distribution means that the elevated soluble calcium ion concentration will rapidly dissipate due to diffusion of the liberated ions into the surrounding volume, as well as due to uptake of the ions by local calcium buffers.

This work explores the spatial resolution of calcium release possible with TPE and reviews the equipment necessary to accomplish this release. We will calculate the quantities of released calcium that can be achieved with a TPE instrument, using previously measured two-photon uncaging action cross sections. Finally, we will explore the temporal behavior of the released calcium concentration distribution as it is influenced by diffusion out of the small TPE volume, as well as by intrinsic and extrinsic calcium buffers. These calculations apply to TPE of any caged compound with substitution of the appropriate diffusion and cellular uptake times and illustrate general principles of the spatial and temporal regimes that can be accessed with TPE. A more detailed treatment of many of these experiments and calculations can be found in the source referenced in footnote 13.

A "user's summary" of relevant calculations and physical data reveals that complete photolysis of calcium cages using TPE is indeed possible. Using 700-nm light from a commercially available Ti:sapphire mode-locked laser (100-fsec pulse length, 80-MHz repetition rate), focused with a 1.3 numerical aperture (NA) lens into a solution containing the calcium cage with the highest known uncaging action cross section, Azid-1,[12] essentially all the cage molecules within an ellipsoidal volume of radial dimension 0.25 μm and axial dimension of 0.86 μm can be photolyzed with a 10-μsec flash of ~7 mW average power at the sample.[13] This highly localized distribution of free calcium ions dissipates rapidly due to diffusion of calcium ions out of the two-photon focal volume and calcium uptake by intrinsic and extrinsic calcium buffers. In the absence of calcium buffers, the elevated calcium concentration at the center of the TPE focal volume will last for ~32 μsec (FWHM), whereas the introduction of buffers can significantly shorten this duration.

---

[12] S. R. Adams, V. Lev-Ram, and R. Tsien, *Chem. and Biol.* **4,** 867 (1997).
[13] E. Brown, J. Shear, S. Adams, R. Tsien, and W. Webb, manuscript submitted (1997).

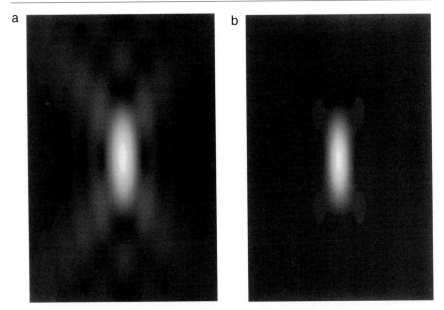

Fig. 1. Spatial distribution of cage activation. (a) One-photon activation. The diffraction limited intensity distribution,[14] with the direction of laser propagation along the z axis (oriented vertically on the page). The effective local uncaging rate per cage–calcium complex via one photon excitation is proportional to the intensity. Note the poor confinement of the distribution in the z dimension. (b) Two-photon activation. The square of the diffraction limited intensity distribution. The effective local uncaging rate per cage–calcium complex via two photon excitation is proportional to the intensity squared. Note the superior confinement of the distribution in the z dimension.

Spatial Resolution of Two-Photon Excitation

The exact representation of the average rate of one- and two-photon excitation (O.P.E. and T.P.E.) for a single cage molecule illuminated by a focused laser beam is

$$\text{O.P.E.} = \sigma \langle I(r, z) \rangle \tag{1}$$
$$\text{T.P.E.} = (1/2)\delta \langle I^2(r, z) \rangle$$

where σ is the one-photon excitation cross section in units of cm^2 and δ is the two-photon excitation cross section in units of $cm^4 \sec^{-1}$ ($10^{-50} cm^4 \sec^{-1}$ equals one Göppert–Mayer, or GM). The bracket symbols, $\langle \ \rangle$, represent a time average, while $I(r, z)$ is the local excitation intensity. The highest spatial resolution possible with a given objective lens and given illumination wavelength is accomplished by uniformly filling the back aperture of the objective with the laser beam, producing the "diffraction limited" intensity

---

[14] D. Sandison and W. Webb, *J. App. Optics* **33**, 603 (1994).

distribution. This diffraction-limited intensity distribution is plotted in Fig. 1. The first power of this distribution (Fig. 1a) is proportional to the one-photon excitation probability and has poor confinement along the z axis, demonstrating the poor three-dimensional resolution of this excitation method. The second power of this distribution (Fig. 1b) is proportional to the two-photon excitation probability and exhibits the spatial confinement intrinsic to two-photon excitation.

The exact form of the diffraction-limited intensity distribution is quite complicated[15] and not generally amenable to use in mathematical analysis. The square of this distribution, however, is closely approximated by the mathematically tractable "ellipsoidal Gaussian"

$$I^2(r, z) = I_0^2 e^{-4r^2/w_r^2} e^{-4z^2/w_z^2} \tag{2}$$

where I_0 is the intensity at the center of the focal spot. The $1/e^2$ characteristic radial and axial dimensions are w_r and w_z, respectively. These characteristic dimensions ultimately dictate the spatial resolution. Values of w_r and w_z can either be measured directly or the following approximations can be used;

$$w_r = \frac{2.6\lambda}{2\pi NA}$$
$$w_z = \frac{8.8n\lambda}{2\pi NA^2} \tag{3}$$

where λ is the wavelength of the excitation light in vacuum, n is the index of refraction of the sample, and NA is the numerical aperture of the illumination optics. For 700-nm light overfilling a 1.3 NA lens, therefore, the two-photon focal volume will have a radial dimension of 0.22 μm and an axial dimension of 0.77 μm. These increase to 0.29 and 1.3 μm for a 1.0 NA lens, demonstrating that the axial resolution of two-photon excitation decreases rapidly with decreasing NA.

One drawback of using an overfilled objective lens to photolyze cages using two-photon excitation is that this method is relatively wasteful of laser power. The transverse intensity profile of lasers operating in the TEM00 mode is a Gaussian with a characteristic width on the order of a millimeter. In order to overfill an objective lens, the width of this profile is greatly expanded using a pair of lenses such that the intensity distribution actually entering the objective lens is essentially uniform. This means that a large fraction of the laser power is rejected by the back aperture stop of the objective lens. This wastefulness is not a significant problem with the common mode-locked tunable Ti:sapphire lasers in use today, with average power outputs of a watt or more. A growing trend in two-photon micros-

---

[15] M. Born and E. Wolf, "Principles of Optics," p. 436. Pergamon Press, Tarrytown, NY, 1980.

copy, however, is the use of single-wavelength, all solid-state lasers that generally have extremely limited power outputs, on the order of 50 mW. Properly overfilling an objective lens with a laser of such limited total output reduces the laser power reaching the sample to levels so small that it becomes problematic to accomplish any significant uncaging. Instead of overfilling the objective, however, the lens can be completely underfilled, allowing the entire laser power (minus the ~10–30% absorption of the objective glass) to reach the sample. The resultant intensity distribution, known as the "Gaussian–Lorentzian" distribution, has significantly lower axial and radial resolution than the diffraction limited intensity distribution. The square of the Gaussian–Lorentzian distribution is

$$I^2(r, z) = \frac{4I_0^2 w_0^4}{\pi w^4(z)} e^{-4r^2/w^2(z)}$$

$$w^2(z) = w_0^2 \left(1 + \left(\frac{z\lambda}{\pi w_0^2}\right)^2\right)$$

(4)

where w_0, the minimum radial beam waist, is dictated by the size of the laser in the back aperture of the objective and the focusing properties of the objective.

This work is concerned with the generation of calcium distributions of the highest spatial and temporal resolutions, and consequently will deal exclusively with diffraction-limited optics, using the intensity profile given by Eq. (2). Most of the results presented here are easily extrapolated to the Gaussian–Lorentzian distribution.

Apparatus for Two-Photon Uncaging

A variety of two-photon photolysis geometries are possible, and two variants using the commercially available two-photon microscopy system sold by Bio-Rad (Hercules, CA) are shown in Fig. 2. In both cases, multiphoton excitation is generated with a mode-locked Ti:sapphire laser, rapid modulation of the near-IR beam is accomplished with a Pockels Cell (Conoptics, Danbury, CT), and the imaging system consists of a Bio-Rad MRC 1024 scanning box and a microscope such as the Zeiss Axiovert 135 (Zeiss, Jena, Germany) or the Olympus BX50WI (Olympus America, Inc., Melville, NY.). The extent to which the objective lens is overfilled can be adjusted with a separate beam expander, whereas dispersion compensation can be accomplished using a pair of SF10 equilateral prisms. Measurement of the two-photon spot size can be accomplished with an objective-mounted piezoelectric translator.

In the first variant, using a multiband dichroic available with the Bio-Rad system, the 488-nm argon laser line is used to generate conventional

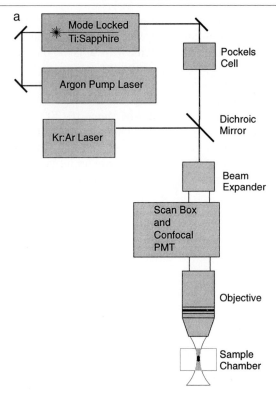

FIG. 2. Cage activation apparatus configuration. (a) The confocal imaging/two-photon photolysis experimental apparatus. (b) The two-photon imaging/two-photon photolysis experimental apparatus.

one-photon confocal images of the target area, whereas photolysis is accomplished using the Ti:sapphire beam directed along the same excitation path. During the imaging process, the near-IR beam is completely attenuated using the Pockels cell, and a confocal image is formed of the target area. The scan mirrors are then used to direct to the target point a high intensity flash of Ti:sapphire light transmitted through the Pockels cell. The subsequent effects of the photolysis are studied using confocal line scans, full images, or other one-photon fluorescence microscopy techniques. The advantage of this technique is that the imaging beam cannot cause unwanted cage photolysis even if the beam dwells in one location for an extremely long time due to slow scan speeds or high zoom factors. The disadvantages of this technique is that the photolysis experiments are limited to the relatively thin specimens available for study with confocal microscopy, and the significant out-of-plane photobleaching caused by one photon excitation during the imaging process will limit the length of cell viability.

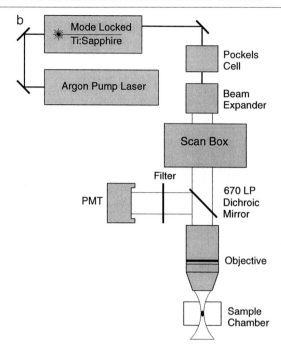

FIG. 2. (*Continued*)

In the second variant, using a simple 670-nm short pass dichroic, the same two-photon beam is used both to image the target area and to photolyze the calcium cage. During imaging, the Pockels cell attenuates the near-IR beam to low powers, sufficient to generate a fluorescent signal but not cause significant photolysis of the calcium cage. The Pockels cell then generates a rapid high power pulse when the scan mirrors have directed the excitation beam to the target area, and various two-photon microscopy techniques (line scans, full images, etc.) can then be used to study the ensuing biological effects. The disadvantage of this technique is that the imaging powers transmitted by the Pockels cell must be carefully chosen to be low enough that no significant uncaging is caused during image formation. The advantage of this technique is that it uses two-photon excitation in the imaging process, with the concomitant benefits this has over confocal microscopy; greater imaging depth and longer sample lifetimes due to reduced photobleaching. The Pockels cell is ideally suited to power modulation of mode-locked lasers because of its rapid response time and wavelength insensitivity, making it superior to shutters and acousto-optical modulators (AOMs) for this purpose. This device works by the application of a high voltage to a set of capacitor plates enclosing a nonlinear optical material. The laser passes through this material, and the plane of polarization of the beam is rotated through an angle dictated by the voltage applied to the plates. An analyzing polarizer is used after the nonlinear material to translate the variation in laser polarization angle into a variation in laser intensity. Pockels cells can easily modulate the transmitted laser power by over a factor of 100 with a response time of a microsecond or less. As will be discussed later, this sort of rapid response time is necessary for TPE generation of calcium concentration elevations with the highest spatial resolution. This temporal requirement eliminates most common mechanical shutters, which have response times of ~1 msec, although pairs of shutters may be used to generate more rapid beam modulation. The pulsed nature of the lasers required for efficient generation of two-photon excitation makes the use of AOMs problematic, although they have sufficiently rapid response times for highly localized generation of calcium. An AOM relies on the diffraction of laser light through the grating generated by an acoustic standing wave in a transparent crystal. A 100-fsec pulse of laser light has a bandwidth of ~15 nm, which can produce undesirable spatial and temporal distortion of the beam as it passes through the diffractive region of the AOM. This leaves Pockels cells as the most viable alternative for rapid modulation of pulsed laser beams.

Dispersion compensation using a pair of SF10 prisms allows the maximum two-photon excitation to be generated with a given average laser power. Group delay dispersion in optical materials causes temporal broadening of transmitted laser pulses, lowering the peak photon densities and thereby reducing TPE. This dispersion arises because the index of refraction of the glass in optical components, such as objective lenses, beam expanders, and tube lenses, varies with wavelength. The speed of light through a material is inversely proportional to this index of refraction, so the speed of light through a material also scales with wavelength. A 100-fsec pulse of light, with ~15 nm of bandwidth, will therefore broaden in time when it passes through an optic because its constituent wavelength components travel at different speeds, with longer wavelengths outrunning shorter ones (positively dispersive) in the Ti:sapphire wavelength range. An appropriately placed pair of prisms will disperse the laser pulses in the opposite, negatively dispersive, sense,[16] effectively giving the long wavelengths a shorter overall optical path length so that the laser pulse will be recombined by passage through the positively dispersive optics and have a pulse width at the sample that is close to its original, nondispersed temporal width. An SF10 prism pair separated by approximately 60 cm provides sufficient

---

[16] R. L. Fork, O. E. Martinez, and J. P. Gordon, *Opt. Lett.* **9,** 150 (1984).

negative dispersion to counter the broadening induced in 100-fsec pulses by a typical high NA objective, a Pockels cell, and a two-lens beam expander.

A method for the measurement of the characteristic axial and radial distances of the two-photon focal volume generated at the sample is highly desirable for direct determination of the spatial resolution of two-photon calcium uncaging. A piezoelectric drive mounted on an objective lens mount with a step size significantly smaller than the expected two-photon axial dimension can be used to directly measure this dimension. The two-photon focal volume can be stepped from within the coverslip to within a fluorescent solution of dye, and the resultant sigmoidal increase in fluorescence can be easily fit to yield w_z, the characteristic axial dimension of the two-photon spot. Likewise, the two photon spot can be scanned transversely across the interface between a fluorescent and nonfluorescent sample using the scanning mirrors of the imaging system in order to determine w_r, the characteristic radial dimension of the two-photon spot. Immobilized subresolution fluorescent beads are also suitable for these measurements.

Two-Photon Uncaging Action Cross Section of Calcium Cages

Once a two-photon uncaging system has been assembled, it is useful to predict the amount of calcium release that is possible under different illumination conditions. However, in order to predict the amount of calcium that can be generated, it is necessary to know the uncaging action cross section of the cage. A convenient method for determining the action cross section of a calcium cage is to focus a photolysis pulse train into a solution containing a calcium cage, calcium, and a high concentration of a rapidly binding calcium indicator dye such as Fluo-3. If a low intensity near-IR laser beam is focused into the sample, a steady fluorescence signal will be generated via TPE of the equilibrium concentration of calcium-bound indicator dye. If the near-IR beam is briefly flashed to a higher intensity, some of the calcium-bound cage within the two-photon focal volume can be photolyzed. This released calcium can then bind to the surrounding indicator dye, resulting in an increase in fluorescence generated by the low-intensity near-IR beam immediately after the photolysis pulse train. This additional fluorescence will dissipate as the newly generated calcium-bound indicator dye diffuses out of the two-photon focal volume. The amplitude of the transient fluorescence increase generated by the photolysis pulse train can be analyzed to determine the two-photon uncaging action cross section of the calcium cage.[13,17]

---

[17] J. Shear, E. Brown S. Adams, R. Tsien, and W. Webb, *Biophys. J.* **70**, A211 (1996).

Neglecting saturation, which will be addressed later, an uncaging pulse train of duration Δt focused very briefly into a sample containing an equilibrium concentration of calcium-loaded cage, $[CCa^{2+}]_0$, will release the following distribution of calcium ions:

$$\Delta[Ca^{2+}(r, z)] = 1 - e^{-(1/2)\delta_u(I_u^2(r,z))\Delta t} [CCa^{2+}]_0 \tag{5}$$

where δ_u is the two-photon action cross section of the calcium cage and consists of the two-photon absorption cross section multiplied by the quantum efficiency of uncaging. $I_u(r, z)$ is the spatial intensity distribution of the focused laser excitation. Equation (5) assumes that Δt, the duration of the uncaging pulse, is significantly shorter than the diffusion time of the cage across the two-photon focal volume. For 700-nm light overfilling a 1.3 NA lens this is satisfied if $\Delta t \sim 10$ μsec. The newly liberated calcium enters a solution with a high concentration of rapid indicator dye. If the rate of uptake of liberated calcium by the indicator dye is significantly faster than the rate of diffusion of the calcium across the two-photon focal volume, then Eq. (5) also represents the concentration distribution of newly calcium-bound indicator dye immediately after the end of the photolysis pulse. For uncaging pulses generated with 700-nm light overfilling a 1.3 NA lens, this requires at least ~ 1 mM of a rapid indicator such as Fluo-3.

The newly generated calcium-bound indicator dye is a nonequilibrium, highly localized concentration distribution superimposed on the equilibrium distribution of calcium-bound indicator. The fluorescence of this concentration distribution can be excited with the same laser, albeit greatly attenuated, that photolyzed the cage. Assuming diffraction limited optics, the fluorescence signal that is emitted by this newly calcium-bound indicator as it dissipates out of the focal volume via diffusion is

$$\Delta F(\tau) = \frac{F_0 (1/2)\delta_u \langle I_u^2 \rangle \Delta t}{2\sqrt{2}} \frac{[CCa^{2+}]_0}{[FCa^{2+}]_0} \frac{1}{(1 + \tau)} \frac{1}{(1 + R\tau)^{1/2}} \tag{6}$$

where F_0 is the steady-state fluorescence signal generated by the attenuated laser beam exciting the equilibrium concentration of calcium-bound indicator dye. The equilibrium concentration of calcium-bound indicator dye is $[FCa^{2+}]_0$ whereas the equilibrium concentration of calcium-bound cage is $[CCa^{2+}]_0$. The quantity τ is a dimensionless "time" equal to $8Dt/w_r^2$, where D is the diffusion coefficient of the fluorescent indicator dye and R is the square of the ratio of the two $1/e^2$ beam waists; $R \equiv (w_z/w_r)^2$. Equation (6) assumes that only a small fraction of the available cage–calcium has been photolyzed ($(1/2)\delta_u \langle I_u^2 \rangle \Delta t \ll 1$) in order to avoid significant depletion of free calcium indicator dye in the focal spot. An example of a transient fluorescent curve generated by two-photon photolysis of the calcium cage

Azid-1 in a solution of Fluo-3, along with the resultant fit using Eq. (6), is shown in Fig. 3.

Three calcium cages have been studied with this technique: the commercially available cages NPEGTA (Molecular Probes, Eugene, OR) and DM-nitrophen (Calbiochem, La Jolla, CA), as well as a new calcium cage, Azid-1.[12] The response of these cages at selected wavelengths between 700

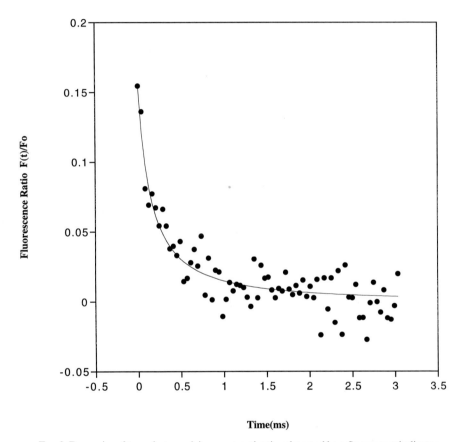

FIG. 3. Dynamics of two-photon calcium cage activation detected by a fluorescent indicator. The fluorescence signal generated by two-photon point photolysis of Azid-1 within a solution of the calcium indicator dye Fluo-3. [Data From E. Brown, J. Shear, S. Adams, R. Tsien, and W. Webb, manuscript submitted (1997).] The bright flash of 700-nm laser light occurs at $t = 0$ on this curve, generating newly calcium-bound Fluo-3, which produces an increase in fluorescence signal $\Delta F(t)$. This fluorescence decays away to equilibrium levels as the new Fluo–calcium diffuses out of the two-photon focal volume generated by the 1.2 NA objective lens. The solid line is a fit to data using Eq. (6). The resultant uncaging action cross section at this wavelength is ~1.4 GM.

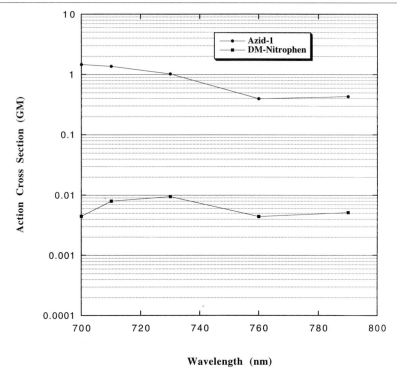

FIG. 4. The two-photon uncaging action on cross sections of Azid-1 and DM-nitrophen with data from 700 to 800 nm. NPEGTA did not produce enough calcium for an uncaging signal to be detected above background noise at any of the wavelengths studied. [Data from E. Brown, J. Shear, S. Adams, R. Tsien, and W. Webb, manuscript submitted (1997).]

and 800 nm were probed using a Ti:sapphire laser source, whereas 622-nm light was provided using a frequency-doubled Cr:Lisgaf (unpublished data). NPEGTA showed no significant reproducible calcium uncaging signal at any of the wavelengths probed, DM-nitrophen displayed a peak uncaging action cross section of ~0.01 GM at 720 nm, and Azid-1 had a peak uncaging action cross section of ~1.4 GM at 700 nm. Data for DM-nitrophen and Azid-1 between 700 and 800 nm is shown in Fig. 4. The uncaging cross section of DM-nitrophen at 622 nm is roughly a factor of six less than the lowest value in the 700- to 800-nm wavelength range, and the cross section of Azid-1 at 622 nm is roughly a factor of 40 lower than in the longer wavelength range. Photolysis experiments using DM-nitrophen and NPEGTA in living cells have qualitatively confirmed the relative efficacy of these two cages.[18]

[18] P. Lipp, J. Kleinle, C. Amstutz, C. Luschner, and E. Niggli, *Biophys. J.* **72,** A44 (1997).

The fact that NPEGTA does not display any uncaging signal using this technique does not necessarily mean that the multiphoton excitation action cross section of this cage is vastly smaller than that of the species with the smallest measurable action cross section at these wavelengths, DM-nitrophen. Equation (6) describes a time-dependent fluorescent signal $F(\tau)$ superimposed on a baseline fluorescent signal F_0 generated by the equilibrium loading of the fluorescent indicator dye. Within the range of uncaging intensities used in these experiments (those that avoid significant dye bleaching), the transient signal $F(\tau)$ is much smaller than the baseline signal F_0. This means that the total noise in the fluorescence amplitude is going to be dominated by the shot noise from the baseline fluorescence F_0. The desired signal is the change in fluorescence due to newly generated Fluo–calcium $\Delta F(\tau)$, therefore the signal-to-noise (S/N) ratio of this technique, assuming that F_0 is shot noise limited, is proportional to

$$\frac{S}{N} \propto \frac{\Delta F(\tau)}{\sqrt{F_0}} \propto \delta_u \frac{[CCa^{2+}]_0}{([FCa^{2+}]_0)^{1/2}} \qquad (7)$$

and is not due solely to the uncaging action cross section, but has important contributions from the equilibrium loading of cage and indicator. Hence, when all other factors are equal, the overall S/N is superior for cages that have lower K_d values. DM-nitrophen has the lowest dissociation constant of the three cages tested here, with a K_d of 5 nM, yielding the most favorable environment for measuring the uncaging peak. The K_d of NPEGTA is 80 nM, more than 10 times greater than that of DM-nitrophen. This factor reduces the S/N ratio for this species by more than threefold compared to DM-nitrophen, making it that much more difficult to measure the uncaging action cross section. The K_d of Azid-1 is ~40-fold greater than that of DM-nitrophen,[12] yielding an equilibrium distribution of calcium that reduces the signal to noise ratio by a factor of six compared to DM-nitrophen. The two-photon action cross section of Azid-1 is 140-fold larger, however, providing sufficient overall signal to noise to measure calcium uncaging. In conclusion, although NPEGTA certainly has a smaller action cross section than DM-nitrophen, the lack of an uncaging signal using this technique can also be contributed to the high equilibrium constant of this species, which contributes to the difficulty in detecting uncaging signal from this cage by decreasing the signal-to-noise ratio by a factor of three.

Total Calcium Yield via Two-Photon Excitation

When the two-photon action cross section of some calcium cages has been measured, the total possible calcium release possible under specific illumination conditions can be determined and the feasibility of two-photon calcium uncaging experiments can be predicted. The total number of calcium ions released by a photolysis pulse train of length Δt can be calculated by integrating Eq. (5) over all space (assuming the total sample volume is much larger than the two-photon focal volume). Assuming a diffraction limited intensity profile, the total number of calcium ions, N_{Ca}, produced by a photolysis pulse that is significantly shorter than the diffusion time of the cage molecules across the two-photon focal volume is then

$$N_{Ca} = -N_A[CCa^{2+}]_0 \frac{\pi^{3/2} w_r^2 w_z}{8} \sum_{n=1}^{\infty} \frac{(-\alpha)^n}{n! n^{3/2}} \quad (8)$$

where N_A is Avogadro's number, $[CCa^{2+}]_0$ again is the equilibrium concentration of calcium–loaded cage, and $\alpha \equiv (1/2)\delta_u \langle I_u^2 \rangle \Delta t$ is the "central uncaging dose," representing in the low intensity limit the average number of uncaging events per molecule at the center of the two-photon spot. Note that in the shallow uncaging limit ($\alpha \ll 1$), the total number of calcium ions N_{ca} does not scale with the NA of the objective lens, just as in the limit of no saturation the total number of fluorescent photons generated by TPE of a fluorophore does not scale with NA. This NA independence of the "total excitation" is a general feature of TPE, whereas increasing the NA will increase the peak intensity squared, $\langle I_u^2 \rangle \sim NA^4$ (assuming a fixed total power), the size of the focal volume will decrease by the same amount, and the two effects exactly cancel out. Once significant amounts of cage–calcium complexes are photolyzed ($\alpha \sim 1$), the total number of calcium ions liberated does indeed scale with NA, albeit in a nonlinear fashion.

Equation (8) contains terms proportional to the equilibrium number of cage–calcium complexes in the two-photon focal volume, as well as terms that describe the fractional release of these ions. This can be simplified greatly by dividing the entire equation by the equilibrium number of cage–calcium complexes in the two-photon volume and expressing the uncaging production as a fractional yield. The two-photon focal volume is a poorly defined region, however, with no specific boundary. A variety of expressions can be chosen to represent it, such as the volume contained within the I_0/e^2 isointensity surface of the excitation beam or the volume contained within the $(1/2)I_0$ isointensity surface. A useful definition of the two-photon volume V arises naturally out of analysis of multiphoton fluorescence correlation spectroscopy, however[19];

$$V \equiv \frac{[\int W^2(r)\,dv]^2}{\int W^4(r)\,dv} \quad (9)$$

[19] J. Mertz, C. Xu, and W. Webb, *Opt. Lett.* **20**, 2532 (1995).

where the unitless intensity profile $W^2(r)$ is derived from Eq. (2) by setting $I_o = 1$. This volume is defined in the low excitation intensity limit without ground-state depletion, or saturation, of the fluorophore undergoing TPE. At high intensities, when ground-state depletion of fluorophores arises, the spatial distribution of two-photon-excited fluorophores loses its three-dimensional Gaussian shape as the fraction excited at the center of the focal volume approaches its maximum value of one. At higher excitation intensities, the spatial distribution of excited fluorophores begins to appear more like a "top hat" with edges that move out from the focal point as the excitation intensity increases. Consequently, Eq. (9) is most representative of the volume of the two-photon excitation spot in the low intensity limit and underestimates the effective excitation volume at high excitation intensities.

The equilibrium number of cage–calcium complexes contained in the two-photon focal volume is just the equilibrium concentration of cage–calcium complexes times the focal volume as defined for our intensity distribution via Eq. (9). We can now define the uncaging ratio R_u as the number of calcium ions generated by a photolysis pulse [Eq. (8)] divided by the number of cage–calcium complexes available to be photolyzed in this two-photon focal volume;

$$R_u = -\frac{\sqrt{2}}{4} \sum_{n=1}^{\infty} \frac{(-\alpha)^n}{n! n^{3/2}} \quad (10)$$

For appropriate central uncaging doses α between zero and 35, this expression can be approximated as

$$R_u \cong 2.54 - 1.967 e^{-0.0543\alpha} - 0.572 e^{-0.405\alpha} \quad (11)$$

which is plotted in Fig. 5. An uncaging ratio of one corresponds to a release of a total number of calcium ions equal to the number of cage–calcium complexes contained in the two-photon excitation volume before the photolysis flash. The uncaging ratio does not asymptotically approach one, however, as $\alpha = (1/2)\delta_u \langle I_u^2 \rangle \Delta t$ grows large because the two-photon volume [Eq. (9)] is defined in the low excitation limit. At high central uncaging doses (large α), significant numbers of cage–calcium complexes that lie outside this equivalent volume V, which is in the lower intensity periphery of the illumination profile, can be photolyzed and, in fact, the uncaging ratio and the effective volume of release grows without bound as the central uncaging dose α approaches infinity.

Equation (10) or (11) can be used to predict the photolysis pulses required to accomplish desired amounts of calcium release. The equilibrium cage–calcium concentration can be set by the cage-loading conditions (the amount of cage–calcium in the loading patch pipette, for example), and

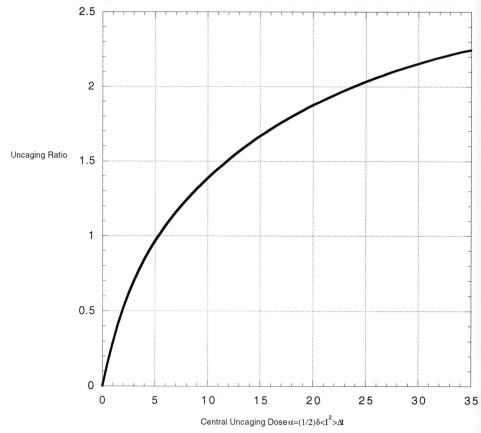

FIG. 5. Excitation dose dependent of calcium cage activation. The total calcium production as a function of the central uncaging dose $\alpha = (1/2)\delta_u \langle I_u^2 \rangle \Delta t$. The total production is expressed as a ratio R_u of the total number of calcium ions released to the total equilibrium number of cage–calcium complexes in the two-photon probe volume and is calculated using Eq. (11).

the two-photon focal volume can be set with the illumination optics [see Eqs. (2), (3), and (9)]. This dictates the total amount of cage–calcium available for photolysis in the two-photon focal volume. All of this caged calcium can then be released via a photolysis pulse that generates an uncaging ratio of one, which can be accomplished with a central uncaging dose of $\alpha \equiv (1/2)\delta_u \langle I_u^2 \rangle \Delta t \sim 6$. Given the reported peak action cross section for Azid-1 of $\sim$1.4 GM at 700 nm, this means that an average power of $\sim$7 mW at the focal volume of an overfilled 1.33 NA objective will achieve an uncaging ratio of one in 10 μsec. DM-nitrophen requires an average power of $\sim$85 mW of 720-nm light to achieve the same results.

The uncaging intensities required for significant, rapid, photolysis of DM-nitrophen are high enough to cause fluorescence saturation, or ground-state depletion, if applied to some fluorescent molecules, so it is conceivable that excitation saturation limits the maximum possible uncaging rate, just as it limits the maximum possible rate of fluorescence.

The kinetics of excitation saturation using pulsed lasers[20] are different from the kinetics using continuous illumination.[21] If commercially available mode-locked Ti:sapphire lasers are exciting common cages, the excited state lifetimes[1] of the cages are generally much longer than the duration of individual laser pulses (nanoseconds versus femtoseconds), with the result that a cage can be excited no more than once per laser pulse. Consequently, the maximum rate of excitation of cages using pulsed lasers is limited to the repetition rate of the laser and will not scale quadratically with the excitation power if a significant fraction of molecules are excited during a single laser pulse. This will limit the maximum rate of calcium release possible with a calcium cage. If the uncaging quantum efficiency of a calcium cage is sufficiently high, however, essentially all of the cage in a given region of the two-photon focal volume will be photolyzed during an uncaging flash with intensities that are lower than intensities required for excitation saturation, and this limit will be irrelevant.

The minimum uncaging quantum efficiency required to render excitation saturation irrelevant for a given experiment can be calculated by modeling the laser pulse train as a sequence of square pulses of intensity I, duration τ, and repetition rate ω. The fraction $F_{\Delta t}$ of cage–calcium complexes photolyzed during an uncaging flash of duration Δt is then

$$F_{\Delta t} = 1 - e^{-(1/2)q\delta I^2 \omega \tau \Delta t} \tag{12}$$

whereas the fraction of cage–calcium complexes excited during a single laser pulse in the uncaging flash, assuming that the intensity is low enough to avoid saturation, is

$$F_\tau = (1/2)\delta I^2 \tau \tag{13}$$

We can therefore solve for the uncaging quantum efficiency q that allows the fraction $F_{\Delta t}$ of cage–calcium complexes photolyzed during an uncaging flash to reach 0.9 (an arbitrarily chosen value that represents nearly complete photolysis) while restricting the fraction F_τ of cage–calcium complexes excited during a single laser pulse to 0.1 (an arbitrarily chosen value that

---

[20] C. Xu and W. Webb, in "Topics in Fluorescence Spectroscopy: Volume 5 of Multi-Photon Excitation and Light Quenching" (J. Lackowicz, ed.). Plenum Press, New York, 1997.

[21] D. Sandison, R. Williams, K. Wells, J. Strickler, and W. Webb, in "Handbook of Biological Confocal Microscopy" (J. Pawley, ed.), p. 39. Plenum Press, New York, 1995.

is certainly below saturation). Setting Eq. (12) equal to 0.9, substituting 0.1 for $\delta I^2 \tau$, and solving for q reveals

$$q \geq \frac{23}{\omega \Delta t} \qquad (14)$$

If the quantum efficiency of uncaging is greater than this value, excitation saturation will not effectively limit the total calcium release possible with a flash of duration Δt because 90% of the cage–calcium complexes will have been photolyzed in a given location at excitation intensities where only 10% of the complexes are excited with each laser pulse, still well below saturation. The calcium yield of DM-nitrophen, with a quantum efficiency[22] of 0.18, will be saturation limited at high intensities only for photolysis pulses that are 1.6 μsec or shorter (assuming a laser repetition rate of 80 MHz). Azid-1, with a quantum efficiency[12] of 1, is even less sensitive to excitation saturation. This demonstrates that excitation saturation will not be a limiting factor in most photolysis experiments, because for uncaging pulses longer than ~2 μsec, the quantum efficiencies of these two calcium cages are high enough that nearly complete photolysis is accomplished at local intensities well below saturation.

Temporal Behavior of Three-Dimensionally Localized Calcium Distributions

The temporal behavior of the calcium concentration distribution released by a multiphoton photolysis pulse is an important factor in the design of uncaging experiments. This behavior can be influenced by the duration of the uncaging pulse, the overall central uncaging dose, the diffusional mobility of the calcium ions, the intrinsic release times of the calcium cage, and the rate of uptake of calcium ions by local buffers, both extrinsic and intrinsic. The influence of buffering on calcium pulses generated via conventional one-photon uncaging has been widely studied.[23,24] However, the extremely localized nature of two-photon uncaging greatly increases the influence of diffusion on the postphotolysis kinetics of the liberated calcium distribution, warranting another look at the problem.

The complex nature of the interplay between buffering and diffusion over the subfemtoliter spatial scales of multiphoton uncaging makes exact calculation of the resultant calcium concentration distributions problematic. Advances in the related problem of buffered calcium influx through mem-

---

[22] J. H. Kaplan and G. Ellis-Davies, *Proc. Natl. Acad. Sci. U.S.A.* **85,** 6571 (1988).
[23] T. Xu, M. Naraghi, H. Kang, and E. Neher, *Biophys. J.* **73,** 532 (1997).
[24] A. Fleet, G. Ellis-Davies, J. Kaplan, and S. Bolsover, *Soc. Neurosci. Abs.* **21,** 775 (1995).

brane channels have involved numerical integration of the relevant equations in order to determine the steady-state distributions of calcium ions in the vicinity of an open calcium channel.[25] In order to gain a physical understanding of the dynamics of the calcium concentration after a photolysis pulse without attempting a numerical solution of the relevant differential equations, we will restrict our attention to the temporal behavior of the calcium concentration at the center of the two-photon focal volume (the peak concentration) and generate a model of this behavior using a few approximations and taking a few limits that greatly simplify the problem. This model will allow us to calculate the concentrations of the three populations that calcium ions flow between during and immediately after an uncaging pulse; they are the population of ground-state cage–calcium complexes, of photoexcited cage–calcium complexes, and of free calcium ions.

The first simplifying approximation is to treat the diffusion of free calcium ions out of the two-photon focal volume as a simple first-order process with a characteristic escape rate k_e given by the reciprocal of the average dwell time. For NA between 0.6 and 1.3 we will approximate this as

$$k_e \cong \frac{1.2 D_{Ca}}{w_r^2} \tag{15}$$

where D_{Ca} is the diffusion coefficient of free (unbuffered) calcium ions and w_r is the $1/e^2$ radius of the excitation volume. The second simplification is to restrict the amount of calcium generated with an uncaging pulse train such that the peak concentration of liberated ions is significantly lower than the local concentration of calcium buffers. In this limit there is no local buffer saturation and we can assign a single characteristic uptake rate constant for intrinsic cellular buffers, k_i, and extrinsic buffers, k_x, where this uptake rate constant is the usual "on rate" of the buffer, with units of $M^{-1} \sec^{-1}$, times the local concentration of the buffer. For freely diffusing calcium buffers with "off rates" similar to Fluo-3 ($k_{off} \sim 240 \sec^{-1}$) based on the equilibrium constant and the "on rate,"[13] the newly calcium-bound buffers will have had ample time to diffuse away from the two-photon probe volume before approaching equilibrium with the new calcium distribution (i.e., for the buffer, $k_e > k_{off}$) and we will therefore assume they act only as calcium sinks, not as calcium sources. We can now assign a single rate constant, $k_o \equiv k_e + k_i + k_x$, to account for all the processes that remove free calcium from the two-photon focal volume.

Two other characteristic rates relevant to this problem are the rate at which ground-state cage–calcium complexes are photoexcited, $(1/2)\delta_u \langle I_u^2 \rangle$, and the rate at which photoexcited cages release calcium, k_r. This intrinsic

---

[25] J. Klingauf and E. Neher, *Biophys. J.* **72**, 674 (1997).

release rate is set by the time required for the internal conformational shifts of a photoexcited calcium–cage complex, which lead to the release of the calcium ion. For the three cages studied here, this rate is on the order of $\sim 10^5$ sec^{-1}.[1,12]

For some calcium cages there is another relevant rate constant; the rate at which photoexcited cage–calcium complexes revert to ground state cage–calcium complexes without releasing calcium. The presence of this extra pathway can add significant complexity to the present calculation. We therefore explicitly consider Azid-1 in these calculations because its quantum efficiency is close to one,[12] suggesting that this pathway can be neglected. The temporal behavior that the following calculations predict is also valid for cages of nonunity quantum efficiency, but the overall amplitude differs by a constant factor.[13]

Using the rates defined earlier, we can now define an extremely simple dynamical relationship between the concentration of the species relevant when a photolysis pulse acts on a solution of caged calcium

$$[CCa^{2+}(0,0;t)] \xrightarrow{(1/2)\delta_u\langle I_u^2\rangle} [C^*Ca^{2+}(0,0;t)] \xrightarrow{k_r}$$

$$\Delta[Ca^{2+}(0,0;t)] \xrightarrow{k_o} \text{Escape} \tag{16}$$

where $[CCa^{2+}(0, 0; t)]$ is the concentration of ground-state cage–calcium complexes at the center of the two-photon spot, $[C^*Ca^{2+}(0, 0; t)]$ is the concentration of photoexcited cage–calcium complexes at the center of the two-photon spot, $\Delta[Ca^{2+}(0, 0; t)]$ is the concentration of liberated calcium ions at the center of the two-photon spot, and "Escape" represents escape of liberated calcium ions from the system either through diffusion out of the two-photon probe volume or chelation by buffers.

We can now form the relatively simple system of differential equations that describe the behavior of these three populations:

$$\frac{\partial}{\partial t}\begin{bmatrix} [CCa^{2+}(0,0;t)] \\ [C^*Ca^{2+}(0,0;t)] \\ \Delta[Ca^{2+}(0,0;t)] \end{bmatrix} = \begin{bmatrix} -(1/2)\delta_u\langle I_u^2\rangle & 0 & 0 \\ (1/2)\delta_u\langle I_u^2\rangle & -k_r & 0 \\ 0 & k_r & -k_o \end{bmatrix} \begin{bmatrix} [CCa^{2+}(0,0;t)] \\ [C^*Ca^{2+}(0,0;t)] \\ \Delta[Ca^{2+}(0,0;t)] \end{bmatrix} \tag{17}$$

In order to generate highly localized calcium distributions of high concentration, rapid uncaging pulses ($\Delta t \ll k_o^{-1}$) of high uncaging rate $(1/2)2\delta_u\langle I_u^2\rangle \gg k_r, k_o)$ will be used. Under these conditions, the free calcium concentration at the center of the two-photon focal volume generated by an uncaging pulse is given by

$$\Delta[Ca^{2+}(0,0;t)] = \frac{k_r}{(k_r - k_o)} (e^{-k_o t} - e^{-k_r t})[CCa^{2+}]_0 \tag{18}$$

where $[CCa^{2+}]_0$ is the equilibrium concentration of cage–calcium complexes. In this high uncaging rate and rapid uncaging pulse limit, the peak calcium concentration generated does not scale with the excitation rate, and the duration of the elevated calcium distribution does not scale with the uncaging pulse duration.

We can now use Eq. (18) to predict the temporal behavior of the peak calcium concentration under a variety of experimental conditions. The simplest case is one in which cage is phytolyzed via two-photon excitation *in vitro* with no extrinsic buffer added. This is also the case that yields the longest possible localized elevations of free calcium with a given illumination pattern. The release rate k_r is assumed to be 10^5 sec^{-1}, and the total rate constant for free calcium loss, k_o, is due only to diffusional escape ($k_o = k_e$). The diffusional escape rate of free calcium ions from the two-photon spot generated with a 1.33 NA lens overfilled with 700-nm light is $k_e \sim 7 \times 10^4$ sec^{-1}, whereas from a 0.5 NA lens it is $k_e \sim 1 \times 10^4$ sec^{-1}. The predicted peak calcium concentration as a function of time after the rapid photolysis pulse is plotted in Fig. 6. We see that the calcium elevations generated with the superior spatial resolution of high NA optics can last up to ~32 μsec FWHM. Both the amplitude and the duration of the free calcium pulse decrease with increasing NA due to the faster diffusional escape time from the smaller focal volume of the higher NA objective. Note, however, that producing similar central uncaging doses while lowering the NA of the objective requires a significant increase in input power.

The next case to consider is two-photon uncaging within a living cell. Prediction of the peak free calcium concentration dynamics now requires knowledge of the intrinsic buffering rate, k_i, extrinsic buffering rate, k_x, and their relation to the characteristic diffusional escape time, k_e. The intrinsic buffering dynamics are is not known for most experimental systems and are, in fact, an attractive subject of study for two-photon calcium uncaging. The extrinsic buffering rate, k_x, is generally known because it is controlled by the quantity and type of calcium buffer added to the system, usually in the form of a fluorescent calcium indicator such as Fluo-3 but also including any calcium-free unphotolyzed cage.[13] In order to explore how buffering interacts with diffusion out of the two-photon spot, we will assume that the total buffering rate in the cellular system ($k_i + k_x$) is equivalent to that exerted by 100 μM of Fluo-3 (a reasonable concentration of indicator that will maintain cell viability). We can then

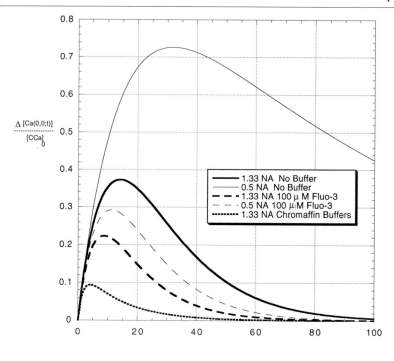

FIG. 6. Buffer and diffusion dependence of the kinetics of the maximum calcium concentration. Free calcium concentration at the center of the two-photon focal volume, as given by Eq. (18). The concentration is expressed as a ratio of the local peak liberated calcium concentration to the equilibrium concentration of cage–calcium complexes before the uncaging pulse. Compare heavy or light lines to observe the effect of calcium buffers, and compare solid or dashed lines for the effect of varying the NA of the objective. The dotted line is calculated using experimentally measured values of the cellular average concentration ($\sim$4 mM) and on rate ($\sim$1 $\times$ 10^8 M^{-1} sec^{-1}) of calcium buffers in bovine chromaffin cells.[23]

calculate the peak free calcium concentration in both the high NA and the low NA case, with results shown in Fig. 6. The presence of calcium buffers obviously greatly accelerates the rate of loss of calcium ions from the two-photon volume, leading to lower free calcium concentrations. This demonstrates the familiar tradeoff between fluorescent signal and biological effect; in order to quantify a dynamical calcium process (such as calcium release by an uncaging pulse), a large amount of fluorescent indicator is desirable to maximize signal. However, the presence of fluorescent indicator dye reduces the amount of free calcium available to induce biological responses.

Experimental measurements have been made of the calcium buffering dynamics in bovine chromaffin cells, and these determined that the cells in question had 4 mM of a calcium buffer with an on rate of $1 \times 10^8 \, M^{-1}$ sec^{-1} [23] for a characteristic intrinsic buffering rate of $k_i \sim 4 \times 10^5$ sec^{-1}, provided that the total calcium release is 400 μM or less. In a cell containing 100 μM of Fluo-3 and this rapid rate of intrinsic calcium buffering, we can estimate that the total effective calcium removal rate constant $k_o = k_i + k_e + k_x = 5.6 \times 10^5$ sec^{-1} for calcium liberated with a 1.33 NA lens. The peak free calcium concentration generated in such an experiment is also plotted in Fig. 6. With such a rapid biochemical mechanism of calcium removal, the rate of loss of free calcium concentration is dominated by calcium buffering over free calcium diffusion.

In conclusion, we find that the longest elevated calcium concentrations, those due to two-photon release of calcium in a nonbuffered environment, are on the order of 32 μsec FWHM for high spatial resolution photolysis using 1.3 NA optics. This duration can be considerably lengthened by reducing the NA of the excitation optics, with the unavoidable consequence of reducing the initial spatial resolution of release. Addition of calcium buffering to the system, whether through intrinsic or extrinsic calcium buffers, can significantly shorten this duration.

Conclusion

This work described equipment for the two-photon excitation of caged calcium and delineated the spatial resolution of calcium release possible with this apparatus. The total calcium release can be calculated easily for any illumination parameters using the equations described here, as can the temporal behavior of the peak calcium concentration generated using rapid uncaging pulses of high release rate.

In practical terms, complete photolysis of caged calcium with high spatial resolution is possible with TPE. Using ~7 mW of average power over a pulse train of 10 μsec, 700-nm light overfilling a 1.3 NA objective can completely photolyze the calcium cage Azid-1 contained within the TPE focal volume. A second calcium cage, DM-nitrophen, requires considerably more excitation power (~85 mW of 720-nm light). The resultant nonequilibrium highly localized distribution of calcium ions will have a radial and axial length scale of 0.25 and 0.86 μm, respectively. The duration of the elevated calcium concentrations produced with TPE will vary depending on the diffusion times and buffering conditions of the released ions. In an unbuffered environment, the aforementioned distribution generated with a high (1.3) NA lens will last for ~32 μsec (FWHM), which is limited by the diffusional escape time of the liberated ions out of the TPE focal volume. This duration can be significantly extended by lowering the spatial resolution of release and can be significantly reduced by the addition of calcium-buffering agents, whether intrinsic cellular calcium buffers or extrinsic calcium indicator dyes.

There are many possible uses for TPE of calcium cages. The competition of buffers for locally released calcium could provide a window into the spatial heterogeneities of the intrinsic calcium-buffering dynamics in living cells. This could be accomplished through utilization of the relative uptake of liberated calcium ions by a fluorescent calcium indicator and intrinsic calcium buffers as a measure of the intrinsic buffering rate and capacity of cells. As has been shown in Fig. 6, for at least one cell type these kinetics should happen significantly faster than the diffusional escape of calcium ions from the TPE focal volume, allowing these buffering properties to be measured with submicron, three-dimensional resolution and thereby extending previous whole cell studies.[23]

These measurements and calculations demonstrate that complete two-photon release of calcium with high three-dimensional spatial resolution is possible and can be an extremely useful tool in the study of a variety of cellular processes.

Acknowledgments

This work was carried out in the Developmental Resource for Biophysical Imaging and Optoelectronics with funding provided by the NSF (Grant DIR 88002787) and NIH (Grant RR07719 and RR04344). EB was supported as a predoctoral trainee under NIH Grant T32GM08267. Laser excitation at 622 nm was supported by Dr. Frank Wise, Xiang Liu, and Lie-Jia Qian of Cornell University Applied and Engineering Physics.

[21] Caged Inositol 1,4,5-Trisphosphate for Studying Release of Ca^{2+} from Intracellular Stores

By NICK CALLAMARAS and IAN PARKER

Introduction

Inositol 1,4,5-trisphosphate ($InsP_3$) is a second messenger that mediates Ca^{2+} release during many physiological processes, including development, gene regulation, secretion, contraction, synaptic transmission, and apoptotic cell death.[1,2] The development of caged-InsP_3 [c-InsP_3: *myo*-inositol 1,4,5-trisphosphate (1,2-nitrophenyl ester)] has provided an elegant and efficient means by which to control levels of this active messenger in the cytoplasm of intact cells, giving an unparalleled degree of temporal and spatial resolution for detailed studies of this ubiquitous intracellular pathway.[3-5] In comparison to other techniques, such as fast-flow perfusion of permeabilized cells[6] or direct introduction from a point source via a pressure or ionophoresis microelectrode,[7] it allows rapid (millisecond), spatially defined elevations in InsP_3 concentration to be achieved without compromising the organization and biochemical composition of the cytoplasm or disrupting the electrical properties of the cell.

The essence of the technique is first to introduce the physiologically inert c-InsP_3 into a cell (e.g., by microinjection or diffusion from a whole cell patch pipette) and allow it to diffuse and equilibrate throughout the cytosol. Subsequent illumination with near-UV light then provides a means to photorelease active InsP_3. Major advantages of this technique are: (1) rapid and reproducible millisecond steps in intracellular concentration of InsP_3 are achieved; (2) the relative concentration of InsP_3 can be precisely regulated by the intensity and duration of illumination; and (3) manipulation of the illumination field allows for homogeneous release of InsP_3 throughout spatially defined regions of the cell.

Several practical points further simplify the use of c-InsP_3 and extend its utility. The caged precursor is extremely stable in the cytosol, and most cells are exquisitely sensitive to InsP_3 (nanomolar range), so that a large "reservoir" of c-InsP_3 may be loaded, allowing numerous photolysis flashes to be delivered, each of which consume only a tiny fraction of the total c-InsP_3. Also, c-InsP_3 is efficiently photolyzed by light with wavelengths <400 nm, thus freeing up the visible spectrum for simultaneous use of long-wavelength indicator dyes to monitor Ca^{2+} liberation induced by InsP_3.

This article describes techniques developed in the authors' laboratory for use of c-InsP_3 in *Xenopus laevis* oocytes and illustrates some of their applications in studying the InsP_3/Ca^{2+} second messenger pathway. The

---

[1] M. J. Berridge, *Nature* **361**, 315 (1993).
[2] T. Michikawa, A. Miyawaki, T. Furuichi, and K. Mikoshiba, *Crit. Rev. Neurobiol.* **10**, 39 (1996).
[3] D. C. Ogden, K. Khodakhah, T. D. Carter, P. T. Gray, and T. Capiod, *J. Exp. Biol.* **184**, 105 (1993).
[4] S. S. Wang and G. J. Augustine, *Neuron* **15**, 755 (1995).
[5] I. Parker, *Neuromethods* **20**, 369 (1992).
[6] T. Meyer, T. Wensel, and L. Stryer, *Biochemistry* **29**, 32 (1990).
[7] M. J. Berridge, *Proc. R. Soc. Lond. B* **238**, 235 (1989).

Xenopus oocyte has long been a common cell type for studies of the InsP_3-mediated Ca^{2+} release system due to its enormous size (>1 mm diameter) and simple geometry, which facilitate many experimental procedures. Additional advantages are the ability of the oocyte to express foreign proteins encoded by microinjected mRNA[8] and a lack of other confounding intracellular Ca^{2+} release channels (i.e., ryanodine receptors). The combined advantages of flash-photolysis of c-InsP_3 and the oocyte system have, for example, permitted high-resolution studies of the kinetics of InsP_3 receptor (InsP_3R) gating in an intact cell system,[9] allowed elucidation of the relationship between InsP_3-evoked Ca^{2+} signals and Ca^{2+}-dependent Cl$^-$ membrane current,[10] and facilitated the resolution of "elementary" Ca^{2+} release events underlying global Ca^{2+} signals.[11–13] Moreover, the ability to evoke reproducible InsP_3 signals has aided pharmacological studies of agents affecting this intracellular messenger pathway.[14,15]

Methodology

Photochemistry of c-InsP$_3$

In the native oocyte, InsP_3 is generated from receptor-mediated breakdown of phosphatidylinositol bisphosphate. InsP_3 then binds to the intracellular InsP_3R to release Ca^{2+} from a subset of the intracellular Ca^{2+} stores and is subsequently metabolized into a bewildering array of other inositol phosphates, some of which retain physiological activity. The flash photolysis of c-InsP_3 avoids the time delays, nonlinearity, and modulation by other messengers associated with ligand-activated InsP_3 generation. This is achieved due to the efficiency of the photochemistry involved in the cleavage of InsP_3 from its protecting nitrophenyl ester group.[16] In brief, the reaction proceeds in two steps: the rapid (nanosecond) absorption of a high energy photon generating active intermediates followed by a slower (millisecond) dark reaction whereby intermediates decay to release InsP_3

---

[8] R. Miledi, I. Parker, and K. Sumikawa, in "Fidia Neuroscience Award Lectures" (J. Smith, ed.), Vol. 3, p. 57. Raven Press, New York, 1989.
[9] I. Parker, Y. Yao, and V. Ilyin, *Biophys. J.* **70**, 222 (1996).
[10] I. Parker and Y. Yao, *Cell Calcium* **15**, 276 (1994).
[11] I. Parker and Y. Yao, *Proc. R. Soc. Lond. B* **246**, 269 (1991).
[12] Y. Yao, J. Choi, and I. Parker, *J. Physiol. (Lond.)* **482**, 533 (1995).
[13] I. Parker, J. Choi, and Y. Yao, *Cell Calcium* **20**, 105 (1996).
[14] I. Parker and I. Ivorra, *J. Physiol. (Lond.)* **433**, 207 (1991).
[15] V. Ilyin and I. Parker, *J. Physiol. (Lond.)* **448**, 339 (1992).
[16] J. W. Walker, J. Feeney, and D. R. Trentham, *Biochemistry* **28**, 3272 (1989).

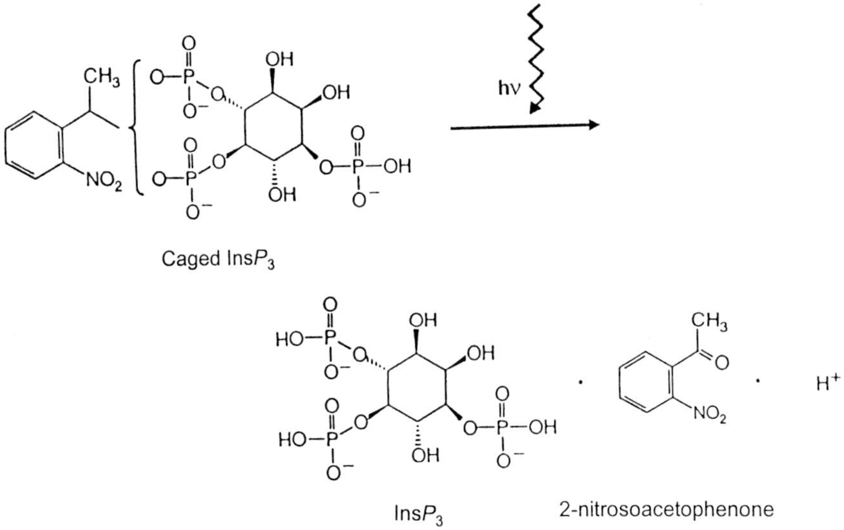

FIG. 1. Schematic photolysis reaction for caged $InsP_3$. In addition to free $InsP_3$, photolysis releases a proton and a by-product, 2-nitrosoacetophenone.

(Fig. 1). The by-products of this reaction, a proton and nitrosoacetophenone, can be toxic at high levels (1 mmol^{-1}), altering the local pH or reacting with -SH groups of proteins, respectively.[17] Fortunately, the concentrations of $InsP_3$ necessary to elicit even maximal responses in oocytes and many other cells are sufficiently small that by-product toxicity appears not to be an issue.

Caged $InsP_3$ can presently be obtained from either Calbiochem (La Jolla, CA) or Molecular Probes (Eugene, OR). In our experience the c-$InsP_3$ available from Molecular Probes is preferable because of lower contamination by active $InsP_3$.

Introduction of c-$InsP_3$ into Cells

Caged $InsP_3$ is a highly charged molecule and thus does not permeate through the cell membrane. For work with single cells, c-$InsP_3$ may be introduced into the cytosol by injection through a micropipette (pneumatic presure injection or ionophoresis) or by diffusion from a whole cell patch pipette. In the latter case, the resulting intracellular concentration should approximate that in the pipette. Pressure injection is convenient for use with large cells, and the final intracellular concentration of c-$InsP_3$ can be

[17] M. G. Hibberd, Y. E. Goldman, and D. R. Trentham, *Curr. Topics Cell. Regul.* **24**, 357 (1984).

estimated from the amount of fluid injected (measured as the diameter of a fluid droplet expelled with the pipette tip in air), the concentration in the injection solution, and the cytosolic volume.

Injection into defolliculated *Xenopus* oocytes[18] is made through broken (tip diameter ca. 5 μm) glass micropipettes. We typically inject about 20 nl of an aqueous solution of 1–5 mM c-InsP_3 (to which Ca^{2+} indicator dyes may also be added), resulting in a final intracellular concentration of about 20–100 μM, assuming a cytosolic volume of 1 μl. It is then necessary to wait for 30 min or longer for diffusional equilibration of c-InsP_3 throughout the cell; this time also allows the cell to metabolize any active free InsP_3 that may be present as a contaminant.[5] Standard room lights, and even halogen fiber-optic illuminators, cause surprisingly little photolysis of c-InsP_3, and we have not found it necessary to take any special precautions with lighting either in handling injection solutions or after loading cells with c-InsP_3.

Clearly, mechanical introduction of c-InsP_3 through pipettes would be impractical for studies on populations of cells. An elegant approach that has recently been developed involves loading c-InsP_3 by extracellular application at a membrane-permeable ester,[19] analogous to the well-known technique for loading indicators such as Fura-2. Alternatively, the plasma membrane may be permeabilized by a variety of techniques[20] to allow access of charged c-InsP_3 to the cytosol.

Linearity and Calibration of Photorelease

The amount of free InsP_3 resulting from a given photolysis flash is expected to be linearly proportional to the starting concentration of c-InsP_3 and the flash intensity and duration. In practice, this linear dependence on flash parameters is followed almost exactly; as illustrated in Fig. 2A, where caged ATP was used to monitor photolysis in place of c-InsP_3 because of the ease of detecting release of ATP by luminescence of a luciferin/luciferase assay system. Thus, an extremely precise control of *relative* amounts of InsP_3 formation within a given cell can be obtained by appropriate adjustment of flash parameters. Our usual method is to adjust the flash intensity using neutral density filters to set an initial working range and then vary the flash duration to control the extent of photorelease. This has advantages in that the duration can be readily altered in very small increments, as opposed to the relatively coarse steps available with filter

---

[18] K. Sumikawa, I. Parker, and R. Miledi, *Methods Neurosci.* **1,** 30 (1989).
[19] W. H. Li, C. Schultz, J. Lopi, and R. Y. Tsien, *Tetrahedron* **53,** 12017 (1997).
[20] K. A. Oldershaw, D. L. Nunn, and C. W. Taylor, *Biochem. J.* **278,** 705 (1991).

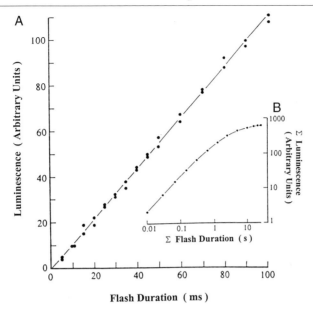

FIG. 2. Linearity of photorelease with flash strength and calibration of extent of photolysis using a luminescence system to assay release of ATP from a caged precursor. (A) Formation of ATP varies linearly with flash duration. Recordings were made from a 20-μl droplet containing 100 μM caged ATP together with undiluted ATP assay reagent (Sigma FL-AAM). The photolysis light was focused by the microscope objective to a small spot within the droplet, and the graph plots the peak amplitude of the ATP-dependent luminescence (monitored by a photomultiplier) evoked following photolysis flashes of identical intensity and varying duration. (B) Estimation of the fractional photolysis of caged ATP resulting from a given photolysis flash. The experiment was performed as in (A), except for use of a smaller droplet that was entirely irradiated by the photolysis spot. A series of repeated flashes of increasing duration were applied, and the peak luminescence recorded following each flash. Double logarithmic plot shows cumulative luminescence (i.e., sum of peak signals evoked by successive flashes) as a function of cumulative duration of exposure to photolysis light. [Reproduced from I. Parker and I. Ivorra, *Am. J. Physiol.* **263,** C154 (1992), with permission.

wheels, and individual flashes within a paired-pulse protocol can be independently controlled. Care should be taken, however, that the longest flashes employed are brief compared to the kinetics of the response under investigation.

For many purposes, control of the *relative* amounts of InsP$_3$ photoreleased in a given experiment is sufficient, but it may also be necessary to estimate the *absolute* concentrations of free InsP$_3$. This is a more difficult and less precise task, for which we have taken two approaches. The first involves an *in vitro* calibration employing caged ATP.[21] The protocol is illustrated in Fig. 2B. A small droplet containing a known amount of ATP together with assay mixture is placed on the microscope stage and is completely irradiated by photolysis flashes of the same intensities as used for physiological experiments. Repeated flashes are applied, and the luminescence signals (linearly proportional to the amount of ATP) are recorded following each flash. After many exposures, the luminescence response declines to zero, indicating complete consumption of the c-ATP (luciferin is present in excess). Thus, the total cumulative luminescence over all responses corresponds to the photorelease of all the ATP, and the fractional amount released by a given light flash can be calculated from the ratio of the luminescence resulting from that flash compared with the total luminescence. This fractional release should then apply equally to photolysis of intracellular c-InsP_3 under the same conditions, with the major caveat that no account is taken of light absorption within the specimen. A second approach is to employ the cell as a bioassay system. In the oocyte, Ca^{2+} waves are initiated at a sharply defined concentration of InsP_3 so that photolysis flashes below threshold evoke little or no Ca^{2+} release or Ca^{2+}-dependent Cl^- current response, whereas just-suprathreshold stimuli give clear responses (cf. Fig. 4). Estimates of the absolute concentration of InsP_3 corresponding to this threshold can then be obtained by injecting increasing amounts of a poorly metabolized but equipotent InsP_3 analog (3-F-InsP_3) to determine the resulting intracellular concentration at which waves are triggered.[12]

Measurements of the InsP_3 concentrations by both of these methods indicate that Ca^{2+} waves are evoked in the oocyte by roughly 50 nM InsP_3, and maximal responses are evoked by concentrations about 20-fold greater.[9] Because the cells can readily be loaded to concentrations of c-InsP_3 as high as 100 μM, flashes giving physiological responses will photolyze only negligible proportions (0.05–1%) of the c-InsP_3. Further, the turbidity of the oocyte cytoplasm limits light penetration to a few tens of μm, so that the greater volume of the cell is effectively shielded from the photolysis light and acts as a virtually inexhaustible reservoir of fresh c-InsP_3. The net result is that it is possible to evoke numerous (hundreds) reproducible responses with repeated flashes, without significant depletion of available c-InsP_3 in the oocyte.

Photolysis Light Sources

The photolysis of caged InsP_3 requires a sufficiently intense source of illumination at wavelengths around 350 nm. Possible light sources capable

---

[21] I. Parker and I. Ivorra, *Am. J. Physiol.* **263**, C154 (1992).

of efficient photolysis in most biological preparations include (1) continuous arc lamps with electronically controlled shutters, (2) xenon flash lamps, and (3) UV lasers. The preceding systems are ordered in terms of increasing energy, complexity, and cost. Each system has relative merits and disadvantages but, fortunately, bigger and costlier systems are not necessarily better.

A continuous UV arc lamp is a stable source of illumination that, when combined with commercially available high-speed electric shutters, provides an inexpensive and highly efficient photolysis system. The stability of the lamp output is very good if the arc is mounted vertically and powered from a high-quality constant current power supply, with a between-flash reproducibility of better than 1%. For use with caged-InsP_3, the limited power of the lamp is not a major consideration because photolysis of only a small percentage of the c-InsP_3 loaded into a cell is sufficient to evoke responses. Either mercury or xenon arc lamps (50–100 W) work well and, indeed, it is usually necessary to attenuate their output with neutral density filters. High-speed electronically controlled shutters provide an affordable and efficient means to produce flashes as brief as 1–2 msec. Durations shorter than this would provide little advantage because the photogeneration of InsP_3 from its caged precursor takes 5–10 msec. Finally, unlike flash lamps or pulsed lasers, it is possible to continuously irradiate the specimen, mimicking physiological InsP_3 generation over several seconds during agonist stimulation. Therefore, a continuous arc lamp is the simplest and most cost-effective choice for most experimental situations. Such systems can be readily constructed from an existing epifluorescence microscope, merely by substituting a UV filter cube to restrict irradiation to wavelengths <400 nm and inserting an electronic shutter into the light path.

Xenon flash lamps[22] provide a much higher (1000 times or greater) energy output in a brief flash than a shuttered arc lamp system but, as noted earlier, this higher energy is not likely to be required when working with c-InsP_3 provided that an efficient optical system is used to channel light to the specimen. Further, flash lamp systems have disadvantages in regard to their greater cost, generation of electrical artifacts, lack of independent control of the duration of each flash in paired-pulse experiments, and an inability to provide continuous illumination of sustained photorelease.

Several types of laser are available that produce high-energy pulses at near-UV wavelengths, including the frequency-doubled ruby laser and the frequency-tripled Nd–YAG (neodymium–ytrrium-aluminum-garnet) laser. In our experience, the only application for which the high cost of these lasers is justified involves focusing the photolysis light to a near-diffraction limited spot in the specimen so as to achieve a virtual point-photorelease

[22] G. Rapp and K. Guth, *Pflugers Arch.* **411**, 200 (1988).

of InsP_3. Because of the perfectly parallel beam from a laser this can be achieved with little loss of light, whereas to achieve the same result with a conventional (arc lamp) source involves insertion of a pinhole aperture with an accompanying drastic loss in light throughput.

Two-photon excitation of caged compounds by femtosecond pulses from mode-locked lasers is a recent development[23] that offers the possibility of even more spatially restricted photorelease because the quadratic dependence of two-photon photolysis on light intensity ensures that this will be restricted only to the plane of focus of the light spot formed by an objective lens. Although this approach has been elegantly employed to map out the sensitivity of neurons to extracellular photorelease of caged neurotransmitter,[24] the possible advantages of this expensive technology for use with c-InsP_3 are unclear and are likely to be limited by the physiology of the InsP_3 pathway. In particular, the rapid intracellular diffusion of InsP_3, together with the long latencies (many milliseconds) before Ca^{2+} release begins, suggests that the spatial localization of InsP_3 will be limited by those factors and not by the minimum volume throughout which photorelease occurs.

Simultaneous Use of c-InsP$_3$ and Fluorescent Calcium Indicators

The primary action of InsP_3 in the cell is to liberate calcium from intracellular stores. One approach to monitor the resulting rise in cytosolic-free calcium concentration is to measure currents through endogenous calcium-activated membrane channels (e.g., the Ca^{2+}-dependent Cl$^-$ channels in the oocyte membrane.[25] This is advantageous in terms of simplicity and ensures that the measurement technique does not perturb the signals of interest. Interpretation may, however, be complicated by uncertainties regarding the steady-state and kinetic relationships between free [Ca^{2+}] and current amplitude,[10] and voltage-clamp recordings of current provide no information regarding the spatial distribution of calcium within the cell. Thus, for many purposes it is desirable to have a more direct measure of cytosolic calcium. The availability of a wide range of fluorescent calcium indicator dyes operating in the visible spectrum makes it possible to combine fluorescence calcium monitoring or imaging together with simultaneous photolysis of c-InsP_3.

Currently, our favorite dyes are the Oregon green family (Molecular Probes), which are optimized for excitation by the 488-nm line of the argon ion laser and fluoresce in the green. The excitation spectra of these dyes are sufficiently separated from that of c-InsP_3 that we find little practical

[23] W. Denk and K. Svoboda, *Neuron* **18**, 351 (1997).
[24] W. Denk, *Proc. Natl. Acad. Sci. U.S.A.* **91**, 6629 (1994).
[25] R. Miledi and I. Parker, *J. Physiol. (Lond.)* **357**, 173 (1984).

difficulty with photolysis of c-InsP_3 resulting from the blue excitation light and, conversely, photolysis flashes at wavelengths <400 nm cause little or no artifacts in fluorescence recordings. Furthermore, the dyes are available as variants with different calcium affinities (e.g., Oregon green 488 BAPTA-1 has an affinity of a few hundred nanomolar, and Oregon green BAPTA-5N has a low affinity of a few tens of micromolar), so that they may be used selectively to examine small calcium signals close to the basal calcium level or large calcium transients evoked by maximal InsP_3 levels. A disadvantage of all currently available long-wavelength indicators is that, unlike the UV-excited dyes Fura-2 and Indo-1, none show shifts in either excitation or emission spectra on binding calcium. Thus, using a single dye, it is not possible to ratio signals at two wavelengths to obtain a calibration of free calcium levels independent of variations in dye loading and path length. A reasonable compromise, however, is to form a "pseudo ratio" signal by expressing calcium-dependent fluorescence signals relative to the resting fluorescence before stimulation.[2]

The basic principle in designing an optical system for simultaneous fluorescence monitoring and photolysis is to "stack" dichroic mirrors so as to split off appropriate parts of the visible and UV spectra. Thus, a UV dichroic mirror placed close to the microscope objective will reflect short wavelengths (<400 nm) from a photolysis light source onto the specimen but transmit longer wavelengths, which may then be further split by an additional dichroic mirror to separate fluorescence excitation and emission wavelengths appropriate for specific dyes. In our earlier optical system[5] employing a Zeiss Universal microscope, this "stacking" was possible by physically mounting one epifluorescence attachment on top of another. More modern microscope designs fail to anticipate the need for such flexibility, and in our present system (described in the following section) a UV dichroic mirror is placed in the regular epifluorescence unit, whereas the fluorescence excitation dichroic mirror is located externally to the microscope with light directed through the video port. Excellent dichroic mirror and filter sets tailored for use with particular dyes are available from Omega Optical Inc. and Chroma Technology Corp. (both at Brattleboro, VT).

Fluorescence calcium monitoring systems may vary in their degree of spatial and temporal resolution (with corresponding increases in cost and complexity) from photomultiplier-based detectors to monitor calcium throughout an entire cell or defined region of a cell,[5] through CCD camera-based wide-field imaging systems[11] to laser-scan confocal microscopes providing millisecond and submicron resolution.[26]

---

[26] I. Parker, N. Callamaras, and W. Wier, *Cell Calcium* **21,** 441 (1997).

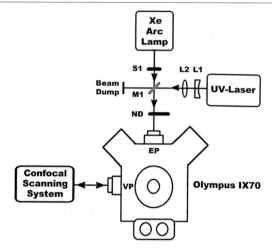

FIG. 3. Schematic diagram of a versatile photolysis and imaging system constructed around an Olympus IX70 inverted microscope. The photolysis system is shown at the top, and interfaces through the epifluorescence port (EP) of the microscope. A confocal scanning system on the left interfaces through the video side port (VP). L1, plano-concave lens; f, −1 cm; L2, biconvex lens; f, 5 cm; S1, electronic shutter; M1, cover glass acting as beam splitter; ND, neutral density filter wheels.

Versatile Photolysis System for Imaging and Electrophysiological Studies

This section describes details of our current photolysis system, which operates through the epifluorescence port of an Olympus IX70 inverted microscope (Lake Success, NY), and can be used either independently or together with a confocal Ca^{2+} imaging system interfaced through the microscope video port (Fig. 3). For further details of the confocal scanner, see Parker et al.,[26] and for more general information pertaining to the principles and applications of biological confocal microscopy, see Pawley.[27]

The photolysis system employs both a continuous arc lamp source (for wide field photorelease) and a pulsed UV laser (for "point" photorelease), with light from both systems being combined by a beam-splitting mirror and directed into the epifluorescence port of the Olympus IX70, which is equipped with a standard UV filter cube and fluor objective lenses. Components are mounted on an optical breadboard, using standard post-mounts (New Focus Inc., Santa Clara, CA). Note that with the exception of the filter cube and holder, no Olympus epifluorescence components are required.

The upper section of Fig. 3 shows the layout of the photolysis light

---

[27] J. B. Pawley, "Handbook of Biological and Confocal Microscopy." Plenum Press, New York, 1990.

paths. All lenses are fused silica for optimal UV transmission, and optical components are postmounted at a height corresponding to the center line of the Olympus epifluorescence port. For experiments where photorelease of caged compound is required over a wide area (10- to 100-μm-diameter spot using a 40× objective), UV light is derived from an arc lamp (75-W xenon lamp, mounted in a Zeiss housing and operated from a stabilized constant-current power supply). An electronic shutter (Uniblitz, Vincent Associates, NY), triggered manually or via TTL input from a Digitimer or computer, controls exposure duration while a set of neutral density wheels (3.0 OD in steps of 0.1 OD, New Focus Inc.) allow control of light intensity. Adjustment of a collector lens in the arc lamp housing provides uniform (Koehler) illumination throughout the photolysis spot in the microscope image plane.

The laser system employs a Mini-Lite frequency-tripled (355 nm) Nd:YAG laser (Continuum, Santa Clara, CA), with the laser head mounted on the optical table by nylon screws and insulating stand-offs to avoid hum loops for electrophysiological recording. Lenses L1 and L2 form a beam expander so that the laser beam fills the back aperture of the objective lens, thus making use of its full numerical aperture. Because the laser beam at full power may crack an objective lens with poor UV transmission, a microscope cover glass is used as a mirror (M4) to reflect only a small percentage of the laser beam into the microscope through the neutral density filter wheels, which allow further attenuation. The invisible beam from the Mini-Lite presents a considerable safety hazard, and appropriate precautions, including use of laser safety goggles, should be taken when the beam is exposed for alignment. A beam dump is placed after M4 to avoid the possibility of the beam passing into the room, and the entire beam path is covered while the laser is operating. The laser spot formed by an objective lens can be viewed using a coverslip marked with a yellow "highlighter" pen, and its position centered by small adjustments of laser position and deflection of M1. Finally, the spot can be brought to a sharp focus by axial adjustment of L2. A UV-blocking filter in the Olympus filter cube, together with an additional long-pass filter (λ >510 nm) inserted in the microscope binocular head, permits safe viewing while the laser is in use. The laser can be operated in single-shot mode (triggered by a push switch or TTL input) or pulsed repeatedly at up to 10 Hz.

Applications of c-InsP$_3$ for Study of Ca^{2+} Signaling in Oocytes

Dose–Response Relation of Ca^{2+} Release

Bath application of calcium-mobilizing agonists to oocytes results in a complex oscillatory Cl$^-$ current, which varies in a highly nonlinear manner

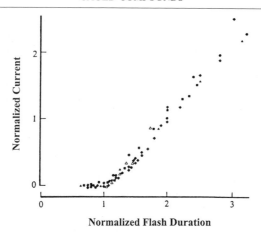

FIG. 4. Dose–response relationship of membrane currents evoked by photorelease of $InsP_3$ by different durations of irradiation. Pooled data from five oocytes (indicated by different symbols) showing the relationship between flash duration and peak size of evoked currents. Flash durations are normalized with respect to that which evoked just-suprathreshold responses in each oocyte, and currents are normalized with respect to currents evoked in each oocyte evoked by a flash of twice threshold duration. In this experiment the entire vegetal hemispheres were exposed to UV light. [Reproduced from I. Parker and I. Ivorra, *Am. J. Physiol.* **263**, C154 (1992), with permission.]

with agonist concentration.[28] The intracellular location of the numerous stages in the signaling pathway between cell surface receptors and activation of the current, however, makes it difficult to determine at which stage the nonlinearity arises. Photolysis of caged $InsP_3$ offers an elegant means to circumvent this problem bypassing earlier stages in the pathway. Indeed, the reproducibility and linearity of photolysis allow intracellular dose–response relationships to be determined with the same ease as bath application of compounds to an extracellular receptor. For example, Fig. 4 illustrates the dependence of Ca^{2+}-activated membrane currents on the strength of photolysis flashes and demonstrates a nonlinear relationship in that a threshold amount of $InsP_3$ is needed before any current is generated. These data also illustrate the reproducibility of the photolysis technique, as evident in the close overlap of measurements from five oocytes when flash strengths are normalized relative to the threshold required to evoke a detectable current in each cell so as to compensate for differing amounts of microinjected c-$InsP_3$.

[28] I. Parker, K. Sumikawa, and R. Miledi, *Proc. R. Soc. Lond. B.* **231**, 37 (1987).

Pharmacological Studies of InsP₃R in Intact Oocytes

The $InsP_3R$ through which calcium liberation occurs is a potential site for modulation by many endogenous messenger compounds and exogenous pharmacological agents. Studies of such effects on calcium mobilization are complicated, however, if extracellular agonists are used to activate the $InsP_3$ pathway because different agents may act on stages between the cell surface receptor and $InsP_3$ formation, as well as on the $InsP_3R$ itself. Again, caged $InsP_3$ provides a means to circumvent these difficulties by allowing the effects of microinjected or (for membrane-permeant substances) bath-applied agents to be monitored on signals evoked by repeated and identical pulses of intracellular $InsP_3$. Two examples are shown in Fig. 5.

The first concerns the possible physiological roles of $Ins(1,3,4,5)P_4$ ($InsP_4$), a higher-order inositol polyphosphate that is formed transiently by phosphorylation of $InsP_3$ during activation of the signaling pathway. As shown in Fig. 5A, evidence that $InsP_4$ acts as a weak agonist at the $InsP_3R$ to potentiate calcium mobilization is obtained by microinjecting $InsP_3$ into oocytes while monitoring calcium-dependent Cl^- currents evoked by $InsP_3$ photoreleased by successive, just-suprathreshold flashes.[14] A strong potentiation of the currents is observed, even with small doses of $InsP_4$ which, themselves, evoke almost detectable responses. This experiment further illustrates a technical point, in that it is necessary to microinject the highly charged $InsP_4$ into the oocyte. The spatial spread of $InsP_4$ in the cell is restricted to a region around the pipette tip because of diffusion and metabolism, and the photolysis light is therefore focused as a small spot centered around the injection pipette. Even though voltage-clamp recordings reflect current from the whole membrane area of the oocyte, the evoked Cl^- currents thereby arise only from the local region exposed to $InsP_4$.

A second example of the use of c-$InsP_3$ for pharmacological studies concerns the actions of caffeine, which is widely used as a tool to discriminate between calcium mobilization mediated through $InsP_3R$ and ryanodine receptors, by virtue of its ability to potentiate ryanodine receptor-mediated responses. Because caffeine readily permeates the cell membrane, we were able to study its actions on $InsP_3$-evoked Cl^- current responses by bath application during trains of repetitive photolysis flashes.[29] This causes a dramatic and reversible reduction in responses (Fig. 5B) and a rightward shift in the dose–response curve for $InsP_3$ (Fig. 5C), suggesting that caffeine may act as a reversible antagonist at the $InsP_3$ receptor. In this and other experiments where calcium liberation is monitored by endogenous calcium-

---

[29] I. Parker and I. Ivorra, *J. Physiol.* (*Lond.*) **433**, 229 (1991).

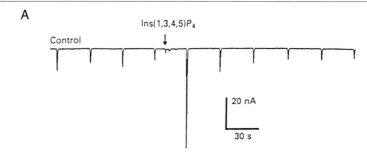

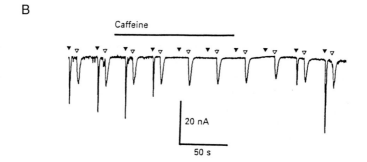

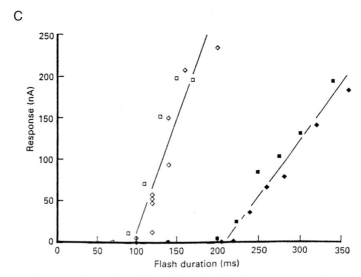

FIG. 5. Utility of c-InsP_3 for pharmacological studies. (A) Responses evoked by photorelease of InsP_3 are facilitated by injection of low doses of InsP_4. Trace shows calcium-dependent membrane current responses (inward currents at a clamp potential of -60 mV) evoked by repetitive, identical photolysis flashes, with a duration just above the threshold to evoke

activated currents it is, however, important to countrol for possible effects on the calcium-activated Cl^{-1} channels themselves. Monitoring of intracellular calcium by means of indicator dyes is one obvious approach, but in the experiment shown in Fig. 5B, intracellular injections of calcium through a micropipette are interspersed with photolysis flashes to directly activate Cl^- currents and serve as a control.

Temporal Control and Kinetics of the Release Process

The photorelease of $InsP_3$ is rapid (a few milliseconds) following a photolysis flash, so that the kinetics of calcium liberation in response to a virtually instantaneous step of $InsP_3$ concentration can be studied. Furthermore, photolysis by a broad light spot results in a homogeneous release of $InsP_3$ throughout a volume of the cell, thus obviating problems of diffusion delays that arise in other techniques such as intracellular microinjection or extracellular applications of $InsP_3$ to permeabilized cells. An example is shown in Fig. 6, where a laser confocal microscope is used to monitor calcium fluorescence signals from a minute volume within the oocyte (ca. 1 fl) in response to $InsP_3$ photoreleased throughout a larger region of the cell surrounding the laser spot. Calcium signals following photolysis flashes of increasing strength begin with progressively shorter latencies and show increasing rates of rise (Fig. 6A). Because the fluorescence signal reflects accumulation of calcium in the cytosol, differentiation of the signal further provides a measure of the rate of calcium efflux (Fig. 6B), allowing a more detailed study of the dependence of the kinetics of opening of the release channels on the concentration of $InsP_3$.[9]

As is evident in Fig. 6B, calcium liberation following a photolysis flash terminates within a few hundred milliseconds, even though the levels of $InsP_3$ remain elevated for several seconds.[21] Thus, the $InsP_3R$ enters a refractory state from which its subsequent recovery can be investigated by applying paired photolysis flashes at varying intervals, in a manner analo-

---

detectable signals. $InsP_4$ (about 1 fmol) was injected into the oocyte when marked by the arrow. [Reproduced from I. Parker and I. Ivorra, *J. Physiol.* (*Lond.*) **433**, 207 (1991), with permission of the Physiological Society.] (B) Bath-applied caffeine inhibits $InsP_3$-evoked membrane current responses, but not the activation of the Ca^{2+}-dependent Cl^- current. Trace shows currents evoked by alternate stimulation by photolysis flashes (filled arrowheads) and intracellular injections of calcium (open arrowheads). Caffeine (5 mM) was bath-applied when marked by the bar. (C) Caffeine increases the threshold amount of $InsP_3$ required to evoke membrane currents. Graph shows amplitudes of currents evoked by photolysis flashes of varying durations in the absence (open symbols) and presence (filled symbols) of 2.5 mM caffeine. [B, C reproduced from I. Parker and I. Ivorra, *J. Physiol.* (*Lond.*) **433**, 229 (1991) with permission of the Physiological Society.]

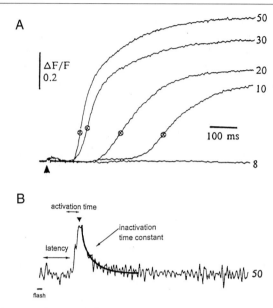

FIG. 6. Calcium transients and kinetics of calcium liberation in response to photorelease of varying amounts of InsP_3. (A) Superimposed fluorescent signals evoked by light flashes of varying durations (indicated in milliseconds at the right of each trace). The arrowhead indicates the time of the flash and circles indicate the point of maximal rate of rise of the signal. The oocyte was loaded with the low-affinity indicator calcium green-5N, together with c-InsP_3. (B) The time differential $[d(\Delta F/F)/dt]$ of the 50-msec trace from (A), indicating the kinetics of Ca^{2+} flux. [Modified from I. Parker, Y. Yao, and V. Ilyin, *Biophys. J.* **70**, 222 (1996), with permission of the Biophysical Society.]

gous to the paired-pulse experiments used to examine the refractory state of voltage-gated channels.[30,31] The onset and recovery from inactivation are determined by delivering paired, identical photolysis flashes at varying intervals and monitoring the calcium-activated currents evoked by the second flash in each pair relative to that evoked by a single flash (Fig. 7). At short intervals the calcium signal is potentiated by a preceding flash, but becomes almost completely suppressed as the interval is lengthened to about 2 sec, and subsequently recovers over several seconds.

Entry into and recovery from the refractory state undoubtedly have an important role in generation of the repetitive calcium spikes observed in various cells during sustained activation by extracellular agonists.[32] A characteristic feature is that the frequency of spiking increases with agonist

---

[30] I. Parker and I. Ivorra, *Proc. Natl. Acad. Sci. U.S.A.* **87**, 260 (1990).
[31] V. Ilyin and I. Parker, *J. Physiol. (Lond.)* **477**, 503 (1994).
[32] M. J. Berridge and A. Galione, *FASEB J.* **2**, 3074 (1988).

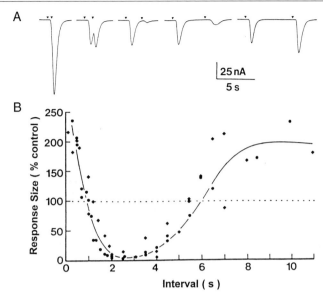

FIG. 7. Paired-flash experiment to investigate the onset and recovery of $InsP_3$-mediated calcium release. (A) Examples of membrane currents evoked by paired, identical flashes delivered at different intervals. Arrowheads mark the times of each flash. (B) Size of the response to the second flash plotted against interval between flashes. Data are shown from two oocytes (different symbols) and are scaled as a percentage of the response evoked by a single flash. [Reproduced from I. Parker and I. Ivorra, *Proc. Natl. Acad. Sci. U.S.A.* **87,** 260 (1990), with permission.]

concentration but, as noted before, interpretation of these findings is complicated by the complex stages in the messenger pathway between the cell surface receptor and the generation of $InsP_3$. An alternative approach is to use sustained illumination with low-intensity UV light to cause a prolonged elevation of intracellular [$InsP_3$] to levels proportional to the light intensity.[33] This results in repetitive spikes in the fluorescence calcium signal during the period of photolysis, which increase in frequency with increasing photorelease but become smaller and superimposed upon a more sustained calcium elevation (Fig. 8).

Spatial Control and Heterogeneity of Calcium Release Sites

The use of light as a stimulus to evoke photolysis of c-$InsP_3$ permits not only control of the magnitude and kinetics of $InsP_3$ formation, but also control of its spatial distribution. Thus, information can be obtained

---

[33] I. Parker and I. Ivorra, *J. Physiol. (Lond.)* **461,** 133 (1993).

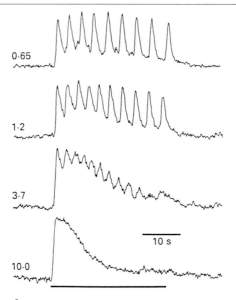

FIG. 8. Repetitive Ca^{2+} spikes during prolonged photorelease of $InsP_3$. Tracaes show point-confocal Ca^{2+} signals (rhod-2 fluorescence) evoked at a single recording spot by various intensities of photolysis light. Horizontal bar indicates the duration of exposure, and numbers next to each trace indicate the intensity of illumination as a percentage of maximum. [Reproduced from I. Parker and I. Ivorra, *J. Physiol.* (*Lond.*) **461,** 133 (1993), with permission of the Physiological Society.]

regarding the spatial aspects of subcellular calcium liberation, either by restricting photolysis to defined regions of the cell or by fluorescence imaging of the patterns of calcium liberation evoked by spatially homogeneous photolysis. The following examples illustrate use of these approaches to study calcium liberation at increasingly fine levels of resolution, from global signals involving the whole cell to elementary calcium events located to within a few microns.

Figure 9 illustrates an experiment in which the regional sensitivity to $InsP_3$ is mapped across the two hemispheres (animal and vegetal) of the polarized oocyte cell. The oocyte was loaded with caged $InsP_3$ and stimulated by identical light flashes focused as a 50×50-μm square. As the photolysis square is moved from the vegetal to the animal pole, the whole-cell Cl^- currents evoked by the flashes grow progressively larger (Fig. 9A), indicating increased sensitivity to $InsP_3$. Thus, even though the voltage-clamp recording summates membrane currents from the whole cell, spatial information is provided by the localized photorelease of $InsP_3$. One complication in such experiments, however, is that the extent of photorelease may

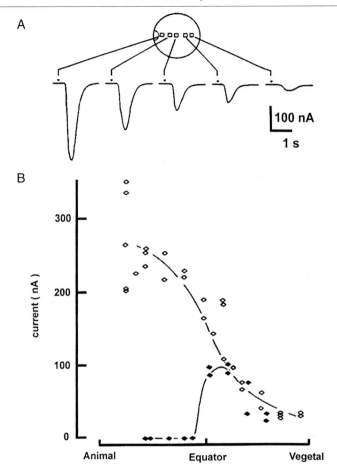

FIG. 9. Spatial variation in currents evoked by $InsP_3$ photoreleased at different locations across the animal/vegetal axis of the oocyte. (A) Traces show currents evoked by identical light flashes, with the photolysis light focused as a small square at different locations on the albino oocyte as denoted in the diagram. The oocyte was oriented with the germinal vesicle (animal pole) to the left. (B) Variation in the peak size of the current at different photolysis positions. Filled symbols are measurements from a pigmented oocyte and open symbols from an albino oocyte. Curves were drawn by eye.

be affected by optical factors in the cell. Wild-type oocytes are strongly pigmented in the animal hemisphere. In these cells, sensitivity to photolysis flashes is greatly reduced in the animal hemisphere (filled symbols, Fig. 9B), presumably because the superficial pigment blocks penetration of photolysis light into the cell. In the experiment illustrated (Fig. 9A and open symbols Fig. 9B), this problem was avoided by use of oocytes from albino frogs, which lack pigmentation.

To examine heterogeneity in sensitivity between different functional calcium release sites at a finer (micrometer) scale,[13] the UV light was focused to a spot of about 2 μm diameter by a pinhole aperture placed in the arc lamp photolysis system, and fluorescence calcium signals were monitored by a confocal laser spot concentric with the photolysis spot (Fig. 10). Calcium liberation at this localized level shows an all-or-none

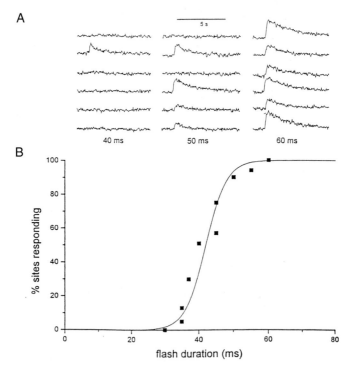

FIG. 10. Local pinhole photorelease of InsP$_3$ reveals little variation in sensitivity to InsP$_3$ at different sites. (A) Traces show Ca^{2+} fluorescence monitored from a stationary confocal spot in response to photorelease of InsP$_3$ evoked by flashes of UV light focused to a 2-μm spot concentric with the confocal spot. Each trace was obtained with the oocyte moved to a new random position, and the columns of traces show responses to three different flash durations as indicated. Note the all-or-none characteristics of the responses. (B) Percentage of sites responding to flashes of varying durations. Data were obtained from traces similar to those in (A). Points indicate the percentage of trials ($n > 20$) with each flash duration in which responses were observed. All responses were from a single albino oocyte, at randomly chosen locations, constrained to fall within a 200 × 200-μm area of the animal hemisphere. [Reproduced by I. Parker, I. Choi, and Y. Yao, *Cell Calcium* **20**, 105 (1996), with permission of *Cell Calcium*.]

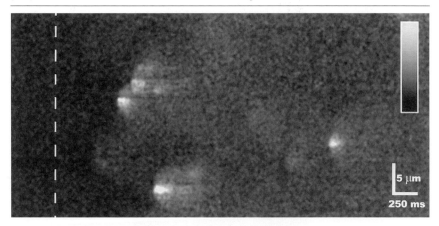

FIG. 11. Uniform photorelease of InsP_3 over a wide area evokes transient, localized, "elementary" calcium release events. The image was obtained by confocal line-scan microscopy, in which the laser spot was repeatedly scanned along a fixed line 50 μm long. Traces from successive lines are stacked from left to right so that the vertical dimension of the image represents distance along the scan line and the horizontal dimension of the image represents distance along the scan line and the horizontal dimension represents time. Increasing Ca^{2+} levels are depicted on a gray scale and are shown as a ratio relative to the resting fluorescence at each pixel before stimulation. A photolysis flash covering a spot approximately 100 μm in diameter around the scan line was applied when indicated by dashed line.

characteristic,[34] so that identical flashes delivered at different, random, locations on the oocyte evoke either no response or responses of similar amplitudes (Fig. 10A). The proportion of sites responding grows progressively as the flash duration is increased. However, this gradation occurs over a very narrow range of flash intensities, suggesting that the microscopic sensitivity to InsP_3 varies only slightly from site to site.

A different approach to obtain spatial information is to image calcium release evoked by homogeneous photolysis of c-InsP_3 over a relatively wide (50–100 μm) area, as illustrated in Fig. 11. In this case, fluorescence calcium images are obtained using a rapid (8.0 msec per line), high-resolution (0.2 μm/pixel) confocal line-scan system. Such studies reveal that homogeneous stimulation gives rise to localized "elementary" release events, originating in a stochastic manner at particular sites that probably represent clusters of InsP_3R.[13]

Finally, line-scan calcium imaging can be combined with localized photorelease induced by a focused spot from a UV laser to visualize calcium

---

[34] I. Parker and I. Ivorra, *Science* **250**, 977 (1990).

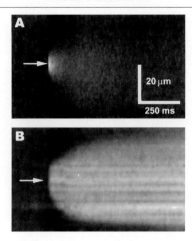

FIG. 12. Examples of UV laser "point" photolysis of caged compounds monitored by confocal line-scan imaging. (A) Control experiment monitoring fluorecsence of fluorecein photoreleased in a droplet of caged precursor. The UV laser flash was delivered at the location and time indicated by the tip of the arrowhead. (B) Ca^{2+} release evoked following photolysis of caged $InsP_3$ by a focused spot from a UV laser. The oocyte was loaded with Oregon green-1 (50 μM) and caged $InsP_3$ (5 μM), and recordings were made at a depth of 7 μm from the surface. The laser spot was positioned in the center of the line scan, and the tip of the arrowhead denotes the location and time of the flash. Line-scan imaging was performed as in Fig. 11. [Reproduced from I. Parker, N. Callamaras, and W. Wier, *Cell Calcium* **21,** 441 (1997), with permission of *Cell Calcium*.]

liberation as $InsP_3$ diffuses from a virtual point source. Figure 12A shows a control experiment in which fluorescein is photoreleased from a droplet of caged precursor (Molecular Probes). The image shows that fluorescein is formed almost immediately following the flash and subsequently spreads rapidly, consistent with diffusion in free solution. However, calcium liberation evoked by "point" photorelease of $InsP_3$ in the oocyte shows more complex characteristics (Fig. 12B). Calcium signals do not begin until after a latency of several tons of milliseconds and are first seen at the site of photorelease. Liberation then occurs at increasing distances from the photolysis spot, giving an indication of the "range of action" of $InsP_3$, reflecting both its diffusion and the requirement for $[InsP_3]$ to exceed a specific threshold to evoke calcium release.

Final Comments

In summary, c-$InsP_3$ is an invaluable tool for the study of intracellular Ca^{2+} signaling due to the degree of control that can be exercised over the intensity, duration, and spatial extent of stimulation. This technique has been used to extend whole cell studies of the InsP_3-mediated Ca^{2+} signaling system in *Xenopus* oocytes. Combination of flash photolysis together with fluorescence Ca^{2+} imaging or membrane current recording provides a means to study the transient kinetics and modulation of spatially complex, dynamic Ca^{2+} signaling events in an intact cell. Moreover, given the flexibility, simplicity, and relative affordability of the basic apparatus necessary to conduct these experiments, the approach should be well within the reach of many investigators.

Acknowledgments

We thank Dr. Jennifer Kahle for editorial help. Financial support was provided by NIH Grant GM48071. Reprint requests and requests for further information should be addressed to Dr. Ian Parker, Laboratory of Cellular and Molecular Neurobiology, Department of Psychobiology, University of California Irvine, CA 92697-4550; e-mail, iparker@uci.edu.

[22] Characterization and Application of Photogeneration of Calcium Mobilizers cADP-Ribose and Nicotinic Acid Adenine Dinucleotide Phosphate from Caged Analogs

By KYLE R. GEE and HON CHEUNG LEE

Introduction

Two independent mechanisms for the mobilization of internal calcium stores have been identified in sea urchin eggs. The calcium-mobilizing metabolites in these mechanisms are the novel nucleotides cyclic ADP-ribose (cADPR, **1**)[1] and nicotinic acid adenine dinucleotide phosphate (NAADP, **2**),[2] derived from NAD and NADP, respectively. Cyclic ADP-ribose can function as a modulator of the calcium-induced calcium response mechanism and it also functions as a calcium messenger itself.[3] In addition to the calcium-releasing activity of cADPR in sea urchin eggs, a variety of mammalian, amphibian, and plant cells have been shown to be responsive to the molecule.[4] The calcium release mechanism activated by NAADP is

[1] H. C. Lee, T. F. Walseth, G. T. Bratt, R. N. Hayes, and D. L. Clapper, *J. Biol. Chem.* **264,** 1608 (1989).
[2] H. C. Lee and R. Aarhus, *J. Biol. Chem.* **270,** 2152 (1995).
[3] H. C. Lee, *Recent Prog. Horm. Res.* **52,** 357 (1996).
[4] H. C. Lee, in "CRC Series on Pharmacology and Toxicology" (V. Sorrentino, ed.), p. 31, CRC Press, Boca Raton, FL, 1995.

very sensitive to low concentrations of the molecule, i.e., nanomolar levels provide sufficient activation. The calcium release mechanisms activated by these molecules are distinct from the one dependent on inositol 1,4,5-trisphosphate (IP_3), e.g., heparin, an antagonist of the IP_3 receptor, has no effect on cADPR or NAADP mechanisms. Additionally, the mechanisms activated by cADPR and NAADP are distinct from each other. The NAADP-dependent calcium release mechanism is not inhibited by high concentrations of magnesium[5] or the specific cADPR antagonist 8-amino-cADPR,[6] distinguishing it from the cADPR-dependent pathway. Also, NAADP and cADPR appear to be operating through distinct receptors; specific binding of [^{32}P]NAADP to sea urchin egg microsomes has been observed, and cADPR has no effect on this binding.[7] A novel property of the NAADP mechanism is that at a subthreshold concentration, NAADP can completely inactivate the release system such that a subsequent challenge with maximal concentrations of NAADP is ineffective.[7]

The cyclic structure of cADPR is formed by linking the N-1 nitrogen of the adenine ring in NAD^+ to the terminal ribose, displacing the nicotinamide group. This cyclization can be performed by treatment of NAD^+ with the ubiquitous enzyme ADP-ribosyl cyclase[1] or by heating solutions of NAD^+ with sodium bromide in dimethyl sulfoxide (DMSO).[8] Nicotinic acid adenine dinucleotide phosphate is formed by replacing the nicotinamide group of NADP with nicotinic acid. This can be accomplished by treatment of NADP with an aqueous base[2] or by ADP-ribosyl cyclase-catalyzed exchange of the nicotinamide in NADP with nicotinic acid.[9]

The use of photolabile, or "caged," precursors of cADPR and NAADP makes it possible to generate these signaling molecules with precise spatial and temporal resolution by UV photolysis when and where desired. This article describes the preparation of cADPR and NAADP caged as 1-(2-nitrophenyl)ethyl (NPE) phosphoesters, and characterization of their photolyses (see Scheme 1).

---

[5] R. Graeff, R. J. Podein, R. Aarhus, and H. C. Lee, *Biochem. Biophys. Res. Commun.* **206**, 786 (1995).
[6] T. F. Walseth and H. C. Lee, *Biochim. Biophys. Acta* **1178**, 235 (1993).
[7] R. Aarhus, D. M. Dicky, R. M. Graeff, K. R. Gee, T. F. Walseth, and H. C. Lee, *J. Biol. Chem.* **271**, 8513 (1996).
[8] Q.-M. Gu and C. I. Sih, *J. Am. Chem. Soc.* **116**, 7481 (1994).
[9] R. Aarhus, R. M. Graeff, D. M. Dickey, T. F. Walseth, and H. C. Lee, *J. Biol. Chem.* **270**, 30327 (1995).

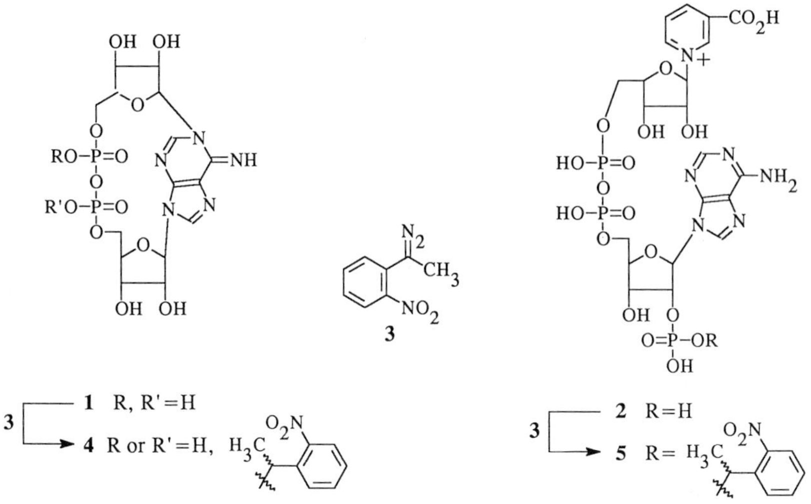

SCHEME 1

Synthesis

Materials

Cyclic ADP-ribose and NAADP are available commercially from Molecular Probes (Eugene, OR). Other reagents are used as received from various suppliers. Thin-layer chromatography (TLC) is performed on aluminum-backed silica gel plates impregnated with a fluorescent (254 nm) indicator, using the indicated solvent systems. Reversed-phase liquid chromatography is performed on Sephadex LH-20 using nanopure water as an eluant. Caging reactions and manipulations are performed under subdued light. 1-(2-Nitrophenyl)diazoethane (3) is prepared according to the method of Walker et al.[10]

Preparation of Caged cADPR

The structure of cADPR presents two obvious sites for attachment of a caging group, i.e., either phosphate of the bridging diphosphate moiety. In theory, a caging group could be attached to the exocyclic amino group of the adenine, but this would likely entail considerable synthetic difficulty.

[10] J. W. Walker, G. P. Reid, J. A. McCray, and D. R. Trentham, *J. Am. Chem. Soc.* **110**, 7170 (1988).

Simple and efficient caging of nucleotide phosphates such as ATP has been previously developed by Walker et al.,[10] and this was the first method used to cage cADPR. It was not known beforehand whether or how modification of the bridging diphosphate moiety would affect the biological activity of the resulting caged cADPR. Fortunately, the synthesis detailed here gives caged cADPR that is biologically inert and, on photolysis, generates cADPR that is active in mobilizing calcium in sea urchin egg microsomes and in live eggs. The procedure for the preparation is similar to that described previously.[11]

Purified cADPR in free acid form (48 mg, 0.092 mmol) is dissolved in 3 ml of ice-cold water. The solution should be acidic at about pH 2.3. To this stirring solution is added the caging reagent 1-(2-nitrophenyl)diazoethane (**3**, 0.28 mmol) in 3 ml of diethyl ether. It was found that low pH is necessary for protonation of the phosphates of cADPR, as deprotonated phosphates are not reactive with the caging reagent. It was also found that a different caging reagent, 1-(4,5-dimethoxy-2-nitrophenyl)diazoethane, at this low pH is unstable and gives a slower reaction with cADPR. The resulting biphasic mixture is vigorously stirred at 0–5° in the dark for 3 hr, during which time the diazoethane solution color changes from amber to pale yellow. The ether layer is drawn off, and the diazoethane/ether treatment is repeated three more times. The aqueous portion is applied to a Sephadex LH-20 column (2 × 20 cm) and eluted with water; 2-ml fractions are collected. The caged product **4** (TLC R_f 0.55, methanol/$CHCl_3$/H_2O/acetic acid, 12.5:10:3.5:0.2) is isolated as a fluffy white powder after lyophilization of the combined product fractions (30 mg, 49%). Unreacted cADPR, which elutes first, is also recovered (R_f 0.13, 15 mg, 31%). The caged product is efficiently photolyzed into free cADPR using a hand-held UV lamp as analyzed by TLC. The caged product represents a mixture of two monocaged isomers, which can be separated by AG MP-1 chromatography. For the unseparated product the molar extinction coefficient at 259 nm is 17,200 using 0.8 mM of the product dissolved in water. The two isomers have similar photolysis efficiency and both are biologically inactive until photolysis. Therefore, for biological applications, the isomers need not be separated.[11] Nuclear magnetic resonance (NMR) and elemental analysis data for caged cADPR indicate monocaging: ^{1}H NMR [$(CH_3)_2$ SO-d_6] δ 9.1-8.6 (m, 2H), 8.0-7.5 (m, 4H), 6.1-5.8 (m, 4H), 5.5 (m, 3H), 4.9 (m, 1H), 4.5-3.8 (m, 10H), 1.65 (m, 2H), 1.50 (dd, J = 18.4, 6.5 Hz, 1H). Anal. calcd for $C_{23}H_{28}N_6O_{15}P_2 \cdot H_2O$: C, 39.05; H, 4.27; N, 11.88. Found: C, 38.42; H, 3.94; N, 10.89. In vitro laser (308 nm) photolysis experiments have determined that the photolysis quantum yield for caged cADPR

---

[11] R. Aarhus, K. Gee, and H. C. Lee, *J. Biol. Chem.* **270**, 7745 (1995).

is 0.11 and that photolysis produces free cADPR with a time constant of 55 msec.[12]

Preparation of Caged NAADP

The synthesis of caged NAADP presents more challenges, as there are several potential sites for attachment of a caging group. Also, it was not known beforehand how modification of these various positions on the molecule would affect biological activity. NAADP is synthesized by incubating NADP (1 mM) at pH 5.0 with the *Aplysia* ADP-ribosyl cyclase (25 ng/ml) in the presence of 30 mM nicotinic acid for several hours at 20–23° and purified by high-performance liquid chromatography (HPLC) using an AG MP-1 column as described previously.[9] The procedure for preparing caged NAADP is similar to that described previously.[13]

NAADP has three phosphate groups and one carboxylate, all of which are potentially reactive with the caging reagent, provided they are protonated during the caging reaction. When the caging reaction is performed with the aqueous phase at pH 4.5, a relatively nonpolar product is formed (R_f 0.87, methanol:chloroform:water:acetic acid, 13:10:3.5:0.2) This product is biologically inactive and is photolyzed to NAADP very slowly. Reducing the pH of the caging reaction to 1.3 results in a caged product that is much more photolabile. NAADP (24 mg, 32 μmol) is dissolved in 2.5 ml of water, and the solution is adjusted to pH 1.3 by addition of dilute aqueous HCl. The diazoethane 3 (0.2 mmol) in 3 ml ether is added and stirred vigorously for 8 hr in darkness. The caging reagent solution is replaced twice and stirred each time for 3 hr. Thin-layer chromatography analysis indicates at this point no remaining NAADP (R_f 0.13) and two major products with R_f values of 0.87 and 0.46. The least polar product (R_f 0.87) coelutes on TLC with the higher pH product described earlier. After the aqueous layer is extracted with ether (2 × 3 ml), the more polar product (R_f 0.46) is separated on a column of Sephadex LH-20 (2 × 8 cm). A pale yellow powder of 8.6 mg (32% yield) of the caged product 5 is obtained, which is much more photolabile than the higher pH product described earlier. Photolysis with a hand-held UV lamp shows significant conversion to free NAADP, as judged by TLC analysis.

High-Performance Liquid Chromatography Analyses of Caged NAADP. HPLC separation of the reaction products is performed with columns packed with AG MP-1 resin (Bio-Rad, Richmond, CA) and eluted with a nonlinear gradient of trifluoroacetic acid. The chromatogram of the

---
[12] T. F. Walseth, R. Aarhus, M. E. Gurnack, L. Wong, H.-G. A. Breitinger, K. R. Gee, and H. C. Lee, *Methods Enzymol.* **280**, 294 (1997).
[13] H. C. Lee, R. Aarhus, K. R. Gee, and T. Kestner, *J. Biol. Chem.* **272**, 4172 (1997).

caged NAADP product mixture using the AG MP-1 column, as well as the gradient of TFA, is shown in Fig. 1 (top). Of the three components, the one indicated with an asterisk is the most effective caged form of NAADP, i.e., most responsive to photolysis. The final purification of the caged NAADP is achieved using a 0.5 × 5-cm Mono Q column (Pharmacia Biotech Inc.). The product is eluted using a gradient of water (solvent A) and 1 M triethylamine bicarbonate (solvent B, pH 8.8): 0–12 min, 0% B; linearly increased to 20% B from 12 to 16 min, linearly increased to 30% B from 16 to 36 min, linearly increased to 100% from 36 to 37 min, and held at 100% B for 3 min before returning to 0% B. A chromatogram

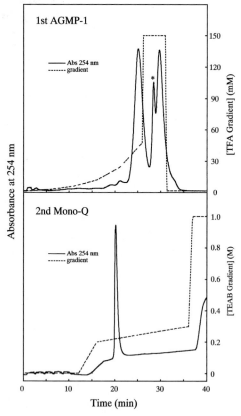

FIG. 1. Purification of caged NAADP (**5**) by HPLC. Sequential chromatographic steps on anion-exchange columns are used to purify a particular form of caged NAADP from a mixture of products generated by the caging reaction. The first step uses an AG MP-1 column (top). The fraction most easily photolyzed is indicated by the asterisk. Final purification is achieved using a Mono-Q column (bottom).

representing the isolation of caged NAADP using this gradient is shown in Fig. 1 (bottom).

Characterization of Caged NAADP. Proton-decoupled ^{31}P NMR spectra of NAADP and caged NAADP are obtained on samples dissolved in a D_2O solution buffered at pH 7.5 with 10 mM HEPES. A phosphoric acid solution in methanol is used as an external (zero) chemical shift standard. The spectrum of NAADP shows two signals in an integration of 2:1, with the upfield resonance at -12.2 ppm and the smaller, downfield signal (2.3 ppm). This downfield signal is absent in the spectrum of NAAD, indicating that the downfield signal in the spectrum of NAADP is due to the 2'-ribosyl phosphate and that the upfield signal originates from the two phosphorus atoms in the internucleotide diphosphate bridge. In the spectrum for caged NAADP, the chemical shift of the upfield signal is essentially unchanged, whereas that corresponding to the 2'-ribosyl phosphate is shifted upfield by about 5 ppm to -2.6 ppm, indicating that this phosphate is the one modified in caged NAADP, as opposed to the diphosphate bridge.

To further confirm the structural characterization, enzymatic degradation of caged and uncaged NAADP with alkaline phosphatase (AP) and nucleotide pyrophosphatase (NP) is performed using the malachite green assay[14] to quantify free phosphate released during reaction. Treatment of NAADP with AP generates ca. 1 mol of free phosphate, by cleavage of the 2'-ribosyl phosphate. Treatment of NAADP with NP generates no free phosphate, but subsequent treatment with AP results in the production of about 3 mol free phosphate (2 mol from the internucleotide diphosphate after cleavage by NP and 1 from the 2'-ribosyl phosphate). In contrast, treatment of caged NAADP with AP results generates no free phosphate. Reaction of caged NAADP with NP, followed by AP, results in the production of 2 mol free phosphate, indicating that the 2-ribosyl phosphate is modified (caged) and thus not available for enzymatic cleavage by AP.

Photolysis of Caged cADPR

Photoactivation of caged cADPR in sea urchin egg (*Strongylocentrotus purpuratus*) homogenates can be achieved in a spectrofluorimeter (Hitachi S-2000) using 350-nm excitation. Every 2 sec the excitation wavelength is alternated between 490 nm for monitoring Fluo-3 fluorescence and then back to 350 nm for further uncaging. Frozen egg homogenates are thawed at 17° for 20 min, then diluted to 5% with a medium containing 250 mM N-methylglucamine, 250 mM potassium gluconate, 20 mM HEPES, 1 mM

---

[14] P. A. Lanzetta, L. J. Alvarez, P. S. Reinach, and O. A. Candia, *Anal. Biochem.* **100**, 95 (1979).

MgCl$_2$, 2 units/ml creatine kinase, 8 mM phosphocreatine, 0.5 mM ATP, and 3 μM Fluo-3, pH 7.2, adjusted with acetic acid. The homogenates are diluted to 2.5% and finally to 1.25% with the medium just described and are incubated at 17° for 1 hr between dilutions. Calcium release is measured in the 1.25% homogenates using an emission wavelength of 535 nm. The measurements are performed in a cuvette maintained at 17°, and the homogenates are stirred continuously. The volume of homogenate is 0.2 ml. Figure 2 shows that adding caged cADPR to egg microsomes does not produce calcium release, but that uncaging with 350-nm light results in calcium mobilization after a brief delay. Addition of the specific cADPR antagonist 8-NH$_2$-cADPR stops the calcium release immediately, indicating that the calcium release is due to uncaged cADPR. Addition of the antagonist before photolysis completely abolishes calcium release during uncaging. A control using ATP caged with the same caging group shows no calcium mobilization.

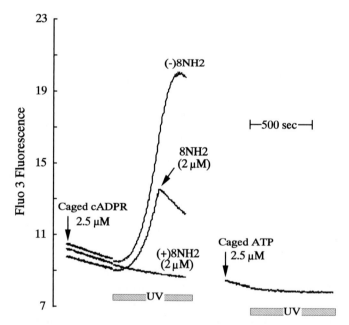

FIG. 2. Calcium release from egg homogenates induced by caged cADPR after photolysis. Photolysis of caged cADPR is performed by alternating the excitation wavelength of the spectrofluorimeter between UV (350 nm) and the monitoring wavelength of 490 nm. Calcium release from egg homogenates is monitored by Fluo-3 fluorescence in the absence [(−)8NH$_2$] or presence [(+)8NH$_2$] of 2 μM of 8-amino-cADPR, a specific antagonist of cADPR. Caged cADPR and caged ATP are added to the final concentrations indicated.

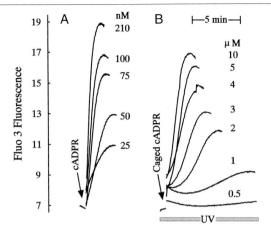

FIG. 3. Efficiency of photolysis of caged cADPR. The efficiency is assessed by comparing the Ca^{2+} release activity of caged cADPR under constant UV illumination with cADPR. Caged cADPR and cADPR are added at the final concentrations indicated. The time scale shown applies to (A) and (B).

The uncaging efficiency can be determined by calibrating the calcium release as a function of caged cADPR concentration versus that obtained using free cADPR. Figure 3 shows the response of egg homogenates to increasing concentrations of caged cADPR on UV photolysis compared to the response obtained using increasing concentrations of cADPR. In this series, uncaging 10 μM caged cADPR induces about the same calcium release as about 100 nM free cADPR, indicating a photolysis efficiency of about 1%; the low number is likely a function of the weak UV source present in the spectrofluorimeter, but it is sufficient for most applications.

Photolysis of intact sea urchin eggs containing caged cADPR also results in calcium release, monitored by Fluo-3, which is coinjected into the eggs with the caged reagent. *Lytechinus pictus* eggs are microinjected by pressure as described previously.[15] Samples of Fluo-3 and caged cADPR are dissolved in an injection buffer containing 0.5 M KCl, 50 μM EGTA, 10 mM HEPES, pH 6.7. Injection volumes are about 1.8–2.5% of the egg volume. For photolysis of caged cADPR in individual eggs, the epifluorescence attachment of a Nikon-inverted fluorescence microscope is modified. A second mercury lamp is attached at 90° to the epifluorescence tube. The light of 300–400 nm for photolysis is selected with a UGI filter (Omega) and reflected 90°, first with a 400 DCLP dichroic filter and then directed toward the objective with a second BCECF Sp dichroic filter. For monitor-

[15] H. C. Lee, R. Aarhus, and T. F. Walseth, *Science* **261,** 352 (1993).

ing Fluo-3 fluorescence, 490-nm light from the first mercury lamp is selected with a 485DF22 filter. The excitation light is passed through the same 400 DCLP dichroic filter and then reflected by the BCECF Sp dichroic filter toward the objective. The Fluo-3 fluorescence is selected by a long-pass filter with a 500-nm cutoff and monitored by a SIT camera. This optical arrangement allows simultaneous measurement of Fluo-3 fluorescence during uncaging photolysis. Figure 4 shows the expected calcium increase on photolysis of caged cADPR in intact eggs. After a slight delay, a threefold increase in Fluo-3 fluorescence is observed consistently in eggs injected

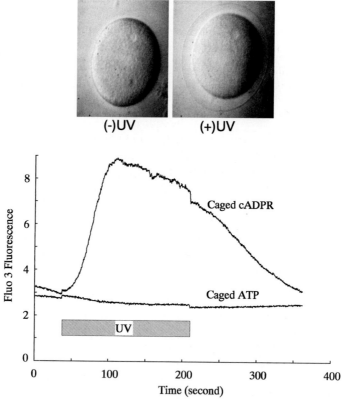

FIG. 4. Calcium release induced by photolysis of caged cADPR in live sea urchin eggs. Calcium release in individual eggs is monitored by Fluo-3 fluorescence. Caged cADPR (~4.9 μm, intracellular) or caged ATP (~72 μm, intracellular) is coinjected with Fluo-3 (~0.25 mM, intracellular) into an egg. The injection volumes are about 1.8–2.5% of the egg. Photolysis is induced by UV light at around 360 nm. The micrographs show an egg loaded with caged cADPR before and after UV exposure. A fertilization membrane surrounding the egg is formed after photolysis.

with 2.3 ± 0.7 μM caged cADPR. Also shown in Fig. 4 is the control experiment using caged ATP. In addition to inducing calcium changes, uncaging also induces the cortical reaction, an index of calcium mobilization. The micrographs in Fig. 4 show an egg injected with caged cADPR before and after UV exposure. After about 20–30 sec of UV exposure, a fertilization membrane is formed surrounding the whole egg. No eggs injected with caged ATP undergo the cortical reaction.

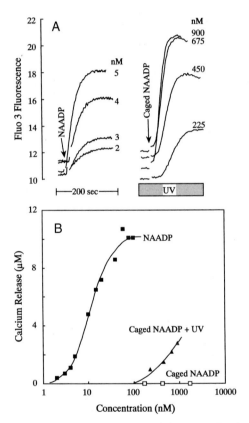

FIG. 5. Efficiency of photolysis of caged NAADP using a spectrofluorimeter. (A) Photolysis was performed by alternating the excitation wavelength of the spectrofluorimeter between 350 nm (UV) and the monitoring wavelength of 485 nm. Calcium release from egg homogenates is monitored by Fluo-3 fluorescence. The calcium release activity of 450 nM caged NAADP during UV is similar to that of 4–5 nM NAADP, indicating a photolysis efficiency of about 1%. (B) Concentration–response curves of NAADP, photolyzed caged NAADP (caged NAADP + UV), and caged NAADP.

Photolysis of Caged NAADP

As for caged cADPR, alternating the excitation wavelength in a spectrofluorimeter between 350 nm for photolysis and 485 nm for monitoring Fluo-3 fluorescence indicates that photolysis of caged NAADP mobilizes calcium in sea urchin egg homogenates. Figure 5 shows that addition of caged NAADP to the egg homogenates with UV excitation results in calcium release. Comparison of the calcium release activity with that induced by NAADP itself shows that about 1% of the caged NAADP is photolyzed. Figure 5B shows the comparison of the concentration response of NAADP and caged NAADP with or without UV photolysis. Because NAADP is effective in releasing calcium at nanomolar concentrations, the low efficiency of photolysis (most likely due to the low intensity in the spectrofluorimeter) does not limit its applications in cell biology.

A novel property of the NAADP-sensitive calcium release is that the release mechanism can be totally inactivated by pretreatment with a sub-

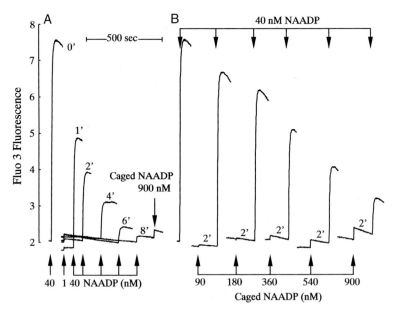

FIG. 6. Assay of the contamination level of NAADP in the caged NAADP preparation. The self-inactivation property of NAADP is used as an assay to estimate its contamination in the caged NAADP preparation. (A) Treatment of sea urchin egg homogenates with 1 nM NAADP for 8 min (A, 8′) effectively inactivates the Ca^{2+} release system such that the response to 40 nM NAADP is essentially eliminated. The extent of inactivation induced by 540 nM caged NAADP (B) is similar to that elicited by 1 nM NAADP (A, trace 2′), indicating that the level of contamination is about 0.1–0.2%. Ca^{2+} release is monitored by Fluo-3.

threshold concentration of NAADP as low as 1 nM.[7] Thus it is critical that purification of caged NAADP results in a concentration of contaminating free NAADP below the self-inactivating levels. Figure 6B shows that addition of purified caged NAADP from 90 to 900 nM to egg homogenates produces no calcium release; the small jumps in Fluo-3 fluorescence at high concentrations of caged NAADP are due to addition artifacts. Figure 6 also shows the results of testing the inactivating effect of caged NAADP. Egg homogenates are pretreated with 1 nM NAADP and subsequently challenged with a maximal concentration (40 nM) of NAADP (Fig. 6A). After 2 min of pretreatment, the response to 40 nM NAADP is reduced substantially (tracing labeled 2′) as compared to no pre-treatment (trace labeled 0′) and is totally eliminated after 8 min of pretreatment (trace labeled 8′). Figure 6B shows that pretreatment of the homogenates with 90 nM caged NAADP for 2 min produces very little inactivation. At 540 nM caged NAADP, the extent of inactivation following the pretreatment is similar to that affected by 1 nM of NAADP (comparing Fig. 6A, 2′, with Fig. 6B, 540 nM). This extent of inactivation by the caged compound can be accounted for if the sample is contaminated with 0.1–0.2% free NAADP. To ensure complete freedom from contaminating NAADP, samples of caged NAADP can be treated with AP immediately before use to remove any free NAADP present. Alkaline phosphatase effectively removes the 2′-phosphate of contaminating NAADP and converts it to NAAD, which is biologically inactive.[13] The caging group attached to the 2′-ribosyl phosphate in caged NAADP protects the caged form from being inactivated by AP.

[23] Applications of Caged Compounds of Hydrolysis-Resistant Analogs of cAMP and cGMP

By U. Benjamin Kaupp, Claudia Dzeja, Stephan Frings, Jürgen Bendig, and Volker Hagen

Cyclic nucleotides, i.e., cAMP and cGMP, control a variety of important cellular processes (for reviews, see Walter[1] and Lincoln and Cornwell[2]). The principal targets of cyclic nucleotides are cAMP- and cGMP-dependent kinases (cAK and cGK),[1,3] cAMP- and cGMP-regulated phosphodiester-

[1] U. Walter, *Rev. Physiol. Biochem. Pharmacol.* **113**, 41 (1989).
[2] T. M. Lincoln and T. L. Cornwell, *FASEB J.* **7**, 328 (1993).
[3] S. S. Taylor, J. A. Buechler, and W. Yonemoto, *Annu. Rev. Biochem.* **59**, 971 (1990).

ases (PDE),[4] cyclic nucleotide-gated (CNG) ion channels,[5-7] and transcription factors.[8] The study of cAMP- and cGMP-signaling pathways has been greatly facilitated by the design of chemical derivatives, dubbed "caged" compounds.[9,10] The popular term *caged* refers to molecules whose biological recognition or activity has been disabled by chemical modification. Photolysis cleaves the modifying group, uncaging and rapidly releasing active molecules. Caged cyclic nucleotides were first synthesized and physiologically tested by Korth and Engels[11] and then successfully employed in photolysis experiments to study a cAMP-dependent slow Ca^{2+} current in the heart.[12,13] Caged cAMP and caged cGMP have also been used to study the activation kinetics of the cGMP-gated channel from rod photoreceptor cells,[14] the signaling pathways in olfactory sensory neurons (OSNs),[15,16] the enhancement of Cl^- currents in ventricular cells,[17] and the motile response of fish melanophores.[18]

In the just-mentioned studies, caged compounds were used that released either cAMP or cGMP. This can limit the experimental application of these compounds as most cells are equipped with powerful PDE whose activity rapidly degrades cyclic nucleotides and thus prevents persistent cellular activation. Further, the low efficiency of photolysis (quantum yield) of many caged compounds limits the size of the concentration step that can be produced inside a cell by pulsed conventional light sources. For many physiological studies it would be desirable to use caged compounds that liberate cAMP or cGMP derivatives that are hydrolysis resistant or have higher biological efficiency. Derivatives that meet both these criteria are 8-Br-cAMP and 8-Br-cGMP. They are poorly hydrolyzable by phosphodies-

---

[4] J. A. Beavo and M. D. Houslay, in "Cyclic Nucleotide Phosphodiesterases: Structures, Regulation and Drug Action." Wiley, Chichester, 1990.
[5] K.-W. Yau and D. A. Baylor, *Annu. Rev. Neurosci.* **12,** 289 (1989).
[6] U. B. Kaupp, *Curr. Opin. Neurobiol.* **5,** 434 (1995).
[7] J. T. Finn, M. E. Grunwald, and K.-W. Yau, *Annu. Rev. Physiol.* **58,** 395 (1996).
[8] A. Kolb, S. Busby, H. Buc, S. Garges, and S. Adhya, *Annu. Rev. Biochem.* **62,** 749 (1993).
[9] S. R. Adams and R. Y. Tsien, *Annu. Rev. Physiol.* **55,** 755 (1993).
[10] J. E. T. Corrie and D. R. Trentham, in "Bioorganic Photochemistry" (H. Morrison, ed.) Vol. 21, p. 243. Wiley, New York, 1993.
[11] M. Korth and J. Engels, *Naunyn-Schmiedeberger's Arch. Pharmacol.* **310,** 103 (1979).
[12] J. Nargeot, J. M. Nerbonne, J. Engels, and H. A. Lester, *Proc. Natl. Acad. Sci. U.S.A.* **80,** 2395 (1983).
[13] J. M. Nerbonne, S. Richard, J. Nargeot, and H. A. Lester, *Nature* **310,** 74 (1984).
[14] J. W. Karpen, A. L. Zimmerman, L. Stryer, and D. A. Baylor, *Proc. Natl. Acad. Sci. U.S.A.* **85,** 1287 (1988).
[15] G. Lowe and G. H. Gold, *J. Physiol.* **462,** 175 (1993).
[16] T. Kurahashi and A. Menini, *Nature* **385,** 725 (1997).
[17] K. Ono, Y. Nakashima and T. Shioya, *Pflügers Arch.* **424,** 546 (1993).
[18] T. Furuta, H. Torigai, M. Sugimoto, and M. Iwamura, *J. Org. Chem.* **60,** 3953 (1995).

terases,[19] and 8-Br-cGMP activates cGK[20] and CNG channels[19,21,22] more effectively than cGMP.

We have described previously the synthesis and properties of 4,5-dimethoxy-2-nitrobenzyl (DMNB) derivatives of 8-Br-cAMP (in the following referred to as DMNB-caged 8-Br-cAMP) and of 8-Br-cGMP (DMNB-caged 8-Br-cGMP), which respectively release 8-Br-cAMP and 8-Br-cGMP after irradiation with ultraviolet light.[23] The following article discusses the properties of these compounds and their usefulness for electrophysiological studies of CNG channels *in situ* using heterologously expressed α subunits of CNG channels from bovine OSNs and cone photoreceptor cells.[24,25]

Design of Caged Derivatives of Hydrolysis-Resistant Cyclic Nucleotides

Cyclic nucleotides such as cAMP or cGMP can be rendered hydrolysis resistant via the addition of substituents to the heterocyclic purine base and caged via esterification at the phosphate moiety with a photolabile group. Carbon C-8 of the purine ring system is the position by which cyclic nucleotides can be modified without compromising their biological efficacy. It is likely that many C-8 substituted congeners are hydrolysis resistant, although this has been demonstrated experimentally for only a few derivatives. Some C-8 substituted derivatives are several orders of magnitude more effective in activating CNG channels than their respective natural agonist.[19,21,22,26,27] Specifically bulky and apolar substituents at C-8 render derivatives very potent agonist. For example, 8-fluoresceinylcarbamoylmethylthio-cGMP activates the rod CNG channel at almost 100-fold lower concentrations than cGMP.[26] In addition, apolar C-8 substituents facilitate permeation across biological membranes. One drawback of this type of modification in combination with apolar caging groups is that it results in

---

[19] A. L. Zimmerman, G. Yamanaka, F. Eckstein, D. A. Baylor, and L. Stryer, *Proc. Natl. Acad. Sci. U.S.A.* **82,** 8813 (1985).
[20] J. D. Corbin, D. Ogreid, J. P. Miller, R. H. Suva, B. Jastorff, and S. O. Døskeland, *J. Biol. Chem.* **261,** 1208 (1986).
[21] K.-W. Koch and U. B. Kaupp, *J. Biol. Chem.* **260,** 6788 (1985).
[22] J. C. Tanaka, J. F. Eccleston, and R. E. Furman, *Biochemistry* **28,** 2776 (1989).
[23] V. Hagen, C. Dzeja, S. Frings, J. Bendig, E. Krause, and U. B. Kaupp, *Biochemistry* **35,** 7762 (1996).
[24] J. Ludwig, T. Margalit, E. Eismann, D. Lancet, and U. B. Kaupp, *FEBS Lett.* **270,** 24 (1990).
[25] I. Weyand, M. Godde, S. Frings, J. Weiner, F. Müller, W. Altenhofen, H. Hatt, and U. B. Kaupp, *Nature* **368,** 859 (1994).
[26] A. Caretta, A. Cavaggioni, and R. T. Sorbi, *Eur. J. Biochem.* **153,** 49 (1985).
[27] R. L. Brown, R. J. Bert, F. E. Evans, and J. W. Karpen, *Biochemistry* **32,** 10089 (1993).

TABLE I
SOLUBILITY OF DMNB ESTERS OF 8-Br-cAMP AND 8-Br-cGMP[a]

Cyclic nucleotide ester	Solubility
DMNB-caged 8-Br-cAMP	
Axial isomer	120
Equatorial isomer	80
DMNB-caged 8-Br-cGMP	
Axial isomer	100
Equatorial isomer	10

[a] In 5% acetonitrile/0.01 M HEPES/KOH buffer, pH 7.2, containing 0.12 M KCl at room temperature. Saturation concentrations in μM 1 hr after dissolution.

reagents that are significantly less soluble (Table I). This disadvantage is offset by a similarly large increase in ligand affinity at most C-8 substituted derivatives.

The rather low quantum yield of the uncaging process limits the concentration step achieved by short flashes from pulsed conventional light sources, such as xenon flash lamps or mercury lamps. Therefore, full activation by a single flash is difficult to achieve. The much higher biological efficacy of C-8 substituted derivatives, in principle, would be also beneficial in this respect. For most purposes, however, it is the contamination by "free" cyclic nucleotides that sets a practical upper limit for the concentration of caged compound rather than its low solubility (see "Optimal Concentrations and Handling of Solutions").

Synthesis

The synthesis of DMNB esters of 8-Br-cAMP and 8-Br-cGMP (Fig. 1, see Hagen et al.[23]) is based on a procedure described for DMNB-cAMP and DMNB-cGMP,[13] but using 4,5-dimethoxy-2-nitrophenyldiazomethane prepared from the hydrazone of 4,5-dimethoxy-2-nitrobenzaldehyde according to Wootton and Trentham.[28]

Briefly, freshly prepared 4,5-dimethoxy-2-nitrophenyldiazomethane is coupled directly to the free acid of 8-Br-cAMP or 8-Br-cGMP by stirring suspensions of the reactants in dimethyl sulfoxide (DMSO) at room temper-

[28] J. F. Wootton and D. R. Trentham, in "Photochemical Probes in Biochemistry" (P. E. Nielsen, ed.), Vol. 272, p. 277. Kluver Academic, The Netherlands, 1989.

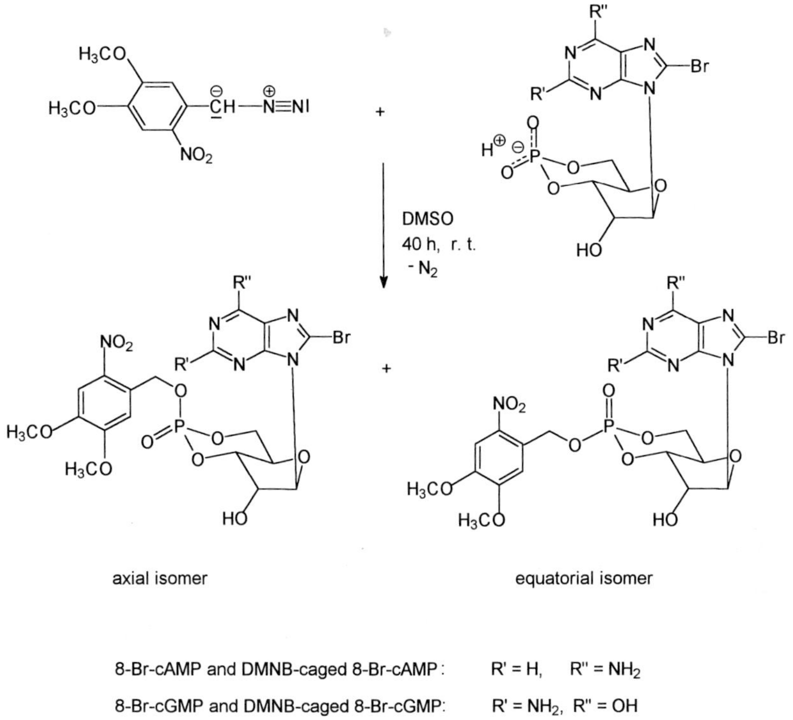

FIG. 1. Structures of diastereoisomers of DMNB-caged 8-Br-cAMP and DMNB-caged 8-Br-cGMP.

ature in the dark for 40 hr. Due to a center of asymmetry on phosphorus, all triesters occur as diastereomeric mixtures. The resulting pairs of diastereomers of the caged compounds are isolated and separated into the axial and equatorial forms (for structures, see Fig. 1) using preparative reversed-phase (RP) high-performance liquid chromatography (HPLC) (for details of the chromatographic procedure, see Hagen et al.[23]). For maximum stability the pure products are lyophilized (see later). Isomeric species are assigned by ^{31}P nuclear magnetic resonance (NMR) in analogy to the corresponding cAMP and cGMP esters.[13,29] The ^{31}P NMR signals at higher field (−5.19 for DMNB-caged 8-Br-cAMP and −5.17 for DMNB-caged 8-Br-cGMP) correspond to the axial isomers, and those at lower field (−3.37 for DMNB-caged 8-Br-cAMP and −4.27 for DMNB-caged 8-Br-cGMP)

[29] J. Engels, *Bioorg. Chem.* **8**, 9 (1979).

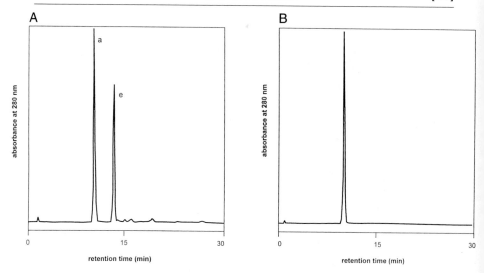

FIG. 2. Reversed-phase HPLC profiles of DMNB-caged 8-Br-cGMP. (A) Mixture of the axial (peak a, retention time 10.2 min) and the equatorial isomer (peak e, retention time 13.2 min) after flash chromatography. The retention time for 8-Br-cGMP was 1.4 min. (B) HPLC profile of the purified axial isomer after preparative HPLC and lyophilisation (PLRP-S column, 120 × 4.6 mm i.d., 8 m; linear gradient of 20–80% B in 40 min; eluent A, water, eluent B, acetonitrile; flow rate: 1 ml/min; detection at 280 nm). [Reprinted with permission from V. Hagen, C. Dzeja, S. Frings, J. Bendig, E. Krause, and U. B. Kaupp. Caged compounds of hydrolysis-resistant analogues of cAMP and cGMP: synthesis and application to cyclic nucleotide–gated channels. *Biochemistry* **35**, 7762–7771 (1996). Copyright (1996) American Chemical Society.].

correspond to the equatorial forms. In RP-HPLC profiles (Fig. 2) the axial isomers show shorter retention times compared to the equatorial isomers. Thin-layer chromatography (silica 60-F_{254}, chloroform/methanol 5:1, v/v) gives R_f values of 0.71 and 0.61 for the axial and the equatorial isomers of DMNB-caged 8-Br-cAMP, respectively, and 0.54 and 0.60 for the axial and the equatorial isomers of DMNB-caged 8-Br-cGMP. The synthesis shows slight (approximately 60%) preferential yield of the caged axial isomer, but all isomers are isolated in acceptable yields and high purities.

Purity

The purity of caged compounds in general, and in particular of hydrolysis-resistant cyclic nucleotides, is of critical importance. The following section discusses two aspects of reagent purity, namely separation of isomer mixtures into their respective diastereomeric components and contamination by the "free," i.e., uncaged, cyclic nucleotide.

On UV illumination the natural cyclic nucleotide is released from both the axial and the equatorial isomer of the caged compound; therefore, in principle, both isomers would be useful for cellular studies. Isomers differ in their solubility and solvolytic stability by at least one order of magnitude (see later); it is therefore advisable to employ pure diastereomers, then, for example, progression of solvolysis inside the cell during an experiment is reduced as well as being more predictable.

Contamination of caged compounds with free cyclic nucleotides is of particular importance with hydrolysis-resistant derivatives because the contaminant is not taken care of by endogenous PDE activity. Free cyclic nucleotide contamination results from two different sources: incomplete caging of the primary product and solvolysis of the triester in aqueous solution. Separation of isomers by RP-HPLC also removes free cyclic nucleotides. Thereby, reagents of 99.9% purity are routinely obtained. Practical hints that help avoid contamination by solvolysis are given later (see "Optimal Concentrations and Handling of Solutions").

Solubility

The successful use of caged compounds in cellular research requires adequate solubilities of the probes in aqueous solutions of moderately high ionic strength. Therefore, the diastereomers of DMNB-caged 8-Br-cAMP and 8-Br-cGMP are tested for their solubilities in 5% acetonitrile/0.01 M HEPES/KOH buffer, pH 7.2, containing 0.12 M KCl at room temperature using RP-HPLC.

When the isomers are dissolved first in pure acetonitrile and then diluted into the buffer, saturation concentrations are obtained as shown in Table I. The greater solubility of axial isomers in aqueous solution, particularly DMNB-caged 8-Br-cGMP, favors the use of pure axial forms in cellular physiology studies.

Photochemical Properties

The DMNB esters of 8-Br-cAMP and 8-Br-cGMP have an absorption peak around 350 nm, a wavelength, that is relatively harmless to cells. Irradiation at 300–400 nm of the axial and the equatorial isomers of DMNB-caged forms of 8-Br-cAMP and 8-Br-cGMP resulted in the liberation of 8-Br-cAMP or 8-Br-cGMP anions, 4,5-dimethoxy-2-nitrosobenzaldehyde, and a proton. The photolysis mechanism is well-known.[10,29,30,31] The effi-

[30] J. H. Kaplan, B. Forbush III, and J. F. Hoffman, *Biochemistry* **17**, 1929 (1978).
[31] J. P. Y. Kao and S. R. Adams, in "Optical Microscopy: Emerging Methods and Applications" (B. Herman and J. L. Lemasters, eds.), p. 27. Academic Press, San Diego, 1993.

ciency of photolysis may be evaluated by both molar extinction coefficients (ε) and quantum efficiencies (Φ) for the disappearance of phosphate esters. Table II summarizes ε values obtained at the long-wavelength absorption maximum and the Φ values of the axial and equatorial forms of DMNB-caged 8-Br-cAMP and DMNB-caged 8-Br-cGMP. Within experimental accuracy, the photochemical properties of the two diastereomers of DMNB-caged 8-Br-cAMP and of DMNB-caged 8-Br-cGMP were identical. The compounds have a relatively high absorbance at 300–400 nm but rather low quantum yields. The quantum yields of the diastereomers of DMNB-caged 8-Br-cAMP are 10-fold larger than of the diastereomers of DMNB-caged 8-Br-cGMP. The photochemical properties are similar to those of DMNB-caged cAMP and DMNB-caged cGMP.[28]

Solvolysis

An ideal caged compound must be hydrolytically stable. With respect to caged derivatives of 8-Br-cAMP and 8-Br-cGMP, hydrolytic stability is crucially important because "free" 8-Br-cAMP or 8-Br-cGMP is a particu-

TABLE II
PHOTOCHEMICAL PROPERTIES FOR DMNB ESTERS OF 8-Br-cAMP AND 8-Br-cGMP[a]

Cyclic nucleotide ester	Long wavelength absorption maximum (nm)	Quantum yield Φ[b]
DMNB-caged 8-Br-cAMP		
Axial isomer	345 (ε 5900)	0.050
Equatorial isomer	345 (ε 5900)	0.045
DMNB-caged 8-Br-cGMP		
Axial isomer	346 (ε 5800)	0.0049
Equatorial isomer	346 (ε 5800)	0.0052[c]

[a] In 5% acetonitrile/0.01 M HEPES/KOH buffer, pH 7.2, containing 0.12 M KCl. Φ values were determined by the relative method [from H. J. Kuhn, S. E. Braslavsky, and R. Schmidt, *Pure Appl. Chem.* **61**, 187 (1989)] using the standard E,E-1,4-diphenyl-1,3-butadiene [Reprinted with permission from V. Hagen, C. Dzeja, S. Frings, J. Bendig, E. Krause, and U. B. Kaupp, Caged compounds of hydrolysis-resistant analogues of cAMP and cGMP: synthesis and application to cyclic nucleotide–gated channels. *Biochemistry* **35**, 7762 (1996). Copyright (1996) American Chemical Society.].
[b] λ_{exc} 333 nm.
[c] In dioxane/0.01 M sodium phosphate buffer, pH 7.2.

larly powerful activator and furthermore their degradation by endogenous phosphodiesterases proceeds at least 500 times slower than that of natural cyclic nucleotides.[19]

In DMNB esters of 8-Br-cAMP and 8-Br-cGMP, as well as of cAMP and cGMP, the electron-releasing dimethoxy groups on the benzene ring increase the stability of the benzyl carbocation, which makes these compounds particularly susceptible to hydrolysis. Half-life values of the pseudo first-order hydrolysis of the diastereomers of DMNB esters of 8-Br-cAMP and 8-Br-cGMP in 5% acetonitrile/0.01 M HEPES/KOH buffer, pH 7.2, containing 0.12 M KCl, are given in Table III. Additionally, half-life values are listed for the DMNB-caged 8-Br-cGMP isomers in different aqueous buffer solutions.

The axial isomers of both caged compounds are sufficiently stable with respect to the time required for photochemical or electrophysiological experiments. In contrast, the respective equatorial isomers are hydrolyzed at pH 7.2 or 4.6 approximately 7–10 times faster than the axial forms. These differences make the axial form much more convenient for physiological experiments than the equatorial form or mixtures of the respective diastereomers. For example, after 0.5 hr in aqueous buffer, the axial form of DMNB-caged 8-Br-cGMP contains approximately 1% 8-Br-cGMP, whereas the equatorial form contains 6%.

The diastereomers of DMNB-caged cAMP and cGMP show similar stabilities in aqueous buffer solutions (V. Hagen, unpublished). In all experiments with these derivatives, we recommend using pure axial isomers.

TABLE III
CALCULATED HALF-LIFE ($t_{1/2}$) FOR DMNB ESTERS OF 8-Br-cAMP AND 8-Br-cGMP[a]

Cyclic nucleotide ester	Aqueous buffer	pH	$t_{1/2}$ (hr)
DMNB-caged 8-Br-cAMP			
Axial isomer	HEPES/KCl	7.2	60
Equatorial isomer	HEPES/KCl	7.2	8
DMNB-caged 8-Br-cGMP			
Axial isomer	HEPES/KCl	7.2	50
Axial isomer	Phosphate	7.2	50
Axial isomer	Phosphate	4.6	50
Equatorial isomer	HEPES/KCl	7.2	5
Equatorial isomer	Phosphate	7.2	6
Equatorial isomer	Phosphate	4.6	5

[a] In aqueous buffer/acetonitrile (95:5).

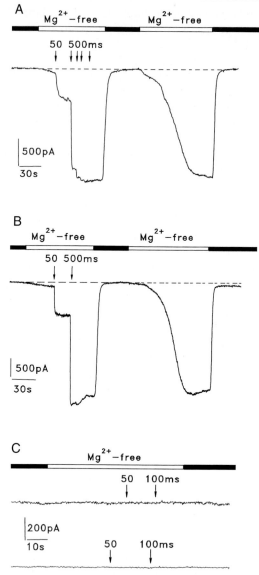

FIG. 3. Whole cell current responses recorded after photolysis of DMNB-caged 8-Br-cGMP (A) and DMNB-caged 8-Br-cAMP (B) from HEK 293 cells expressing the olfactory α subunit. (A) Current was activated by 50- and 500-msec flashes of UV light (arrows). Black and white bars indicate superfusion of the cell with Mg^{2+}-containing and Mg^{2+}-free solutions, respectively. Recording pipette contained 6.25 μM DMNB-caged 8-Br-cGMP. The membrane voltage was held at $V_m = -70$ mV. The initial Mg^{2+}-sensitive current was 106.5 pA and the maximal current was 2154 pA. This corresponds to a 8-Br-cGMP contamination of 41 nM or 0.66%.

"Optimal" Concentrations and Handling of Solutions

The following section discusses criteria to choose the optimal concentration of caged cyclic nucleotides. This concentration is determined by (1) ligand affinity, (2) the contamination by free cyclic nucleotides, and (3) the tolerable level of "background" activation. For example, the mean constant of half-maximal activation $K_{1/2}$ and the Hill coefficient, n (at -50 mV), are 0.13 ± 0.03 μM and 2.6, respectively, for the activation of the olfactory CNG channel α subunit by 8-Br-cGMP (see Fig. 5B). If 10% of the maximal current is considered tolerable as a background current, and if the caged compound is contaminated with 0.5% of free 8-Br-cGMP, then up to 11.2 μM of DMNB-caged 8-Br-cGMP can be used in the pipette. If experiments require a lower background current, either the DMNB-caged 8-Br-cGMP must be further purified or its concentration must be decreased accordingly. However, in order to produce the same incremental increase in Δ[8-Br-cGMP] at lower concentrations of the caged compounds a stronger (or longer) light flash is needed. Therefore, a practical lower limit of the concentration of caged cyclic nucleotides is set by the energy output of the actinic light source and possibly also by deleterious effects of strong UV illumination on the cell. The $K_{1/2}$ value of activation of the olfactory α subunit by 8-Br-cAMP is roughly 10-fold larger compared to 8-Br-cGMP; accordingly, 10-fold higher concentrations of DMNB-caged 8-Br-cAMP may be employed (Fig. 3B). In summary, an "optimal" concentration of caged compound is the highest concentration that can be employed, given a tolerable level of background activation by free cyclic nucleotides. This "optimal" concentration may be estimated from the $K_{1/2}$ value of channel activation and the level of contamination.

The solution may also become contaminated with free cyclic nucleotide during an experiment due to solvolysis in aqueous solution. Solvolysis can be kept at a minimum by dissolving small aliquots of lyophilized reagent in water-free DMSO. Final concentrations are adjusted by preparing aqueous dilutions of this stock solution shortly before the experiment.

(B) Flash-induced whole cell current responses with DMNB-caged 8-Br-cAMP. The recording pipette contained 100 μM DMNB-caged 8-Br-cAMP. All other conditions were as in (A), except $V_m = -50$ mV. (C) Control experiments with transfected HEK 293 cells and 100 μM caged ATP in the recording pipette (upper trace) and nontransfected HEK 293 cells with 6.25 μM DMNB-caged 8-Br-cGMP in the recording pipette (lower trace); $V_m = -50$ mV. [Reprinted with permission from V. Hagen, C. Dzeja, S. Frings, J. Bendig, E. Krause, and U. B. Kaupp. Caged compounds of hydrolysis-resistant analogues of cAMP and cGMP: synthesis and application to cyclic nucleotide–gated channels. *Biochemistry* **35,** 7762–7771 (1996). Copyright (1996) American Chemical Society.].

Cyclic Nucleotide-Gated Channel Activation by Photolysis of Caged Cyclic Nucleotides

DMNB-caged 8-Br-cGMP and DMNB-caged 8-Br-cAMP were tested in HEK 293 cells expressing the α subunit of the CNG channel from bovine OSNs.[24,32]

Figure 3A shows a whole cell recording of membrane current evoked by photolysis of DMNB-caged 8-Br-cGMP. The cell was initially bathed in a solution containing 10 mM Mg^{2+}, which effectively blocks currents activated by 8-Br-cGMP during equilibration of the cell with the caged compound.[32] Shortly before illumination by UV light, the perfusion was switched to a solution that contained 10 mM EGTA. In most experiments, a small current of several tens of picoampères developed after removal of Mg^{2+}. The small Mg^{2+}-sensitive currents are not due to the caged compound itself, but are completely accounted for by the activation of CNG channels with contaminating 8-Br-cGMP. No, or negligible, Mg^{2+}-sensitive currents were observed in nontransfected cells or in cells that have been equilibrated with caged ATP (Fig. 3C). This Mg^{2+}-sensitive current progressively increased during an experimental session due to solvolytic hydrolysis of DMNB-caged 8-Br-cGMP.

On irradiation of the cell with flashes of UV light (50 or 500 msec), large currents were recorded (Fig. 3A). The following control experiments indicate that these currents were activated by liberation of 8-Br-cGMP from the caged compound: (i) in the absence of the caged compound or in nontransfected HEK 293 cells, there was no current response to a light flash (Fig. 3C); (ii) the light-evoked current responses were saturable. A 500-msec flash activated ≥90% of the maximal current; subsequent flashes caused much smaller increments of current or no changes at all (see also Fig. 5A). (iii) The flash-induced current was almost entirely and reversibly suppressed by extracellular Mg^{2+} (Fig. 3A) and (iv) the nitrosobenzaldehyde photolysis product did not change the membrane conductance because photolysis of caged ATP (100 μM), which has the same caging group as DMNB-caged 8-Br-cGMP, was ineffective (Fig. 3C). We conclude that the light-evoked current was activated by 8-Br-cGMP produced by the photolysis of DMNB-caged 8-Br-cGMP.

Figure 4 shows a photolysis experiment using either NPE-caged cGMP or DMNB-caged 8-Br-cGMP in the pipette. Because the $K_{1/2}$ constant of the olfactory channel for cGMP is ≈12-fold larger than that for 8-Br-cGMP, the pipette solution contained 75 μM NPE-caged cGMP. The flash-evoked currents differ from each other in that the rapid rise produced by NPE-

---

[32] S. Frings, R. Seifert, M. Godde, and U. B. Kaupp, *Neuron* **15,** 169 (1995).

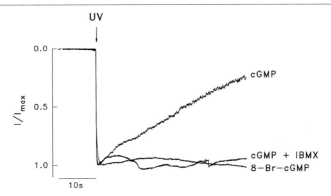

FIG. 4. Comparison of the time courses of flash-induced whole cell currents produced by either NPE-caged cGMP or DMNB-caged 8-Br-cGMP. Concentrations in the pipette were 75 μM for NPE-caged cGMP, 6.25 μM for DMNB-caged 8-Br-cGMP, and 0.3 mM for IBMX. Flash-induced currents measured from three different cells were normalized to the peak of the response shortly after the flash. Maximal whole cell currents produced by cGMP were −1.2 nA (−IBMX), −2.3 nA (+IBMX), and −2.1 nA produced by 8-Br-cGMP; V_m = −50 mV, 50-msec flashes. [Reprinted with permission from V. Hagen, C. Dzeja, S. Frings, J. Bendig, E. Krause, and U. B. Kaupp. Caged compounds of hydrolysis-resistant analogues of cAMP and cGMP: synthesis and application to cyclic nucleotide–gated channels. *Biochemistry* **35**, 7762–7771 (1996). Copyright (1996) American Chemical Society.].

caged cGMP is followed by a slower decay, whereas the current produced by DMNB-caged 8-Br-cGMP remains constant. The time of half-maximal decay of the cGMP-induced current was $\tau_{1/2} \approx 34.1 \pm 15.6$ sec (9 experiments), range 14.9–65.6 sec. The current decline was almost completely abolished by 0.3 mM of the PDE inhibitor 3-isobutyl-1-methyl-xanthine in the pipette solution. We interpret these results to indicate that cGMP released by light is hydrolyzed by endogenous PDE activity, whereas 8-Br-cGMP is not.

Calibration

The photolysis-induced step in cyclic nucleotide concentration generated inside the cell can be accurately calibrated *in situ*. It requires two different measurements. First, the dose–response relationship of the CNG channel must be determined in excised inside-out patches. Figure 5B shows dose–response relationships of 8-Br-cAMP and 8-Br-cGMP for the olfactory α subunit. Second, the ratio I_{flash}/I_{max} between the flash-induced current (I_{flash}) and the maximal current (I_{max}) at saturating concentrations of cyclic nucleotides must be determined. I_{max} can be determined either by illuminating cells with repetitive light flashes until whole cell currents saturate or by a single bright flash. From the I_{flash}/I_{max} ratio and the dose–response

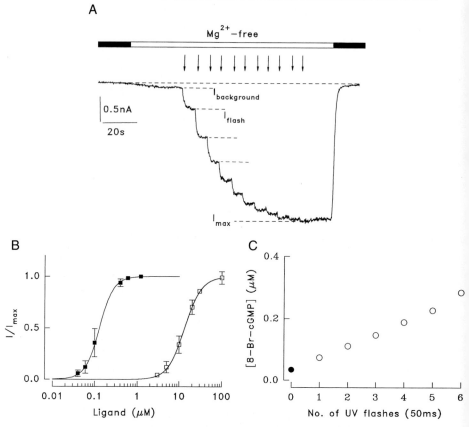

Fig. 5. Calibration of flash-induced changes in intracellular 8-Br-cGMP. (A) Current responses to successive flashes of UV light. Flash duration was 50 msec. The recording pipette contained 6.25 μM DMNB-caged 8-Br-cGMP. The initial concentration of uncaged 8-Br-cGMP in the pipette solution was 34 nM, corresponding to a contamination of 0.54%. The concentration after the fifth flash was 226 nM. The average flash-induced increment of the 8-Br-cGMP concentration was 38.4 ± 2.4 nM (5); $V_m = -50$ mV. (B) Dose–response relation of the heterologously expressed α subunit of the olfactory CNG channel for 8-Br-cGMP (■) and 8-Br-cAMP (□). Steady-state cyclic nucleotide-activated currents I measured from inside-out patches at −50 mV were normalized by the maximal current I_{max} at saturating concentrations of the respective nucleotide. Data were obtained from five experiments for 8-Br-cAMP and from nine experiments for 8-Br-cGMP. The curves represent a least-squares fit to the Hill equation. Mean values ± SD of the constant of half-maximal activation $K_{1/2}$ and the Hill coefficient n were 13.3 ± 1.7 μM and 2.2 ± 0.1 for 8-Br-cAMP and 0.13 ± 0.03 μM and 2.6 ± 0.1 for 8-Br-cGMP. (C) Increase of 8-Br-cGMP concentration with number of UV flashes, calculated from the dose–response relation of (A) and the ratios of $I_{h\nu}/I_{max}$ determined from the current recording of (B). [Reprinted with permission from V. Hagen, C. Dzeja, S. Frings, J. Bendig, E. Krause, and U. B. Kaupp. Caged compounds of hydrolysis-resistant analogues of cAMP and cGMP: synthesis and application to cyclic nucleotide–gated channels. *Biochemistry* **35**, 7762–7771 (1996). Copyright (1996) American Chemical Society.].

relationship, the concentration step per flash can be calculated. Figure 5A shows current recordings after repetitive irradiation with 50-msec flashes. The current amplitude increased incrementally with the number of flashes until it saturated after roughly 10 flashes. A concentration jump of 38.4 ± 2.4 nM 8-Br-cGMP/flash was calculated from the first five flashes. This experiment also demonstrates that the amount of cyclic nucleotide released per flash is roughly constant, i.e., the [8-Br-cGMP] increases linearly with the flash number (Fig. 5C). This conclusion is only valid when the incremental increase in cyclic nucleotide concentration is much smaller than the concentration of caged compound (i.e., Δ[8-Br-cGMP] ≪ [DMNB-caged 8-Br-cGMP]).

Instead of applying successive flashes, light flashes of longer duration may be used. Figure 6A shows such an experiment with flashes of 10, 50,

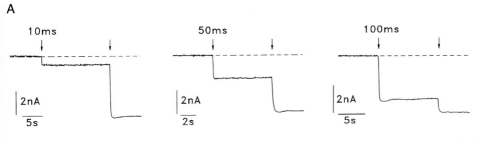

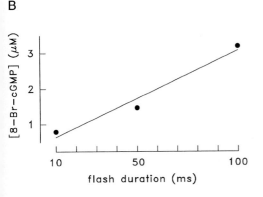

Fig. 6. Amplitude of flash-induced whole cell currents recorded from HEK 293 cells expressing the α subunit of bovine cone photoreceptors. (A) Currents evoked by flashes of 10 (left), 50 (middle), and 100 (right) msec duration. The pipette contained 15 μM of DMNB-caged 8-Br-cGMP; V_m = −50 mV. The saturating light flash was 1 sec in duration. (B) Incremental increase of 8-Br-cGMP concentration with flash duration, calculated from the dose–response relation and the ratios of I_{hv}/I_{max} (A).

and 100 msec duration. The incremental change in intracellular [8-Br-cGMP] is linearly related to the duration of the flash (Fig. 6B).

Poorly Hydrolyzable vs Hydrolyzable Analogs of Cyclic Nucleotides

The principal advantage of C-8 substituted derivatives of cyclic nucleotides is their hydrolysis-resistant nature. Caged derivatives of these compounds therefore allow persistent elevation of cyclic nucleotides *in situ*. The decay of the flash-induced concentration step on a time scale of several minutes is probably determined by slow dialysis of the cell with the pipette solution and diffusion of liberated C-8 substituted cyclic nucleotides across the cell membrane rather than by PDE hydrolysis. The hydrolysis-resistant nature along with the high biological efficacy of most C-8 substituted cyclic nucleotides require reagents of extremely high purity. For example, a contamination of DMNB-caged 8-Br-cGMP (100 μM) with 1% of the free (8-Br-substituted) cyclic nucleotide would result in the activation of $\geq$90% of the olfactory CNG channel in a typical transiently expressing HEK 293 cell.

In this respect, caged derivatives of the natural cyclic nucleotides are more convenient to use than their C-8 substituted congeners. Residual free cyclic nucleotides either become degraded by endogenous PDE activity while the cell is equilibrated with the caged compound or it can be removed beforehand by treatment of solutions with trypsin-activated PDE.[14] In addition, due to hydrolysis of the flash-generated cyclic nucleotide, the experiment can be repeated many times on the same cell as long as it contains sufficient amounts of caged compound. However, considering the modest, if not poor, purity of commercially available caged cyclic nucleotides, that advantage may be elusive because unintentional activation of the cellular process under study may occur before endogenous PDE had time to remove free cyclic nucleotide. For this reason, it is mandatory to check chromatographically the purity of each batch, even of hydrolyzable cyclic nucleotides.

Acknowledgments

We thank Dr. J. Bradley (Paris) for critical reading and A. Eckert for preparing the manuscript. This work was supported by a grant from the Deutsche Forschungsgemeinschaft (to S.F.). Claudia Dzeja is the recipient of a fellowship from the Studienstiftung des Deutschen Volkes. We are also grateful to the Fonds der Chemischen Industrie for financial support.

[24] Caged Probes for Studying Cellular Physiology: Application of o-Nitromandelyloxycarbonyl (Nmoc) Caging Method to Glutamate and a Ca^{2+}-ATPase Inhibitor

By Francis M. Rossi, Michael Margulis, Robert E. Hoesch, Cha-Min Tang, and Joseph P. Y. Kao

The release of bioactive molecules from inert, photosensitive precursors by flash photolysis is an increasingly useful technique in cellular experimentation.[1,2] The precursors, wherein some functional group that is essential for bioactivity has been chemically modified by a photolabile masking group to abolish activity temporarily, are often referred to as "caged" molecules.[3] This article focuses on a newly developed caging group, two caged molecules based thereon, as well as biological applications of these two new caged reagents.

o-Nitromandelyloxycarbonyl (Nmoc) Caging Group

The newly introduced photosensitive o-nitromandelyloxycarbonyl (Nmoc) caging group offers several advantages. First, flash photolysis of Nmoc-caged species releases bioeffector molecules on the millisecond time scale, with quantum yields of ~0.1. Second, the Nmoc cage is synthesized with relative ease. Third, Nmoc-imidazole (Fig. 1), a chemically activated form of the Nmoc cage, reacts smoothly with hydroxyl and amino groups to form, respectively, carbonate and carbamate linkages that are highly resistant to spontaneous cleavage in the absence of light. These characteristics combine to make Nmoc an attractive caging group for biomolecules bearing hydroxyl or amino groups.

The active caging reagent, Nmoc-imidazole, is prepared in two versions: the methyl and *tert*-butyl ester forms [Fig. 1, R = CH_3 and $C(CH_3)_3$, respectively]. For caging hydroxyl groups in biomolecules, the methyl ester form of Nmoc-imidazole is quite satisfactory. This form readily reacts with hydroxyl groups to form stable carbonate ester linkages that remain intact while the methyl group on the cage is removed by base hydrolysis to liberate

---
[1] S. R. Adams and R. Y. Tsien, *Ann. Rev. Physiol.* **55**, 755 (1993).
[2] J. P. Y. Kao and S. R. Adams, *in* "Optical Microscopy: Emerging Methods and Applications" (B. Herman and J. J. Lemasters, eds.), p. 27. Academic Press, San Diego, 1993.
[3] J. H. Kaplan, B. Forbush III, and J. F. Hoffman, *Biochemistry* **17**, 1929 (1978).

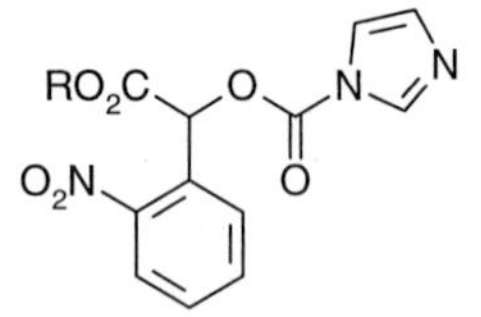

1-Nmoc-imidazole
R = methyl, or *tert*-butyl

FIG. 1. Structure of Nmoc-imidazole.

the final caged molecule. Nmoc-DBHQ,[4] a reagent for photoreleasing DBHQ, an inhibitor of SR/ER Ca^{2+}-ATPase,[5–7] was prepared using methyl Nmoc-imidazole.[8] For caging amino groups, however, the *tert*-butyl ester form of Nmoc-imidazole should be used. When the methyl ester form of Nmoc-imidazole is used for caging primary amino groups in biomolecules (e.g., the α-amino group in amino acids), the major product is not the desired caged molecule, but rather a cyclic oxazolidinedione (Fig. 2). This undesirable side reaction can be suppressed by replacing the methyl ester in Nmoc-imidazole with the *tert*-butyl ester. The *tert*-butyl form of Nmoc-imidazole readily couples with free amino groups to form stable carbamate linkages that are unaffected by the acidic conditions required for removing the *tert*-butyl group to generate the final caged molecule. Nmoc-Glu, a caged version of the neurotransmitter glutamate, was prepared with *tert*-butyl Nmoc-imidazole.

Syntheses

Synthetic procedures for methyl Nmoc-imidazole, Nmoc-DBHQ, and the membrane-permeant acetoxymethyl (AM) ester of Nmoc-DBHQ have been published.[8] Preparative procedures for *tert*-butyl-Nmoc-imidazole and Nmoc-Glu are outlined in Fig. 3 and are described below.

---

[4] APV, DL-2-amino-5-phosphonovaleric acid; CNB, α-carboxy-2-nitrobenzyl; DBHQ, 2,5-di(*tert*-butyl)hydroquinone; EGTA, ethylene glycol bis(2-aminoethyl)ether-*N,N,N',N'*-tetraacetic acid; GluR, glutamate receptor; HEPES, *N*-[2-hydroxyethyl]piperazine-*N'*-[2-ethanesulfonic acid]; NMDA, *N*-methyl-D-aspartate; NMR, nuclear magnetic resonance; NPE, 1-(2-nitrophenyl)ethyl; PIPES, piperazine-*N,N'*-bis(2-ethanesulfonic acid); SERCA, sarcoplasmic/endoplasmic reticulum Ca^{2+}-ATPase; TTX, tetrodotoxin.

[5] G. A. Moore, D. J. McConkey, G. E. Kass, P. J. O'Brien, and S. Orrenius, *FEBS Lett.* **224**, 331 (1987).

[6] G. E. Kass, S. K. Duddy, G. A. Moore, and S. Orrenius, *J. Biol. Chem.* **264**, 15192 (1989).

[7] D. Thomas and M. R. Hanley, *Methods Cell Biol.* **40**, 65 (1994).

[8] F. M. Rossi and J. P. Y. Kao, *J. Biol. Chem.* **272**, 3266 (1997).

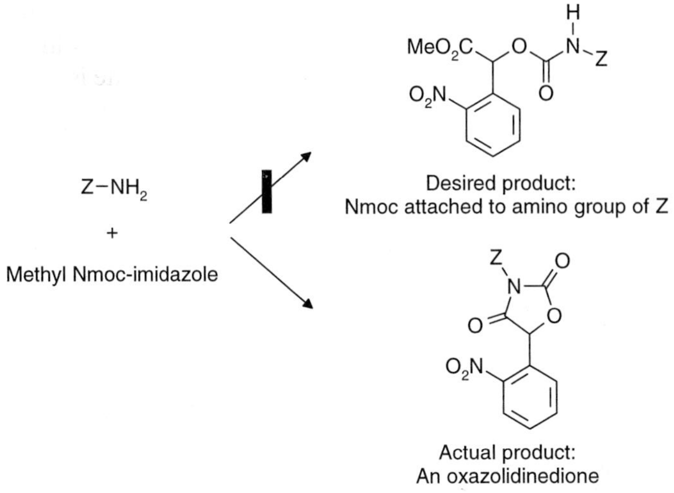

FIG. 2. Side-product formation in reaction of methyl Nmoc-imidazole with amines.

tert-Butyl-2-acetoxy-2-(2-nitrophenyl)acetate (2)

Step a of Fig. 3: 40 mmol (7.88 g) o-nitromandelic acid (**1**)[8] is refluxed for 45 min in 60 ml acetic anhydride. After the reaction mixture cools to room temperature, it is diluted with 50 ml each of tetrahydrofuran and water. The mixture is stirred for 2 hr. The organic layer is separated from the aqueous layer, diluted with 100 ml toluene, and extracted twice with 100 ml water. The organic phase is dried over $MgSO_4$ and evaporated to yield a brown oil that is directly used in step b of Fig. 3.

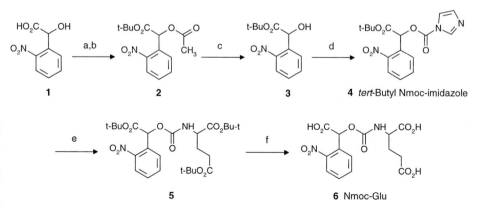

FIG. 3. Scheme outlining syntheses of *tert*-butyl Nmoc-imidazole and Nmoc-Glu.

Step b of Fig. 3: 40 mmol (8.74 g) *tert*-butyl trichloroacetimidate is added to a solution of the product of step a in 50 ml benzene. After 1 hr of stirring, the reaction mixture is filtered to remove solids. The filtrate is concentrated on a rotary evaporator and then chromatographed with ethyl acetate/hexane (1:5) to yield 7.41 g (63%) of compound **2** as an oil. ^{1}H NMR: 8.02 (d, J = 8.1 Hz, 1H), 7.65 (d, J = 3.9 Hz, 1H), 7.56–7.51 (m, 2H), 6.75 (s, 1H), 2.21 (s, 3H), 1.40 (s, 9H).

tert-Butyl-2-hydroxy-2-(2-nitrophenyl)acetate (3)

Step c of Fig. 3: To a stirred solution of compound **2** (6.63 g, 22.4 mmol) in 50 ml methanol, 1.12 mmol Cs_2CO_3 (0.365 g) is added. After 1 hr, the reaction mixture is diluted with 100 ml ethyl acetate and filtered through a plug of 15 g silica gel. Concentration of the filtrate on a rotary evaporator yields 5.36 g (94%) of compound **3** as an oil. ^{1}H NMR: 7.96 (dd, J = 1.2, 6.8 Hz, 1H), 7.72–7.60 (m, 2H), 7.51–7.45 (m, 1H), 5.84 (d, J = 3.7 Hz, 1H), 3.67 (br d, J = 4.4 Hz, 1H), 1.38 (s, 9H).

*tert-Butyl-Nmoc-imidazole [tert-butyl-2-(2-nitrophenyl)-2-
(oxycarbonylimidazole)acetate] (4)*

Step d of Fig. 3: Compound **3** (8.00 mmol, 2.02 g) and carbonyldiimidazole (8.00 mmol, 1.30 g) are dissolved in 25 ml dichloromethane. After 1 hr of stirring, the reaction mixture is extracted three times with 25 ml water, dried over $MgSO_4$, concentrated by rotary evaporation, and chromatographed with ethyl acetate/hexane (2:3) to yield 2.15 g (77%) of compound **4** as an oil. ^{1}H NMR: 8.20 (s, 1H), 8.11 (dd, J = 1.3, 6.6 Hz, 1H), 7.73–7.63, (m, 3H), 7.48 (s, 1H), 7.11 (s, 1H), 6.88 (s, 1H), 1.43 (s, 9H).

*N-[tert-Butyl-2-[2-nitrophenyl]-2-oxycarbonyl acetate]-L-glutamic Acid,
Di-tert-butyl Ester (5)*

Step e in Fig. 3: L-Glutamic acid di-*tert*-butyl ester hydrochloride (2.49 mmol, 0.736 g) and compound **4** (2.49 mmol, 0.866 g) are dissolved in 10 ml dichloromethane. After 0.35 ml triethylamine is added, the reaction mixture is stirred for 40 hr. The residue obtained after the solvent evaporation is chromatographed with ethyl acetate/hexane (1:5) to yield 0.858 (64%) of compound **5**. ^{1}H NMR: 8.00, (d, J = 7.82 Hz, 1H), 7.65–7.62 (m, 2H), 7.53–7.47 (m, 1H), 6.72 (s, 0.5H), 6.68 (s, 0.5H), 5.64–5.58 (m, 1H), 4.29–4.21 (m, 1H), 2.32–2.08 (m, 4H), 1.93–1.87 (m, 2H), 1.57 (s, 4.5H), 1.54 (s, 4.5H), 1.47 (s, 4.5H), 1.44 (s, 4.5H), 1.43 (s, 4.5H), 1.42 (s, 4.5H).

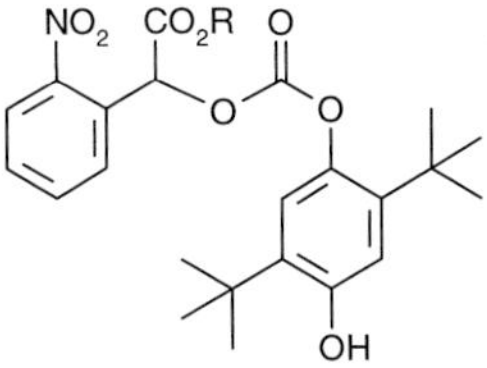

Nmoc-DBHQ: R = H
Nmoc-DBHQ/AM: R = CH$_2$O$_2$CCH$_3$

FIG. 4. Structures of Nmoc-DBHQ and the corresponding acetoxymethyl (AM) ester.

Nmoc-Glu (N-[2-[2-nitrophenyl]-2-oxycarbonyl acetic acid]-L-glutamic acid) **(6)**

Step f of Fig. 3: To a solution of 5 ml trifluoroacetic acid in 5 ml dichloromethane is added 0.882 mmol (0.475 g) of compound **5**. After 2 hr, the solution is concentrated by rotary evaporation. After azeotropic removal of residual acid with several small volumes of benzene, the yield of Nmoc-Glu (compound **6**) is 0.32 g (98%), in the form of a colorless glass. This material is dissolved directly in physiologic buffers for biological experiments. High-resolution mass spectrometry (electron ionization): calculated for C$_{14}$H$_{14}$N$_2$O$_{10}$[M$^+$] m/z = 370.0649, observed 370.0635.

Nmoc-DBHQ: Photomodulation of SR/ER Ca^{2+}-ATPase (SERCA) Activity

The structures of Nmoc-DBHQ and the corresponding AM ester are shown in Fig. 4. Because the AM ester can be passively loaded into cells by incubation, it is the most convenient form of the reagent for typical cellular experiments. Loading is accomplished by incubating cells with 10–40 μM of the AM ester in aqueous medium for 30–60 min at room temperature. In order to monitor the effects of DBHQ photorelease on intracellular Ca^{2+} dynamics, it is often desirable to load cells with a fluorescent indicator such as Fluo-3. For such applications, a few micromolar of the AM ester of the indicator can also be present in the loading medium. A small amount of the surfactant Pluronic F-127 (final concentration <0.02% w/v) may be used to aid dispersion of the hydrophobic AM esters into aqueous medium.[9]

---

[9] J. P. Y. Kao, *Methods Cell Biol.* **40,** 155 (1994).

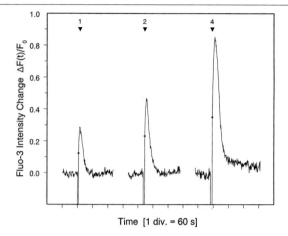

FIG. 5. Effect of DBHQ photorelease on $[Ca^{2+}]_i$. A REF52 rat embryo fibroblast was loaded with Nmoc-DBHQ and Fluo-3 Ca^{2+} indicator by incubation with 40 and 5 μM of the corresponding AM esters in DMEM at room temperature for 60 min. The experiment was conducted in HBSS. Indicator fluorescence data are presented as relative changes in intensity, $\Delta F(t)/F_0$, where $\Delta F(t)$ is the change of intensity at time t relative to the average initial intensity, F_0, which is measured with the cell at rest. Arrowheads and associated numbers indicate the times and durations (sec), respectively, of photolysis light flashes. Corresponding dips in the traces result from temporary interruptions of fluorescence data acquisition by an electromechanical shutter to avoid potential damage to the fluorescence photomultiplier tube that might be caused by leakage from the high-intensity photolysis flash.

Figure 5 illustrates the effects of transiently and reversibly inhibiting SERCA pumps by photoreleasing DBHQ in rat fibroblasts loaded with Nmoc-DBHQ. Changes in $[Ca^{2+}]_i$ are monitored with the Ca^{2+} indicator Fluo-3. Photolysis is performed as previously described.[8] A stable resting $[Ca^{2+}]_i$ is the result of a dynamic balance between active pumping processes that remove Ca^{2+} from the cytosol and passive leakage of Ca^{2+} into the cytosol. Photoreleased DBHQ, which blocks the SERCA Ca^{2+} pumps and thus disrupts the pump–leak balance, causes transient, dose-dependent rises of $[Ca^{2+}]$ in the cytosol. The $[Ca^{2+}]_i$ rise is transient because SERCA inhibition by DBHQ is reversible and because DBHQ is an uncharged, membrane-permeant molecule that easily diffuses out of the cell. As soon as photoreleased DBHQ is cleared from the cell by diffusion into the extracellular medium, the SERCA pumps are no longer inhibited, and the resting $[Ca^{2+}]_i$ becomes reestablished.

Because DBHQ is fairly water soluble and is membrane permeant, one might wonder whether delivery of DBHQ by rapid-flow superfusion would be as efficacious as by photorelease. A comparison of the two delivery

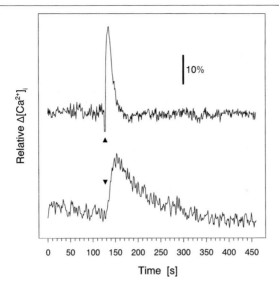

FIG. 6. Comparison of the kinetics of changes in [Ca^{2+}]$_i$ in response to DBHQ delivered by photorelease and by superfusion. (Upper trace) Response of REF52 cells to 1-sec photorelease of DBHQ (marked by upward arrowhead). Loading and experimental conditions are exactly as in Fig. 5. (Lower trace) Response of REF52 cell to 1-sec pulse of 60 μM DBHQ delivered by superfusion (marked by downward arrowhead). Cell was loaded with Fura-2 indicator by incubation with 1 μM of the AM ester in DMEM for 60 min at room temperature. Experiment was conducted in HBSS. Results are presented as changes of [Ca^{2+}]$_i$ (Δ[Ca^{2+}]$_i$) relative to resting [Ca^{2+}]$_i$ (typically 70–90 nM, as determined by separate calibrations). Rise and decay of the response to DBHQ photorelease are clearly much faster than for delivery by superfusion. Time constants ($t_{1/e}$) for rise and decay are 1.9 ± 0.1 and 9.7 ± 0.3 sec, respectively, for photorelease and 17 ± 5 and 82 ± 4 sec, respectively, for superfusion.

techniques is shown in Fig. 6, where cellular responses to 1-sec DBHQ application by photorelease and by superfusion are presented. Data in Fig. 6 show that both the rise and the decay of the response to photorelease are much faster than for superfusion delivery. More quantitatively, the rising phases of the photorelease and superfusion responses are characterized by time constants of 1.9 and 17 sec, respectively; the decay phases have corresponding times constants of 9.7 and 82 sec, respectively. These results demonstrate that even for molecules that are quite water soluble and membrane permeant, photorelease still affords much better temporal and kinetic control in cellular experimentation than conventional flow techniques.

Figure 7 shows that DBHQ photorelease can be used to probe dynamic cellular phenomena, such as [Ca^{2+}]$_i$ oscillations. In REF52 cells (rat embryo fibroblasts), activation of the inositol trisphosphate signaling pathway coupled with depolarization-induced Ca^{2+} influx causes persistent and regular

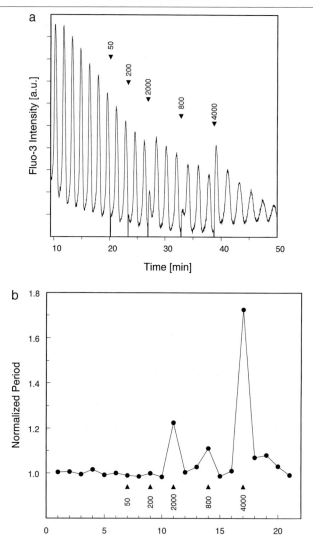

FIG. 7. Effect of transient SERCA Ca^{2+} pump inhibition on $[Ca^{2+}]_i$ oscillations in rat embryo fibroblasts. (a) $[Ca^{2+}]_i$ oscillations induced in a REF52 cell by application of 500 nM each of R^8-vasopressin, which activates $Ins(1,4,5)P_3$ production, and gramicidin D, an ionophore that depolarizes the cell by facilitating Na^+ influx. Arrowheads and associated numbers mark the times and durations (msec) of photolytic light flashes. Cell loading and experimental conditions are exactly as in Fig. 5. (b) Data from (a) analyzed to reveal alterations in the period between oscillation peaks caused by photoreleased pulses of DBHQ. Periods are normalized to the unperturbed period length. Arrowheads mark the oscillation cycles during which photorelease occurred; the corresponding numbers indicate the duration of the light flash used.

oscillations in $[Ca^{2+}]_i$.[10] The mechanism underlying these oscillations requires positive cooperativity between $[Ca^{2+}]_i$ rises and $Ins(1,4,5)P_3$ production.[11,12] The period between oscillation spikes is dependent on the ability of intracellular Ca^{2+} stores to actively sequester Ca^{2+} from the cytosol.[13] Nmoc-DBHQ makes it possible to study the effect of attenuating the ability of intracellular Ca^{2+} stores to sequester Ca^{2+}. The behavior of $[Ca^{2+}]_i$ oscillations in response to photomodulation of SERCA activity is shown in Fig. 7a. The results are presented more quantitatively in Fig. 7b, which shows that photoreleased DBHQ causes a dose-dependent lengthening of the period of that oscillation cycle during which photorelease occurred. This type of period lengthening is just what is expected from modeling studies.[13] Such an experiment thus provides a direct test of the role of SERCA pumps in the $[Ca^{2+}]_i$ oscillation mechanism.

Nmoc-Glu: Caged Glutamate for Studying Cellular Neurophysiology

Nmoc-Glu is a new caged glutamate that photoreleases free glutamate on the one-to-few millisecond time scale, at a quantum yield of 0.11. The principal advantage of Nmoc-Glu over γ-*O*-CNB-Glu,[14] a commonly used caged glutamate with rapid photorelease kinetics, is resistance to spontaneous hydrolysis under physiological conditions. Hydrolytic instability is detrimental for two reasons. First, non-NMDA glutamate receptors (GluRs) require more than 1 m*M* glutamate to activate fully and yet are significantly desensitized by less than 10 μ*M* glutamate.[15,16] Second, micromolar levels of glutamate are sufficient to activate NMDA GluRs.[17] Thus, achieving full activation of non-NMDA GluRs by photolysis of caged glutamate requires the caged reagent concentration to approach 10 m*M* in the medium. Any tendency to hydrolyze spontaneously would cause a caged glutamate to liberate free glutamate gradually, even in the absence of light. Even if free glutamate contamination is insignificant by the standards of analytical chemistry, it may still be significant biologically—0.1% spontaneous hydro-

---

[10] A. T. Harootunian, J. P. Y. Kao, and R. Y. Tsien, *Cold Spring Harb. Symp. Quant. Biol.* **53**, 935 (1988).

[11] A. T. Harootunian, J. P. Y. Kao, S. Paranjape, S. R. Adams, B. V. L. Potter, and R. Y. Tsien, *Cell Calcium* **12**, 153 (1991).

[12] A. T. Harootunian, J. P. Y. Kao, S. Paranjape, and R. Y. Tsien, *Science* **251**, 75 (1991).

[13] T. Meyer and L. Stryer, *Proc. Natl. Acad. Sci. U.S.A.* **85**, 5051 (1988).

[14] R. Wieboldt, K. R. Gee, L. Niu, D. Ramesh, B. K. Carpenter, and G. P. Hess, *Proc. Natl. Acad. Sci. U.S.A.* **91**, 8752 (1994).

[15] L. O. Trussell and G. D. Fischbach, *Neuron* **3**, 209 (1989).

[16] C.-M. Tang, M. Dichter, and M. Morad, *Science* **243**, 1474 (1989).

[17] M. L. Mayer and G. L. Westbrook, *J. Physiol.* **354**, 29 (1984).

lysis in 10 mM caged glutamate yields 10 μM free glutamate, a concentration sufficient to desensitize non-NMDA GluRs and to activate NMDA GluRs.

Glutamate photorelease from Nmoc-Glu is represented in Fig. 8. An important feature is that photolysis generates an intermediate species, the carbamate of glutamate, which then decarboxylates to yield free glutamate and carbon dioxide. The decarboxylation step is rate limiting under physiological conditions; therefore, photorelease of free glutamate from Nmoc-Glu is slower than from γ-O-CNB-Glu. The decarboxylation step is also accelerated at lower pH. The foregoing suggests that the actual rate of free glutamate generation by photolysis can be increased in two ways: by increasing the starting concentration of Nmoc-Glu in the medium and by decreasing pH.

For photolysis, an argon ion laser (Coherent I90-5, Santa Clara, CA) with an output of ~400 mW in the 351- to 364-nm region is used. The laser beam is steered into a Nikon Diaphot inverted microscope through the epifluorescence port and is directed through the back aperture of the objective (Fluor ×40, N.A. 1.3, Nikon, Columbia, MD) by a 400-nm long-pass dichroic mirror. A divergent fused silica lense with a focal length of −150 mm positioned 20 cm in front of the dichroic mirror allows imaging of the laser beam into a ~50-μm-diameter spot in the image plane. The photolysis spot can be further delimited in diameter by closing the diaphragm of the microscope objective. Final intensity in the image plane is estimated at ~70 μWμm^{-2}. Light pulse duration is controlled by a laser shutter (Unblitz LS2, Vincent Associates, Rochester, NY). Electrophysiological measurements are performed in the whole cell voltage-clamp mode with a patch-clamp amplifier (Dagan 3900, Minneapolis, MN). Borosilicate glass electrodes with resistances of 3–5 MΩ are used for whole cell recordings. Electrode series resistance is compensated 80–90% during whole cell recordings. Signals are filtered at 2 kHz, sampled at 5 kHz, and collected and analyzed with pClamp software (Axon Instruments, Foster City, CA). Extracellular solutions contain (in mM) 150 NaCl, 3 KCl, 2 CaCl$_2$, 1 MgCl$_2$, and 10 HEPES (or PIPES for pH < 7), adjusted to pH 7.3 with NaOH.

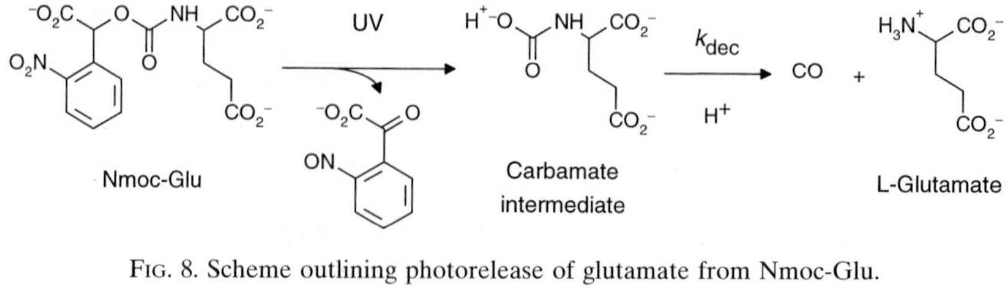

FIG. 8. Scheme outlining photorelease of glutamate from Nmoc-Glu.

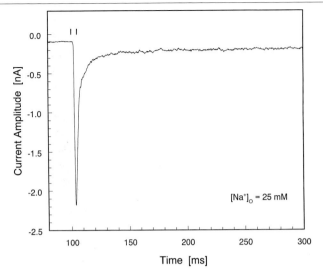

FIG. 9. Glutamate photorelease from Nmoc-Glu evokes robust inward current in a hippocampal neuron. Whole cell current response of a hippocampal neuron[18] to a 5-msec photorelease from 10 mM Nmoc-Glu, pH 6.3, at room temperature. To attenuate the peak current amplitude, the extracellular Na$^+$ concentration was reduced to 25 mM and photolysis was restricted to a 30-μm-diameter spot over the soma. The 10–90% rise time of the response is 0.98 msec. The double bars above the trace delimit the duration of the UV light pulse. Pipette and extracellular solution compositions were as given in the text.

Where extracellular [Na$^+$] is reduced, Na$^+$ is replaced with choline. TTX (1 μM) and APV (100 μM) are added to external solutions. The internal pipette contains (in mM) 140 CsCl, 5 EGTA, 1 MgCl$_2$, 2 CaCl$_2$, and 10 HEPES, titrated to pH 7.3 with CsOH.

An experiment where glutamate photorelease from Nmoc-Glu is used to stimulate a whole cell voltage-clamped rat hippocampal neuron[18] is shown in Fig. 9. Pulse photorelease of glutamate evokes a robust inward current with a 10–90% rise time of 0.98 msec. The current declines rapidly as the GluR channels desensitize.

Figure 10 illustrates experiments that demonstrate the high chemical stability of Nmoc-Glu. These experiments show the effect of applying freshly prepared 1 mM solutions of two different caged glutamates, *without photolysis*, to hippocampal neurons. Whereas γ-O-CNB-Glu[19] evokes significant inward current even in the absence of light, Nmoc-Glu is essentially inert. By repeating such paired tests on fresh cells, it was observed that

---

[18] C.-M. Tang, M. Margulis, Q.-Y. Shi, and A. Fielding, *Neuron* **13**, 1385 (1994).
[19] Purchased from Molecular Probes, Inc., Eugene, OR.

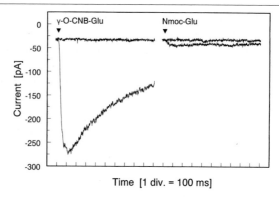

FIG. 10. Prephotolysis activity of γ-O-CNB-Glu and Nmoc-Glu. Solutions of the caged compounds were pressure delivered, in the absence of light, to each cell tested. The upper trace in each pair of experiments depicts the baseline current. Arrowheads mark the times of reagent delivery. Nmoc-Glu was inert, whereas γ-O-CNB-Glu definitely evoked an inward current. These traces were collected within 30 min of preparing 1 mM solutions of the reagents from solid samples. The solutions were kept on ice and in the dark until use. Pipette and external solution compositions were as given in the text.

the prephotolysis activity of the γ-O-CNB-Glu solution increases with time (the average evoked whole cell current increases approximately threefold, from ~120 to ~330 pA, over the period from 30 to 90 min after the γ-O-CNB-Glu solution was prepared). The Nmoc-Glu solution, however, remains inert, evoking whole cell inward currents of only ~10 pA. These results suggest that the intrinsic lability of γ-O-CNB-Glu causes gradual liberation and accumulation of free glutamate under physiological conditions, even in the absence of photolysis. In contrast, Nmoc-Glu appears to be stable in physiologic solution indefinitely.

In the absence of an ideal caged glutamate, deciding which available reagent is most appropriate depends on the demands of a particular experiment. For example, if the rapid gating kinetics of non-NMDA GluR channels are of primary interest, then γ-O-CNB-Glu, which exhibits very fast photorelease, would be appropriate. However, Nmoc-Glu has utility in applications where its principal advantages, low prephotolysis activity and high hydrolytic stability, are most useful. Such applications include (1) experiments where caged glutamate needs to be in physiologic solution for extended periods of time, (2) experiments where background activation of neurons must be minimized, (3) experiments where desensitization of non-NMDA GluRs must be minimized, and (4) experiments where the photoreleased free glutamate concentration must be high (i.e., in the millimolar range), which requires that the caged glutamate be present initially at high concentration. Thus, potential applications for Nmoc-Glu include quantitative mapping of the spatial distribution of functional non-NMDA GluRs on dendrites and in the use of photostimulation to analyze brain circuitry.[20,21]

In summary, the advantage of the Nmoc caging method is threefold: (1) ease of preparation and use of the active caging reagent, Nmoc-imidazole; (2) stable caging of hydroxyl and amino groups in bioactive molecules; and (3) moderately fast photorelease kinetics.[22–24] The illustrative examples described in this article show that the Nmoc approach can have wide use in developing new reagents for cellular physiological studies.

Acknowledgments

This work was supported by NIH Grant GM-46956 to J.P.Y.K.

[20] E. M. Callaway and L. C. Katz, *Proc. Natl. Acad. Sci. U.S.A.* **90**, 7661 (1993).
[21] L. C. Katz and M. B. Dalva, *J. Neurosci. Methods* **54**, 205 (1994).
[22] The Nmoc-caged molecules described here photorelease on the millisecond time scale, comparable to NPE-caged ATP and Ins(1,4,5)P_3, which have time constants ($t_{1/e}$) of release of 4–5 and 12 msec, respectively (Refs. 23 and 24).
[23] J. W. Walker, P. R. Gordon, J. A. McCray, and D. R. Trentham, *J. Am. Chem. Soc.* **110**, 7170 (1988).
[24] J. W. Walker, J. Feeney, and D. R. Trentham, *Biochemistry* **28**, 3272 (1989).

[25] Development and Application of Caged Ligands for Neurotransmitter Receptors in Transient Kinetic and Neuronal Circuit Mapping Studies

By GEORGE P. HESS and CHRISTOF GREWER

Neurotransmitter receptor-mediated reactions on the surface of some 10^{12} cells in the central, sympathetic, and peripheral mammalian nervous systems play a central role in the way in which environmental information is received, transmitted, transduced, encoded, and stored.[1,2] Malfunctions of these reactions are implicated in a number of diseases (e.g., Huntington's disease, Parkinson's disease, epilepsy) and mediate the effects of many clinically important compounds (e.g., tranquilizers, antidepressants) and of

[1] F. Crick, "The Astonishing Hypothesis: The Scientific Search for the Soul," Charles Scribner & Sons, New York, 1994.
[2] E. R. Kandel, J. H. Schwartz, and T. M. Jessell, "Essentials of Neural Science and Behavior," Appleton & Lange, Norwalk, CT, 1995.

abused drugs (e.g., cocaine).[2,3] Transient kinetic techniques suitable for investigating the mechanism of neurotransmitter receptor-mediated reactions on cell surfaces have been developed[4,5] and are the subject of this article.

Here we describe the instrumentation and theory involved in transient kinetic investigations of neurotransmitter receptor-mediated reactions on the surface of nervous system cells and in the membranes of *Xenopus laevis* oocytes in which many receptors have been expressed.[6] Photolabile, biologically inert precursors of neurotransmitters (caged neurotransmitters) that can be photolyzed to the neurotransmitters in the microsecond time region form an integral part of the transient kinetic techniques. The design, synthesis, and characterization of caged excitatory[7–9] and inhibitory[10–12] neurotransmitters that have been used in transient kinetic investigations are reviewed by Gee *et al.*[13]

Six structurally related excitatory and inhibitory neurotransmitter receptors, and many isoforms, have been identified using recombinant DNA technology.[14–16] These cell surface proteins, on binding chemical signals (neurotransmitters) released from an adjacent cell, transiently form open transmembrane channels (Fig. 1A). Excitatory receptors (for acetylcholine and glutamate) form cation-specific channels and inhibitory receptors [for γ-aminobutyric acid (GABA) and glycine] anion-specific ones. If the sign and amplitude of the resulting transmembrane voltage change reach a critical threshold value of about -40 mV, the cell will transmit an electrical signal along its long projection, an axon. When this signal reaches the end

---

[3] A. G. Gilman, T. W. Randal, A. S. Neis, and P. Taylor, "The Pharmacological Basis of Therapeutics," 8th Ed. Macmillan, New York, 1990.
[4] N. Matsubara, A. P. Billington, and G. P. Hess, *Biochemistry* **31**, 5507 (1992).
[5] G. P. Hess, L. Niu, and R. Wieboldt, *Ann. N.Y. Acad. Sci.* **757**, 23 (1995).
[6] D. Langosch, B. Laube, N. Rundstrom, V. Schmieden, J. Borman, and H. Betz, *EMBO J.* **13**, 4223 (1994).
[7] T. Milburn, N. Matsubara, A. P. Billington, J. B. Udgaonkar, J. W. Walker, B. K. Carpenter, W. W. Webb, J. Marque, W. Denk, J. A. McCray, and G. P. Hess, *Biochemistry* **29**, 49 (1989).
[8] R. Wieboldt, K. R. Gee, L. Niu, D. Ramesh, B. K. Carpenter, and G. P. Hess, *Proc. Natl. Acad. Sci. U.S.A.* **91**, 8752 (1994).
[9] L. Niu, K. R. Gee, K. Schaper, and G. P. Hess, *Biochemistry* **35**, 2030 (1996).
[10] D. Ramesh, R. Wieboldt, L. Niu, B. K. Carpenter, and G. P. Hess, *Proc. Natl. Acad. Sci. U.S.A.* **90**, 11074 (1993).
[11] K. R. Gee, R. Wieboldt, and G. P. Hess, *J. Am. Chem. Soc.* **116**, 8366 (1994).
[12] L. Niu, R. Wieboldt, D. Ramesh, B. K. Carpenter, and G. P. Hess, *Biochemistry* **35**, 8136 (1996).
[13] K. R. Gee, B. K. Carpenter, and G. P. Hess, *Methods Enzymol.* **291**, [2] (1998) (this volume).
[14] H. Betz, *Neuron* **5**, 383 (1990).
[15] R. M. Stroud, M. P. McCarthy, and M. Shuster, *Biochemistry* **29**, 11009 (1990).
[16] G. Mandel and D. McKinnon, *Annu. Rev. Neurosci.* **16**, 323 (1993).

of an axon, neurotransmitter is released near receptors on the surface of an adjacent cell[2] (Fig. 1A).

The transient kinetic techniques described here were developed to measure the rate and equilibrium constants that determine the concentration of open receptor channels as a function of neurotransmitter concentration and time in the physiologically relevant time region of microseconds to a few milliseconds. The concentration of open receptor channels, and their conductance and ion specificity, determine the electrical signals in the membrane of the cells that trigger signal transmission to another cell.[17,18,19]

The minimum reaction scheme for a receptor-mediated reaction is shown in Fig. 1B. It is based on experiments with the excitatory nicotinic acetylcholine receptor[27a,28] about which most information is available. It is the only neurotransmitter receptor about which we have three-dimensional structural information.[17,18] The conductance, lifetime, and ion specificity of the open receptor channel are conveniently determined by the single-channel current-recording technique.[19a] The new techniques described here were developed to solve another problem. To account for the integration of excitatory and inhibitory signals arriving at one cell and the resulting transmembrane voltage changes, one needs to know the receptor-controlled rate at which inorganic cations and anions cross the membrane through the receptor channels. The rates of the cation and anion fluxes across a semipermeable membrane are related to the transmembrane voltage.[20,21] They depend not only on the conductance of the open receptor channels, but also on their concentration in the cell membrane.[22] To determine the concentration of open channels as a function of neurotransmitter concentration and time, transient kinetic techniques are required. One needs to determine (i) the reaction path starting with free receptor and neurotransmitter and leading to the formation of neurotransmitter: receptor complexes and open receptor channels, (ii) the reaction path leading to desensitized (transiently inactivated) receptor forms, and (iii) the rate and equilibrium constants associated with individual steps in the reaction. The reactions steps outlined in Fig. 1B, like those of many enzyme-catalyzed reactions, occur in the

---

[17] M. Planck, *Ann. Physik u. Chem* **40**, 561 (1890).
[18] D. E. Goldman, *J. Gen. Physiol.* **27**, 37 (1943).
[19] G. P. Hess, H.-A. Kolb, P. Läuger, E. Schoffeniels, and W. Schwarze, *Proc. Natl. Acad. Sci. U.S.A.* **81**, 5281 (1984).
[19a] E. Neher and B. Sakmann, *Nature* **260**, 799 (1976).
[20] N. Unwin, *J. Mol. Biol.* **229**, 1101 (1995).
[21] N. Unwin, *Nature* **373**, 37 (1995).
[22a] C. L. Weill, M. G. McNamee, and A. Karlin, *Biochem. Biophys. Res. Commun.* **61**, 997 (1974).
[22b] J. A. Reynolds and A. Karlin, *Biochemistry* **17**, 2035 (1978).

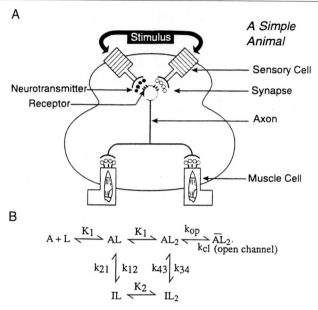

Fig. 1. (A) Two sensory cells are located on the surface of an animal. Activation of the sensory cell on the left leads to release of inhibitory neurotransmitters (●). Activation of the sensory cell on the right leads to release of excitatory neurotransmitters (○). Inhibitory neurotransmitters (e.g., glycine and GABA) on binding to their specific receptors on the cell surface of neurons located inside the animal, transiently open anion-conducting transmembrane channels. Excitatory neurotransmitters (e.g., acetylcholine and glutamate) transiently open cation-conducting transmembrane channels. If the transmembrane voltage change resulting from the activation of excitatory and inhibitory neurotransmitters is of appropriate sign and amplitude (~−40 mV), an electrical signal is elicited that travels along the axon of the neuron and results in the release of neurotransmitters adjacent to receptors in other cells, thus repeating the signal transmission process. (B) A minimum mechanism for the nicotinic acetylcholine receptor from the *Electrophorus electricus* electroplax and BC_3H1 muscle cells. The reaction scheme is consistent with the available structural[17,18,18a,18b] and kinetic[5,28,52] information. The reaction is initiated when a neuron secretes neurotransmitter L, which then binds to the active, nondesensitized receptor form, A, in the membrane of an adjacent cell. The chemical reaction contains several ligand-binding steps, characterized by the dissociation constant K_1, and leads to receptor–ligand complexes, where the subscript represents the number of ligand molecules bound. The binding of two ligand molecules before the acetylcholine receptor channel opens was first suggested by Katz and Thesleff.[28] Several transitions in protein conformation result: One change in the submillisecond time region leads to the formation of a transmembrane channel ($\overline{AL_2}$), characterized by the rate constants for channel opening (k_{op}) and closing (k_{cl}). In the presence of neurotransmitter, receptor proteins are also rapidly (15–200 msec) converted to inactive forms, I, with altered ligand-binding properties, characterized by the dissociation constant K_2. Katz and Thesleff[28] observed a slow (seconds) desensitization process in the muscle acetylcholine receptor. It is now known that a slow desensitization process is associated with a minor receptor form in BC_3H1 cells that contain muscle acetylcholine receptors.[33] It is assumed that a single K_1 value pertains to

micro- to millisecond time domain and must, therefore, be investigated with techniques that have a micro- to millisecond time resolution. Such techniques exist for investigations of reactions mediated by soluble proteins.[23-27] The function of neurotransmitter receptors, however, must be studied in intact cells or vesicles with the receptor protein embedded in a membrane separating two solutions of different ionic composition. Kinetic techniques for studying cell surface receptors in the micro- to millisecond time region and over a wide range of reactant concentrations only recently became available.[4,5,7,53] We shall give a few examples of the application of the new techniques using caged neurotransmitters in studies of (i) the channel-opening process of neurotransmitter receptors, (ii) drug–receptor interactions, (iii) integration by a cell of inorganic cation and anion transmembrane fluxes mediated by excitatory and inhibitory receptors, respectively, and (iv) locating and elucidating the neurotransmitters and receptors that control signal transmission in known circuits of neurons controlling measurable responses.

Rapid Chemical Reaction Techniques for Studying Reactions on Cell Surfaces

Cell-Flow Technique

Rapid receptor desensitization (the reversible inactivation of a receptor by its specific neurotransmitter) provided the impetus for developing techniques for rapidly equilibrating receptors with the neurotransmitter in solutions flowing over cells. Desensitization of neurotransmitter receptors was first discovered in studies with the nicotinic acetylcholine receptor and appeared to be slow, occurring in the second time region.[28] When it was

---

[23] M. Eigen, in "Fast Reactions and Primary Processes in Chemical Kinetics: Nobel Symp. 5" (S. Claesson, ed.), p. 333. Interscience, New York, 1967.
[24] G. G. Hammes, "Enzyme Catalysis and Regulation." Academic Press, New York, 1982.
[25] A. Fersht, "Enzyme Structure and Mechanism." Freeman, New York, 1985.
[26] K. A. Johnson, in "The Enzymes," Vol. 20, p. 1. Academic Press, New York, 1993.
[27] H. Gutfreund, "Kinetics for the Life Sciences: Receptors, Transmitters and Catalysts." Cambridge University Press, UK, 1995.
[27a] G. P. Hess, D. J. Cash, and H. Aoshima, *Annu. Rev. Biophys. Bioeng.* **12**, 443 (1983).
[28] B. Katz and S. Thesleff, *J. Physiol. (Lond.)* **138**, 63 (1957).

---

the dissociation of L from the AL and AL_2 forms. The direct conversion of A, or $\overline{AL_2}$ to the inactive I forms is neither required nor excluded by data available so far.[27a,52] The equations used to evaluate the rate and equilibrium constants of the reaction scheme using rapid chemical reaction techniques are listed in Table I.

TABLE I
Equations for Reaction Scheme Shown in Fig. 1B

Equation	Equation number
$(\overline{AL_2})_o = \dfrac{\overline{AL_2}}{A + AL + AL_2 + \overline{AL_2}} = \dfrac{L^2}{(L+K_1)^2 \Phi + L^2} = P_o$	(I-A)
$I_A = I_M R_M (\overline{AL_2})_o$	(I-B)
$[I_M R_M (I_A)^{-1} - 1]^{1/2} = \Phi^{1/2} + \Phi^{1/2} K_1 [L_1]^{-1}$	(I-C)

Eq. (I-A): $(\overline{AL_2})_o$ represents the fraction of receptor molecules in the open channel form. $\Phi^{-1} = k_{op}/k_{cl}$ is the channel-opening equilibrium constant, L represents the molar concentration of activating ligand, and P_o is the conditional probability, determined in single-channel current recordings, that the receptor is in the open-channel form.[30] All other symbols have been defined (Fig. 1B).

Eq. (I-B): I_A is the current due to open receptor channels in the cell membrane corrected for receptor desensitization. I_M is the current due to 1 mol of open receptor channels and R_M represents the number of moles of receptors in the cell membrane.

Eq. (I-C): A linear version of Eq. (I-B)[27a]

$(I_{obs})_t - (I_{obs})_{t\infty} = [(I_{obs})_{t=0} - (I_{obs})_{t\infty}]e^{-\alpha t}$	(II-A)
$\alpha = \Phi \left[\dfrac{L k_{43} + 2 K_2 k_{21}}{(L + 2 K_2) \Phi} + \dfrac{(L k_{34} + 2 K_1 k_{12}) L}{L^2(1 + \Phi) + 2 K_1 L \Phi + K_1^2 \Phi} \right]$	(II-B)
$\alpha = \dfrac{k_{34} \Phi L^2}{(L + K_1)^2 \Phi + L^2}$	(II-C)

Eq. (II-A): α represents the rate coefficient for receptor desensitization obtained from the falling phase of the current in cell-flow experiments.[33] I_{obs} represents the current during the falling phase. The subscripts t and $t = 0$ refer to the time of measurement, and $t = \infty$ to the time when an equilibrium between active and desensitized forms has been reached.

Eq. (II-C): When k_{34} (Fig. 1B) is the dominant rate constant, a simplified equation is obtained[33]

$I_A/I'_A = 1 + I_o/K_I$	(III-A)
$\dfrac{1}{K_I} = \dfrac{F_A}{(K_I)_1} + \dfrac{F_{AL}}{(K_I)_2} + \dfrac{F_{AL_2}}{(K_I)_3} + \dfrac{F_{\overline{AL_2}}}{(K_I)_4}$	(III-A')
$\dfrac{I_A}{I'_A} = 1 + \dfrac{I_o}{K_I} \dfrac{(2K_1 L + K_1^2)\Phi}{(L + K_1)^2 \Phi + L^2}$	(III-B)
$\dfrac{I_A}{I'_A} = 1 + \dfrac{I_o}{K_I} (\overline{AL_2})_o$	(III-C)
$\dfrac{I_A}{I'_A} = 1 + \dfrac{I_o}{K_I} + \dfrac{II_o}{K_{II}}$	(III-D)

TABLE I (Continued)

Equation	Equation number
$\dfrac{I_A}{I'_A} = 1 + \dfrac{I_o}{K_I} + \dfrac{II_o}{K_{II}} + \dfrac{I_o}{K_I}\dfrac{II_o}{K_{II}} = 1 + \dfrac{I_o}{K_I} + \dfrac{II_o}{K_{II}}\left(\dfrac{K_I + I_o}{K_I}\right)$	(III-E)

Eq. (III-A): The dissociation constant of the inhibitor from the nondesensitized receptor can be determined by both cell-flow and photolysis measurements. In order to simplify the equations, we use a ratio method, I_A/I'_A, where I_A and I'_A represent the current maxima corrected for receptor desensitization in the absence and presence of inhibitor, respectively. I_o represents the inhibitor concentration and K_I the observed dissociation constant

Eq. (III-A'): The relationship between the observed inhibitor dissociation constant K_I and the inhibitor dissociation constant for A, AL, AL_2, and $\overline{AL_2}$ receptor forms. F_A, F_{AL}, F_{AL_2}, and $F_{\overline{AL_2}}$ represent the fraction of receptors in forms A, A, AL_2, and $\overline{AL_2}$

Eq. (III-B): For a competitive inhibitor $1/K_I$ is multiplied by the fraction of receptor molecules in the A and the AL form

Eq. (III-C): An inhibitor binding only to the open-channel form

Eq. (III-D): Two inhibitors I_o and II_o binding to the same receptor site

Eq. (III-E): Two inhibitors I_o and II_o binding to two different receptor sites

$I_t = I_{max}[1 - \exp(-k_{obs}t)]$	(IV-A)
$k_{obs} = k_{cl} + k_{op}\left(\dfrac{L}{L + K_1}\right)^2$	(IV-B)
$k_{obs} = k_{cl}\dfrac{K_I}{K_I + I_o} + k_{op}\left(\dfrac{L}{L + K_1}\right)^2$	(IV-C)
$k_{obs} = k_{cl} + k_{op}\left(\dfrac{L}{L + K_1}\right)^2\dfrac{K_I}{K_I + I_o}$	(IV-D)

Eq. (IV-A): In the laser-pulse photolysis experiments with BC$_3$H1 cells containing nicotinic acetylcholine receptors, the current rise time was observed to follow a single exponential rate law over 85% of the reaction.[4] I_t represents the observed current at time t and I_{max} represents the maximum current

Eq. (IV-B): The relationship between the observed rate constant for the current rise k_{obs} and k_{op}, k_{cl}, K_1 of the reaction scheme (Fig. 1B)

Eq. (IV-C): k_{obs} in the presence of an inhibitor that binds only to the open-channel form of the receptor

Eq. (IV-D): k_{obs} in the presence of an inhibitor that binds only to the closed-channel form of the receptor. If the inhibitor binds both to the open- and closed-channel forms a combination of Eqs. (IV-C) and (IV-D) is obtained

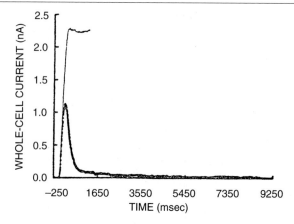

FIG. 2. Cell-flow measurement with a BC$_3$H1 muscle cell containing nicotinic acetylcholine receptors, pH 7.4, 23°, −60 mV transmembrane potential. A solution of 200 μM acetylcholine emerged from the cell-flow device (Fig. 3A) at a rate of 1 cm sec^{-1}. The thick solid line represents the observed current, and the thinner line represents the calculated current corrected for receptor desensitization using Eq. (1). [From G. P. Hess, J. B. Udgaonkar, and W. L. Olbricht, *Annu. Rev. Biophys. Biophys. Chem.* **16**, 507 (1987)].

discovered, using rapid reaction techniques[29] and subsequently single-channel recording measurements,[30] that this receptor desensitizes in the millisecond time region, an elegant device for flowing solutions containing neurotransmitter over a single cell was developed.[31] The current due to opening of receptor channels can then be recorded by the whole cell current recording technique.[32] Equilibration of ligands in the flowing solutions with cell surface receptors, however, can be slow compared to receptor desensitization, and the amplitude of the current recorded must be corrected to take into account desensitization that occurs during the equilibration of the neurotransmitter in the flowing solutions with the cell surface receptors.[33,34] The thick solid line in Fig. 2 is the observed whole cell current recorded when a BC$_3$H1 muscle cell containing acetylcholine receptors was equilibrated with 200 μM acetylcholine in a solution flowing over the cell. The current first rises due to cations moving through receptor channels that open in

[29] G. P. Hess, D. J. Cash, and H. Aoshima, *Nature* **282**, 329 (1979).
[30] B. Sakmann, J. Patlak, and E. Neher, *Nature* **286**, 71 (1980).
[31] O. A. Krishtal and V. I. Pidoplichko, *Neuroscience* **5**, 2325 (1980).
[32] O. Hamill, A. Marty, E. Neher, B. Sakmann, and F. J. Sigworth, *Pfluegers Arch.* **391**, 85 (1981).
[33] J. B. Udgaonkar and G. P. Hess, *Proc. Natl. Acad. Sci. U.S.A.* **84**, 8758 (1987).
[34] G. P. Hess, J. B. Udgaonkar, and W. L. Olbricht, *Annu. Rev. Biophys. Biophys. Chem.* **16**, 507 (1987).

response to the neurotransmitter. The falling phase of the current is due to receptor desensitization in which the receptor channels close. The thin line in Fig. 2, which reaches an amplitude about twice that reached in the recorded current trace, is a calculated line [Eq. (1)] and is the current after correction for receptor desensitization that occurs during the equilibration of the receptors with neurotransmitters in the flowing solution.[33,34]

Equation (1) was derived[33,34] to calculate the current I_A, which is associated with the active nondesensitized receptor form A (Fig. 2). To obtain the value of I_A from measurements of the observed current I_{obs}, the current time course is divided into constant (1–5 msec) time intervals to take into account the equilibration time of small segments of the cell surface with ligand. The current is then corrected for the desensitization occurring during each time interval Δt. After n constant time intervals ($n\Delta t = t_n$), during each of which the current $(I_{obs})_{\Delta t}$ is measured, the corrected current is given by[33,34]

$$(I_A)_{t_n} = (e^{\alpha \Delta t} - 1) \sum_{i=1}^{n} (I_{obs})_{\Delta t_i} + (I_{obs})_{\Delta t_n} \tag{1}$$

where $(I_{obs})_{\Delta t_i}$ is the observed current during the ith time interval and t_n is equal to or greater than the current rise time, the time it takes the current to reach a maximum value. Under conditions of laminar flow of the neurotransmitter solution over the cell, the value of I_A was found to be independent of the solution velocities used and could be determined with good precision ($\pm 10\%$).[33] Equation (1) is based on the theory of the flow of solutions over spherical objects.[35,36] BC$_3$H1 cells are nearly spherical when suspended by the current-recording electrode in solutions bathing the cell. In experiments with central nervous system neurons, which are firmly attached to their stratum, it has been possible to obtain spherical, tightly sealed receptor-containing membrane vesicles of about 15 μm diameter attached to the recording electrode[37a] (which can be suspended in the fluid stream emerging from the flow device).

A diagram of a cell-flow device is shown in the lower left-hand corner of Fig. 3A. Solutions flow at rates of less than 2 cm sec^{-1} from a 150–250-μm-diameter porthole of the flow device over a single cell[32,34] (or vesicle of 10–15 μm diameter. The cell (or vesicle) is suspended from a

---

[35] V. G. Landau and E. M. Lifshitz, "Fluid Mechanics," p. 219. Pergamon, Oxford, 1959.
[36] A. G. Levich, "Physicochemical Hydrodynamics." Prentice-Hall, Engelwood Cliffs, NJ, 1962.
[37] K. M. Walstrom and G. P. Hess, *Biochemistry* **33**, 7718 (1994).
[37a] W. Sather, S. Dieudonne, J. F. MacDonald, and P. Ascher, *J. Physiol.* (*London*) **450**, 643 (1992).

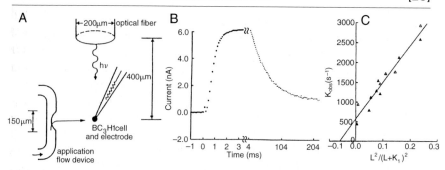

FIG. 3. (A) BC$_3$H1 cell carrying nicotinic acetylcholine receptors, of ~20 μm diameter, attached to an electrode for whole-cell current recording was equilibrated with caged carbamoylcholine.[4] The side view of the cell-flow device[31] used to equilibrate the cell surface with ligands is shown on the left. A Candela SLL500 dye laser is used. Rhodamine 640 or sulforhodamine 640 laser dye, together with a second harmonic generator, produces wavelengths of 320 and 328 nm, respectively. The laser beam is introduced from an optical fiber of 200 μm diameter. The fiber is adjusted to be ~400 μm away from the cell so that the area illuminated around the cell has a diameter of 300–400 μm. The energy of the laser pulse emerging from the fiber is ~500 μJ, and the pulse length is 600 nsec. By projecting visible light through the optical fiber, the cell is illuminated and the fiber properly positioned. The wavelength of 328 nm was chosen to avoid cell damage at lower wavelengths and too low a product yield at higher wavelengths. Current is recorded in the whole-cell configuration.[32] Data were low-pass filtered (Krohn-Hite 3322) with a 1- to 10-kHz cutoff frequency (−3 dB point) and then digitized at a 2- to 20-kHz sampling frequency using a PDP 11/23 minicomputer; data were then transferred to a Convex C210 computer where the constants for the rising and decaying phases of the whole-cell current were analyzed using the data analysis program PLOT. (B) Whole-cell current induced by 200 μM released carbamoylcholine at pH 7.4, 22–23°, and −60 mV.[4] The points represent digitized current data [Table I, Eqs. (IV-A) and (IV-B)]. An observed first-order rate constant of 2140 sec^{-1} was evaluated from the rise-time. (C) Determination of k_{op}, k_{cl}, and K_1 for the opening of acetylcholine receptor-channels in BC$_3$H1 cells at pH 7.4, 23°, and −60 mV.[4] The values of k_{obs} determined from experiments shown in (B) are plotted according to $k_{obs} = k_{cl} + k_{op} L^2/(L + K_1)^2$ [Table I, Eq. (IV-A) and (IV-B)]. The values of k_{op}, k_{cl}, and K_1 are given in Table IV.

current-recording electrode[32] and is about 50–100 μm away from the porthole. Because the orientation of the porthole of the flow device with respect to the cell is critical, we modified the device[38] first designed by Krishtal and Pidoplichko[31] in which the porthole is stationary and mixing of solutions occurs before leaving the porthole. A flow device that enables one to flow solutions of different composition over the cell has been described in detail by Niu *et al.*[38] As many as 20 measurements have been made with the same BC$_3$H1 muscle cell. At the flow rates used the current must be corrected for desensitization (Fig. 2). The time resolution of the cell-flow techniques

[38] L. Niu, C. Grewer, and G. P. Hess, *Tech. Protein Chem.* **VII**, 139 (1996).

is about 10 msec.[33] To obtain a better time resolution, the flow rate of the neurotransmitter solution could be increased. However, if one increases the flow rate of the solutions over the cell, turbulent flow and mixing occurs between the neurotransmitter solution and the buffer solutions surrounding the cell. Consequently, the concentration of neurotransmitter in the solution equilibrating with the cell surface receptors is not known.

Laser-Pulse Photolysis Technique

In contrast to the cell-flow technique, the time resolution of a laser-pulse photolysis technique is two to three orders of magnitude better. For this method, biologically inert photolabile precursors of neurotransmitters (caged neurotransmitters) suitable for investigations, in the microsecond time region, of the excitatory acetylcholine and glutamate receptors and the inhibitory γ-aminobutyric acid and glycine receptors have been developed.[3,7–12,39] A cell attached to a current-recording electrode[31] is preequilibrated with a caged neurotransmitter. At zero time a pulse of laser light photolyzes the caged compound in the microsecond time region. The liberated neurotransmitter binds to receptors on the cell surface and initiates the formation of transmembrane channels; the resulting current is measured (Fig. 3B). The apparatus used is illustrated in Fig. 3A. The time resolution of the method is governed by the rate with which the neurotransmitter is liberated by a light pulse. Photolysis rates with $t_{1/2}$ values between 1 and 45 μsec have been obtained with the compounds listed in Tables II and III. Caged neurotransmitters that are not suitable for transient kinetic measurements because they either photolyze too slowly, in the millisecond time region, or are not biologically inert[39] have also been synthesized[40–43] but are not listed on Tables II and III. Several caged phenylephrine derivatives exist[44–46] suitable for investigations of the α_1-adrenergic receptor. These are also not reviewed here.

[39] K. R. Gee, L. Niu, K. Schaper, and G. P. Hess, *J. Org. Chem.* **60**, 4260 (1995).
[40] M. Wilcox, R. W. Viola, K. W. Johnson, A. P. Billington, B. K. Carpenter, J. A. McCray, A. P. Guzikowski, and G. P. Hess, *J. Org. Chem.* **55**, 1585 (1990).
[41] A. P. Billington, K. M. Walstrom, D. Ramesh, A. P. Guzikowski, B. K. Carpenter, and G. P. Hess, *Biochemistry* **31**, 5500 (1992).
[42] J. E. T. Corrie, A. DeSantis, Y. Katayama, K. Khodakhah, J. B. Messenger, D. C. Ogden, and D. R. Trentham, *J. Physiol. (Lond.)* **465**, 1 (1993).
[43] R. Wieboldt, D. Ramesh, B. K. Carpenter, and G. P. Hess, *Biochemistry* **33**, 1526 (1994).
[44] S. M. Muralidharan, G. M. Mayer, W. B. Boyle, and J. M. Nerbonne, *Proc. Natl. Acad. Sci. U.S.A.* **90**, 5199 (1993).
[45] J. W. Walker, H. Martin, F. R. Schmitt, and R. J. Barsotti, *Biochemistry* **32**, 1338 (1993).
[46] S. Muralidharan and J. M. Nerbonne, *J. Photochem. Photobiol. B* **27**, 123 (1995).

In kinetic measurements the concentration of neurotransmitter generated by photolysis of a caged precursor is measured by using the cell-flow technique.[32,33] Before photolysis, a known concentration of free neurotransmitter is applied to the cell and the amplitude of the resulting current, together with a dose–response curve (e.g., Fig. 4), is used to calibrate the concentration of neurotransmitter generated in the photolysis experiment. Cell-flow experiments are also used at the end of the experiment to check if the laser pulse has damaged the cell or the receptors. The cell-flow device (Fig. 3A)[38] (see Fig. 1 in Niu et al.[38] for details) is used both to remove the neurotransmitter from the culture dish and to add caged neurotransmitter. The results obtained by laser-pulse photolysis and the single-channel current-recording technique[19a] can be compared only at low neurotransmitter concentrations at which receptor desensitization is relatively slow. As can be seen in Fig. 4 there is good agreement among the results obtained by laser-pulse photolysis of caged neurotransmitters, the single channel current-recording technique,[19a] and the cell-flow technique.[33]

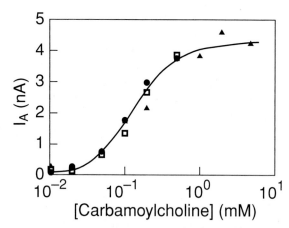

FIG. 4. Concentration dependence of the current amplitude corrected for receptor desensitization, I_A. BC$_3$H1 muscle cells, pH 7.4, 22–23°, and −60 mV transmembrane potential. Data are from J. B. Udgaonkar and G. P. Hess, *Proc. Natl. Acad. Sci. U.S.A.* **84,** 8758 (1987). (○, single-channel current recordings; ▲, cell-flow measurements) and N. Matsubara, A. P. Billington, and G. P. Hess, *Biochemistry* **31,** 5507 (1992) (□, laser-pulse photolysis). The line indicating the concentration of open receptor channels was calculated from the constants pertaining to the channel-opening process determined in laser-pulse photolysis experiments (Table IV).

Photochemical Properties of Caged Neurotransmitters

Attempts to make chemical kinetic studies of acetylcholine receptors on cell surfaces were initiated in innovative studies by Bartels et al.,[46a] who synthesized the *photoisomerizable* 3,3'-bis(trimethyl aminomethyl)azobenzene (Bis Q) and reported that the *trans*, but not the *cis*, form was an activating ligand for the acetylcholine receptor. Lester then used photoisomerization of Bis Q to study the kinetics of the opening of the acetylcholine receptor channel.[46b] In our development of a rapid reaction method using light activation, we foresaw the need to measure adverse effects of caged neurotransmitters and their photolysis products; it became one of the strategic considerations. Using rapid mixing techniques we had developed for use with vesicles, we found that "inactive" *cis*-Bis Q causes receptor desensitization and *trans*-Bis Q becomes a receptor inhibitor even at low concentrations.[47] We, therefore, began to consider the synthesis of photolabile inert precursors of neurotransmitters.

Photocleavable protecting groups for biologically important compounds have many uses in cell biology. This is particularly true of cases where access of a compound to its reaction partner is slow but the induced reaction is fast.[48–53] The most frequently used photocleavable protecting group is the 2-nitrobenzyl group, with various substituents. The photochemically induced transfer of oxygen in 2-nitrobenzyl derivatives by which the *ortho* CH group becomes C-OH and the NO_2 group NO on irradiation was first described by De Mayo[54] in 1960. Barltrop et al.[55] then protected the carboxy group of glycine and leucine using a 2-nitrobenzyl derivative. Subsequently, Patchornik et al.[56,56a] used the 2-nitrobenzyloxycarbonyl group as a photosensitive blocking reagent in peptide synthesis. The use of this protecting group with biologically important phosphates was pioneered by Kaplan *et*

[46a] E. Bartels, N. H. Wasserman, and B. F. Erlanger, *Proc. Natl. Acad. Sci. U.S.A.* **68**, 1820 (1971).
[46b] H. A. Lester and J. M. Nerbonne, *Annu. Rev. Biophys. Bioeng.* **11**, 151 (1982).
[47] A. H. Delcour and G. P. Hess, *Biochemistry* **25**, 1793 (1986).
[48] J. H. Kaplan, B. Forbush, and J. F. Hoffman, *Biochemistry* **17**, 1929 (1978).
[49] J. A. McCray and D. R. Trentham, *Proc. Natl. Acad. Sci. U.S.A.* **77**, 7237 (1980).
[50] R. S. Adams and R. Y. Tsien, *Annu. Rev. Physiol.* **55**, 755 (1993).
[51] J. E. T. Corrie and D. R. Trentham, "Bioorganic Photochemistry 2" (H. Morrison, ed.). Wiley, New York, 1993.
[52] G. P. Hess, *Biochemistry* **32**, 989 (1993).
[53] J. M. Nerbonne, *Curr. Opin. Neurobiol.* **6**, 379 (1996).
[54] P. DeMayo, *Adv. Org. Chem.* **2**, 367 (1960).
[55] J. A. Barltrop, P. J. Plant, and P. Schofield, *J. Chem. Soc. Chem. Commun.* 822 (1966).
[56] P. Patchornik, B. Amit, and R. B. Woodward, *J. Am. Chem. Soc.* **92**, 6333 (1970).
[56a] P. Kuzmic, L. Pavlickova, and M. Soucek, *Czech. Chem. Commun.* **51**, 1293 (1986).

al.[48] and McCray and Trentham[49] and led to widespread use of "caged compounds" in which the 2-nitrobenzyl group with and without substituents is the protecting group. When the α-amino group of glutamate was blocked by this reagent, the photolysis rate was 17 sec^{-1} and the quantum yield 0.65.[51] Several other caging groups that can be used to protect the amino group or carboxyl group of neurotransmitters have been reported, e.g., the 3′,5-dimethoxybenzoin group[57–60] and the p-hydroxyphenacyl group.[61] For transient kinetic investigations of neurotransmitter receptors on the surface of muscle cells and neurons, caged neurotransmitters must have a number of properties that have not yet been examined with these compounds.

Caged neurotransmitters suitable for transient kinetic experiments, in addition to being photolyzed in the microsecond time region so that photolysis is not rate limiting, must also be (i) photolyzed to give the neurotransmitter with sufficient quantum yield to allow kinetic investigations to be made over a wide range of neurotransmitter concentration, (ii) water soluble and sufficiently stable in aqueous solution before photolysis, (iii) photolyzed at a wavelength greater than 335 nm to avoid cell damage, and (iv) last but not least neither the caged compound nor the photolysis products, with the exception of the liberated neurotransmitter, may modify the receptor-mediated reaction being studied.

All these criteria were kept in mind when we initiated the synthesis of photolabile inert precursors of neurotransmitters. 2-Nitrobenzyl derivatives were used to protect the amino group of carbamoylcholine,[7,62] an analog of acetylcholine, and of the inhibitory neurotransmitters glycine[63] and GABA.[64] For the acetylcholine receptor, carbamoylcholine is used because it is a stable and well-characterized analog of acetylcholine and, unlike acetylcholine, it contains a functional group, an amino group, to which the protecting group can be attached. The introduction of the α-carboxy-2-nitrobenzyl protecting group (α-CNB)[7] improved the photolysis rate of caged carbamoylcholine; the carbamate is photolyzed with a rate of 17,000 sec^{-1} and a quantum yield of 0.8.[7] It also improved the photolysis rate of the caged amino groups compared to the caging group used previously[44] for amino groups. However, although the N-protected photolabile precur-

---

[57] J. C. Sheehan, R. M. Wilson, and A. W. Oxford, *J. Am. Chem. Soc.* **93,** 7222 (1971).
[58] K. R. Gee, L. Niu, K. Schaper, and G. P. Hess, *Biochemistry* (in press).
[59] G. Papageorgiou and J. E. T. Corrie, *Tetrahedron* **53,** 3917 (1997).
[60] G. G. Hammes and C. W. Wu, *Annu. Rev. Biophys. Biophys. Chem.* **16,** 507 (1974).
[61] C. H. Park and R. S. Givens, *J. Am. Chem. Soc.* **119,** 2453 (1997).
[61] D. L. Purich, *Enzymes* **249,** 3 (1995).
[62] J. W. Walker, J. A. McCray, and G. P. Hess, *Biochemistry* **25,** 1799 (1986).
[63] A. P. Billington, K. W. Johnson, D. Ramesh, A. P. Guzikowski, B. K. Carpenter, and G. P. Hess, *Biochemistry* **31,** 5500 (1992).
[64] R. Wieboldt, D. Ramesh, B. K. Carpenter, and G. P. Hess, *Biochemistry* **33,** 1526 (1994).

sors of glycine[63] and GABA,[64] and we presume of other amino group-containing compounds, are photolyzed 35 times faster than the protecting groups previously used for amino groups,[51] they are photolyzed 30 times slower than the caged carbamoylcholine we synthesized.[7]

When we used the α-CNB group to protect the carboxyl group of neurotransmitters, we obtained compounds that meet all the criteria listed earlier for transient kinetic investigations of the excitatory glutamate[8] and kainate[9] receptors and the inhibitory GABA[11] receptors. The photochemical properties of neurotransmitters caged with the α-CNB group[7] are listed in Table II.

The 2-methoxy-5-nitrophenyl group (MNP)[56a] was used to cage the inhibitory neurotransmitter glycine.[10] Although this derivative is photolyzed in the 1-μsec time region, it is not stable in aqueous solution. β-Alanine also activates the glycine receptor[65] and MNP-caged β-alanine has all the desirable properties for transient kinetic investigations of the glycine receptor.[9] The photochemical properties of the MNP caged compounds are listed in Table III.

Determination of Rate and Equilibrium Constants of Channel-Opening Mechanism

The nicotinic acetylcholine receptor in BC$_3$H1 muscle cells will be used as an example. Figure 3B shows some results obtained in a laser-pulse photolysis experiment. The cell was equilibrated with caged carbamoylcholine. At zero time the caged carbamoylcholine was photolyzed, liberating free carbamoylcholine in the microsecond time region. The time resolution of the technique allows one to observe three distinct phases of the reaction: a rising phase of the current reflecting the opening of acetylcholine receptor channels, a maximum current amplitude, a measure of the concentration of open receptor channels, and on a different and slower time scale, the falling phase of the current reflecting receptor desensitization. In experiments with the excitatory acetylcholine, glutamate and kainate receptors and the inhibitory GABA and glycine receptors[4,7–12] the rise time of the current follows a single exponential rate equation over 85% of the reaction. In Fig. 3C the dependence of the first-order rate constant for the current rise time is plotted versus carbamoylcholine concentration (Fig. 3C) according to Eq. (IVB) (Table I). The ordinate intercept of the line gives the value of the channel-closing rate constant, k_{cl}, and the slope allows evaluation of the channel-opening rate constant, k_{op}. From the dependence of the maximum current amplitude on the concentration of carbamoylcho-

---

[65] D. Choquet and H. Korn, *Neurosci. Lett.* **84**, 329 (1988).
[65a] M. Amador and J. A. Dani, *Synapse* **7**, 207 (1991).

TABLE II
PHOTOLYTIC PROPERTIES OF BIOLOGICALLY INERT, PHOTOLABILE DERIVATIVES OF
NEUROTRANSMITTERS CAGED WITH α-CARBOXY-2-NITROBENZYL GROUP[a]

α-Carboxy-2-nitrobenzyl group: 2-nitrobenzyl-CH(COOH)—

Caged neurotransmitter	Group caged	Photolysis $t_{1/2}$ (μsec)	Product quantum yield	Target receptor	Ref.
I. Carbamoylcholine NO_2-C$_6$H$_4$-CH(COOH)-NH-C(=O)-O-(CH$_2$)$_2$-$\overset{+}{N}$(CH$_3$)$_3$	Carbamate	45	0.8	Acetylcholine	Milburn et al., (1989)[7]
II. Glutamate NO_2-C$_6$H$_4$-CH(COOH)-O-C(=O)-(CH$_2$)$_2$-CH(COOH)-NH$_2$	γ-Carboxyl	21	0.14	Glutamate	Wieboldt et al. (1994)[8]
III. γ-Aminobutyric acid (GABA) NO_2-C$_6$H$_4$-CH(COOH)-O-C(=O)-(CH$_2$)$_2$-NH$_2$	γ-Carboxyl	19	0.16	GABA	Gee et al. (1994)[11]
IV. Kainate NO_2-C$_6$H$_4$-CH(COOH)-O-C(=O)- [kainate moiety with H$_3$C-C(=CH$_2$)- and -NH-CH(COOH)- ring]	γ-Carboxyl	45	0.37	Kainate	Niu et al. (1996)[a,9]

[a] pH 7.5, 22°.

TABLE III
PHOTOLYTIC PROPERTIES OF BIOLOGICALLY INERT, PHOTOLABILE DERIVATIVES OF NEUROTRANSMITTERS CAGED WITH 2-METHOXY-5-NITROPHENYL GROUP[a]

2-Methoxy-5-nitrophenyl group: (2-methoxy-5-nitrophenyl)–O–

Caged neurotransmitter	Photolysis $t_{1/2}$ (μsec)	Product quantum yield	Ref.
I. Caged glycine (2-methoxy-5-nitrophenyl)–O–C(=O)–CH$_2$–NH$_2$	<1	0.2	Ramesh et al. (1993)[a,10]
II. Caged β-alanine (2-methoxy-5-nitrophenyl)–O–C(=O)–(CH$_2$)$_2$–NH$_2$	<1	0.2	Niu et al. (1996)[a,12]

[a] pH 7.5, 22°C.

line, one can obtain the value of the channel-opening equilibrium constant, $\Phi = k_{cl}/k_{op}$, and the value of K_1, the dissociation constant of the receptor site controlling channel opening (Fig. 1B), by using Eq. (I-C) (Table I). The value of K_1 and Φ ($\Phi = k_{cl}/k_{op}$) obtained from the effect of ligand concentration on the current amplitude can be compared to the values of K_1, and k_{cl}, and k_{op} obtained from the effect of ligand concentration on the observed rate constant, k_{obs}, for the rise time of the current [Fig. 3C; Eq. (IV-B) in Table I].

The falling phase of the current gives information about the rate of receptor desensitization. Receptor desensitization is slow compared to channel opening and occurs in a different time scale. It is investigated more conveniently by the cell-flow method.[33]

Comparison of Values Obtained for Rate and Equilibrium Constants of Channel-Opening Process Using Independent Techniques

The values of k_{op}, k_{cl}, and K_1 (Fig. 1B) obtained for the acetylcholine receptor in BC_3H1 muscle cells by the laser-pulse photolysis technique and two independent techniques (cell-flow and single-channel current recordings) are compared in Table IV. (i) The cell-flow technique is a chemical kinetic method, even though its time resolution is at least two orders of magnitude less than that of the laser-pulse photolysis technique. It does not allow one to determine k_{op} and k_{cl}. It does, however, allow one to determine the current I_A corrected for desensitization.[33,34] I_A is a measure of the concentration of open receptor channels. The relationship between I_A and the constants of the mechanism in Fig. 1B (K_1, $\Phi = k_{cl}/k_{op}$) and neurotransmitter concentration is given in Eqs. (I-A) and (I-B) (Table I). (ii) In the single-channel current-recording technique[19a] the receptor is equilibrated with neurotransmitter before the measurements are made. At low concentrations of neurotransmitter, at which the receptor is not completely desensitized, the technique allows one to determine the ion specificity, conductance, and lifetime of the open receptor channel[19] where the lifetime is a measure of k_{cl}. Under conditions where the receptor site density is low and the recording electrode ideally records the current from a single receptor channel it is possible to determine P_o.[29] P_o is the conditional probability that the channel is open while the receptor is in an active, nondesensitized state. The relationship between P_o, the fraction of nondesensitized receptors in the open-channel form, and the constants of the mechanism in Fig. 1B is given by Eq. (I-A) (Table I).

The values of the constants for the acetylcholine receptor in BC_3H1 cells determined by the three techniques are listed in Table IV. The relationship between the concentration of open acetylcholine receptor channels and carbamoylcholine concentration, obtained by laser-pulse photolysis, the cell-flow,[33] and the single-channel current-recording[19a] technique, is given in Fig. 4. As can be seen from Table IV, when the single-channel current-recording and cell-flow techniques are used at appropriate ligand concentrations and in the appropriate time domain, there is excellent agreement between the results obtained with these techniques and with the laser-pulse photolysis technique.

Investigations of Mechanism of Receptor Inhibition

Effects of Inhibitors on k_{op} and k_{cl}. The nicotinic acetylcholine receptor in BC_3H1 muscle cells will be taken as an example. Inspection of Fig. 3B and 3C and of Eq. (IV-A) and (IV-B) (Table I) indicates that the laser-pulse photolysis technique allows one to determine the rate constant of

TABLE IV
Constants for Reaction Steps in Opening of Nicotinic Acetylcholine Receptor Channel in BC$_3$H1 Muscle Cells[a]

Constant	Method	Value of constant
k_{op}	Laser–pulse photolysis	12,000 sec^{-1}
	Previous techniques (22 studies)[b]	450–40,000 sec^{-1}
k_{cl}	Laser–pulse photolysis	500 ± 100 sec^{-1}
	Single channel current	400 ± 130 sec^{-1}
Fraction of open channels	Laser–pulse photolysis	0.94
	Cell flow	0.84
K_1	Laser–pulse photolysis	210 ± 90 μM
	Cell flow	240 μM

[a] pH 7.4, at 22–23°, and −60 mV. From N. Matsubara, A. P. Billington, and G. P. Hess, *Biochemistry* **31**, 5507 (1992).

[b] See Jackson, *J. Physiol. (Lond.)* **397**, 555 (1988); reviewed Madsen and Edeson, *Trends Pharm. Sci.* **9**, 315 (1988); Liu and Dilger *Biophys. J.* **60**, 424 (1991).

the current rise indicative of channel opening and, therefore, to study the effects of inhibitors on k_{cl} [Eq. (IV-C), Table I] and on k_{op} [Eq. (IV-D), Table I] independently of one another. Before the laser-pulse photolysis technique was developed, it was possible to determine the effects of inhibitors only on the channel-closing rate, using the single-channel current-recording technique.[65,66] It was noticed that in the presence of the noncompetitive inhibitor procaine, the acetylcholine receptor channel rapidly opened and closed (flickered) while it was open. This was ascribed to the inhibitor rapidly entering and leaving the open channel, thereby blocking and unblocking it. This inhibition mechanism was generally accepted before the introduction of transient kinetic techniques for investigations of receptor–inhibitor interactions (e.g., see Lena and Changeux[67]). An alternative mechanism is not, however, excluded by the measurements. In an alternative mechanism, the inhibitor binds to a regulatory site on the receptor before, and after, the channel opens. In this mechanism, binding of a compound to the receptor per se may not cause receptor inhibition, but the subsequent conformational change does. Such mechanisms are well documented in enzymes regulating metabolism.[24–27]

How can the two mechanisms be distinguished? In both mechanisms, the rate constant for channel closing is predicted to decrease as the inhibitor concentration is increased. However, a mechanism in which an inhibitor binds to the receptor before the channel opens makes a clear-cut prediction

[66] E. Neher and J. H. Steinbach, *J. Physiol. (London)* **277**, 779 (1978).

[67] C. Lena and J. P. Changeux, *Trends Neurosci.* **16**, 181 (1993).

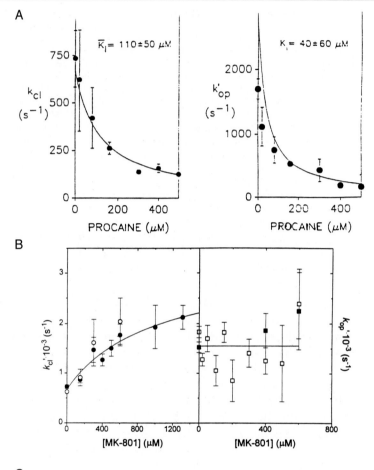

that is not made by the channel-blocking mechanism. It predicts that the channel-opening rate will change as the inhibitor concentration is increased. The two mechanisms can, therefore, be distinguished, providing one can measure the effects of inhibitors not only on the channel-closing rate, but also on the channel-opening rate. This can now be done using the laser-pulse photolysis method.[68]

Laser-pulse photolysis experiments in which the mechanism of inhibition of the acetylcholine receptor can be studied are illustrated in Fig. 5A. The graph on the left shows the effect of the inhibitor procaine on the channel-closing rate, which decreased as the procaine concentration increased. This decrease in the channel-closing rate is predicted by both mechanisms and agrees with the results obtained in single-channel current recordings. The graph on the right shows the effect of procaine on the channel-opening rate [Eq. (IV-D), Table I]. The channel-opening rate also decreases as the procaine concentration is increased. This observation is consistent with a mechanism in which the inhibitor can bind and exert its effect both before and after the channel opens.

Additional details of the inhibition mechanism have been obtained in laser-pulse photolysis experiments with the noncompetitive inhibitors cocaine[69] and MK-801.[70] MK-801 is a noncompetitive inhibitor of the excitatory acetylcholine[65] and N-methyl-D-aspartic acid receptors[79] and has anticonvulsant properties.[72] The results in Fig. 5B demonstrate that MK-801 increases the channel-closing rate without affecting the channel-opening

[68] L. Niu and G. P. Hess, *Biochemistry* **32**, 3831 (1993).
[69] L. Niu, L. G. Abood, and G. P. Hess, *Proc. Natl. Acad. Sci. U.S.A.* **92**, 12008 (1995).
[70] C. Grewer, L. Niu, and G. P. Hess, *Biochemistry* submitted (1998).
[71] J. E. Huettner and B. P. Bean, *Proc. Natl. Acad. Sci. U.S.A.* **85**, 1307 (1998).
[72] E. H. F. Wong, J. A. Kemp, T. Priestley, A. R. Knight, G. N. Woodruff, and L. L. Iversen, *Proc. Natl. Acad. Sci. U.S.A.* **83**, 7104 (1996).

FIG. 5. Laser-pulse photolysis measurements with the acetylcholine receptor in BC$_3$H1 cells, pH 7.4, 22–24°, and −60 mV. (A) The effect of procaine on k'_{cl} and k'_{op}. The concentration of carbamoylcholine released from caged carbamoylcholine is 20 and 115 μM in the experiments on the left and right, respectively. (B) The effect of MK-801 on k'_{cl} and k'_{op}. The concentration of photoreleased carbamoylcholine is 25 and 160 μm in the experiments on the left and right, respectively. Solutions of caged carbamoylcholine without and with MK-801 flowed over the cell for 200 msec before the laser pulse.[70] (C) Tentative scheme for the reaction of inhibitors with the receptor based on the experiments shown. A, AL_2, and IA, IAL_2 represent the closed form of the receptor without and with inhibitor bound, respectively. $\overline{AL_2}$ and $\overline{IAL_2}$ represent the open-channel form of the receptor without and with inhibitor bound respectively, K_I and K'_I are the receptor–inhibitor dissociation constants, and $IA*L_2$ represents the inhibited closed-channel form of the receptor.[70]

rate. The resulting decrease in the channel-opening equilibrium constant accounts for receptor inhibition in the millisecond time region. A subsequent and slower process characterized by the observed rate constants k'_b and k'_f leads to complete inhibition of the receptor (Fig. 5C).[70]

Evaluation of Receptor:Inhibitor Dissociation Constants. The time resolution of the kinetic methods just described allows one to determine the maximum current amplitudes and, therefore, the concentration of open receptor channels before desensitization occurs. This allows one to determine the effects of inhibitors on the current amplitudes at low concentrations of neurotransmitter, when the receptor is mainly in the closed-channel form, and at high concentrations, when the receptor is mainly in the open-channel form.[68–70] These measurements allow one to differentiate between noncompetitive and competitive inhibitors [Eqs. (III-A) and (III-B), Table I] and to determine whether two noncompetitive inhibitors bind to two different sites [Eq. (III-E), Table I] or to the same site [Eq. (III-D), Table I]. An example of a plot of the maximum current amplitudes obtained before receptor desensitization occurs, in the absence and presence of inhibitors, I_A and I'_A respectively, is shown in Fig. 6. In these experiments,

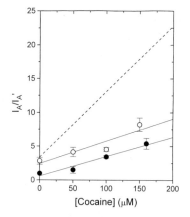

FIG. 6. Cell-flow experiments, pH 7.4, 22–23°, and −60 mV. Effect of cocaine concentration on the ratio of current amplitudes, I_A/I'_A, in the presence of 20 μM carbamoylcholine and the absence or presence of MK-801. I_A is the current amplitude in the absence of inhibitor, and I'_A is the current amplitude in the presence of inhibitor. ●, 20 μM carbamoylcholine, and varying cocaine concentrations, in the absence of MK-801; $K_I(app) = 35$ μM [Eq. (III-A), Table I]. ○, 20 μM carbamoylcholine, and varying cocaine concentrations, in the presence of 100 μM MK-801 ($K_{I(app)}$ for MK-801 is 53 μM); $K_{I(app)}$ for cocaine evaluated by use of Eq. (III-D) (Table I) is 30 μM and is similar to the value (35 μM) obtained in the absence of MK-801. The dotted line was calculated for the case in which two inhibitors do not compete for one site [Eq. (III-E), Table I] using the dissociation constants for the inhibitors determined in experiments in which only one or the other inhibitor is present.

cocaine and MK-801 were the inhibitors used. The slopes of the lines, proportional to the apparent cocaine–receptor dissociation constant K_I, are the same when the concentration of cocaine is varied, but 100 μM MK-801 is absent (●) or present (○). Equation (III-D) (Table I) indicates that this is expected when two noncompetitive inhibitors bind to the same site. The dashed line is calculated for the case in which cocaine and MK-801 bind to two different noncompetitive binding sites, using Eq. (III-E) (Table I) and using the apparent dissociation constants for cocaine and MK-801 from experiments in which only one of the inhibitors is present.

Integration by a Single Cell of Signals Elicited by Excitatory and Inhibitory Neurotransmitters

In classical experiments, Hubel and Wiesel[73,74] showed that individual nerve cells in the visual pathway respond to specific environmental signals. For instance, the direction from which a light dot moves across the visual field of a monkey and the time interval between the movements can be discerned from the electrical recordings of a single cell.[74] Many different and complicated theories involving complex circuits of cells have been proposed to explain how electrical signals can arise in a single cell that are indicative of both the direction and the time interval between environmental signals. None of the theories take into account the chemical mechanism of receptor-mediated reactions because the reaction mechanisms were not known. The results in Fig. 7 suggest that when the reaction mechanism and the associated rate and equilibrium constants of the receptor-mediated reactions are considered, a relatively simple interaction of excitatory and inhibitory receptors can explain electrical signals in single cells, which are indicative of the direction from which a signal comes and the time interval between the signals. Assume that the sensory cells in Fig. 1A contain photoreceptors and that activation of the photoreceptor on the left leads to the release of the inhibitory neurotransmitter GABA and that activation of the photoreceptor some distance away on the right leads to the release of the excitatory neurotransmitter glutamate. Figure 7 shows the current oscillation expected when (i) the glutamate and GABA receptors are activated at the same time (Fig. 7B), (ii) the GABA receptor is activated first and then the glutamate receptor (Fig. 7A), and (iii) the glutamate receptor is activated first and then the GABA receptor (Fig. 7C). Figures 7A and 7C are based on calculations in which the currents due to the GABA and glutamate receptors are subtracted from each other. In the central panel

[73] D. H. Hubel and T. N. Wiesel, *Proc. Roy. Soc.* (*London*) (*Biology*) **198,** (1977).
[74] D. Hubel, "Eye, Brain, and Vision." Scientific American Library, New York, 1988.

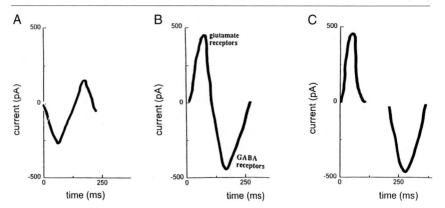

FIG. 7. Calculated integration of currents by a single cell arising from activation of inhibitory (anion-conducting) GABA receptors and excitatory (cation-conducting) glutamate receptors. The simulations are based on experiments with hippocampal neurons equilibrated with 500 μM caged GABA or 500 μM caged glutamate. In each case the neurotransmitter was released by a 308-nm, 10-nsec laser pulse of ~500 μJ. The concentration of the released neurotransmitters was assumed to remain unchanged. The observed current versus time trace was normalized to the same maximum current amplitude. The current due to the opening of GABA- and glutamate-activated receptor channels was then subtracted from each other. (A) The time interval between activation of first the inhibitory GABA receptors and then the excitatory glutamate receptors was taken as ~40 msec. (B) Simultaneous activation of glutamate and GABA receptors. (C) The time interval between activation of first the glutamate and then the GABA receptors was taken as ~200 msec.

(Fig. 7B) the concentrations of the excitatory glutamate and inhibitory GABA were chosen such that the amplitudes of the current flowing through excitatory and inhibitory receptor channels in a single cell were the same. The current flows through the receptors with opposite polarity (Fig. 7B). We would, therefore, not expect to detect a signal. This is not what happens. When glutamate and GABA arrive at their respective receptor sites at the same time, the current due to the excitatory glutamate receptor rises and is not neutralized by the current due to the inhibitory GABA receptor. Later the current trace falls to negative values, due to the inhibitory GABA receptor. The currents do not neutralize each other because the channel-opening rates, and the rates of desensitization, are quite different for the two receptors. Both rates are much faster for the glutamate receptor than for the GABA receptor. The repeated simultaneous release of glutamate and GABA would lead to oscillation of the electrical signal. Such oscillations are frequently observed in central nervous system neurons.[68a] Figure 7C shows what happens when the glutamate receptors are activated first and then the inhibitory GABA receptors. The signals due to the excitatory glutamate and inhibitory GABA receptor are now separated by a clearly discernible time interval. Figure 7A shows what happens when the GABA receptors are activated first and then the glutamate receptors. The electrical signals are quite different from those shown in the other panels of Fig. 7C. The amplitudes of the excitatory and inhibitory signals have decreased because the slower GABA receptor is activated first and the faster glutamate receptor somewhat later. As can be seen from Fig. 7, depending on which receptor type is activated first and which second, and the time interval between the activations, the resulting electrical signals are different. This difference is accounted for by the underlying mechanism of the receptor-mediated reactions and its associated constants. We have assumed here that the concentration of the released neurotransmitter remains constant for 125 msec. The conclusions reached, however, are valid as long as channel opening is fast compared to receptor desensitization or removal of neurotransmitter from the receptors. Under these conditions it takes only two receptor types in a single cell to indicate the direction and time interval between stimuli that contain spatial and temporal information.

Use of Caged Neurotransmitters to Determine How Neuronal Circuits Function

Caged neurotransmitters have been used in identifying and locating receptors in cells[75] and tissue slices.[76,77] The use of caged neurotransmitters has also made it possible to identify receptors and cells that secrete the neurotransmitter in a known circuit of cells controlling a measurable response in the nematode *Caenorhabditis elegans*.[78] Contraction of the *C. elegans* pharynx is controlled by 20 neurons.[79,80] Analogous to the mammalian heart, the *C. elegans* pharynx pumps liquid in and out of the organism. The positions of the 20 neurons controlling pharyngeal pumping have been identified by electron microscopy.[79] Among the important questions to be answered are which cells secrete which neurotransmitters and which cells contain the responding neurotransmitter receptors? Because a tough cuticle surrounds the worm, and because the animal is under turgor pressure, electrical recordings from single cells cannot be made routinely. Electrical recordings from the intact pharynx are, however, possible (electropha-

---

[75] W. Denk, K. R. Delaney, A. Gelperin, D. Kleinfeld, B. W. Strowbridge, D. W. Tank, and R. Yuste, *J. Neurosci. Methods* **54**, 151 (1994).
[76] E. M. Callaway and L. C. Katz, *Proc. Natl. Acad. Sci. U.S.A.* **90**, 7661 (1993).
[77] L. C. Katz and M. B. Dalva, *J. Neurosci. Methods* **54**, 205 (1994).
[78] H. Li, L. Avery, W. Denk, and G. P. Hess, *Proc. Natl. Acad. Sci. U.S.A.* **94**, 5912 (1997).
[79] D. G. Albertson and J. N. Thomson, *Phil. Trans. R. Soc. Lond B* **275**, 299 (1976).
[80] D. M. Raizen and L. Avery, *Neuron* **12**, 483 (1994).

ryngeograms, EPGs[80–82]) (Fig. 8A). Laser ablation studies in which a single known neuron is eliminated by a laser pulse[83,84] have shown that two of the neurons (M3s) control the time interval between contraction and relaxation of the pharyngeal muscle[78] (Fig. 8A). The M3 neurons are also necessary for the small negative spikes in the EPGs; the spikes disappear when M3 neurons have been destroyed[80,81] (Fig. 8A).

It was found that in nematodes without the M3 neurons, application of the glutamate to the pharynx via photolysis of caged glutamate restored the negative spikes in the EPG and shortened the period between contraction and relaxation of the pharynx. Photolytic release of carbamoylcholine or GABA from the corresponding caged compound had no effect on the EPG. Cells responding to glutamate were identified[78] by photolyzing caged glutamate on the pharynx over areas 7 μm in diameter. Responses were found on muscle cells pm4, pm5, and pm6 in the metacorpus, isthmus, and the front region of the terminal bulb, respectively (Fig. 8B). Connections between motor neurons M3 and muscle cells pm4 and pm5 have been identified in electron micrographs.[79] The experiments suggest that these cells and pm6 contain receptors that respond to glutamate and that cell M3 secretes a neurotransmitter (presumably glutamate) that activates these glutamate receptors. The experiments also indicate that the M3 neurons regulate the time interval between contraction and relaxation of the pharynx by activating glutamate receptors and that the glutamate receptors in muscle cell pm6 in the terminal bulb, innervated by a yet unidentified neuron, determine the relaxation of the pharynx.

Neurotransmitter Receptors that Can Now Be Investigated in the Microsecond Time Region Using Caged Neurotransmitters

Suitable photolabile precursors for transient kinetic investigations in the microsecond time region of the excitatory acetylcholine receptor have been available for a number of years.[7] Caged neurotransmitters suitable for transient kinetic investigations in the microsecond time region have now become available for investigations of the excitatory glutamate,[8] and kainate[9] receptors, and the inhibitory GABA[11] and glycine receptors[12] (Tables II and III). Some representative laser-pulse photolysis experiments using these new caged neurotransmitters are shown in Fig. 9. They demon-

---

[81] L. Avery, D. M. Raizen, and S. Lockery, *Methods Cell Biol* **48**, 251 (1995).
[82] L. Avery, *J. Exp. Biol.* **175**, 283 (1993).
[83] J. E. Sulston and J. G. White, *Dev. Biol.* **78**, 577 (1980).
[84] C. I. Bargmann and L. Avery, *Methods Cell Biol.* **48**, 225 (1995).

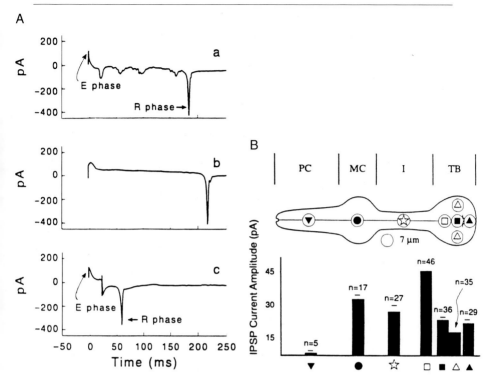

Fig. 8. (A) (a) A typical electropharyngeogram (EPG)[80,81] recorded from wild-type (N2) *C. elegans*.[78] This portion of the EPG corresponds to a single pumping action of the pharynx, a contraction followed by a relaxation. It contains three phases. The E or excitation phase, indicative of muscle contraction, is the positive spike in the EPG. The P or plateau phase is defined as the period between the E phase and R phase; the large negative spike, the R phase, is indicative of muscle relaxation. In the P phase, there are small negative spikes (inhibitory postsynaptic potentials, IPSPs).[80,81] (b) An EPG from a nematode whose M3 neurons were killed by laser ablation.[80,84] All IPSPs disappeared in this trace and the pump duration, the time interval between the E and R phases, increased when compared to the wild type. (c) An EPG from the same nematode as in (b). The pharynx was equilibrated with 0.5 mM caged glutamate from which free glutamate was generated by photolysis 20 msec after the E spike. (B) 0.5 mM caged glutamate was equilibrated with the pharynx of a nematode in which the M3 neurons were ablated and no IPSPs were generated. The caged glutamate was photolyzed in areas 7 μm in diameter, indicated by circles in the diagram of the pharynx; the circles are drawn in correct proportion to the length of the pharynx. The procorpus (PC), metacorpus (MC), and isthmus (I) of the pharynx each contain one type of muscle cell.[78] The bar graph below the drawing of the pharynx gives the IPSP current amplitudes obtained from various areas of the pharynx upon photolysis of 0.5 mM caged glutamate. Error bar is SEM. [Copyright (1996) National Academy of Sciences, U.S.A.]

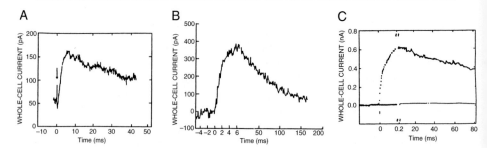

Fig. 9. Current recorded from mouse neurons at pH 7.4, 22–23°, −60 mV. In each experiment a neuron was incubated for 200 msec with caged compound in physiological saline. The concentrations of the caged compound were 500 μM caged glutamate in the experiment on the left, 500 μM caged GABA in the experiment in the middle, and 620 μM caged glycine in the experiment on the right. At time 0 the cell was exposed to a 0.5-mJ flash of 343-nm light generated by a Candella SLL500 pulsed dye laser and delivered through a fiber optic. The release of the neurotransmitter in the microseconds time region led to activation of the receptors on the cell surface. The concentration of open receptor channels was determined by a whole-cell current-recording method.[32]

strate that techniques are now available to investigate the mechanism of the channel-opening process, and of receptor activation and inhibition by clinically used compounds and abused drugs, for the excitatory glutamate and inhibitory GABA and glycine receptors, in addition to the nicotinic acetylcholine receptor.

Transient Kinetic Experiments with Neurotransmitter Receptors Expressed in *Xenopus* Oocytes

Xenopus laevis oocytes are used extensively to express neurotransmitter receptors from various regions of the nervous system, from different species, and of varying subunit composition and also to express receptors altered by site-specific mutations, nervous system diseases, or exposure to abused drugs. The chemical kinetic techniques suitable for use with cells of 15 to 20 μm diameter are not suitable for use with oocytes,[85] the diameter of which is 50 to 70 times larger. One reason for this is that flow techniques are required in conjunction with laser-pulse photolysis experiments to calibrate the concentration of the photoreleased neurotransmitters and to assess possible receptor or membrane damage by the laser pulse. An approach to overcome these difficulties has been developed[85] using electrical

---

[85] L. Niu, R. W. Vazquez, G. Nagel, T. Friedrich, E. Bamberg, R. E. Oswald, and G. P. Hess, *Proc. Natl. Acad. Sci. U.S.A.* **93,** 12964 (1996).

recordings from giant (~20 μm diameter) membrane patches.[86,87] The experiments in Fig. 10 illustrate that receptors expressed in *Xenopus* oocytes can now be investigated, by using giant oocyte membrane patches, with the same time resolution as receptors in muscle cells and neurons. The measurements[85] shown were made with oocytes expressing the muscle receptor from BC_3H1 cells.[88] The maximum current amplitudes observed in the experiments were normalized to each other. On the left (solid line, Fig. 10A) is an experiment with an oocyte membrane patch and the experiment on the right (dashed line, Fig. 10A) is an experiment with a whole oocyte using the two-electrode voltage clamp. In the experiment on the left the time it takes the current to reach a maximum value is 2 msec. This current rise time of 2 msec reflects the equilibration time of cell surface receptors with ligand. The current decay, indicative of receptor desensitization, occurs in two phases: ~70% of the current decays with a $t_{1/2}$ value of ~14 msec and the remaining current decays with a $t_{1/2}$ value of ~65 msec. In contrast, when the measurements were made with the whole oocyte using the two-electrode voltage-clamp method (the experiment on the right, dashed line, Fig. 10A), the current rise time is 2 sec. The two rapid desensitization processes seen in the experiment on the left (solid line) can no longer be seen. They are expected to have gone to completion during the 2-sec equilibration process between cell surface receptors and the receptor-activating ligand carbamoylcholine. From the difference in surface area between the oocyte membrane patch and the whole oocyte, the current observed in the whole oocyte experiment that desensitizes slowly may reflect the property of only a small fraction of receptors present in the oocyte membrane that desensitize relatively slowly. The properties of the major forms of the receptors present, which desensitize rapidly and which can be observed in the experiment on the left, are not observed in the experiment on the right. Similar results have been obtained in experiments with the neuronal α_7-acetylcholine receptor.[85] Such results are expected in all investigations with receptors expressed in oocytes in which the two-electrode voltage-clamp method is used and in which the current rise time is slow compared to receptor desensitization. An example of a laser-pulse photolysis measurement with a giant oocyte membrane patch is shown in Fig.

[86] D. W. Hilgemann, *Pfluegers Arch.* **415,** 247 (1989).
[87] J. Rettinger, L. A. Vasilets, S. Elsner, and W. Schwartz, in "The Sodium Pumps" (E. Bamberg and W. Schoner, eds.), p. 553. Steikopff, Darmstadt, Germany, 1994.
[87a] D. Bertrand, E. Cooper, S. Valera, D. Rungger, and M. Ballivet, *Methods Neurosci.* **4,** 174 (1991).
[87b] W. Stühmer and A. B. Parekh, in "Single-Channel Recording" (B. Sakmann and E. Neher, eds.), 2nd Ed., p. 341, Plenum, New York, 1995.
[88] D. Schubert, A. J. Harris, E. E. Devine, and S. Heinemann, *J. Cell Biol.* **61,** 398 (1974).

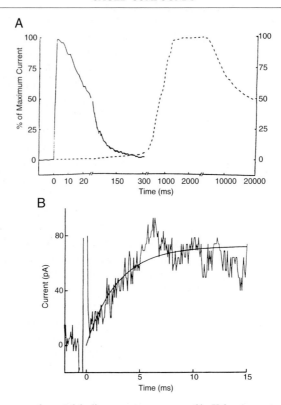

FIG. 10. Mouse muscle acetylcholine receptors expressed in *X. laevis* oocytes; measurements made at pH 7.4, 22–23°, and −60 mV.[77] (A) The solid line (on the left) represents the current response obtained using a giant outside-out oocyte membrane patch with a flow technique. The dotted line on the right represents the response obtained with the whole oocyte two-electrode voltage-clamp technique.[87a,87b] The maximum current amplitude of the response in the experiment on the left (solid line) is 4.0 nA and for the experiment on the right (dotted line) 17.7 μA. (B) Current induced by flash photolysis of 700 μM caged carbamoylcholine preequilibrated with a giant outside-out oocyte membrane patch. The light source for photolysis was a flash lamp (Chadwick-Helmuth, El Monte, CA; Model 278). The wavelength was in the 300- to 390-nm range because of the cutoff by the lens and dichroic mirror used. The pulse energy was ~700 μJ and the pulse length 300 μsec. The spikes at time zero are instrument artifacts.[85] [Copyright (1996) National Academy of Sciences, U.S.A.]

10B. It may, therefore, be of interest to reexamine the many interesting observations that have been reported in experiments with oocytes in which inadequate time resolution techniques have been employed.

Conclusions

Techniques have been described, using caged neurotransmitters, that enable transient kinetic investigations to be made of cell surface receptors expressed in muscle cells, neurons, and *Xenopus* oocytes with the same time resolution previously possible only with soluble proteins. The use of these techniques was illustrated in investigations of the formation of transmembrane channels, receptor–drug interactions, and integration of excitatory and inhibitory signals by a single central nervous system neuron that have temporal and spatial information. The use of caged neurotransmitters in identifying cells that contain a specific receptor and cells that secrete the neurotransmitter in a circuit of cells controlling a measurable response has also been illustrated.

The techniques described have been developed recently so that only a few examples could be given. Many receptors and their isoforms from various regions of the nervous system, from different species, and from animals that exhibit defects in receptor-mediated reactions are now available. Hundreds of therapeutically useful compounds and abused drugs have been identified that affect receptor-mediated reactions, and new therapeutic agents are continuously being discovered. By analogy with the use of transient kinetics in the investigations of reactions mediated by soluble proteins, which rapidly (millisecond) interconvert between various forms with different ligand-binding properties and biological activities,[23–27] and the few examples available in investigations of cell surface receptors,[5] it appears likely that the new transient kinetic techniques, and caged compounds, will be useful for studying the many interesting and unexplored areas of cell surface receptor function that have already been identified and those that will come to light in the future.

Acknowledgments

Work reviewed here was supported by grants (GM04842 and NS08527) awarded to G.P.H. by the National Institutes of Health Institute of General Medical Sciences and the National Institutes of Health Institute of Neurological Diseases and Stroke, respectively, and by a grant (922061) awarded to G.P.H. by the National Science Foundation. C.G. was supported by a Feodor Lynen Fellowship of the Alexander von Humboldt Foundation.

[26] Caged Plant Growth Regulators

By ANDREW C. ALLAN, JANE L. WARD, MICHAEL H. BEALE, and ANTHONY J. TREWAVAS

Introduction

Although plant growth regulators (plant hormones) have been the focus of intense research since their first isolation early this century, is it only now becoming possible to elucidate how and where these small molecules are first perceived by the plant cell. Most natural plant growth regulators are weak permeable acids, making it unlikely that outwardly facing plasma membrane receptors are the exclusive site for hormone perception. Although such receptors have indeed been inferred [e.g., in the perception of gibberellin (GA) by aleurone protoplasts[1]], as the hormone diffuses past these receptors many internal sites will also experience the signal. For example, it has been shown that part of the growth response to auxin resides in the action of internal receptors.[2]

To answer the question of the functional localization of plant hormone perception, studies have been made of the effects of injecting natural regulator molecules [e.g., abscisic acid (ABA)[3,4] and GA[4]] into plant cells. However, this technique allows for no temporal separation between the injection event (being quite traumatic for the plant cell due to piercing the cell wall) and subsequent exposure to the signal. We have developed UV-activated caged plant growth regulators to allow intracellular release of these compounds. "Caging" provides the ideal way to release the hormone precisely, after the cell has recovered from injection. It also allows internal release of plant growth regulators whose permeability does not allow for easy loading into the cell.

This article describes the synthesis of caged plant growth regulators, and the implications of our finding that cell responses to plant growth regulators, mediated by pH and calcium signals, can be elicited after release within the plasma membrane.

---

[1] R. Hooley, M. H. Beale, and S. J. Smith, *Planta* **183**, 274 (1991).
[2] M. Claussen, H. Lüthen, and M. Böttger, *Physiol. Plant* **98**, 861 (1996).
[3] B. E. Anderson, J. M. Ward, and J. I. Schroeder, *Plant Physiol.* **104**, 1177 (1994).
[4] S. Gilroy and R. L. Jones, *Plant Physiol.* **104**, 1185 (1994).

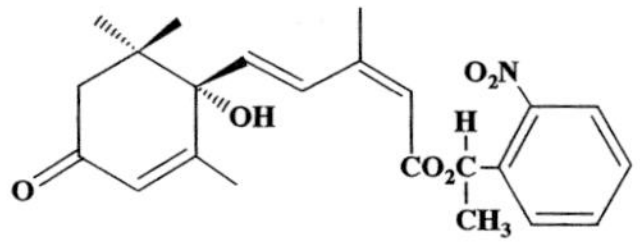

FIG. 1. The structures of caged plant growth regulators and of caged fluorescein.

Abscisic acid 1-(2-nitrophenyl)ethyl ester (ABA-2NPE)

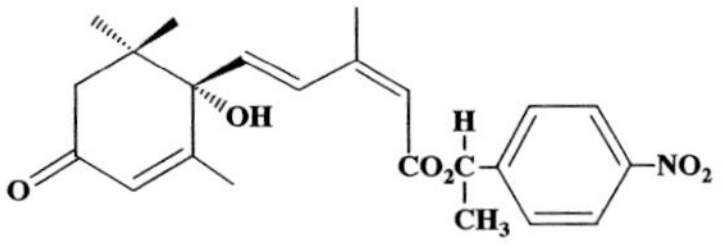

Abscisic acid 4-nitrobenzyl ester (ABA-4NBE)

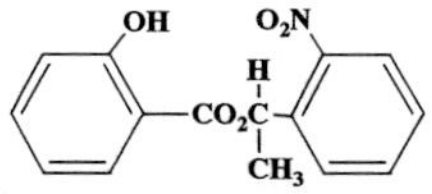

Salicylic acid 1-(2-nitrophenyl)ethyl ester (SA-2NPE)

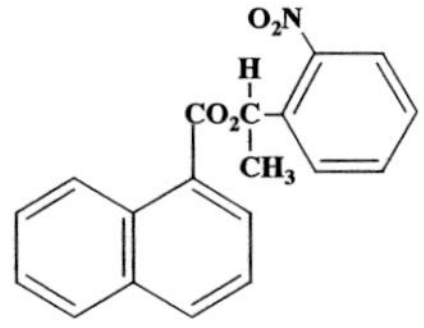

1-Naphthylacetic acid 1-(2-nitrophenyl)ethyl ester (1-NAA-2NPE)

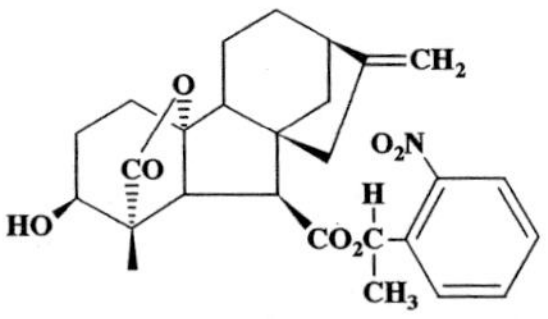

Gibberellin A$_4$ 1-(2-nitrophenyl)ethyl ester (GA$_4$-2NPE)

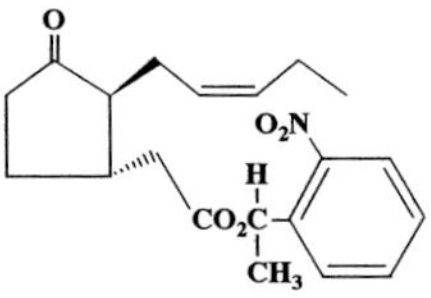

Jasmonic acid 1-(2-nitrophenyl)ethyl ester (JA-2NPE)

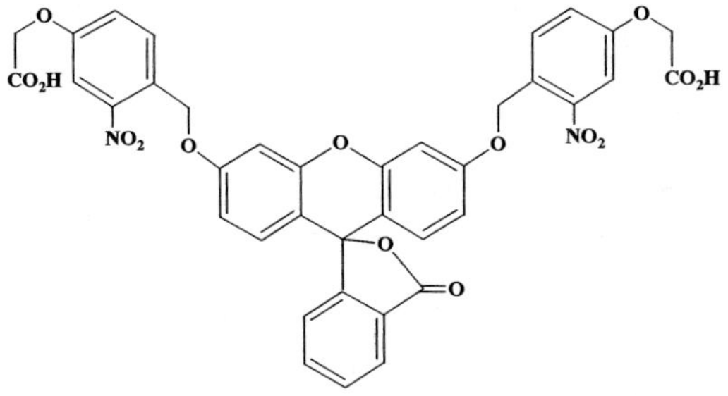

Fluorescein bis(5-carboxymethoxy-2-nitrobenzyl) ether

Synthesis of Caged Plant Hormones[5]

Those plant growth regulators that are carboxylic acids are relatively easy to cage as photocleavable esters (Fig. 1). These are formed by coupling to appropriate derivatives of 2-nitrobenzyl alcohol. Although 2-nitrobenzyl esters themselves are suitable, more efficient photolysis is achieved when

[5] J. L. Ward and M. H. Beale, *Phytochemistry* **38,** 811 (1995).

the benzyl carbon is substituted as in 1-(2-nitrophenyl)ethyl esters.[6] In our work, we also make use of the corresponding 4-nitrobenzyl esters, which are nonphotocleavable and, as such, serve as excellent controls in microinjection experiments.

Many methods are available to activate carboxylic acids for coupling to alcohols. For GA, AGA, jasmonic acid (JA), and 1-naphthylacetic acid (1-NAA), we have used[5] the mixed anhydride method because this requires no protection of other functional groups on the hormone. This method is not applicable to acids such as salicylic acid (SA), which contain a reactive hydroxy group. In this case, we found that activation by formation of N-hydroxysuccinimidyl esters was convenient.

Procedure for Preparation of Caged ABA

Commercial ABA (Lancaster Synthesis, Morecambe, Lancashire, UK) (150 mg), dissolved in anhydrous tetrahydrofuran (distilled from calcium hydride) (3 ml), is treated with isobutylchloroformate (80 μl) and triethylamine (80 μl) at room temperature until complete formation of a less polar product (TLC monitoring) has been observed (ca. 10 min). The solution is then diluted in ethyl acetate containing 5% (w/v) aqueous citric acid. The organic layer is dried (magnesium sulfate) and evaporated *in vacuo* to yield ABA-isobutyloxycarbonyl anhydride (189 mg, 91%), ^{1}H NMR (CDCl$_3$) δ: 7.82 (1H, d, J = 16Hz, H-5), 6.32 (1H, d, J = 16Hz, H-4), 5.95 (1H, s, H-3′), 5.76 (1H, s, H-2), 4.06 [2H, d, J = 7Hz, OCH_2CH(CH$_3$)$_2$], 2.48 and 2.32 (2H, 2d, J = 17Hz, H$_2$-5′), 2.07 and 1.92 (6H, 2s, H$_3$-7′ and H$_3$-6); 2.05 [1H, m, CH$_2$CH(CH$_3$)$_2$], 1.12 and 1.04 (6H, 2s, H$_3$-8′ and H$_3$-9′), 0.99 [6H, d, J = 7Hz, OCH$_2$CH(CH_3)$_2$].

The anhydride (68 mg) in acetonitrile (2 ml) is then treated with 1-(2-nitrophenyl)ethanol (63 mg) and 4-dimethylaminopyridine (30 mg) for 1 hr at room temperature. After addition of ethyl acetate and 5% aqueous citric acid, and recovery as described earlier, the product is purified by flash chromatography on silica gel (BDH, 40–63 μm) eluting with hexane : ethyl acetate (3 : 1) to give ABA-1-(2-nitrophenyl)ethyl ester (caged ABA, ABA-2NPE) (60 mg, 78%) ^{1}H NMR (CDCl$_3$) δ: 7.98 (1H, dd, J = 8 and 1Hz, ArH), 7.80 (1H, d, J = 16Hz, H-5); 7.65 (2H, m, ArH$_2$), 7.45 (1H, dt, J = 8 and 1Hz, ArH); 6.40 (1H, q, J = 7Hz, OCH(CH$_3$)Ar); 6.18 (1H, d, J = 16Hz, H-4); 5.90 (1H, br s, H-3′); 5.80 (1H, br s, H-2); 2.45 and 2.29 (2H, 2d, J = 17Hz, H$_2$-5′); 2.10 and 2.02 (6H, 2s, H$_3$-7′ and H$_3$-6); 1.67 (3H, d, J=7Hz, CHCH_3Ar); 1.10 and 0.98 (6H, 22s, H$_3$-8′ and H$_3$-9′). MS *m/z*

---

[6] J. W. Walker, *in* "Cellular and Molecular Biology: A Practical Approach" (J. Chad and H. Wheal, eds.), p. 179. IRL Press, Oxford, 1991.

(relative intensity) 413 ([M⁺], 0.4), 383(1), 263(5), 190(49), 162(19), 150(100), 135(33), 120(43) and 91(65).

This procedure is also used to prepare ABA-4-nitrobenzyl ester (ABA-4NBE; see Fig. 1) and the corresponding gibberellin A_4 (GA_4-2NPE and GA_4-4NBE), JA (JA-2NPE and JA-4NBE), and 1-NAA (1-NAA-2NPE and 1-NAA-4NBE) esters.[5]

Procedure for Preparation of Caged Salicylic Acid

Salicylic acid (2.5 g) in acetone (75 ml) is treated with *N*-hydroxysuccinimide (2.08 g). Dicyclohexylcarbodiimide (3.73 g) is added slowly, and the mixture is stirred for 3 hr, giving a white precipitate. The mixture is filtered and the filtrate evaporated *in vacuo* to yield salicylic acid *N*-hydroxysuccinimidyl ester (SA-NHS ester; 4.41 g), which is used in subsequent coupling reactions without further purification.

SA-NHS ester (100 mg) in dry acetonitrile (10 ml) is treated with 1-(2-nitrophenyl)ethanol (71 mg) and 4-dimethylaminopyridine (30 mg). The solution is stirred under nitrogen at room temperature for 1 hr. Ethyl acetate (50 ml) and 5% citric acid (50 ml) are added, and the product, recovered from the organic layer, is purified by flash chromatography using hexane–ethyl acetate mixtures to give salicylic acid 1-(2-nitrophenyl)ethyl ester (caged SA, SA-2NPE) (92 mg). ^{1}H NMR (CDCl$_3$) δ: 7.99 (1H, dd, $J = 8$ and 1Hz, ArH), 7.95 (1H, dd, $J = 8$ and 1.5Hz, ArH), 7.71 (1H, dd, $J = 8$ and 1.5Hz, ArH), 7.64 (1H, dt, $J = 8$ and 1Hz, ArH), 7.46 (2H, m, ArH$_2$); 6.96 (1H, dd, $J = 8$ and 0.7Hz, ArH), 6.92 (1H, dt, $J = 7$ and 0.8 Hz, ArH), 6.61 [1H, q, $J = 7$Hz, OCH(CH$_3$)Ar]; 1.80 (3H, d, $J = 7$Hz, CHCH_3Ar). MS m/z (relative intensity) 287 (M⁺, 3), 219(3), 151(11), 150(100), 138(31), 120(82), 103(17) and 92(64). This procedure is also used to prepare SA-4-nitrobenzyl ester (SA-4NBE).

The caged plant growth regulators are made up as a 10 mM stock in methanol, vacuum dried, and frozen until use.

Injection of Plant Cells and Flash Photolysis

Although direct loading of cells by these caged molecules is possible, injection is preferred as a means to deliver the cage into the cell. Therefore, the choice of plant tissue used for study is usually a compromise between what is possible to inject (the optimum being a single layer of cells with little autofluorescence) and what has a measurable response to the plant growth regulator. Guard cells, protoplasts, pollen tubes, and epidermal cells have been studied. These are mounted on thin coverslips and placed on the stage of a Nikon Diaphot TMD inverted microscope (Nikon, UK Ltd.,

Shropshire, UK) to enable the micromanipulators (Narashige NT-88 4D) of the microinjection unit to approach from above. Microelectrodes with tip diameters of approximately 0.3 μ were pulled from GC120 filamented electrode glass (Clark Electromedical Instruments, Reading, UK) and back-filled with 100 μM Calcium Green-1 or SNARF-1 (Molecular Probes, Eugene, OR) and 200 μM caged probe. The remainder of the electrode was back-filled with 3 M KCl. Care was taken to keep the cage-filled electrodes out of direct light; filling was possible under diffuse light in an otherwise darkened room.

For data discussed in this study, guard cells of *Commelina communis* were injected. These have the advantage of having visual responses to many phytohormones (a change in pore aperture), although they are harder to successfully inject than neighboring epidermal cells. The fluorescent probe and the cage were coinjected by ionotophoresis using a current of 0.1–0.2 nA for approximately 3 sec. At least 15 min was then allowed for the cell to recover from the injection event. Cells with visible signs of damage (e.g., collapse, condensation of the nucleus, irregularities in chloroplast arrangement) were discarded at this stage. Calcium or pH measurements were then made using fluorescence microscopy with an attached photometry system (Newcastle Photonic Systems, Newcastle-upon-Tyne, UK) or a Bio-Rad MRC-600 laser scanning confocal microscope.[7,8]

We have often used Calcium Green-1 (Molecular Probes),[9] which has high affinity and selectivity for free Ca^{2+}. Although Calcium Green-1 has the disadvantage that it is a single wavelength dye, we chose to use it for these studies, as it may be visualized using an argon laser-scanning confocal microscope. In addition, experiments with ratiometric calcium dyes, such as Fura-2 and Indo-1, are not possible as they are usually excited at 340–350 nm, overlapping with the wavelengths used for photolysis of caged probes. However, SNARF-1, used to measure intracellular pH, has the advantage of being both ratiometric and excited by visual wavelengths, which avoid release of the caged probe (excitation around 510 nm, and emissions at 580 and >610 nm).

For *in vivo* photolysis of the injected probe, we have used two flash release systems with success. First, a "pulse" of up to 30 sec, focused through the Plan Apo 40× objective, uses the epifluorescence lamp (Xenon 75-W) of the microscope via a 360 ± 10-nm interference filter under computer control (Newcastle Photonic Systems). Such exposures could be easily

---

[7] A. C. Allan, M. D. Fricker, J. L. Ward, M. H. Beale, and A. J. Trewavas, *Plant Cell* **6**, 685 (1994).

[8] V. E. Franklin-Tong, B. K. Drøbak, A. C. Allan, P. A. C. Watkins, and A. J. Trewavas, *Plant Cell* **8**, 1305 (1996).

[9] R. P. Haugland, "Handbook of Fluorescent Probes and Research Chemicals." Molecular Probes, Inc., Eugene, OR, 1992.

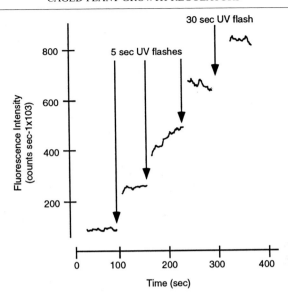

FIG. 2. Calibrating photolysis by the use of caged fluorescein. An individual guard cell was injected with the caged probe and then exposed to successive UV pulses. Fluorescence intensity was monitored using microphotometry (excitation, 480 nm; emission, 525 nm).

calibrated to a relative release by using caged fluorescein (see Fig. 2). For example, we injected individual epidermal cells with caged fluorescein; after 10 sec UV exposure, fluorescence increased to 400,000 counts. Aqueous solutions of fluorescein in cell-sized droplets (within an oil emulsion on the microscope) were used to calibrate this release to approximately 1 μM. One million counts correlated with a fluorescein concentration of 7 μM. Generally, 80% of the injected cage was released following a 20-sec exposure.

The second system for cage release that we have used successfully[10] requires purchase of a flash photolysis "gun." This system produces a 1-msec flash of UV light (300 mJ in the 330- to 380-nm band) from a XF-10 flash photolysis system (Hi-Tech Scientific, Salisbury, UK). Exposures are therefore much faster, but photometry (as shown in Figs. 2 and 4) cannot be performed with this setup because the epifluorescent lamp (needed for excitation of Calcium Green-1 or SNARF-1) is replaced by the flash gun. However, it is ideal for studies using laser-scanning confocal microscopy.

The viability of cells after photolysis events is always checked using noninjected controls or cells injected with compounds such as caged IP_3

[10] R. Malhó and A. J. Trewavas, *Plant Cell* **8,** 1935 (1996).

(Molecular Probes). Another good control molecule for our studies are the 4-nitrobenzyl esters (e.g., ABA-4NBE) of the caged plant growth regulators. Changing the position of the nitro group from the *ortho* to the *para* position makes ABA-4NBE nonphotolabile. The 4-nitrobenzyl esters therefore control against any activity of caged compounds without release or any nonspecific UV activation.

Caged Plant Growth Regulators: Biological Function after Photolysis

We have found that release of plant growth regulators in the cytosol of cells, following photolysis of the cage, results in biological activity as measured by changes in stomatal pore aperture. For example, a decrease in stomatal aperture occurs following a 30-sec UV photolysis of ABA-2NPE in the injected cell without affecting the uninjected companion guard cell (Fig. 3). Closure could be observed within 2 min of ABA release inside the cell. Measurements of closure rate can be made using either an adjustable Nikon Filar micrometer 10× eyepiece with an accuracy of ±0.2 μm or on the computer after image capture. We have photolyzed ABA-2NPE in a total of over 50 cells and observed closure in every case. The guard cell could be incubated after injection of ABA-2NPE for up to 1 hr with no deterioration in subsequent closing activity after photolysis. Closure can also result from damage due to microinjection. However, this possibility can be excluded using subsequent treatment with fusicoccin (10

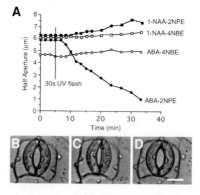

Fig. 3. The effect of caged ABA or caged 1-NAA on stomatal aperture after photolytic release in the cytosol of the guard cell. (A) Apertures of individual guard cells injected with cages or their nonphotolysible analogs. (B) Guard cell injected with caged ABA (right cell of the pair) before photolysis. (C) Thirty minutes after UV photolysis. (D) One hour after subsequent perfusion with 10 μM fusicoccin. Bar: 10 μm. [Modified from A. C. Allan, M. D. Fricker, J. L. Ward, M. H. Beale, and A. J. Trewavas, *Plant Cell* **6**, 685 (1994).]

TABLE I
EFFECTS OF INTRACELLULAR RELEASE OF PLANT GROWTH REGULATORS ON CYTOPLASMIC pH, Ca^{2+}, AND FINAL APERTURE RESPONSE OF *Commelina* STOMATAL GUARD CELL

Caged plant growth regulator	Effect of release on		Effect on stomatal aperture
	pH_{cyt}	$[Ca^{2+}]_{cyt}$	
1-NAA-2NPE	Acidification, sometimes oscillatory	None	Slight opening
ABA-2NPE	Slight acidification	Increase dependent on plant growth	Closure
JA-2NPE	Little change	Slight increase	None
SA-2NPE	Acidification	None	None

μM), which opens guard cells (by activating the plasma membrane H^+-ATPase[11]). Opening of the guard cell resulted from the release of 1-NAA within the guard cell (Fig. 3), after photolysis of 1-NAA-2NPE. Ultraviolet exposure of cells injected with the nonphotolyzable analogs, ABA-4NBE and 1-NAA-4NBE, resulted in no significant change in guard cell aperture.

We have also studied the effect of plant growth regulator release within the cell on cytoplasmic Ca^{2+} concentration and pH. Table I summarizes these results. Figure 4 shows how photometry experiments can provide a high temporal resolution to hormone-induced pH changes, as monitored by SNARF-1 fluorescence. To calibrate changes in the ratio of SNARF-1 fluorescence mixtures of isobutyric acid and ammonium hydroxide were added at the end of each experiment to verify a fixed pH reading. Although these results are only preliminary (our major study has involved caged ABA and calcium measurements[7]), they provide an indication that auxin and ABA have effects on secondary messengers, leading to changes in aperture when released *within* the cell. Other plant growth regulators, particularly salicylic acid, also have interesting effects worthy of future study. Gibberellin A_4 was not tested in this study.

Concluding Remarks

The use of caged phytohormones has already helped in our understanding of the basis of Ca^{2+}-induced guard cell closure.[7] We have found that the release of ABA inside cells results in guard cell closure. Controlled release of ABA inside the cells allowed for reduction in variation (loading, effects of differing cell populations) seen in previous experiments involving

---

[11] E. Marré, *Annu. Rev. Plant Physiol.* **30**, 273 (1979).

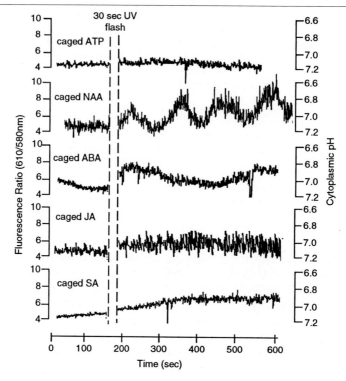

FIG. 4. The effect of photolysis of caged plant growth regulators on the cytosolic pH of *Commelina* guard cells, as measured by SNARF-1 ratio photometry.

exogenously applied hormones. We were therefore able to conclude that Ca^{2+}-mediated events are up-regulated in plants that have experienced elevated temperatures.

The finding that release of a plant hormone within the plant cell produces a rapid response indicates that our ideas of tightly controlled receptors pointing out into the apoplast, perceiving extracellular ligands, should be adjusted for the plant world. It may be that phytohormone-binding proteins in the cytoplasm or nucleus directly perceive the primary signal and so transduce it to a response in an entirely intracellular system. We can not exclude the possibility that a fraction of the cytosolically released hormone leaks out after the photolysis, thereby activating outward-facing receptors. However, the rapid perfusion of these cells, and the observation that the release of approximately 1 μM ABA induces closure at the same rate as exogenously applied ABA, at the same concentration, suggests that interior receptors can also transduce this signal. Future studies with caged plant growth regulators, perhaps complexed to dextrans, may provide us with information about the functionality of cytosolic receptors for plant responses.

[27] Use of Caged Compounds in Studies of the Kinetics of DNA Repair

By R. A. MELDRUM, R. S. CHITTOCK, and C. W. WHARTON

Introduction

DNA is susceptible to a wide range of damaging influences that can cause severe problems for a cell if it is not repaired with high fidelity. Oxygen radicals, alkylating agents, and radiation are perhaps the three most potent types of damaging agents. The fidelity of the genome in living cells is maintained by a complex series of processes that repair damaged DNA. The repair system constantly scans the genome for damage and affects repair of a wide range of types of damage on a continuous basis. The predominant forms of DNA damage are base mismatches, UV-induced pyrimidine dimers, base hydroxylation/alkylation, and single- or double-strand breaks induced by ionizing radiation. There are several pathways for repair, but probably the most important are nucleotide excision repair (e.g., UV-induced pyrimidine dimers) and base excision repair (e.g., ionizing radiation-induced single-strand breaks, where base damage and/or sugar damage occurs). These two pathways are illustrated in outline in Fig. 1, which shows that several sequential steps are involved in each process (for an excellent general reference to DNA repair, see Friedburg *et al.*[1]). In each case repair is completed by polymerization using the undamaged strand as the template to fill the gap created by the repair process, followed by ligation to form continuous strands.

Elegant genetic experiments have delineated the sequence of events that take place and the enzymes involved in the catalysis of each step. Relatively little information is available concerning the details of the kinetics of repair *in vivo* as severe perturbations (i.e., added inhibitors) have to be applied if specific steps are to be interrogated. Almost nothing is known

---

[1] E. C. Freidburg, G. C. Walker, and W. Siede, "DNA Repair and Mutagenesis." ASM Press, Washington, DC, 1995.

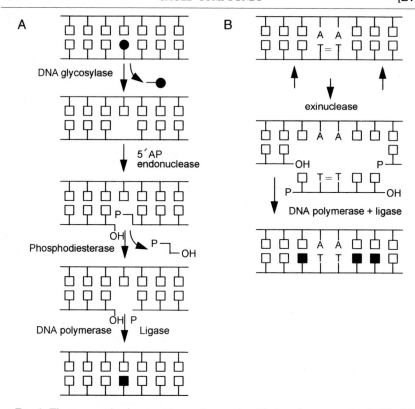

FIG. 1. The two predominant pathways for repair of lesions in mammalian DNA. (A) Pathway of base excision repair, which is responsible for the repair of oxidative and ionizing radiation (strand break) damage. Only one to five bases are replaced in this process. (B) Pathway for nucleotide excision repair. In this process, which is responsible for the repair of UV-induced pyrimidine dimers, some 30 nucleotides are replaced in the repair process. Solid square symbols show bases that have been replaced in the repair process and so can be used to monitor repair if radiolabeled nucleotides are used in repair experiments.

about events that occur within a few seconds of damage, and it is to this aspect that the application of caged nucleotides has been addressed.

Although overall repair processes take minutes or hours, it is extremely unlikely that cells will take a leisurely attitude to damage assessment and the requirement for emergency repairs. For a multicellular organism the danger of mutation (and hence the chance of carcinogenesis) is much more serious than cell death so significant differences will be expected, and are seen to occur, between repair in single cells, e.g., bacteria, and multicellular cells, e.g., mammalian. It is thus important that we know what criteria operate in the selection and kinetics of repair pathways in the face of various types of DNA damage.

Against this somewhat uncertain background the application of caged nucleotides to the study of DNA repair in mammalian cells carries a number of advantages as well as some disadvantages.[2-5] Perhaps the main advantage is that a clean start can be made to studies of repair, as the reagent can be decaged with a laser flash at a time point that is well-defined relative to the damaging event. Use of readily permeable nucleoside bases would blur the start point of the experiment due to the requirement for (tri)phosphorylation before incorporation into DNA. Also, the start point can be rendered indistinct by incorporation of nucleotides by replication synthesis prior to the application of DNA damage.

The damaging event itself can be applied by using lasers (either UV- or laser-induced X-rays), so both the damage and the trigger for monitoring of the subsequent repair kinetics can be well defined.[2-5] The caged nucleotide is radiolabeled so that after quenching of the repair process, nucleotide incorporation into DNA can be followed, although this requires the tedious procedure of DNA isolation (or, at the least, separation from unincorporated nucleotides). Because the method monitors the common polymerization step, both the nucleotide excision repair pathway and the base excision pathway are assayed. Differentiation of the pathways will require a more sophisticated approach, although the type of damage applied directs the pathway that is predominantly utilized. With the exception of the radiochemical assay for incorporation of nucleotides into DNA, the experiment readily lends itself to automation and this has been achieved as will be described later.

The method necessarily involves one procedure that can be regarded as a potential disadvantage and this stems from the impermeability of the mammalian cell wall to nucleotides, even when in caged form. In order to gain access for the caged reagents, electroporation has been used to load the cells. This process, although kinetically advantageous compared with, for example, detergent-induced permeabilization, which requires minutes rather than milliseconds, it does induce damage in DNA and the nuclear envelope. Although it has proved possible to optimize the electrical parameters for electroporation so as to minimize DNA damage, it seems that the lamins of the nuclear envelope are disrupted by voltages lower than those

---

[2] R. A. Meldrum, S. Shall, D. R. Trentham, and C. W. Wharton, *Biochem. J.* **266**, 885 (1990).
[3] R. A. Meldrum, S. Shall, and C. W. Wharton, *Biochem. J.* **266**, 885 (1990).
[4] C. W. Wharton, R. A. Meldrum, C. Reason, J. Boone, and W. Lester, *Biochem. J.* **293**, 825 (1993).
[5] W. S. Meaking, J. Edgerton, C. W. Wharton, and R. A. Meldrum, *Biochim. Biophys. Acta* **1264**, 357 (1995).

required to achieve poration. It is possible to introduce a refractory period in which cellular functions fully recover prior to application of DNA damage. An important fact is that electroporation-induced DNA damage is not repaired on the time scale of repair experiments with caged nucleotides, although it may influence the kinetics and fidelity of the repair response. This aspect is discussed in more detail later.

The synthesis and application of the 2-hydroxyethyl o-nitrophenyl derivatives of [methyl-^{3}H]thymidine triphosphate, [α-^{32}P]thymidine triphosphate, and [α-^{32}P]dideoxyadenosine triphosphate (ddATP) are described. The latter is a chain terminator and so when this is employed only a single label can be inserted at a given repair site. Ideally, dideoxythymidine should be used, as thymidine is exclusive to DNA; however, it was not possible to obtain this material in ^{32}P-labeled form. A variety of unlabeled caged nucleotide triphosphates have also been used; these were prepared using the standard method.[6] The only serious problem that was encountered in the caging reaction was when [methyl-1',2'-^{3}H]thymidine triphosphate was used. The additional tritium atoms on the ribose ring caused double caging to occur, one presumably placed on the ribose hydroxyl group. This is assumed to arise from a tritium isotope effect relayed to the 3'-OH reaction center. This material was very hydrophobic and could not be electroporated into cells, although it decaged satisfactorily.

Synthesis of High Specific Activity [α-^{32}P]Dideoxyadenosine Triphosphate and [methyl-^{3}H]Thymidine Triphosphate

The preparative procedure[2] is based on using 1 mCi [α-^{32}P]dideoxyadenosine triphosphate labeled at 5000 Ci/mmol. Due to the very high specific activity the mass of ddATP is very small and it proved necessary to use a higher concentration of 1-(2-nitrophenyl)diazoethane than the 80 mM used in the standard preparation.[6] Because there is also almost no buffering power available in such a dilute solution, ATP was added to provide buffering at pH 4.0. Accordingly, 50 μl of 5 mM ATP, adjusted to pH 4.0, was added to 1 mCi ddATP (in 100 μl aqueous solution as delivered by Amersham International). To this solution, contained in a vial that could be tightly stoppered, was added 250 μl 1 M 1-(2-nitrophenyl)diazoethane in chloroform, prepared as described.[2,6,7] Stirring of the solution proved only partly effective so the reaction vial, wrapped in aluminum foil, was *vigorously* shaken for 24 hr.

[6] J. W. Walker, G. P. Reid, J. A. McCray, and D. R. Trentham, *J. Am. Chem. Soc.* **110**, 7170 (1988).
[7] J. W. Walker, G. P. Reid, and D. R. Trentham, *Methods Enzymol.* **172**, 288 (1989).

Safety Note

Considerable pressure can build up during the shaking process so it is essential to release this pressure before opening the vial. This is achieved by inserting a hypodermic needle through the plastic lid of the vial before removing the aluminum foil. Both the shaking and the venting should be performed in a well-protected fume hood.

To the reaction mixture was added 1 ml water, followed by 2 ml chloroform. The mixture was agitated by drawing a stream of air through the solution and then the chloroform was removed with a pipette and discarded to a ^{32}P store vessel. This procedure was repeated twice more before traces of chloroform were removed by drawing a stream of air through the solution for 15 min.

Purification of Caged [α-^{32}P]ddATP

This was achieved by using high-performance liquid chromatography (HPLC) with a reversed-phase column (Waters μBondapak C-18). The column was eluted isocratically with 10 mM phosphate buffer, pH 5.5, containing 25% methanol at a flow rate of 1 ml/min. The ATP added to act as buffer served to perform another important function in the purification stage. It acts as a carrier to prevent the severe tailing that occurred when ATP was not added to the reaction mix. ATP eluted essentially coincidentally with uncaged ddATP whereas caged ATP eluted well before caged ddATP. The elution of ATP and caged ATP was detectable using the photometric detector set at 260 nm, whereas the elution of the ^{32}P-labeled materials was detected by scintillation counting of small aliquots of the 1-ml samples collected during elution. Caged ATP elution is sufficiently well separated from caged ddATP elution that essentially complete removal of caged ATP is achieved on two passages through the HPLC column. In practice, the presence of a small quantity of caged ATP in the caged ddATP should have no adverse effects on DNA repair experiments, but because of the potential cytotoxic effects of the nitrosoketone photolysis product, material to be used in experiments with cells was always passed through the column twice.

Analysis of Purified Caged [α-^{32}P]ddATP

Analytical HPLC, performed using the same conditions as for the preparative runs, showed that the radiolabeled material was purified to 96–98% in terms of caging after two passages through the HPLC column, whereas

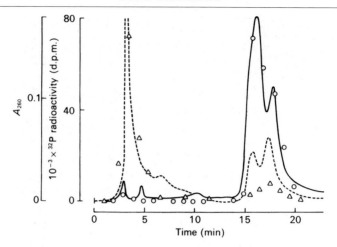

Fig. 2. Photolysis and HPLC analysis of purified ^{32}P-caged ddATP. Purified unlabeled compound (10 nM) was added to act as a carrier. A Waters μBondapak C-18 reversed-phase column was eluted at 1 ml/min with 10 mM phosphate buffer, pH 5.5, containing 25% (v/v) methanol. Open circles and line, unphotolyzed compound; triangles and dashed line, 3-min photolysis using a 366-nm chromatography lamp. Samples were exposed in a 3.5-cm cell culture dishes.

the total yield was >70%. The results of a typical run are shown in Fig. 2. Note that poor results with severe tailing occurred if a carrier was not added. In this case unlabeled ddATP was used. Similar experiments showed that if ATP was not added as carrier and buffer in the synthesis of caged material the yield was 30–40% and the purity in terms of caging was 70% after two passages through the HPLC column. The presence of the two expected diastereoisomers of the caged compounds was observed for all compounds tested, except caged ara-CTP.

Photolysis of Caged Compounds

We have not made kinetic measurements of caged compound photolysis and the release of uncaged material, but we have assessed the efficiency of the decaging process using UV difference spectroscopy. When caged compounds are photolyzed, a difference spectrum appears at 310 nm and has a differential extinction coefficient of 4040 M^{-1} cm^{-1} for all the compounds analyzed. This can be used to follow the decaging process, provided the complete experiment is completed within 10 min. A slow decay of the difference spectrum occurs, which clearly has to be avoided in experiments where quantitative measurements of decaging are required. During analytical procedures, an Hg (nom. 366 nm) chromatography lamp has been used to induce photolysis, but photolysis has also been measured using a 351-nm XeF excimer laser (10-nsec flashes, 200 mJ) as used in DNA repair experiments. No significant differences were found in decaging behavior, despite the very large difference in illumination intensity. Figure 3 shows the results of a typical photolysis experiment.

[^{3}H]Thymidine triphosphate (40–80 Ci/mmol) was caged using this procedure, although a carrier proved unnecessary for HPLC purification.

Stability of Caged Compounds in Mammalian Cells

Serum-supplemented medium proved to contain one or more enzymes capable of causing rapid (16% in 2 hr) degradation of caged compounds, and the degradative activity was not eliminated by heat treatment.[2] It is therefore necessary to use a serum-free medium or phosphate-buffered saline for cellular experiments with caged compounds. When [^{32}P]ddATP was incubated with thoroughly washed mouse (PY3T3) fibroblasts, there was very little degradation in 30 min, but extensive degradation was observed after 18 hr. Similar experiments in which permeabilized or lysed cells were used gave very similar results, suggesting that a cell surface enzyme may be involved. It thus seems that caged nucleotides appear to be sufficiently stable in the cellular milieu to permit short term kinetic experiments up to 30 min without the complication of the degradation of caged nucleotides. It is unlikely that the nature of the nucleotide will have

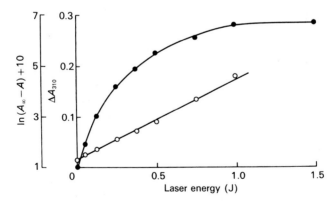

FIG. 3. Photolysis of caged ddTTP with 351-nm excimer laser irradiation. ddTTP (69 μM) in 30 mM HEPES buffer, pH 7.0, was irradiated in a 1-ml quartz microcuvette with pulses of 25 mJ. The absorbance at 310 nm was measured by removal from the target position, and the experiment was completed in less than 10 min. The open symbols represent a semilogarithmic plot of the data, which shows that it is a first-order process.

a large effect on the degradative process as phosphodiesterases, presumed to be responsible for the hydrolysis, tend to be very nonspecific.

DNA Repair Experimental Protocol

The overall scheme of the experiments involves the following steps, as shown in Fig. 4. The caged nucleotides are loaded into cells by electroporation (see later). An interval of a few seconds or longer is introduced to allow some recovery and reequilibration of the cells following the electric shock. Damage to the DNA is applied in the form of a 248-nm laser pulse (or a pulse of soft X-rays generated by a laser-induced plasma). A second variable delay allows repair to start, which is followed by photoactivation of the caged nucleotides with a pulse of 351-nm UV laser light. A further variable incubation period allows radiolabeled nucleotides to be incorporated at repair sites before the incubation is quenched and the cells are lysed with sodium hydroxide. There are two possible ways to collect data. The caged nucleotides may be photoactivated immediately after the DNA damaging pulse and a variable interval allowed before quenching with NaOH. This approach leads to profiles of the accumulation of radiolabeled bases in the DNA known as a *progress curve*. The second approach is to vary the interval between the DNA damaging laser pulse and the pulse that photoactivates the caged reagent. This gives rise to profiles of the *rate* of incorporation of radiolabeled bases in the DNA or *rate profiles*.

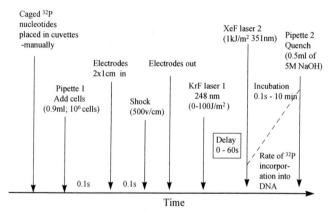

FIG. 4. The experimental scheme for the study of early events in the repair of UV and soft X-ray-induced DNA damage in mammalian cells. See text for experimental details.

Design of Computer-Controlled Apparatus

The features of the apparatus are shown in Fig. 5. The turntable has eight stations that accommodate 1-ml quartz microcuvettes, which have all four faces polished. The 248-nm light is applied on the larger face of the cuvette (1 × 2 cm) as the dose (e.g., <100 J/m^2) is much less than that required from 351-nm laser and the power loss from expansion of the beam is inconsequential. Also, because the 248-nm radiation is more strongly absorbed than the 351-nm photoactivating light, it is advantageous to arrange it so that it has to penetrate only the short dimension of the cuvette (4 mm), where attenuation is about 20% when 10^6 cells/ml are used. Custom-made cuvettes are used in which the side walls are fabricated from high-quality quartz in order to minimize the attenuation of the 248-nm light. Control measurements ensure the nominal dose reaches the center of the cuvette. Although similar considerations apply to the application of the 351-nm light, the absorption/scattering is much less. The 351-nm UV laser beam traverses 1 cm, the normal path length of the cuvette. The apparatus is interfaced with a microcomputer that allows flexibility in programming so that a wide range of experiments can be conducted with this apparatus.

Following an experiment, the lysed cells were treated with trichloroacetic acid, and the precipitated material was collected on glass fiber filters and extensively washed prior to scintillation counting. In some experiments

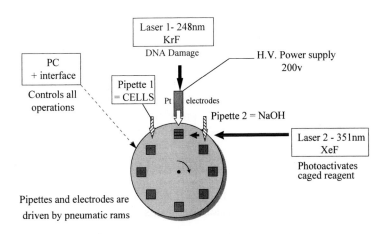

Turntable device for automatic dispensing,
laser exposure, permeabilization and quenching samples

FIG. 5. The computer-controlled apparatus used in DNA repair experiments with caged nucleotides.

the DNA was isolated and it was shown that >90% of the radioactivity incorporated into precipitated material was in the DNA.

Electroporation

Electroporation is a widely used technique for introducing molecules into cells to which the cell membrane is naturally impermeable. The most common use and thus the best assessed for its efficiency is to introduce foreign DNA into cells (i.e., transfection). It is the most suitable method to introduce caged nucleotides into cells because it is rapid and potentially reversible. These properties minimize leakage of cellular components in permeabilized cells. When the process is used for transfection of DNA, it is only important that a sufficient proportion of the cells into which the foreign DNA has integrated remain viable and that the surviving cells maintain a clonogenic potential. For experiments with caged compounds, it is important that permeation of the whole cell population is optimal. It is also important that the cellular process being studied, and for which the caged compounds are intermediates, is not disturbed by the effects of electroporation. Since we have studied the repair of DNA damage using caged nucleotides, we have examined, in some depth, the possibility that electroporation may itself induce DNA damage and how this can be minimized.

The electroporation methods and apparatus used follows closely that of Chu *et al.*[8] The permeabilization buffer used consists of 20 mM HEPES, pH 7.05, 137 mM NaCl, 5 mM KCl, 0.7 mM Na$_2$HPO$_4$, 6 mM dextrose, pH 7.05. Two different capacitor sizes were used: 1160 and 690 μF. The $t_{1/2}$ for discharge of a capacitor of 1160 μF is 10 msec and for a capacitor of 690 μF is 5.8 msec. HL60 cells (0.9 ml, 10^6/ml in 1-ml microcuvettes, as earlier) in permeabilization buffer were subjected to an electrical discharge at varying voltages and at two different values of capacitance.

The DNA from the cells was then analyzed by the alkaline elution method for breaks. Figure 6 shows the relationship between the electric charge (coulombs) and the strand scission factor (SSF), which is a standard parameter related to strand breaks.[9] When these results were combined with those of an experiment that estimated the sizes of molecules that permeated the cells at different field strengths, it was fortunately apparent

---

[8] G. Chu, H. Hayakawa, and P. Berg, *Nucleic Acids Res.* **15,** 1311 (1987).

[9] C. Baumstark-Kahn, S. Griensenbach, and R. R. Weichselbaum, *Free Rad. Res. Commun.* **16,** 381 (1992).

[10] J. S. Clegg and S. A. Jackson, *Biochem. J.* **255,** 335 (1988).

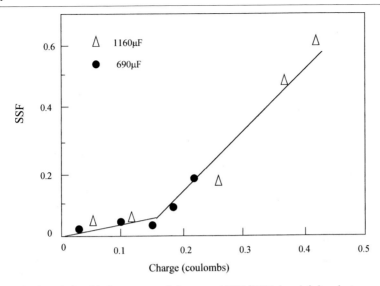

Fig. 6. The relationship between total charge and SSF (DNA breaks) for electroporation of HL60 cells. The total charge passed through the cell suspension is Q (coulombs) = C (capacitance in Farads) × V (voltage discharge across the electrodes). Triangular symbols represent points at which the capacitance was 1160 μF, and circular symbols represent points at which the capacitance was 690 μF; the voltage was varied at both levels of capacitance.

that caged nucleotides would permeate the cell membrane under electroporation conditions that would not induce DNA damage, i.e., below a charge of 0.15 coulombs. Cells were electroporated at different voltages in the presence of various sizes of fluorescein isothiocyanate (FITC)-labeled dextrans, spun through silicon oil to separate the cells from the surrounding aqueous medium, lysed, and the fluorescence measured. Figure 7 shows profiles of the permeability of two different sized molecules into the cells at different field strengths. The caged nucleotide is of intermediate size between these two molecules, and optimum permeation of a molecule of this size can take place at a field strength of 500 V/cm without inducing DNA damage. At this field strength, it is estimated from these experiments that between 5000 and 10,000 caged nucleotides gain access to the cytosol in a single cell when the external concentration of caged molecules is 0.02 μM.

Notably there was no apparent repair of electroporation-induced DNA breaks in a 2-hr period after application of the electric pulse. We assume the breaks induced are double stranded and that these must severely compromise the survival of those cells in which they occur.

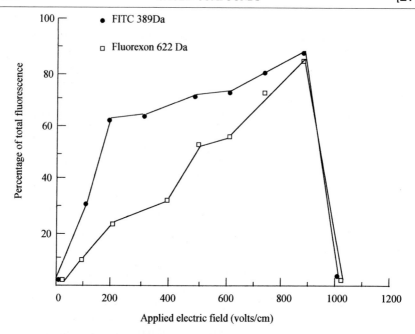

FIG. 7. FITC-dextrans were added to electroporation buffer at concentrations that standardized the fluorescence per unit volume. HL60 cells were electroporated in these solutions, and the cells were then separated from the aqueous medium by centrifugation through oil and lysed in 1% Triton X-100. The fluorescence of each sample is expressed as the percentage of total possible fluorescence if 100% of the cells were permeabilized and took up the dye. The capacitance used in these experiments was 1160 μF.

Results of DNA Repair Experiments

An unexpected result was found in studies of the early stages of DNA repair using the apparatus and methods described earlier.[11] For both 248-nm UV light *and* 1-keV soft X-rays, a transient incorporation of nucleotides into DNA occurred. Maximal incorporation occurs after ca. 30 sec after which most of the radiolabel is removed, the lowest level of incorporation being seen after ca. 1 min. With thymidine triphosphate there is then a steady incorporation following the initial transient, but with ddATP all nucleotides are removed and there is only minor further incorporation. The maximal value of incorporation in the transient phase is approximately equal for each nucleotide, which indicates that only a single base is incorporated at each repair site as ddATP is a chain terminator. Two main deductions have been made from these data.

[11] R. A. Meldrum, W. Meaking, and C. W. Wharton, *Nucleic Acids Res.* **22,** 1234 (1994).

The first is that a common form of damage, caused by both UV laser and soft X-ray irradiation, is being "repaired" in the transient phase, which we have named the "panic" response. Based on measurements of strand breaks induced by each type of radiation, it seems most likely that *direct breaks* in DNA are being "repaired." The high intensity of the UV laser radiation induces ionizing radiation-like direct breaks and so this is proposed to be the common feature. Pyrimidine dimers, normally the primary photoproduct of UV irradiation, do not seem to play a role in the transient response.

That TTP behaves very similarly compared with ddATP suggests that the transient phase represents low fidelity base incorporation and that the fidelity is sufficiently low that almost all bases need to be removed before "correct" repair can start. It has been suggested that this response may be related to the need to attain strand continuity (to maintain the reading frame) in circumstances where severe damage may be imminent. Single-strand breaks in close proximity on either strand may look like double-strand breaks to the enzymes that survey the DNA for integrity.

An analogy for these surprising results has been proposed which likens the panic response to the emergency repairs a dweller in a hurricane zone might attempt to make to the roof of their house *during* the hurricane. Exactness would not be important at this time, whereas a more leisurely and purposeful repair could be made after the passage of the hurricane. Whatever the explanation, it is clear that much remains to be done to achieve a clear insight into the biological role of the panic response. Could, for example, it be related to a switch into apotosis?

Author Index

Numbers in parentheses are footnote reference numbers and indicate that an author's work is referred to although the name is not cited in the text.

A

Aarhus, R., 403, 404, 406, 407, 407(9), 411, 415(7, 13)
Abood, L. G., 30, 32(13), 463, 464(69)
Abrahams, J. P., 308
Acerbis, G., 152
Adachi, S., 256, 277
Adams, P. J. H. M., 83
Adams, R. S., 455
Adams, S. R., 30, 77, 78, 95, 117, 155, 223, 323, 324, 357, 358, 365, 365(13), 367, 368, 369(12), 374(12), 376(12), 416, 421, 431, 439
Adhya, S., 416
Afanas'eva, G. B., 69
Agrawal, S., 153
Albers, W., 291, 294(13), 297(13), 298(13), 305(12)
Albertson, D. G., 467, 468(79)
Alfano, R. R., 179
Allan, A. C., 474, 478, 480
Allen, P. M., 123
Allen, T. S., 310, 318
Allin, C., 223
Almers, W., 221, 241, 357
Almo, S. C., 95, 216, 217(24), 256, 259(12), 308
Altendorf, K., 305(46), 306
Altenhofen, W., 417
Alvarez, L. J., 409
Alves, A. M., 136
Amador, M., 457
Amit, B., 31, 98, 155, 455
Ämmälä, C., 220
Amos, A. A., 21
Amstutz, C., 368
Andersen, M. L., 10
Anderson, B. E., 474
Anderson, J. C., 8

Andersson, I., 266
Andres, V., 278
Andrews, M. A. W., 327
Andrews, P. C., 78
Andrieux, C. P., 10
Anson, M., 323
Antar, Y. M., 192
Aoki, T., 335, 337(76), 340(76)
Aoshima, H., 31, 46, 447, 448(27a), 450, 460(29)
Apell, H.-J., 204, 291, 305(11), 305(47), 305(48), 305(49), 305(50), 306
Applied Biosystems, 145, 153(38)
Arbuckle, T., 269
Arner, A., 215
Arnost, M. J., 69
Asami, K., 279
Ashley, C. C., 310, 323, 357
Athey, P. S., 4, 5(16, 17), 7(16, 17), 8(16, 17), 17(16, 17), 18(16, 17), 19(16, 17), 321, 323
Augustine, G. J., 122, 381
Augustinsson, K.-B., 265
Avery, L., 30, 467, 468, 468(78, 80), 469(78, 80, 81, 83)
Axelsen, P. H., 265, 273(6)

B

Baas, P. W., 353
Bagshaw, C. R., 215, 335
Baldwin, J. E., 8
Ballivet, M., 471, 472(87a)
Balogh-Nair, V., 236
Bamberg, E., 32, 289, 290, 291, 291(2), 293(8, 10), 294(9), 298(9), 299(8), 300, 301, 305(5, 6, 8–10, 25, 44, 46, 52), 306, 470, 471(85), 472(85)

Barabas, K., 293, 296(17)
Barany, F., 153
Bargmann, C. I., 468, 469(83)
Barltrop, J. A., 31, 455
Barman, T., 116
Barnekow, F., 137
Barner, J., 10, 13(31), 24(31), 25(31)
Barnes, J., 15, 32, 37(32)
Barsotti, R. J., 80, 88(4), 240, 301, 310, 311, 318, 323, 325, 328, 328(37), 340, 357, 376(1)
Barth, A., 122, 226, 227, 231, 234(15), 235(15), 237(18), 240(15), 241(18)
Bartlett, W., 10, 13(32), 15, 25, 25(32)
Barz, W. Z., 24
Baumstark-Kahn, C., 492
Bauschke, E., 136
Bax, A., 90
Bayer, E. A., 135, 135(1), 139, 152(1)
Bayley, H., 35, 96, 116(17), 117, 118, 119, 120, 121, 122, 122(21), 123, 123(17, 21, 22), 128, 129, 131, 132, 133(21), 134, 134(30), 155, 266
Baylor, D. A., 176, 416, 417, 430(14)
Beale, M. H., 474, 475, 476(5), 478, 480
Beaugé, L., 301
Beavo, J. A., 416
Bechem, M., 222
Becker, A., 233, 300, 305(25)
Becker, P. L., 219
Bendig, J., 415, 417, 419, 419(23), 420, 422, 425, 427, 428
Benning, M. M., 308
Berchtold, H., 308
Berendzen, J., 256
Berg, P., 492
Berger, C. L., 318, 335
Bergot, B. J., 153
Berovic, N., 247, 250(2, 3)
Berridge, M. J., 381, 396
Bert, R. J., 417
Bertrand, D., 471, 472(87a)
Betz, H., 444
Beynon, T. D., 247, 250(2, 3)
B'hlen, P., 147
Billington, A. P., 30, 31, 32(11, 20), 41(11, 20, 56), 48, 48(11, 20), 139, 147, 150(30, 39), 155, 266, 269(18b), 270(18b), 272(18b), 447(4), 449(4), 452(4), 453(7), 453(7), 454, 456, 456(7), 457(7), 461, 468(7)

Black, M. M., 353
Blättler, W. A., 96, 118, 141, 155, 156(13), 175(13)
Blaustein, M. P., 301
Blay, G., 54
Block, S. M., 335, 337(79)
Boerner, W. M., 192
Bokvist, K., 220
Boldac, J. M., 277
Bolsover, S., 374
Bon, S., 265, 272, 273(25)
Bond, M., 312
Bonhoeffer, T., 176
Boone, J., 485
Borgstahl, G. E. O., 256
Borisy, G. G., 75
Borlinghaus, R., 291, 305(11, 50), 306
Borman, J., 444
Born, M., 179, 190(20), 194(20), 360
Böttger, M., 474
Bottiroli, G., 152
Bouet, F., 265, 273(6)
Bourgeois, D., 256, 277
Bourne, Y., 265, 273(9)
Boyd, A. E., 77
Boyer, P. D., 225
Bradford, M. M., 103, 112(31)
Braha, O., 134
Braslavsky, S. E., 421
Bratt, G. T., 403, 404(1)
Braxmeir, T., 169
Brenner, B., 311, 313
Breter, H., 136
Brinton, C. C., 136
Brintzinger, H., 232
Brown, A. M., 220
Brown, E., 203, 204(3), 356, 358, 365, 365(13), 367, 368, 375(13), 377(13)
Brown, R. L., 417
Brubaker, M. J., 252
Bruhnke, J., 118, 255, 266
Brundage, R. A., 95
Brune, M., 318, 332
Brustovetsky, N., 300, 305(25)
Buechler, J. A., 415
Burch, J. M., 179
Burgess, W. H., 89, 91(10)
Burke, P. M., 256
Burmester, C., 95, 215, 251
Burtnick, L. D., 98
Bus, H., 416

AUTHOR INDEX

Busby, S., 416
Butt, H.-J., 289
Buu-Hoi, N. P., 20

C

Cabib, E., 269
Calder, G. M., 357
Callamaras, N., 380, 389, 390(26), 402
Callaway, E. M., 30, 443, 467
Callender, R., 236
Cameron, J. F., 100, 155
Candia, O. A., 409
Cannon, A. M., 152
Cantor, C. R., 135
Capiod, T., 221, 381
Cardona, M. L., 54
Caretta, A., 417
Carpenter, B. K., 30, 31, 32, 32(20), 35(22), 41(31), 48(20, 22), 122, 139, 147, 150(30, 39), 155, 266, 269(18b), 270(18b), 272(18b), 439, 444, 453(7, 10, 12), 456, 456(7), 457(7, 8, 10), 468(7, 8)
Carraway, R. E., 94
Carter, T. D., 219, 381
Cash, D. J., 46, 447, 448(27a), 450, 460(29)
Cavaggioni, A., 417
Cepus, V., 223, 243, 244(47)
Cerny, R. L., 137, 153(24)
Chageux, J. P., 461
Chaisson, E., 180
Chanal, P., 265
Chang, C., 35
Chang, C.-Y., 117, 118, 121, 123(17, 22), 129, 132, 155, 266
Chao, J. L., 233
Chapman, A. C., 229
Chase, P. B., 313
Chatonnet, A., 265
Chattopadhyay, A., 46
Cheley, S., 123, 134
Chen, C. X., 236
Chen, Y., 256
Cheney, R. E., 307
Chillingworth, R. K., 318, 332
Chimilio, L., 11
Chittock, R. S., 245, 247, 250(2, 3), 483
Choi, I., 400, 400(13)
Choi, J., 382, 386(12)
Choquet, 457, 461(65)

Chorongiewski, H., 223, 233(2)
Chow, R. H., 221
Christensen, B., 305(52), 306
Christensen, L. M., 153
Christy, R. W., 179
Chu, G., 492
Chung, S., 463, 464(68), 466(68a)
Cifuentes, F., 241
Clapper, D. L., 403, 404(1)
Clark, R. B., 220
Clark-Lewis, I., 134
Claussen, M., 474
Clegg, J. S., 492
Clore, G. M., 90
Cocuzza, A. J., 136
Cody, W. L., 78
Coherent, Inc., 178
Cohn, M., 127
Cole, N. B., 78
Cole, P. A., 133
Collins, A., 301
Colthup, N. B., 227, 242(16, 17)
Conibear, P. B., 215, 335
Conklin, E. J., 135, 135(3), 152(3)
Connolly, B. A., 225
Cook, C. S., 333
Cook, S. N., 134
Cooke, R., 308, 309, 313, 324(22), 343
Cooper, E., 471, 472(87a)
Corbin, J. D., 417
Cornwell, T. L., 415
Corrie, J. E. T., 3, 7, 30, 101, 122, 139, 204, 220, 221, 223, 226, 227(5), 231(5), 234(15), 235(25), 240(15), 252, 260, 263(19), 306, 318, 321, 323, 324(52), 328(52), 330(52), 332, 343(52), 416, 421(10), 455, 456, 456(52), 457(52)
Coull, J. M., 137, 150(25)
Courtney, K. D., 278
Craik, J. S., 101, 318
Cram, D. J., 10
Crevel, I. M.-T. C., 310
Crick, F., 443
Crivici, A., 90
Cross, R. A., 310
Cruickshank, D. W. J., 265, 277(11)
Cubitt, A. B., 77
Cummings, J. P., 68
Currie, K. P., 222
Cusack, E., 123
Cuzikowski, A. P., 155

D

Dairman, W., 147
Dalva, M. B., 30, 443, 467, 472(77)
Daly, J. W., 227, 242(16)
Dani, J. A., 457
Danley, L. E., 134
Dantzig, J. A., 95, 214, 225, 307, 309, 318, 319, 320(21), 323, 324, 324(21), 325(54), 328(54), 330(20), 331(21), 339(21)
Davis, B. A., 20
Dawson, B. A., 137, 153(22, 23)
Dawson, P. E., 134
Deacon, G. B., 241
De Boni, U., 349
DeGroot, C., 135
Delaney, K. R., 467
Delcour, A. H., 455
DeLuca, M., 245
DeMayo, P., 455
Denk, W., 30, 31, 32(20), 41(20), 48(20), 139, 147, 150(30, 39), 176, 179, 266, 269(18b), 270(18b), 272(18b), 357, 388, 444, 453(7), 456(7), 457(7), 467, 468(7, 78), 469(78)
de Pont, J. J. H. H. M., 290, 305(5)
Desai, S., 135, 135(3), 152(3)
Desantis, A., 220
Desel, H., 114
Devine, E. E., 471
de Weer, P., 295, 298
Dicky, D. M., 404, 407(9), 415(7)
Diffee, G. M., 241
Dilger, 461
DiPolo, R., 301
Ditmar, R. M., 233
Dolphin, A. C., 222
Doppler, T., 141
Dose, K., 305(52), 306
Døskeland, S. O., 417
Drechsel, D., 335
Drexel, D., 72
Drøbak, B. K., 357, 478
Dröse, S., 305(46), 306
Drummond, R. M., 94
Duddy, S. K., 432
Dudley, G., 10, 13(31), 15, 24(31), 25(31), 32, 37(32)
Duke, E. M. H., 217, 219(27)
Duong, T. T., 139
Dupont, Y., 241

Durden, D. S., 20
Durrant, I., 136
Dyer, D. H., 252, 277
Dzeja, C., 415, 417, 419, 419(23), 420, 422, 425, 427, 428

E

Eccleston, J. F., 417
Eckstein, F., 225, 417
Edelson, R., 122
Edeson, 461
Edge, M. D., 136
Edgerton, J., 485
Edidin, M., 78
Edman, K. A. P., 343
Ehret-Sabatier, L., 265, 273(6), 277
Eichinger, L., 107
Eigen, M., 447, 473(23)
Eiler, J. J., 229
Einstein, A., 179
Eisenberg, D., 101
Eisenberg, E., 110, 309, 324
Eisenrauch, A., 301, 305(44), 306
Eismann, E., 417, 426(24)
Eliasson, L., 220
Elliott, G., 134
Ellis-Davies, G. C. R., 219, 221, 240, 241, 241(36), 253, 301, 310, 318, 323, 324, 331, 340, 357, 374, 376(1)
Ellman, G. L., 278
Ellman, J. A., 118, 134, 266
Elsner, S., 471
Engels, J., 58, 59, 416, 419, 421(29)
Engert, F., 176
Epel, D., 278, 282, 284, 286, 286(6), 287, 288
Epstein, W. W., 8
Escobar, A., 241, 357
Evan, G. I., 78
Evans, F. E., 417
Evans, M. R., 136
Eysselein, V. E., 3

F

Fabiato, A., 325, 347(56)
Fabiato, F., 325, 347(56)
Fawcett, M.-C., 117
Fay, F. S., 78, 94, 95

AUTHOR INDEX

Fearon, K. L., 153
Featherstone, M. R., 278
Feeney, J., 221, 382, 443
Feldmeyer, D., 222
Fendler, K., 289, 290, 291, 293(8, 10), 294(9, 13), 297(13), 298(9, 13), 299(8), 304, 305(6, 8–10, 12, 40, 46), 306
Ferenczi, M. A., 254, 318, 321, 323, 324(52), 328(52), 330(52), 332, 334(70), 343, 343(52)
Fernandez, J. M., 357
Ferraretto, A., 152
Fersht, A., 447, 473(25)
Ferster, D., 463, 464(68), 466(68a)
Fetscher, C. A., 247
Feuerstein, J., 225
Fidler-Lim, N., 241
Fielding, A., 325, 441
Fill, M., 340, 357
Finer, J. T., 335
Fink, R. H. A., 214
Finkelstein, A., 136
Finn, F. M., 136
Finn, J. T., 416
Fischbach, G. D., 46, 439
Fischmeister, R., 220
Fister, T., 321, 323
Fleet, A., 374
Fletterick, R. J., 308
Fodor, G., 24
Fodor, S. P. A., 98
Foldman, A., 265
Forbush, B. III, 1(4), 3, 130, 131(37), 134(37), 155, 168(2), 176, 233, 240(33), 247, 290, 291(4), 293, 299(4), 307, 421, 431, 455
Fork, R. L., 178, 364
Forster, A. C., 136
Frace, A. M., 220
Franken, S. M., 260
Franklin-Tong, V. E., 357, 478
Franks-Skiba, K., 343
Franzini-Armstrong, C., 310, 318, 357
Frazier, A. W., 230
Fréchet, J., 100, 155
Fricker, M. D., 478, 480
Friedburg, E. C., 483
Friedrich, T., 32, 289, 291(2), 305(51), 306, 470, 471(85), 472(85)
Frings, S., 415, 417, 419, 419(23), 420, 422, 425, 426, 427, 428

Fritz-Wolf, K., 308
Froehlich, J. P., 291, 294(13), 297(13), 298(13), 304, 305(12, 40)
Frolow, F., 265, 273(5)
Frunder, H., 279
Fryer, M. W., 340
Fujimori, T., 312
Fujiwara, A., 279
Fukui, K., 112
Fuldner, H. H., 225
Funakoshi, T., 348, 350, 352(7), 354(7)
Funatsu, T., 318, 333, 335
Furman, R. E., 417
Furtak, T. E., 179
Furuichi, T., 381, 389(2)
Furuta, T., 50, 51, 52, 60(7), 61(7), 63(7, 9)
Futura, T., 8

G

Gabriel, D., 95, 97, 155, 158(14), 162(14)
Gadsby, D. C., 295, 298
Gait, M. J., 136
Galione, A., 396
Garcia, M. B., 54
Garges, S., 416
Garrossian, M., 8
Gärtner, W., 291, 294(9), 298(9), 305(9)
Gasparro, F., 122
Gayen, S. K., 179
Gee, K. R., 10, 13(31), 15, 24(31), 25(31), 30, 31, 32, 34(27), 35(22), 37(32), 39, 42(21, 23), 43(21), 45(23), 48(21, 22), 49(21, 27), 63, 96, 104(19), 110(18), 111(18), 122, 403, 404, 406, 407, 415(7, 13), 439, 444, 453(8, 9, 11), 456, 457(8, 9, 11), 458, 468(8, 9, 11)
Gelfand, V. I., 75
Gelperin, A., 467
Genick, U. K., 256
Georg, H., 227, 237(18), 241(18)
Gerisch, G., 109
Gerrard, A., 179
Gerwert, K., 223, 232, 233(2), 234, 241, 243, 244(47)
Ghosh, A. K., 139
Ghung-Yu, C., 96, 116(17)
Gibbons, I. R., 308

Gilbert, S. H., 94, 95
Gildea, B. D., 137, 150(25)
Giles, W. R., 220
Gilman, A. G., 444, 453(3)
Gilroy, S., 474
Ginsburg, S., 267
Givens, R. S., 1, 2, 4, 5(16, 17), 6, 7(16, 17), 8, 8(14–17), 10, 11, 13(31, 32), 15, 17(16, 17), 18(16, 17), 19(16, 17), 20(27), 21(27), 22(27), 24, 24(31), 25, 25(31, 32), 28(27), 32, 37(32), 50, 117, 122, 134(27), 306, 321, 323, 456
Godde, M., 417, 426
Godt, R. E., 327
Goeldner, M., 265, 266, 267, 269(17, 19), 270(17, 19, 20, 22), 271, 272(17, 19, 20), 273(6, 22), 274, 275, 275(22), 276(22), 277, 277(17)
Golan, R., 117
Gold, G. H., 416
Goldenberg, S. S. S., 349
Goldmacher, V. S., 118
Goldman, A., 265, 273(5)
Goldman, D. E., 445
Goldman, Y. E., 95, 96, 176, 214, 225, 307, 309, 310, 311, 312, 312(35), 318, 319, 320(21, 27), 323, 324, 324(21, 32), 325, 325(54), 327(32), 328(54), 330, 330(20, 32), 331, 331(21), 332, 333, 334, 334(69), 339(21), 341(32), 343, 357, 383
Goldschmidt-Clermont, P. J., 74(17), 78, 98
Goldstein, L. S. B., 308
Goody, R. S., 95, 176, 204, 214, 215, 216, 217(23, 24), 243, 244(46, 47), 251, 253, 256, 257, 259(12), 260, 261(18), 263(19), 308, 318, 319
Göppert-Mayer, M., 357
Gordon, D. J., 101
Gordon, J. P., 181, 364
Gordon, P. R., 443
Gorenstein, G., 58
Grace, M. R., 133
Graeff, R. M., 404, 407(9), 415(7)
Gravel, D., 155
Gray, P. T., 381
Greene, L. E., 309
Gregoriou, V. G., 233
Grell, E., 220, 290, 291, 293(8, 10), 299(8), 305(5, 8, 10, 44, 52), 306

Grewer, C., 30, 32, 41(4), 43(4), 46(4), 48(4), 96, 443, 446(38), 452, 454(38)
Griensenbach, S., 492
Griffiths, P. J., 357
Gronenborn, A. M., 90
Gross, L. A., 77
Grunwald, M. E., 416
Grynkiewicz, G., 323, 324
Gu, Q.-M., 404
Guesdon, J. L., 135
Gunther, J., 279
Guo, X. Q., 219
Gurnack, M. E., 407
Gurney, A. M., 3, 214
Gurney, E. L., 230
Gutfreund, H., 447, 473(27)
Güth, K., 209, 212(7), 214, 253, 320, 387
Gutweiler, M., 305(52), 306
Guzikowski, A. P., 31, 456
Györke, S., 340, 357

H

Haas, A. L., 137, 153(23)
Hackney, D. D., 225, 309, 310
Hadfield, A. T., 217, 219(27), 252, 256(2)
Hagen, C., 423, 425, 427, 428
Hagen, V., 415, 417, 419, 419(23), 420
Hajdu, J., 252, 256(2), 266
Hamill, O., 450, 451(32), 452(32), 454(32), 460(32), 470(32)
Hamm, H. E., 308
Hammes, G. G., 447, 456, 473(24)
Hammill, O. P., 44(52), 46
Hancock, D. C., 78
Hanley, M. R., 432
Hansen, H. J., 141
Hao, J. P. Y., 323
Harada, I., 231, 232(25), 236(25), 240(25), 244(25)
Harada, Y., 333, 335, 337(76), 340(76)
Hardie, R. C., 220
Harel, M., 265, 273(5–7)
Harootunian, A. T., 439
Harris, A. J., 471
Hartshorne, D. J., 91, 93(14)
Hartung, K., 289, 301, 302(33), 303(33), 304, 304(33), 305(40), 306
Hartzell, H. C., 220

Hasselbach, W., 306
Hata, Y., 217
Hatayama, M., 52, 63(9)
Hatch, M. A., 267
Hatchard, C. G., 26
Hatt, H., 417
Haubs, M., 291, 293(8), 299(8), 305(8)
Haugland, R. P., 67(5), 68, 69(5), 71(5), 77(5), 118, 478
Haugwitz, M., 101
Hauser, K., 122, 226, 234(15), 235(15), 240(15)
Hayakawa, H., 492
Hayat, M. A., 356
Hayes, R. N., 403, 404(1)
He, Z.-H., 318, 332
Heathcock, C. H., 22
Heidecker, M., 95, 96, 97(18), 104(18), 109, 112, 114(18), 118, 121(18)
Heim, R., 77
Heinemann, C., 221
Heinemann, S., 471
Helliwell, J. R., 265, 277(11)
Hellrung, B., 6
Henderson, R., 277
Hendrix, J., 214
Herbette, L., 175, 176, 293
Herman, T., 137, 153(20–23)
Herrmann, C., 116
Herzog, K., 229
Hess, B., 232
Hess, G. P., 30, 31, 32, 32(11–13, 20), 34(27), 35(22), 41, 41(4, 11, 20, 31, 56), 42(21, 23), 43, 43(4, 21), 44(42, 45), 45(23, 30, 42, 57), 46, 46(4, 42, 45), 47(42, 45), 48, 48(4, 6, 11, 20–22), 49(21, 27), 96, 122, 147, 150(30, 39), 155, 266, 269(18a; 18b), 270(18a; 18b), 272, 272(18a; 18b), 439, 443, 444, 445, 446(5, 38, 53), 447, 447(4, 5, 33, 53), 448(27a; 33), 449(4), 450, 451, 451(33, 34), 452, 452(4, 33), 453(7–12), 454, 454(33, 38), 455, 456, 459(33), 460(29, 33), 467, 468(7–9, 11, 12, 78), 469(78), 470, 471(85), 472(85), 473(5)
Hessling, B., 223, 233(2), 234
Heyse, S., 305(48), 306
Hibberd, M. G., 95, 96, 176, 307, 309, 311, 318, 319, 323, 324, 324(32), 325(54), 327(32), 328(54), 330(20, 32), 331, 332, 334(69), 341(32), 383

Higuchi, H., 95, 214, 307, 318, 319, 330, 333, 334, 335, 336, 337(81), 340(81), 343(81)
Hijden, H. T. W. M. v. d., 290, 305(5)
Hilgemann, D. W., 289, 295(1), 298, 301, 305, 471
Hilgenfeld, R., 308
Hiraoki, T., 241
Hirokawa, N., 348, 349, 350, 350(3), 352(3, 7, 8), 353(3), 354(7)
Hirose, K., 318
Hirth, C., 265, 273(6), 277
Hoard, D. E., 22
Hobbiger, F., 265
Hobbs, A., 291, 294(13), 297(13), 298(13), 305(12)
Hodota, C., 50
Hoesch, R. E., 431
Hoffman, J. F., 1(4), 3, 130, 131(37), 134(37), 155, 168(2), 176, 233, 240(33), 247, 290, 291(4), 299(4), 307, 421, 431, 455
Hoffman, K., 136
Hoffmann, H., 222
Hoffmann, W., 215
Holden, H. M., 308
Holland, D., 136
Holman, J. P., 217
Holmes, K. C., 308
Holzbaur, E. L. F., 308
Homsher, E., 214, 225, 309, 310, 318, 319, 320(21), 323, 324(21), 331(21), 332, 334(70), 339(21)
Homsher, H., 318
Hood, L., 153
Hooley, R., 474
Horiuti, K., 214, 311, 312, 330
Hoskins, B. K., 357
Houslay, M. D., 416
Hryschko, L. V., 302
Hubel, D., 465
Hursthouse, M. B., 7, 260
Hyman, A. A., 72, 335

I

Ikebe, M., 78, 90, 93(14), 94
Ikebe, R., 90, 91
Ikura, M., 90
Ilyin, V., 219, 382, 395(9), 396
Inoue, Y., 335, 336, 337(81), 340(81), 343(81)

Ireland, R. E., 21
Irisawa, H., 301
Irving, M., 318
Ishibashi, M., 50
Ishihara, A., 39
Ishijima, A., 318, 333, 335
Ishikawa, T., 50
Ito, K., 59
Iversen, L. L., 463
Iverson, R. M., 287
Ivorra, I., 95, 382, 385, 386, 392, 393, 393(14), 395, 395(21), 396, 397, 398, 401
Iwamura, H., 50
Iwamura, M., 8, 50, 51, 52, 60(7), 61(7), 63(7, 9), 416

J

Jack, W. E., 134
Jackson, 461
Jackson, J. B., 247, 250(2, 3)
Jackson, S. A., 492
Jacobsen, K., 39
Jaffe, E. K., 127
Jager, A., 59
Jagla, A., 215
Jakes, K. S., 136
Jaruschewski, S., 291, 294(13), 297(13), 298(13), 305(12)
Jastorff, B., 417
Jessel, T. M., 443, 445(2)
Johaszova, M., 301
John, J., 215, 217(23), 225, 243, 244(46), 257, 261(18)
Johns, R. B., 34
Johnson, K. A., 447, 473(26)
Johnson, K. W., 31, 155, 456
Johnson, L. N., 217, 219(27), 265, 277(11)
Johnston, L. J., 10
Jones, R. L., 474
Jonges, G. N., 278
Jovin, T. M., 157, 168(16), 171(16)
Jung, A. H., 1, 10, 11, 13(32), 15, 24, 25, 25(32)
Juorio, A. V., 20

K

Kabsch, W., 95, 216, 217(24), 256, 259(12), 308
Kagawa, K., 214
Kahn, M., 33
Kaiser, R., 153
Kalow, W., 278
Kamimura, A., 31
Kanai, Y., 350, 352(8)
Kandel, E. R., 443, 445(2)
Kang, H., 374, 378, 380(22)
Kao, J. P. Y., 323, 324, 357, 421, 431, 432, 433(8), 435, 436(8), 439
Kaplan, J. H., 1(4), 3, 130, 131(37), 134(37), 155, 168(2), 176, 233, 240, 240(33), 241, 241(36), 247, 253, 290, 291(4), 299(4), 301, 307, 310, 318, 323, 324, 331, 340, 357, 374, 376(1), 421, 431, 455
Kappl, M., 301, 302(33), 303(33), 304(33)
Karlin, A., 445, 446(18a; 18b)
Karpen, J. W., 176, 416, 417, 430(14)
Kass, G. E., 432
Katayama, Y., 220, 323
Kato, H., 217, 218
Katsube, Y., 217
Katz, B., 46, 446(28), 447
Katz, L. C., 30, 443, 467, 472(77)
Kaupp, U. B., 415, 417, 419, 419(23), 420, 422, 425, 426, 426(24), 427, 428
Kawai, M., 313
Kellogg, D., 72, 335
Kemp, B. E., 91, 93(14), 123
Kemp, D. S., 50
Kemp, J. A., 463
Kent, S. B. H., 134
Kestner, T., 407, 415(13)
Keszthelyi, L., 293, 296(17)
Keyes, R. W., 182
Khan, S., 219
Khodakhah, K., 219, 220, 381
Kibak, H., 288
Kihara, T., 175, 293
Kimura, A., 217
Kimura, J., 301
Kinosita, K., 96, 117, 155, 156(10), 158, 158(10), 161(17), 162(17)
Kirschner, M. W., 348
Kirstgen, R., 168
Kitazawa, T., 312
Klee, C. B., 90
Klein, M. V., 179
Kleinfeld, D., 467
Kleinle, J., 368
Kleywegt, G. J., 265, 273(7)
Klingauf, J., 375

AUTHOR INDEX

Klingenberg, M., 298, 300, 305(25)
Knaupp, U. B., 416
Knight, A. R., 463
Knight, P., 162
Knölker, H. J., 169
Kobayashi, S., 312
Koch, K.-W., 417
Kodama, T., 112
Koechner, W., 178, 179(13), 195(13), 198(13)
Koenigs, P., 118, 218, 255, 266
Koga, N., 50
Kogelnik, H., 188
Kojima, H., 318, 333, 335
Kolb, A., 416
Kolb, H.-A., 445
Kometani, K., 112
Kono, E., 318
Korn, 457, 461(65)
Korn, E. D., 101
Korth, M., 416
Koshland, D. E., 277
Koshland, D. E., Jr., 252
Koster, H., 137, 150(25)
Kovacs, O., 24
Koyama, Y., 50
Kozlowski, R. Z., 220, 301
Krafft, G. A., 68
Krämer, R., 298
Krause, E., 417, 419, 419(23), 420, 422, 425, 427, 428
Krejci, E., 265
Krengel, U., 308
Kreutz, W., 227, 231, 237(18), 241(18)
Krishnasastry, M., 134
Krishtal, O. A., 46, 450, 452(31), 453(31)
Kron, S. J., 110, 335
Krzymanska-Olejnik, E., 135, 138, 154, 156
Kubert, M., 153
Kucherov, V. F., 20
Kueper, L. W. III, 2, 4, 5(16, 17), 7(16, 17), 8(15-17), 10, 13(31), 15, 17(16, 17), 18(16, 17), 19(16, 17), 24(31), 25(31), 32, 37(32), 117, 321, 323
Kuhn, H. J., 421
Kulba, M., 20
Kull, F. J., 308
Kumler, W. D., 229
Kurahashi, T., 416
Kurzchalia, T. V., 136
Kushmerick, M. J., 313

Kusianowicz, J., 134
Kuzmic, P., 32, 455
Kyogoku, Y., 229

L

Lacktis, J., 309, 318, 319, 320(21), 323, 324(21), 331(21), 339(21)
Ladouceur, G., 155
La Du, B. N., 277
Laflamme, M. A., 219
Lamb, G. D., 340
Lambert, J. M., 96, 118, 141, 155, 156(13), 175(13)
Lamm, G. M., 136
Lancet, D., 417, 426(24)
Landau, V. G., 47, 451
Landegren, U., 153
Lando, L., 221
Langer, P. R., 136
Langosh, D., 444
Lännergren, J., 215
LaNoue, K., 298
Lanzetta, P. A., 409
Lappi, S., 273
Lattman, E. E., 308
Lau, R., 308
Lau, W., 89, 91(10)
Laube, B., 444
Läuger, P., 291, 305(11, 12, 47), 306, 445
Lautwein, A., 260, 263(19)
Lavit, D., 20
Lea, T. J., 310, 323
Lederer, W. J., 3, 301
Lee, A. K., 221
Lee, H. C., 403, 404, 404(1, 2), 406, 407, 407(9), 411, 415(7, 13)
Lee, J., 39
Lefever, E., 137, 153(21)
LeGrand, A., 256
Leikauf, E., 137
Leimbruger, W., 147
Lenart, T. D., 310, 318, 357
Lend, C., 461
Lentfer, A., 95, 216, 217(24), 256, 259(12)
Leonard, D. M., 78
Leslie, A. G. W., 308
Lester, H. A., 3, 214, 416, 418(13), 419(13)
Lester, W., 485
Leterrier, F., 272, 273(25)

Levich, A. G., 47, 451
Levitsky, D. O., 302
Lev-Ram, V., 358, 369(12), 374(12), 376(12)
Levy, J., 137, 153(20)
Lewis, S. M., 3
Lewitzki, E., 220
Leytus, S. P., 170
Li, H., 30, 467, 468(78), 469(78)
Li, T., 188
Lidzey, D. G., 247, 250(2, 3)
Lien-Vien, D., 227, 242(17)
Lifshitz, E. M., 47, 451
Lim, S. S., 75
Lin, M., 343
Lincoln, T. M., 415
Lindberg, U., 108
Lindsay, H. A., 278
Ling, N., 318
Lipp, P., 301, 368
Lippincott-Schwartz, J., 78
Liu, 461
Lockery, S., 468, 469(81)
Lockhart, A., 310
Lockridge, O., 265, 277
Locktis, J., 225
Long, L. M., 140
Long, W., 10
Longhurst, R. S., 179
Lorentzon, T., 290
Lough, J., 137, 153(22, 23)
Loughrey, H., 152
Lowe, G., 416
Lu, A. T., 98
Lu, Z., 3, 310
Ludwig, J., 417, 426(24)
Lukas, T. J., 89, 91(10)
Luo, Y., 313
Luschner, C., 368
Lüthen, H., 474
Lutter, R., 308
Lwebuga-Mukasa, J., 273
Lymn, R. W., 309

M

Madsen, 461
Maeda, Y., 214, 253, 318, 319
Maher, L. J. III, 137, 153(24)
Maiman, T. H., 182

Makhatadze, G. I., 216
Malhó, R., 478
Mallavarapu, A., 63, 96, 104(19), 110(18), 111(18)
Mandel, G., 444
Mandelkow, E., 215
Mandveno, A., 215
Mangel, W. F., 170
Mann, D., 98
Mannherz, H. G., 308
Manor, D., 236
Mantele, W., 122, 226, 227, 231, 234(15), 235(15), 237(18), 240(15), 241(18)
Marchot, P., 265, 273(9)
Marell, A. E., 228
Margalit, T., 417, 426(24)
Margulis, M., 431, 441
Markham, K. R., 34
Marque, J., 31, 32(20), 41(20), 48(20), 139, 147, 150(30, 39), 266, 269(18b), 270(18b), 272(18b), 444, 453(7), 456(7), 457(7), 468(7)
Marré, E., 480
Marriott, G., 95, 96, 97(16, 18), 98(16), 99(16), 104(18), 109, 110(15), 112, 114(18), 117, 118, 121(9, 18), 156(10), 155, 157, 157(12), 158, 158(10, 14), 161(17), 162(14, 17), 168(12, 16), 171(12, 16), 173(12), 175(12)
Marston, S., 324
Martell, A. E., 325
Martin, H., 80, 88(4), 311, 328, 328(37)
Martinez, O. E., 364
Marty, A., 450, 451(32), 452(32), 454(32), 460(32), 470(32)
Marty, E., 44(52), 46
Maruyama, J., 59
Marx, A., 215
Masserini, M., 152
Massoulié, J., 265, 272, 273(25)
Mathias, L. J., 36, 71(16), 78
Mathivanan, N., 10
Matsubara, I., 333
Matsubara, N., 30, 31, 32(11, 20), 41(11, 20, 56), 48, 48(11, 20), 139, 147, 150(30, 39), 266, 269(18b), 270(18b), 272(18b), 444, 447(4), 452(4), 453(7), 454, 456(7), 457(4, 7), 461, 468(7)
Matsuoka, S., 301, 302
Matsuura, M., 90

Mattson, G., 135, 135(3), 152(3)
Matuszewski, B., 4, 5(16, 17), 7(16, 17), 8(14, 16, 17), 17(16, 17), 18(16, 17), 19(16, 17), 50, 321, 323
Maughan, D. W., 327
Maydell, M. R. D., 214
Mayer, M. L., 439
Mazid, M. A., 7, 260
McCarthy, M. P., 444
McConkey, D. J., 432
McConnaughie, A. W., 8
McCray, J. A., 3, 31, 32(20), 33, 41(20), 43, 43(33), 46(33, 43), 48(20), 88, 95, 96, 130, 131(38), 134(38), 139, 147, 150(30, 39, 40), 155, 175, 176, 203, 204(1, 2), 218(1), 219, 223, 225, 226, 226(9), 231(4, 9), 233(4), 241, 247, 253, 266, 269(18a; 18b), 270(18a; 18b), 271, 272(18a; 18b), 293, 296(16), 307, 320, 323, 324(46), 325, 347(56), 405, 406(10), 443, 444, 453(7), 455, 456, 456(7), 457(7), 468(7), 486
McDonald, K. S., 357
McElroy, W. D., 245
McInnes, J. L., 136, 152(8)
McKee, S. P., 139
McKinnon, D., 444
McMillan, S., 180
McNamee, M. G., 46, 445, 446(18a)
Meaking, W. S., 485
Means, A. R., 91, 93(14)
Meldrum, R. A., 245, 247, 279, 483, 485, 486(2), 489(2)
Melhado, L. L., 170
Melzer, W., 222
Mendel, D., 118, 266
Menetret, J. F., 215
Menini, A., 416
Mercer, T., 24
Mertz, J., 370
Mery, P. F., 220
Messenger, J., 220
Meyer, B., 114
Meyer, T., 381, 439
Michikawa, T., 381, 389(2)
Mikoshiba, K., 381, 389(2)
Milburn, T., 31, 32(20), 41(20), 48(20), 139, 147, 150(30, 39), 266, 269(18b), 270(18b), 272(18b), 444, 453(7), 456(7), 457(7), 458, 468(7)
Miledi, R., 382, 384, 388, 392

Milford, F. J., 179
Millar, N. C., 214, 225, 309, 310, 318, 319, 320(21), 323, 324(21), 331(21), 339(21)
Millard, P. J., 39
Miller, J. P., 417
Miller, W. T., 117, 266
Minta, A., 323, 324
Misiura, K., 136, 150(12)
Mitchison, T. J., 63, 64, 66, 68(4), 72, 74(17), 75, 78, 95, 96, 98, 104, 104(19), 110(18), 111(18), 155, 335, 348, 350(1)
Mitra, A., 33
Miyata, H., 96, 117, 155, 156(10), 158, 158(10), 161(17), 162(17)
Miyawaki, A., 381, 389(2)
Mizani, S. M., 298
Moffat, K., 256, 277
Moisescu, D. G., 327, 347
Mokrin, S. C.
Moloney, M. G., 8
Momotake, A., 52, 63(9)
Montibeller, J. A., 136
Moore, G. A., 432
Mooseker, M. S., 307
Mooser, G., 273
Morgensen, S., 135, 135(3), 152(3)
Morrat, K., 277
Morrison, H., 117, 121(1), 124, 134(1)
Moss, R. L., 3, 241, 310, 311, 313, 357
Muir, T. W., 134
Mulhern, S. A., 110
Müller, F., 417
Mulligan, I. P., 310, 323
Muralidharan, S., 222
Murata, H., 231, 232(25), 236(25), 240(25), 244(25)
Murray, J. M., 318, 357
Murray, S., 155
Muslin, A. J., 123
Muto, E., 319, 335, 336, 337(81), 340(81), 343(81)

N

Nagel, G., 32, 289, 291, 291(2), 293(10), 305(10, 51), 306, 470, 471(85), 472(85)
Naim, M., 117
Nair, H. K., 265, 269
Nara, M., 241

Naraghi, M., 374, 378, 380(22)
Nargeot, J., 416, 418(13), 419(13)
Neher, E., 44(52), 46, 48, 48(48), 221, 374, 375, 378, 380(22), 445, 448(30), 450, 454(19), 460(19)
Neis, A. S., 444, 453(3)
Nerbonne, J. M., 3, 30, 117, 121(4), 214, 222, 416, 418(13), 419(13), 455
Neveu, D., 221
Ng, K., 256
Niblack, B., 35, 96, 116(17), 117, 118, 123(17), 129, 155, 266
Nicoll, D. A., 302
Nielander, G. W., 135, 135(3), 152(3)
Niggli, E., 3, 301, 368
Nishioka, T., 217
Niu, L., 30, 31, 32, 32(12, 13), 34(27), 35(22), 41(31), 42(23), 45(23, 30), 48(22), 49(27), 122, 439, 444, 446(5, 36), 447(4, 5), 449(4), 452, 453(8, 9, 10, 12), 454(38), 456, 457(8–10, 12), 458, 459, 463, 464(69), 468(8, 9, 12), 470, 471(85), 472(85), 473(5)
Noegel, A. A., 101, 107
Noel, J. P., 308
Noma, A., 301
Noren, C. J., 134
Nosek, T. M., 327
Nunn, D. L., 384
Nunn, D. S., 31

O

Oberhauser, A. F., 357
O'Brien, P. J., 432
O'Brien, R. D., 278
Ochoa, E. L. M., 46
O'Connor, T. P., 348
Oda, J., 217
Oefner, C., 265, 273(5)
Oesterhelt, D., 291, 294(9), 298(9), 305(9)
Offer, G., 162
Ogden, D. C., 219, 220, 221, 381
Ogreid, D., 417
O'Hare, P., 134
Okabe, S., 348, 349, 350, 350(3), 352(3), 352(8), 353(3)
Olbrich, A., 168
Olbricht, W. L., 43, 44(45), 46(45), 47(45), 450, 451(34)

Oldershaw, K. A., 384
Olejnik, J., 135, 138, 154, 156
Olsen, S. R., 123
Ono, K., 416
Ordoukhanian, P., 117
O'Reilly, N. J., 78
Orrenius, S., 432
Osawa, T., 50, 51
Ostap, E. M., 3, 318, 335
Oswald, R. E., 32, 470, 471(85), 472(85)
Otis, T., 32
Ott, D. G., 22
Ottl, J., 95, 155, 158(14), 162(14)
Oxford, A. W., 7, 31, 40(16)

P

Pabst, R., 305(52), 306
Pai, E. F., 95, 215, 216, 217(23, 24), 243, 244(46), 256, 257, 259(12), 260, 261(18), 263(19), 308
Palmer, R. A., 233
Pan, P., 117, 119, 120, 122(21), 123(21), 128, 133(21), 266
Panchal, R., 121, 123, 123(22), 132
Papageorgiou, G., 456
Paranjape, S., 439
Pardee, J. D., 73, 101
Parekh, A. B., 471, 472(87b)
Park, C.-H., 1, 10, 13(32), 24, 25, 25(32), 122, 134(27), 456
Park, S.-H., 8, 15, 20(27), 21(27), 22(27), 28(27)
Parker, C. A., 26
Parker, I., 95, 219, 380, 381, 382, 384, 385, 386, 386(12), 388, 388(10), 389, 389(5, 11), 390(26), 392, 393, 393(14), 395, 395(9, 21), 396, 397, 398, 400, 400(13), 401, 402
Parsons, T. D., 221, 241, 357
Pashkevich, K. I., 69
Patchornik, A., 31, 98, 117, 155
Patchornik, P., 455
Pate, E., 309, 313, 324(22), 343
Patel, J. R., 241, 357
Patel, R., 219
Patlak, J., 46, 48(48), 448(30), 450
Patton, C., 325

Patzlaff, M., 24
Paulsen, P. J., 242
Paulus, G. G., 176
Pavlickova, L., 32, 455
Pawley, J. B., 390
Pearson, R. B., 123
Peck, A. W., 265
Peckham, M., 318
Pedro, J. P., 54
Peeters, J. M., 83
Peng, L., 265, 266, 267, 269(17, 19), 270(17, 19, 20, 22), 271, 272(17, 19, 20), 273(22), 274, 275, 275(22), 276(22), 277(17)
Petratos, K., 95, 216, 217(24), 218, 255, 256, 259(12)
Petricdvic, V., 179
Petsko, G. A., 95, 118, 217(24), 218, 255, 256, 259(12), 308
Pfaff, E., 298
Philipson, K. D., 302
Phillips, G. N., Jr., 256
Phillips, R. J., 241
Pidoplichko, V. I., 46, 450, 452(31), 453(31)
Pieles, U., 136
Pillai, V. N. R., 1, 98
Pirrung, M. C., 7, 8(21), 31, 98
Piston, D. W., 179
Planck, M., 445
Plant, P. J., 31, 455
Plunkett, S. E., 233
Podein, R. J., 404
Pohl, B., 222
Pollard, T. D., 308
Pon, R. T., 136, 151(10)
Poole, K. J. V., 204, 214, 253, 318, 319
Porter, N. A., 118, 218, 255, 266
Portnoy, D. A., 74(17), 78, 98
Postovskii, I. Y., 69
Potma, E. J., 332
Potter, B. V. L., 439
Potter, J. D., 215
Powell, T. A., 220, 301
Pradervand, C., 256, 277
Prassler, J., 109
Pratt, A. J., 8
Prendergast, F. G., 89, 91(10)
Priestley, T., 463
Privalov, P. L., 216
Purich, D. L., 456

Q

Qiu, X. Q., 136
Quinn, D. M., 265, 267(2), 269, 273(2)
Quirk, S., 308

R

Raaben, R. J., 83
Rademacher, J., 273
Raible, A. M., 153
Raizen, D. M., 467, 468, 469(61, 80, 81)
Rakowski, R. F., 295, 298
Rameriz, F., 19, 28(35)
Ramesh, D., 31, 32, 35(22), 41(31), 45(30), 48(22), 122, 439, 444, 453(8, 10, 12), 456, 457(8, 10), 459, 468(8)
Rammelsberg, R., 223, 233(2)
Randal, T. W., 444, 453(3)
Rapoport, T. A., 136
Rapp, G., 95, 202, 204, 209, 212(7), 214, 215, 216, 217(23, 24), 243, 244(46), 253, 256, 257, 258(18a), 259(12), 261(18), 318, 319, 320, 387
Rapp, R., 357
Ratcliffe, R., 22
Ravelli, R. B. G., 265, 273(7)
Rayment, I., 308
Read, J. L., 98
Reason, C., 485
Recek, J., 50
Rees, B. B., 288
Reese, C. B., 8
Reese, T. S., 335
Regnier, M., 318
Rehwinkel, H., 168
Reich, J. G., 279
Reid, G. P., 3, 7, 33, 43(33), 46(33), 59, 68, 86, 88, 130, 131(38), 134(38), 147, 150(40), 176, 225, 226, 226(9), 231(9, 13), 247, 260, 263(19), 293, 296(16), 318, 321, 323, 324(52), 325, 328(52), 330(52), 343(52), 347(56), 405, 406(10), 486
Reinach, P. S., 409
Reiser, C. O. A., 308
Reisler, E., 110, 112(36)
Reitz, J. R., 179
Ren, Z., 256, 277
Renström, E., 220
Reshetnikova, L., 308

Rettinger, J., 471
Reynolds, J., 445, 446(18b)
Rhodes, Ch. K., 179
Richard, S., 416, 418(13), 419(13)
Ried, G. P., 271
Rieger, D., 101
Ringe, D., 118, 218, 255
Robinson, I. M., 357
Rodionov, V. I., 75
Rogalski, A. A., 135
Romberg, L., 310
Root, D. D., 110
Rorsman, P., 220
Rosch, P., 225
Rosenblatt, J., 74(17), 78, 98
Roskoski, R., 130
Rossi, F. M., 431, 432, 433(8), 436(8)
Rothschild, K. J., 135, 138, 154, 156
Rottenberg, H., 298
Roudna, M., 204
Rozycki, M., 108
Ruf, H., 220
Rundstrom, N., 444
Rungger, D., 471, 472(87a)
Russo, M. J., 123, 134(30)
Rypniewski, W. R., 308

S

Sablin, E. P., 308
Sabry, J., 348
Sachs, G., 290
Sakmann, B., 44(52), 46, 48, 48(48), 445, 448(30), 450, 451(32), 452(32), 454(19, 32), 460(19, 32), 470(32)
Sakoda, T., 330
Sakuma, K., 50
Sakurada, K., 335, 337(76), 340(76)
Salmon, E. D., 75
Salser, S., 72, 335
Sanchez-Llorente, A., 260, 263(19)
Sanders, J., 153
Sandison, D., 373
Sano, T., 135
Savage, M. D., 135, 135(3), 152(3)
Savéant, J.-M., 10
Sawatari, A., 30
Sawin, K. E., 63, 64, 72, 96, 104(19), 110(18), 111(18), 335
Sawyer, D. T., 242
Sawyer, T. K., 78
Schafer, F. P., 179
Schalk, I., 265, 273(6), 277
Schaper, K., 31, 32, 34(27), 42(23), 45(23), 49(27), 444, 453(9), 456, 457(9), 468(9)
Schawlow, A. L., 182
Scheidig, A., 215, 243, 244(47), 251, 260, 263(19)
Schildkamp, W., 256, 277
Schiltz, E., 225
Schlaeger, E. J., 58
Schlechtingen, G., 169
Schleicher, M., 101, 107
Schlichting, I., 95, 215, 216, 217(23, 24), 225, 243, 244(46), 256, 257, 259(12), 261(18)
Schmid, H., 141
Schmidt, C. F., 335, 337(79)
Schmidt, R., 421
Schmidt-Bäse, K., 308
Schmieden, V., 444
Schmirmer, N. K., 308
Schmitt, F., 80, 88(4)
Schnabel, W., 40, 43(40, 41), 46(40, 41)
Schnapp, B. J., 335, 337(79)
Schneider, S., 226, 237(14)
Schnyder, T., 308
Schoenberg, M., 324
Schoffeniels, E., 445
Schofield, P., 455
Schröder, R. R., 215
Schroeder, J. I., 474
Schubert, D., 471
Schulman, H., 273
Schultz, P. G., 118, 134, 266
Schupp, H., 40, 43(40, 41), 46(40, 41)
Schutt, C. E., 108
Schwartz, J. H., 443, 445(2)
Schwartz, S., 39
Schwartz, W., 471
Schwarze, W., 445
Sciaky, N., 78
Scott, R. H., 222
Scott, W. G., 277
Seifert, R., 426
Self, C. H., 117
Self, S. A., 188
Senter, P. D., 96, 118, 141, 155, 156(13), 175(13)
Seravalli, J., 269
Serebryakov, E. P., 20

Shai, R., 117, 134(8)
Shall, S., 247, 279, 485, 486(2), 489(2)
Shao, Y., 134
Shaw, A. S., 123
Shaw, P. J., 357
Shear, J., 358, 365, 365(13), 367, 368, 375(13), 377(13)
Sheehan, J. C., 7, 31, 40(16), 456
Sheidig, A., 95
Shi, Q.-Y., 441
Shim, S. B., 8
Shimanouchi, T., 229
Shimkus, M., 137, 153(20, 21)
Shioya, T., 301
Shofield, P., 31
Shoger, R. L., 279
Shuey, S. W., 7, 8(21)
Shuster, M., 444
Siebert, F., 227, 233, 237(18), 241(18)
Siede, W., 483
Siegmen, A. E., 179
Sigler, P. B., 308
Sigman, D. S., 273
Sigworth, F. J., 44(52), 46, 450, 451(32), 452(32), 454(32), 460(32), 470(32)
Sih, C. I., 404
Silman, I., 265, 267, 270(22), 273(5–7, 22), 274, 275, 275(22), 276(22)
Simmons, R. M., 311, 312(35), 333, 335, 343
Singer, P., 277
Singer, S. J., 135
Skingle, D. C., 136
Slatin, S. L., 136
Sleep, J. A., 116, 254, 321
Smirnoff, P., 117
Smith, C. L., 78, 135
Smith, R., 308
Smith, R. M., 228, 325
Smith, S. J., 474
Snape, W. J., Jr., 3
Solas, D., 98
Somlyo, A. P., 214, 311, 312
Somlyo, A. V., 214, 311, 312
Sonar, S., 138
Soria, M. R., 152
Soucek, M., 32, 455
Soukup, G. A., 137, 153(24)
Souvignier, G., 232, 234
Spoors, J. A., 117
Sprinzl, M., 308

Sproat, B. S., 136
Spudich, J. A., 73, 101, 110, 335
Spudich, J. L., 219
Srajer, V., 256, 277
Sreekumar, R., 78, 94
Steffen, P., 72, 335
Steger, E., 229
Steglich, W., 168
Stein, T., 147
Stempel, K. E., 225
Stengelin, M., 290, 305(6)
Stephenson, D. G., 340
Stepinska, M., 91, 93(14)
Stiekema, F., 135
Stienen, G. J. M., 332
Still, W. C., 33
Stocker, S., 109
Stoddard, B. L., 118, 218, 252, 255, 277
Strickler, J., 176, 357, 373
Stroud, R. M., 444
Strowbridge, B. W., 467
Stryer, L., 98, 176, 381, 416, 417, 430(14), 439
Stühmer, W., 471, 472(87b)
Stults, J. T., 153
Stürmer, W., 305(47, 48), 306
Suck, D., 308
Sudati, F., 152
Sugimoto, M., 8, 50, 52, 60(7), 61(7), 63(7, 9), 416
Sulston, J. E., 468
Sumikawa, K., 382, 384, 392
Suslova, L. M., 20
Sussman, J. L., 265, 267, 270(22), 273(5–7, 22), 274, 275, 275(22), 276(22)
Sutton, W. R., 68
Suva, R. H., 417
Svelto, O., 179
Svensson, E. C., 318, 319, 335
Svoboda, K., 335, 337(79), 388
Swain, K. E., 64
Sweet, R. M., 256, 277
Swezey, R. R., 280, 282, 283(5), 284, 286, 286(6), 288
Symons, R. H., 136, 152(8)

T

Takeda, S., 350, 352(7), 354(7)
Takeuchi, H., 231, 232(25), 236(25), 240(25), 244(25)

Tallec, A., 10
Tanaka, D., 31
Tanaka, H., 220
Tanaka, J. C., 417
Tanaka, M., 90
Tang, C. I., 178
Tang, C.-M., 431, 439, 441
Tank, D. W., 467
Tanner, J. W., 123, 330
Tanokura, M., 241
Tansey, M. J., 96, 141, 155, 156(13), 175(13)
Taran, C., 233
Tardivel, R., 10
Tardy, C., 10
Tasumi, M., 241
Taylor, A. W., 230
Taylor, C. W., 384
Taylor, E. W., 309
Taylor, J.-S., 117
Taylor, P., 265, 273, 273(9), 444, 453(3)
Taylor, S. S., 415
Telford, J. N., 278
Temsamani, J., 153
Teng, T.-Y., 256, 277
Ten Kortenaar, P.B.W., 83
Terasaki, M., 78
Tesser, G. I., 83
Theriot, J. A., 63, 64, 66, 68(4), 74(17), 78, 95, 96, 98, 104(19), 110(18), 111(18), 155, 348
Thesleff, S., 46, 446(28), 447
Thieleczek, R., 347
Thio, L. I., 46
Thirlwell, H., 318, 321, 324(52), 328(52), 330(52), 343(52)
Thirlwell, T., 254
Thomas, D., 432
Thomas, D. D., 3, 318, 319, 330, 335, 337(76), 340(76)
Thomas, D. T., 3
Thomas, J. L., 117
Thomas, P., 221
Thomason, D. B., 247
Thompson, S., 117
Thompson, W. J., 139
Thomson, J. N., 467, 468(79)
Thorson, J., 325
Tibor, R. S., 417
Tijssen, P., 136
Till, U., 279
Tilney, L. G., 78

Tipler, P. A., 180
Tirrell, D. A., 117
Toker, L., 265, 273(5)
Tokuda, K., 50
Tolman, R. C., 179
Tomchick, D. R., 308
Toner, M., 123, 134(30)
Torigai, H., 8, 50, 51, 52, 60(7), 61(7), 63(7, 9), 416
Townes, C. H., 181, 182
Toyoshima, Y. Y., 335
Travers, F., 116
Trentham, D. R., 3, 7, 30, 33, 43, 43(33), 46(33, 43, 44), 59, 68, 86, 88, 95, 96, 122, 130, 131(38), 134(38), 139, 147, 150(40), 155, 175, 176, 203, 204, 204(1), 218(1), 219, 220, 221, 223, 225, 226, 226(9), 227(5), 231(4, 5, 9, 13), 233(4), 234(15), 235(15), 240(15), 247, 252, 260, 271, 293, 296(16), 307, 309, 311, 312, 318, 319, 320, 321, 323, 324, 324(32, 46, 52), 325, 325(54), 328(52, 54), 330(20, 32, 52), 331, 332, 334(69, 70), 341(32), 343(52), 347(32, 56), 382, 383, 405, 406(10), 416, 418, 421(10), 422(28), 443, 455, 456(52), 457(52), 486
Trewavas, A. J., 474, 478, 480
Tripathi, R. K., 278
Troullier, A., 223, 241
Troutman, H. D., 140
Trussell, L. O., 32, 46, 439
Tschisgale, M., 279
Tsien, R. Y., 30, 77, 78, 95, 117, 155, 223, 323, 324, 358, 365, 365(13), 367, 368, 369(12), 374(12), 375(13), 376(12), 377(13), 384, 416, 431, 439, 455
Tsuboi, M., 229
Tsukita, S., 318
Tsutsumi, A., 241
Twist, V. W., 220

U

Udenfriend, S., 147
Udgaonkar, J. B., 31, 32(20), 41, 41(20), 43, 44(42, 45), 45(42), 46(42, 45), 47(42, 45), 48(20), 139, 147, 150(30, 39), 266, 269(18b), 270(18b), 272(18b), 444, 447(33), 448(33), 450, 451(33, 34),

452(33), 453(7), 454, 454(33), 456(7), 457(7), 459(33), 460(33), 468(7)
Uhl, R., 114
Uhler, M. D., 123
Uhmann, W., 233
Ulbrich, C., 223
Ulman, A., 117
Umeyama, T., 350, 352(8)
Umezawa, J., 7
Unoue, Y., 319
Unwin, N., 445, 446(17, 18)
Ursby, T., 256, 277
Uyeda, T. Q. P., 335

V

Valdemanis, J. A., 178
Vale, R. D., 308, 310
Valera, S., 471, 472(87a)
Vallee, R. B., 308
Vanaman, T. C., 98
Van Duk, B. G., 83
Van Noorden, C.J.F., 278
Vasilets, L. A., 471
Vazquez, R. W., 32, 470, 471(85), 472(85)
Veigel, C., 214
Velez, P., 357
Vergara, J., 357
Vergara, J. L., 241
Vigny, M., 272, 273(25)
Viktorova, T. S., 69
Vilaseca, R., 179
Vinson, C., 308
Viola, R. W., 31, 155

W

Wagner, N., 305(52), 306
Wakatsuki, S., 217, 219(27)
Walba, D. M., 21
Waldrop, A. A., 136
Walker, B., 35, 96, 116(17), 118, 123(17), 129, 134, 155
Walker, G. C., 483
Walker, J. E., 308
Walker, J. M., 3, 88, 88(4)
Walker, J. W., 3, 31, 32(20), 33, 41(20), 43(33), 46(33), 48(20), 59, 68, 78, 80, 86, 94, 130, 131(38), 134(38), 139, 147, 150(30, 39, 40), 176, 225, 226, 226(9), 231(9, 13), 247, 266, 269(18a; 18b), 270(18a; 18b), 271, 272(18a; 18b), 293, 296(16), 310, 323, 325, 347(56), 382, 405, 406(10), 443, 444, 453(7), 456, 456(7), 457(7), 468(7), 476, 486
Wallimann, T., 308
Wallmark, B., 290
Walmsley, I. A., 178
Walseth, T. F., 403, 404, 404(1), 407, 407(9), 411, 415(7)
Walsh, C. T., 133
Walstrom, K. M., 451
Walter, U., 415
Wang, S.-H., 122
Wang, S. S., 84, 381
Ward, D. C., 136
Ward, J. L., 474, 475, 476(5), 478, 480
Ward, J. M., 474
Warmuth, R., 220
Watkins, P. A. C., 478
Watterson, D. M., 89, 91(10)
Wayner, D. D. M., 10
Webb, M. R., 318, 332, 334(69)
Webb, W. W., 31, 32(20), 41(20, 56), 48, 48(20), 139, 147, 150(30, 39), 176, 178, 179, 203, 204(3), 266, 269(18b), 270(18b), 272(18b), 356, 357, 358, 365, 365(13), 367, 368, 370, 373, 375(13), 377(13), 444, 453(7), 456(7), 457(7), 468(7)
Weber, J., 10, 13(32), 15, 25(32)
Weber, J. F. W., 1, 11, 25
Weiboldt, R., 444, 447(4, 5)
Weichselbaum, R. R., 492
Weigele, M., 147
Weill, C. L., 445, 446(18a)
Weiner, J., 417
Weiss, C. O., 179
Weiss, J. N., 302
Wells, K., 373
Weng, G., 236
Wensel, T., 381
Wesenberg, G., 308
Westbrook, G. L., 439
Weyand, I., 417
Wharton, C. W., 245, 247, 250(2, 3), 279, 483, 485, 486(2), 489(2)
Wheat, H., 463, 464(68), 466(68a)

Whitaker, M., 219
White, D. C. S., 325
White, H. D., 318, 335
White, J. G., 468
Wiberley, S. E., 227, 242(16, 17)
Wieboldt, R., 31, 32, 35(22), 41(31), 42(21), 43(21), 45(30), 48(21, 22), 49(21), 96, 122, 439, 444, 446(5), 453(8, 10–12), 456, 457(8, 10–12), 458, 468(8, 11, 12), 473(5)
Wiedmann, M., 136
Wiegand-Steubing, R., 214
Wier, W., 389, 390(26), 402
Wiesel, 465
Wilchek, M., 135, 135(1), 139, 152(1)
Wilcox, M., 31, 155
Wilding, M., 219
Willenbucher, R. F., 3
Williams, R., 373
Willner, I., 117, 134(8)
Wilson, I. B., 267, 269
Wilson, K., 95, 216, 217(24), 256, 259(12)
Wilson, R. M., 7, 31, 40(16)
Winkelmann, D. A., 308
Wirz, J., 6, 267, 269(19), 270(19, 20), 272(19, 20)
Wise, F. W., 178, 179
Wittinghofer, A., 95, 215, 216, 217(23, 24), 225, 243, 244(46), 256, 257, 259(12), 260, 261(18), 263(19), 308
Wolf, E., 179, 190(20), 194(20), 360
Wolf, M. R., 357
Wong, E. H. F., 463
Wong, J. G., 221
Wong, W. K., 40, 43(40), 46(40)
Wood, S. W., 136
Woodruff, G. N., 463
Woodward, R. B., 31, 455
Wootton, J. F., 43, 46(44), 221, 222, 323, 418, 422(28)
Wordeman, L., 72, 335
Wormmeester, J., 135
Wright, E. M., 219
Wu, C. W., 456
Wuddel, I., 305(47–49), 306
Wulff, M., 256, 277

X

Xie, Y. N., 3
Xiong, X., 134
Xu, C., 370, 373
Xu, T., 374, 378, 380(22)
Xue, J., 321, 323
Xue, J.-y., 4, 5(16, 17), 7(16, 17), 8(16, 17), 17(16, 17), 18(16, 17), 19(16, 17)

Y

Yamada, K., 214
Yamaguchi, H., 217
Yamakawa, M., 311
Yamanaka, G., 417
Yanagida, T., 319, 330, 333, 335, 336, 337(76, 81), 340(76, 81), 343(81)
Yang, J., 247
Yan-Marriott, Y., 112
Yanovsky, V. P., 179
Yao, Y., 219, 382, 386(12), 388(10), 389(11), 395(9), 396, 400, 400(13)
Yariv, A., 196, 198
Yasumasu, I., 279
Yau, K.-W., 416
Yazawa, M., 241
Yonemoto, W., 415
Yu, L. C., 311
Yuste, R., 467

Z

Zahn, D., 279
Zang, R., 215
Zechel, K., 157, 168(16), 171(16)
Zehavi, U., 98, 117, 155
Zeiger, H. Z., 181
Zernike, F., 197
Zhang, S., 32
Zhao, J., 215
Zhao, Y., 313
Zhu, G., 90
Zhu, Q. Q., 40, 43(41), 46(41)
Ziegler, P., 21
Zimmerman, A. L., 176, 416, 417, 430(14)
Zimmermann, W., 136
Zöllner, P., 222
Zoramski, C. Z., 46
Zucker, R. S., 205, 212(6), 221, 301
Zwierzak, A., 20

Subject Index

A

A23187, 1-(4,5-dimethoxy-2-nitrophenyl) ethyl carboxylate synthesis, 39–40
AAC, see ADP, ATP carrier
Abscisic acid, caged
    biological function after photolysis, 480–483
    calcium response measurement with fluorescent probes, 478, 481–482
    injection into plant cells, 477–478
    photolysis conditions, 478–480
    structure, 475
    synthesis, 475–477
Acetic acid, caged compound
    pH changes using photolysis, 246, 248–249
    synthesis, 247–248
Acetylcholinesterase
    caged ligands
        carbamylcholine, 266–267, 269, 273–276
        choline, 266–267, 269, 273–276
        criteria for ideal compounds, 273
        inhibition analysis, 272–274
        kinetics of photolysis, 273
        noracetylcholine, 266–267, 269
        norbutyrylcholine, 267, 269
        photochemical properties, 269–272
        photolysis conditions, 277–278
        recovery of activity following photolysis, 275–276, 278
    physiological function, 265
AChE, see Acetylcholinesterase
Actin, see also Muscle fiber
    cross-linking with photocleavable reagents
        caged G-actin complex preparation, 173–175
        cross-linking with 4-bromomethyl-3-nitrobenzoic acid succinimide ester, 159–162
        labeling with tetramethylrhodamine iodoacetamide, 157–158
        labeling with thiol-reactive caged rhodamine reagent, 170–171, 173
    G-actin, 6-nitroveratryloxycarbonyl chloride caged conjugate
        fluorescence photoactivation microscopy, 104–105
        functional analysis of caged and decaged proteins, 105–107
        labeling ratio, 101–102
        photoactivation conditions, 104
        polymerization competence, 102–103
        preparation, 101
        tetramethylrhodamine labeling of caged protein, 103
    labeling with fluorescent caged compounds, 72–74
ADP, p-hydroxyphenacyl caged compound synthesis, 22–24
ADP, ATP carrier
    electrophysiological measurements with caged ATP
        bilayer system, 298–299
        time-resolved measurements, 300–301
        transport modes, 300
    physiological function, 298
Affinity labeling, see Biotin affinity labeling; Photocleavable affinity tags
β-Alanine
    2-methoxy-5-nitrophenyl carboxylate caged compound
        kinetics of photolysis, 40–41
        quantum yield determination, 43, 45
        synthesis, 38–39
    receptor, see Neurotransmitter receptor
γ-Aminobutyric acid
    α-Carboxy-o-nitrobenzyl carboxylate caged compound
        kinetics of photolysis, 43
        synthesis, 35–37

desyl caged compound, 10–11
p-hydroxyphenacyl caged compound, 10–11
receptor, see Neurotransmitter receptor
Aminodextran, thiolation with 2-iminothiolane, 157
ATP, caged
  ADP, ATP carrier, electrophysiological measurements with caged ATP
    bilayer system, 298–299
    time-resolved measurements, 300–301
    transport modes, 300
  calcium affinity, 321, 324, 346–347
  dideoxy ATP in DNA repair kinetic analysis
    analytical high-performance liquid chromatography, 487–488
    apparatus design for studies, 491–492
    DNA damage and analysis, 490, 494
    panic response, 494–495
    photolysis conditions, 488
    purification, 487
    stability in mammalian cells, 489–490
    synthesis of high specific activity α-^{32}P compound, 486–487
  efficiency of photoconversion, 341–342
  p-hydroxyphenacyl caged compound
    photolysis conditions, 28–29
    properties and applications, 8
    stability testing, 29
  luciferase activation for bioelectronic imaging, 246–248, 250
  magnesium affinity, 321, 324, 346–347
  muscle fiber photolysis studies
    actomyosin studies
      ATPase assays, 332–333
      ATP binding, 330–332
      caged calcium studies, 332
      cross-bridge detachment, 330, 334, 344–346
      relaxation of tension, 330, 334
      muscle shortening velocity, inhibition before photolysis, 343–346
    photochemical conversion, quantification, 338–341
  sodium, potassium-ATPase, electrophysiological measurements with caged ATP
    bilayer experiments, 291–295

  binding of caged compounds, 293
  patch clamp studies, 295–298
  photolysis conditions, 291, 293
  time-resolved measurements, 293–294, 297–298
Azid-1, calcium uncaging efficiency by two-photon photolysis, 367–369, 379

B

Beam radius, Gaussian beam optics, 187
Benzoin phosphate
  limitations in photochemical applications, 7
  photolysis properties, 6
Biotin affinity labeling, see also Photocleavable affinity tags
  applications, 135–136, 152
  avidin affinity, 136
  biotinylation of macromolecules, 136
  release of target molecule, 136–137
Birefringence, Gaussian beam optics, 194–195
Bovine serum albumin, thiolation
  2-iminothiolane reaction, 157
  thiol content, determination, 157
8-Bromo-cyclic AMP
  4,5-dimethoxy-2-nitrobenzyl caged compound
    calibration of flash-induced changes, 427, 429–430
    concentration optimum determination, 425
    design, 417–418
    gated channel activation studies, 426–427
    handling of solutions, 425
    photochemical properties, 421–422
    purity requirements, 420–421, 430
    solubility, 418, 421
    solvolysis, 422–423
    synthesis, 418–420
  phosphodiesterase resistance, 416–417
8-Bromo-cyclic GMP
  4,5-dimethoxy-2-nitrobenzyl caged compound
    calibration of flash-induced changes, 427, 429–430

concentration optimum determination, 425
design, 417–418
gated channel activation studies, 426–427
handling of solutions, 425
photochemical properties, 421–422
purity requirements, 420–421, 430
solubility, 418, 421
solvolysis, 422–423
synthesis, 418–420
phosphodiesterase resistance, 416–417
4-Bromomethyl-3-nitrobenzoic acid succinimide ester
photochemical properties, 159
reactivity, 159
synthesis, 156–159
BSA, *see* Bovine serum albumin
BuChE, *see* Butyrylcholinesterase
Butyrylcholinesterase
caged ligands
carbamylcholine, 266–267, 269, 273–276
choline, 266–267, 269, 273–276
criteria for ideal compounds, 273
inhibition analysis, 272–274
kinetics of photolysis, 273
noracetylcholine, 266–267, 269
norbutyrylcholine, 267, 269
photochemical properties, 269–272
photolysis conditions, 277–278
recovery of activity following photolysis, 275–276, 278
physiological function, 265

C

Caffeine, calcium release studies with caged inositol 1,4,5-triphosphate, 393, 395
Caged compounds, *see also specific compounds and functional groups*
advantages in substrate release studies, 1, 30, 64, 95–96
criteria in design, 1–2, 64, 120
signal transduction studies, *see* Calcium
Calcium
affinity for caged compounds, 321, 324, 346–347
ATPase, *see* Sarcoplasmic reticulum/endoplasmic reticulum calcium ATPase
caged, *see* Calcium, caged
fluorescent probe measurement
apparatus for simultaneous electrophysiological measurement, 388–391
response to calcium release by caged compounds
cyclic ADP-ribose, 410–413
2,5-di(*tert*-butyl)hydroquinone caged compound, 435–437, 439
inositol 1,4,5-triphosphate, 388–391, 395, 403
nicotinic acid adenine dinucleotide phosphate, 414–415
plant hormones, 478, 481–482
signaling, *see* Cyclic ADP-ribose; Inositol 1,4,5-triphosphate; Nicotinic acid adenine dinucleotide phosphate; Sarcoplasmic reticulum/endoplasmic reticulum calcium ATPase
Calcium, caged
flash lamp photolysis studies with caged compounds, 219–222
sodium, calcium-exchanger, DM-nitrophen caged calcium studies
electrophysiological measurements, 302–306
photolysis kinetics, 301
rate constant determinations, 305–306
regulatory site studies, 301–302, 304–305
photochemical conversion, quantification, 339–340
two-photon photolysis
apparatus, 361–365
applications, 356–357, 380
calcium yield determination, 358, 369–374
temporal behavior of calcium concentration distribution, 374–379
uncaging action cross section determination, 365–369
cAMP, *see* Cyclic AMP
Carbamylcholine
caged compound studies with crystallized cholinesterases, 266–267, 269, 273–276
receptor, *see* Neurotransmitter receptor

α-Carboxy-*o*-nitrobenzyl carboxylates
  γ-aminobutyric acid
    kinetics of photolysis, 43
    synthesis, 35–37
    applications, 31–32
    suitability for neurotransmitter study with cell-flow, 48–50
Cell-flow, caged neurotransmitter characterization
  channel opening, rate and equilibrium constant determination, 460
  instrumentation, 46–47, 451–453
  quantification of photolysis, 454
  suitability of compounds, determination, 48–49
  theory, 47, 447, 450–451
  validation, 47–48
cGMP, *see* Cyclic GMP
Choline, caged compound studies with crystallized cholinesterases, 266–267, 269, 273–276
CNB, *see* α-Carboxy-*o*-nitrobenzyl carboxylates
4-Coumarinylmethyl group, phosphate compound caging
  cyclic AMP derivatives
    4-(7-acetoxycoumarinyl)methyladenosine derivative
      properties, 51–52
      synthesis, 51–54, 58
    biological activities, 62–63
    4-(7-hydroxycoumarinyl)methyladenosine derivative
      properties, 51–52
      synthesis, 51–53, 56, 58
    4-(7-methoxycoumarinyl)methyladenosine derivative
      properties, 51–52
      synthesis, 50–54, 58
    photochemical and spectral properties, 60–62
    4-(7-propionyloxycoumarinyl)methyladenosine derivative
      properties, 51–52
      synthesis, 51–53, 55, 58
    purification, 59
    silver(I) oxide promoted condensation, 51, 56–58
    stability in dark, 60
  structural determination, 59
  fluorescent properties, 50–51
Cross-linking, heterobifunctional photocleavable reagents
  actin
    caged G-actin complex preparation, 173–175
    cross-linking with 4-bromomethyl-3-nitrobenzoic acid succinimide ester, 159–162
    labeling with tetramethylrhodamine iodoacetamide, 157–158
    labeling with thiol-reactive caged rhodamine reagent, 170–171, 173
  applications, 155–156, 175
  bovine serum albumin
    thiolation with 2-iminothiolane, 157
    thiol content, determination, 157
  4-bromomethyl-3-nitrobenzoic acid succinimide ester
    photochemical properties, 159
    reactivity, 159
    synthesis, 156–159
  criteria in design, 162
  1-(3,4-dimethoxy-6-nitrophenyl)-2,3-epoxypropyl chloroformate
    reactivity, 168
    synthesis, 165
  1-(3,4-dimethoxy-6-nitrophenyl)-2,3-epoxypropyl hydroxide
    absorption spectroscopy, 167–168
    reactivity with thiol groups, 168
    synthesis, 165
  N-1-([3,4-dimethoxy-6-nitrophenyl)-2,3-epoxypropyl]oxy]carbonyl)4-(*N*,*N*-dimethylamino)pyridinium, 3-oxo-2,5-diphenyl-2,3-dihydrothiophene 1,1-dioxide 4-hydroxide
    caged G-actin complex preparation, 173–175
    solubility, 169
    synthesis, 166
  1-(3,4-dimethoxy-6-nitrophenyl)-2,3-epoxypropylsuccinimidyl carbonate, synthesis, 166–167
  1-hydroxyl-1-(3,4-dimethoxy-6-nitrophenyl)-2-propene, synthesis, 164
  4-[1-(3,4-dimethoxy-6-nitrophenyl)-2,3-epoxypropyl-1-oxycarbonyloxy]-3-oxo-2,5-diphenyl-2,3-dihydrothio-

phene 1,1-dioxide
  reactivity, 168–169
  rhodamine, preparation of thiol-reactive caged compound, 169–170
  synthesis, 165–166
Cyclic ADP-ribose
  caged 1-(2-nitrophenyl)ethyl compound
    calcium release analysis with fluorescent probes, 410–413
    photolyis conditions, 409–410
    synthesis, 405–407
    uncaging efficiency determination, 411
  calcium release in sea urchin eggs, 403–404
  synthesis, 404
Cyclic AMP
  4-coumarinylmethy derivatives
    4-(7-acetoxycoumarinyl)methyladenosine derivative
      properties, 51–52
      synthesis, 51–54, 58
    biological activities, 62–63
    4-(7-hydroxycoumarinyl)methyladenosine derivative
      properties, 51–52
      synthesis, 51–53, 56, 58
    4-(7-methoxycoumarinyl)methyladenosine derivative
      properties, 51–52
      synthesis, 50–54, 58
    photochemical and spectral properties, 60–62
    4-(7-propionyloxycoumarinyl)methyladenosine derivative
      properties, 51–52
      synthesis, 51–53, 55, 58
    purification, 59
    silver(I) oxide promoted condensation, 51, 56–58
    stability in dark, 60
    structural determination, 59
  desyl caged compound
    photolysis conditions, 27
    properties and applications, 4–6
    synthesis, 19–20
  hydrolysis-resistant analogs, see 8-Bromocyclic AMP
  targets in signaling, 415–416

Cyclic AMP-dependent protein kinase
  caging at essential cysteine, 129–130
  thiophosphorylation of peptides, 125
Cyclic GMP
  hydrolysis-resistant analogs, see 8-Bromocyclic GMP
  targets in signaling, 415–416
Cysteine, caged protein conjugates
  caging of *in vitro* translated protein, 128–129
  cyclic AMP-dependent protein kinase, 129–130
  engineering for site modification, 118, 134
  heavy meromyosin
    functional analysis
      ATPase assay, 110, 114
      in vitro motility assay, 110, 114–116
    photoactivation
      duration of irradiation, 112–114
      transient kinetics analysis, 113–116
    preparation
      4,5-dimethoxy-2-nitrobenzyl bromide conjugation, 111–112, 114–115
      materials, 97
      purification of protein, 111–112
  α-hemolysin R104C
    modification with 2-nitrobenzyl group, 121, 128–129
    photolysis conditions, 130–133
  kemptide
    modification with 2-nitrobenzyl group, 124–125
    photolysis conditions, 130–132
  mass spectrometry analysis, 119, 124
  2-nitrobenzyl reagents, efficiency in preparation, 120–121

D

Desyl group
  carboxylate ester compound caging
    γ-aminobutyric acid, 10–11
    applications, 13–14, 16
    L-glutamate
      properties and applications, 10–11, 13–14

synthesis, 25–26, 37–38
photochemistry, 11
phosphate compound caging
cyclic AMP
photolysis conditions, 27
properties and applications, 4–6
synthesis, 19–20
desyl dihydrogen phosphate synthesis, 17–19
limitations, 7
photolysis, 4, 6
photolysis of caged compounds, 26–27
2,4′-Dihydroxyacetophenone, synthesis, 24
4,5-Dimethoxy-2-nitrobenzyl bromide, see Cysteine, caged protein conjugates
4,5-Dimethoxy-2-nitrobenzyl carboxylate
applications, 31
1-(3,4-Dimethoxy-6-nitrophenyl)-2,3-epoxypropyl chloroformate
reactivity, 168
synthesis, 165
1-(3,4-Dimethoxy-6-nitrophenyl)-2,3-epoxypropyl hydroxide
absorption spectroscopy, 167–168
reactivity with thiol groups, 168
synthesis, 165
N-1-([3,4-Dimethoxy-6-nitrophenyl)-2,3-epoxypropyl)oxy]carbonyl)4-(N,N-dimethylamino)pyridinium, 3-oxo-2,5-diphenyl-2,3-dihydrothiophene 1,1-dioxide 4-hydroxide
caged G-actin complex preparation, 173–175
solubility, 169
synthesis, 166
4-[1-(3,4-Dimethoxy-6-nitrophenyl)-2,3-epoxypropyl-1-oxycarbonyloxy]-3-oxo-2,5-diphenyl-2,3-dihydrothiophene 1,1-dioxide
reactivity, 168–169
rhodamine, preparation of thiol-reactive caged compound, 169–170
synthesis, 165–166
1-(3,4-Dimethoxy-6-nitrophenyl)-2,3-epoxypropylsuccinimidyl carbonate, synthesis, 166–167
1-(4,5-Dimethoxy-2-nitrophenyl)ethyl carboxylate
A23187, synthesis, 39–40
applications, 31

2,2′-Dinitrobenzhydryl carboxylate
applications, 32
N-methyl-D-aspartate, synthesis, 33–35
DMNB, see 4,5-Dimethoxy-2-nitrobenzyl carboxylate
DM-nitrophen
calcium uncaging efficiency by two-photon photolysis, 367–369, 379
cation chelation, 240–241
Fourier transform infrared spectroscopy difference spectra of photolysis, 241–242
photolysis reaction, 241
sodium, calcium-exchanger, caged calcium studies
electrophysiological measurements, 302–306
photolysis kinetics, 301
rate constant determinations, 305–306
regulatory site studies, 301–302, 304–305
DMNPE, see 1-(4,5-Dimethoxy-2-nitrophenyl)ethyl carboxylate
DNA repair
caged radioactive nucleotides in kinetic studies
advantages and disadvantages, 485
apparatus design for studies, 491–492
dideoxy ATP
analytical high-performance liquid chromatography, 487–488
high specific activity α-^{32}P compound synthesis, 486–487
purification, 487
DNA damage and analysis, 490, 494
electroporation of compounds, 485–486, 492–493
panic response, 494–495
photolysis conditions, 488
stability in mammalian cells, 489–490
thymidine triphosphate synthesis, 489
pathways, overview, 483
single cell versus multicellular organisms, 484–485
DNB, see 2,2′-Dinitrobenzhydryl carboxylate
Dynamic bleaching
conditions causing bleaching, 184
critical intensity estimation, 184
theory, 182–183

Dynein
  caged compounds in photolysis studies
    calcium chelators, 321, 324
    nucleotides, 321, 325
    purity requirements, 321, 324–325
    table of properties, 322–323
  mechanism of motion, 308, 310
  photolysis applications, 310

E

Electron microscopy, caged fluorescein-labeled tubulin, 352–353, 355–356
Electroporation, caged substrates, 283–285
Enkephalin, see Leucine-enkephalin

F

Flash lamp photolysis
  applications
    calcium flux in response to caged compounds, 219–222
    chemotaxis, serine effects on bacteria, 219
    microtubule dynamics, 215
    muscle contraction, 214–215, 320
  driving circuit, 207–209
  energy of pulse, measurement, 213–214
  lamp selection, 204–205, 207
  optics
    lenses, 210–212
    light guides, 213
    mirrors, 209–210
    refractive optics versus ellipsoidal mirrors, 212–213
  pulse duration, 222
  X-ray crystallography
    heat damage of crystals by flash, 216–218
    measuring compounds released by photolysis, 219
    Ras GTPase studies in crystals, 215–216
Flash photolysis, see Flash lamp photolysis; Laser photolysis
Fluorescein, caged compounds
  caging group selection, 67
  practical considerations for use, 67
  protein labeling, 72–74
  synthesis, 68
  tubulin, see Tubulin
  types and structures, 64–66, 475
Fluorescence microscopy, see Fluorescence photoactivation imaging
Fluorescence photoactivation imaging
  cameras, 74
  G-actin studies, 104–105
  green fluorescent protein, 77–78
  light sources, 74–75
Fourier transform infrared spectroscopy
  caged GTP photolysis
    band assignment by isotopic labeling, 237–240
    difference spectra, 231
    dithiothreitol effects on spectra, 233–234
    intermediates in decay, 234–235
    magnesium effects on spectra, 232
    photolysis reaction, 231
    synthesis of compounds, 225
  caged phosphate photolysis
    band assignment by isotopic labeling, 227–229
    difference spectra, 226–229
    magnesium effects on spectra, 229–231
  DM-nitrophen photolysis
    cation chelation, 240–241
    difference spectra, 241–242
    photolysis reaction, 241
  instrumentation for photolysis studies, 223–224, 244
  Ras studies with caged GTP
    data collection, 224–225
    difference spectra, 243–245
  time-resolved measurement
    data collection techniques, 232–233
    temporal resolution, 223
Frequency doubling, theory, 196–198, 201
FTIR, see Fourier transform infrared spectroscopy

G

G-actin, see Actin
GABA, see γ-Aminobutyric acid
Gaussian beam optics
  beam radius, 187
  birefringence, 194–195
  focusing, 187
  Rayleigh range, 189

reflection, 192–193
refraction, 194
Self's equations, 188–190
waist position, 188, 190
waist radius magnification, 189
wave polarization, 190–191
GFP, see Green fluorescent protein
Gibberellin, caged
  calcium response measurement with fluorescent probes, 478, 481–482
  injection into plant cells, 477–478
  photolysis conditions, 478–480
  structure, 475
  synthesis, 475–476
Glucose 6-phosphate, caged compound studies in vivo
  electroporation of caged substrate into sea urchin eggs, 283–285
  glucose 6-phosphate dehydrogenase assay, in vivo, 285–286
  fertilization of sea urchin eggs, effect on activity, 286–288
  inertness testing of caged compound, 282–283
  photolysis conditions, 285
  synthesis of caged compound
    2-nitroacetophenone hydrazone preparation and reaction with glucose 6-phosphate, 281
    radiolabeling, 280
L-Glutamate
  desyl caged compound
    properties and applications, 10–11, 13–14
    synthesis, 25–26, 37–38
  p-hydroxyphenacyl caged compound
    properties and applications, 10–11, 14, 16
    synthesis, 24–25
  o-nitromandelyloxycarbonyl caged compound
    electrophysiology studies of glutamate receptor effects, 440–443
    photochemistry, 440
    synthesis, 434–435
  receptor, see Neurotransmitter receptor
Glycine receptor, see Neurotransmitter receptor
Glycogen phosphorylase, photolysis studies in crystals, 252–253

Green fluorescent protein, fluorescence photoactivation imaging, 77–78
GTP, caged
  Fourier transform infrared spectroscopy of photolysis
    band assignment by isotopic labeling, 237–240
    difference spectra, 231
    dithiothreitol effects on spectra, 233–234
    intermediates in decay, 234–235
    magnesium effects on spectra, 232
    photolysis reaction, 231
    Ras studies, 224–225, 243–245
    synthesis of compounds, 225
  synthesis of compounds, 225

H

Heavy meromyosin, caged protein conjugate
  functional analysis
    ATPase assay, 110, 114
    in vitro motility assay, 110, 114–116
  photoactivation
    duration of irradiation, 112–114
    transient kinetics analysis, 113–116
  preparation
    4,5-dimethoxy-2-nitrobenzyl bromide conjugation, 111–112, 114–115
    materials, 97
    purification of protein, 111–112
α-Hemolysin R104C
  modification with 2-nitrobenzyl group, 121, 128–129
  photolysis of caged protein, 130–133
HMM, see Heavy meromyosin
1-Hydroxyl-1-(3,4-dimethoxy-6-nitrophenyl)-2-propene, synthesis, 164
p-Hydroxyphenacyl group
  carboxylate ester compound caging
    γ-aminobutyric acid, 10–11
    applications, 13–14, 16
    L-glutamate
      properties and applications, 10–11, 14, 16
      synthesis, 24–25
    peptides, 11–13
    photochemistry, 11
  phosphate compound caging

ADP coupling in synthesis, 22–24
ATP
  photolysis conditions, 28–29
  properties and applications, 8
  stability testing, 29
  p-hydroxyphenacyl phosphate
    photolysis, 28
    synthesis, 20–22
  photolysis reactions, 8–10
photolysis of caged compounds, 26–27

I

Infrared spectroscopy, see Fourier transform infrared spectroscopy
Inositol 1,4,5-triphosphate
  caged compound
    applications, 381–382
    calcium fluorescent probe measurement with electrophysiological measurement, 388–391, 395, 403
    kinetics of calcium release, 395–397
    light sources for photolysis, 386–388
    linearity and calibration of photorelease, 384–386
    photochemistry, 382–383
    spatial control and heterogeneity of calcium release sites, 397–402
    Xenopus oocyte studies
      caffeine studies, 393, 395
      dose–response relationship of calcium release, 391–392
      InsP_4 agonist studies, 393
      microinjection, 383–384
    mediation of calcium release, 380–381, 404
Ion pump, see also Calcium; Sodium, calcium-exchanger; Sodium, potassium-ATPase
  electrophysiological measurements with caged substrates, 289–290
  overview of caged substrate studies, 290, 305–306

J

Jasmonic acid, caged
  calcium response measurement with fluorescent probes, 478, 481–482
  injection into plant cells, 477–478
  photolysis conditions, 478–480
  structure, 475
  synthesis, 475–476

K

Kainate receptor, see Neurotransmitter receptor
Kemptide
  modification with 2-nitrobenzyl group, 124–125
  photolysis of caged peptide, 130–132
  thiophosphorylation and modification, 125–126
Kinesin
  ATP-binding core, 308
  caged compounds in photolysis studies
    calcium chelators, 321, 324
    nucleotides, 321, 325
    purity requirements, 321, 324–325
    table of properties, 322–323
  mechanism of motion, 307–310
  motility assay
    apparatus for combined photolysis studies, 336–337
    bead assay for force measurement, 338
  photolysis applications, 310
  preparation from bovine brain, 335
  translocation inhibition by caged ATP, 346

L

Laser photolysis
  caged carboxylates, 40
  characteristics of laser types, 203, 319
  comparison to flash lamp systems, 203–204
  components of basic laser, 185–186
  dynamic bleaching
    conditions causing bleaching, 184
    critical intensity estimation, 184
    theory, 182–183
  frequency doubling, theory, 196–198, 201
  Gaussian beam optics
    beam radius, 187
    birefringence, 194–195
    focusing, 187
    Rayleigh range, 189
    reflection, 192–193

refraction, 194
Self's equations, 188–190
waist position, 188, 190
waist radius magnification, 189
wave polarization, 190–191
group velocity dispersion, 199–200
mode locking, theory, 198–199
Q-switching, production, 195
safety
  fail-safe protection, 186
  goggles, 183
  voltage supply, 186
single-photon photolysis
  development, 175–176
  laser systems, 176, 201–202
  spatial resolution, 359
  stimulated emission, theory, 179–182
two-photon photolysis, see Two-photon photolysis
Laue diffraction, see X-ray crystallography, photolysis studies
Leucine-enkephalin, photocleavable biotin labeling
  avidin affinity purification, 144–145, 148–149
  conjugation, 144
  fluorescamine assay, 145
  photocleavage kinetics, 146–148
LSM1
  caged tyrosine peptide
    design of inactive caged peptide, 88–90
    kinetics of photolysis, 87–88
    myosin light-chain kinase inhibition, 93–94
    spectral and photochemical properties, 85–87
    synthesis, 80–85
  structure and function, 79–80
Luciferase
  activation by caged ATP, 246, 248, 250
  bioelectronic imaging and pattern recognition applications, 246–247, 250
  immobilization in agarose gels, 249–250
  inhibition by caged acetate pH effects, 246, 248–249
  reaction catalyzed, 245
Lysine, caged protein conjugates
  G-actin
    fluorescence photoactivation microscopy, 104–105

    functional analysis of caged and decaged proteins, 105–107
    labeling ratio, 101–102
    photoactivation conditions, 104
    polymerization competence, 102–103
    preparation of 6-nitroveratryloxycarbonyl chloride conjugate, 101
    tetramethylrhodamine labeling of caged protein, 103
  inactivation of proteins, 98–99
  photolytic mechanism, 98
  preparation of 6-nitroveratryloxycarbonyl chloride conjugates
    4-dimethylaminopyridine salt preparation, 100
    materials, 97
    quantification of conjugation, 99–101
  profilin
    functional analysis of caged protein, 108–110
    preparation of 6-nitroveratryloxycarbonyl chloride conjugate, 108
    recombinant protein purification, 107–108

M

Magnesium
  affinity for caged ATP, 321, 324, 346–347
  effects on caged GTP infrared spectra, 229–231
Mass spectrometry, 2-nitrobenzyl caged protein analysis, 119, 124
Meromyosin, see Heavy meromyosin
2-Methoxy-5-nitrophenyl carboxylate β-alanine
  kinetics of photolysis, 40–41
  quantum yield determination, 43, 45
  synthesis, 38–39
  applications, 32
N-Methyl-D-aspartate, 2,2'-dinitrobenzhydryl carboxylate synthesis, 33–35
Microtubule, see also Tubulin
  motility assay
    apparatus for combined photolysis studies, 336–337
    gliding assay, 337
    preparation from tubulin, 335

MK-801, neurotransmitter receptor inhibition
  dissociation constant determinations, 464–465
  effects on opening and closing rates, 463–464
MNP, see 2-Methoxy-5-nitrophenyl carboxylates
Mode locking, theory, 198–199
Muscle fiber, see also Actin; Heavy meromyosin; Myosin
  apparatus for photolysis experiments
    light sources, 315, 319–320
    setups, 313, 315
  biophysical signals recorded in conjunction with photolysis, 315–317
  caged ATP
    actomyosin studies
      ATPase assays, 332–333
      ATP binding, 330–332
      caged calcium studies, 332
      cross-bridge detachment, 330, 334, 344–346
      relaxation of tension, 330, 334
    muscle shortening velocity, inhibition before photolysis, 343–346
  caged compounds in photolysis studies
    calcium chelators, 321, 324
    nucleotides, 321, 325
    purity requirements, 321, 324–325
    table of properties, 322–323
  motility assays, 334–335
  photochemical conversion of caged compounds, quantification
    ATP, 338–341
    calcium, 339–340
    phosphate, 339
  photolysis
    applications, 214–215, 253–254, 310
    uniformity in sample chamber, 341–342
  preparation for photolysis experiments
    detergent treatment, 311
    glycerol extraction, 311
    mounting, 312–313
  solutions for photolysis experiments
    buffer, 327
    cations, 326
    changes following photolysis, 328–329
    computer analysis, 325–326
    free radical scavengers, 327–328
    ionic strength, 325–327
    nucleotide depletion, 328
Myosin, see also Muscle fiber
  ATP-binding core, 308
  caged compounds in photolysis studies
    calcium chelators, 321, 324
    nucleotides, 321, 325
    purity requirements, 321, 324–325
    table of properties, 322–323
  mechanism of motion, 307–309
  photolysis applications, 310

N

NAADP, see Nicotinic acid adenine dinucleotide phosphate
Neurotransmitter receptor
  caged compounds
    apparatus for photolysis, 453
    α-carboxy-2-nitrophenyl neurotransmitters, photolytic properties, 453, 455–458
    cell-flow quantification of photolysis, 454
    channel opening, rate and equilibrium constant determination, 457, 459–460
    inhibitor studies
      dissociation constant determinations, 464–465
      effects on opening and closing rates, 460–461, 463–464
    2-methoxy-5-nitrophenyl neurotransmitters, photolytic properties, 453, 457, 459
    microsecond electrophysiology measurements, 468, 470
    neuronal circuit function analysis, 467–468
    Xenopus oocyte studies, 470–472
  cell-flow studies
    apparatus, 451–453
    channel opening, rate and equilibrium constant determination, 460
    principle, 447, 450–451
  classification, 444–445
  malfunctions in disease, 443–444
  minimum reaction scheme for mediated reactions

characteristics, 445–446
equations for analysis, 448–449
single cell excitatory and inhibitory neurotransmitter signal integration, 465–467
Nicotinic acetylcholine receptor, *see* Neurotransmitter receptor
Nicotinic acid adenine dinucleotide phosphate
  caged 1-(2-nitrophenyl)ethyl compound
    calcium release analysis with fluorescent probes, 414–415
    high-performance liquid chromatography, 407–409
    photolyis conditions, 414
    structural characterization, 409
    synthesis, 407
  calcium release in sea urchin eggs, 403–404
2-Nitrobenzyl group
  applications in caged compounds, 3–4, 64, 118–119
  limitations in photochemical applications, 3–4
  mass spectrometry analysis of modified proteins, 119, 124
  photolysis reaction, 2–3, 226
  rate of photolysis, 122
  scavenging free radicals after photolysis, 134, 233, 328
o-Nitromandelyloxycarbonyl group
  2,5-di(*tert*-butyl)hydroquinone caged compound
    calcium studies with fluorescent indicators, 435–437, 439
    cell loading, 435
    synthesis, 432
  glutamate caged compound
    electrophysiology studies of glutamate receptor effects, 440–443
    photochemistry, 440
    synthesis, 434–435
  imidazole
    reactivity, 431–432
    synthesis, 431–434
2-Nitrophenethyl group
  limitations in photochemical applications, 3–4
  photolysis reaction, 2

6-Nitroveratryloxycarbonyl chloride conjugate, *see* Lysine, caged protein conjugates
NMDA, *see* *N*-Methyl-D-aspartate
Nmoc, *see* *o*-Nitromandelyloxycarbonyl group
Noracetylcholine, caged compound studies with crystallized cholinesterases, 266–267, 269
Norbutyrylcholine, caged compound studies with crystallized cholinesterases, 267, 269
NPEGTA, calcium uncaging efficiency by two-photon photolysis, 367–369

O

Oligonucleotide, photocleavable biotin labeling
  affinity purification, 145, 150–151
  high-performance liquid chromatography, 149–150
  phosphorylation at 5′-end, 153
  photocleavage, 145–146, 150–151
  polyacrylamide gel electrophoresis analysis, 151–152
  synthesis, 145, 149, 153

P

Peptide synthesis, caged tyrosine peptides, 83–85
Phosphate, caged
  caging groups, *see* 4-Coumarinylmethyl group; Desyl group; *p*-Hydroxyphenacyl group
  Fourier transform infrared spectroscopy
    band assignment by isotopic labeling, 227–229
    difference spectra, 226–229
    magnesium effects on spectra, 229–231
  photochemical conversion, quantification, 339
Photobleaching, limitations compared to fluorescence photoactivation, 63

Photocleavable affinity tags
  applications, 137–138
  design, 138
  photocleavable biotins
    applications, 152–154
    criteria in design, 152
    leucine-enkephalin
      avidin affinity purification, 144–145, 148–149
      conjugation, 144
      fluorescamine assay, 145
      photocleavage kinetics, 146–148
    oligonucleotide labeling
      affinity purification, 145, 150–151
      high-performance liquid chromatography, 149–150
      phosphorylation at 5'-end, 153
      photocleavage, 145–146, 150–151
      polyacrylamide gel electrophoresis analysis, 151–152
      synthesis, 145, 149, 153
    structures, 138–139
  synthesis
    materials, 139–140
    photocleavable biotin NHS carbonate, 140–142
    photocleavable biotin phosphoramidite, 142–144, 153
Photolysis rate, calculation, 131
Photoprotecting compounds, see Caged compounds
PKA, see Cyclic AMP-dependent protein kinase
Profilin, 6-nitroveratryloxycarbonyl chloride caged conjugate
  functional analysis of caged protein, 108–110
  preparation, 108
  recombinant protein purification, 107–108

Q

Q-switching, production, 195
Quantum yield
  caged carboxylates, 43, 45–46
  caged tyrosine peptides, 85–87
  calculation, 131

R

Raman spectroscopy
  caged versus uncaged GTP, 235–237
  instrumentation for photolysis studies, 225
Ras
  caged GTP studies in crystals
    caged GTP affinity, 257, 259–260
    data collection, 259
    flash lamp photolysis, 95, 215–216, 257, 261
    rapid freezing followed by data collection, 261
    Ras mutation for thermostability, 260
    resolution of data, 259–260, 262
  Fourier transform infrared spectroscopy studies with caged GTP
    data collection, 224–225
    difference spectra, 243–245
Rayleigh range, Gaussian beam optics, 189
Reflection, Gaussian beam optics, 192–193
Refraction, Gaussian beam optics, 194
Resorufin, caged compounds
  caging group selection, 67
  practical considerations for use, 67–69
  protein labeling, 72–74
  synthesis, 68–69
  types and structures, 65–66
Rhodamine, caged compounds
  applications, 75, 77
  caging group selection, 67, 72
  practical considerations for use, 67, 69, 71–72
  protein labeling, 72–74
  synthesis, 69
  thiol-reactive caged compound
    actin labeling and cross-linking, 170–171, 173–175
    preparation, 169–170
  types and structures, 66
RS-20
  caged tyrosine peptide
    calmodulin binding analysis, 90–93
    design of inactive caged peptide, 88–90
    kinetics of photolysis, 87–88
    spectral and photochemical properties, 85–87
    synthesis, 80–85
    structure and function, 79–80

S

Salicylic acid, caged
  calcium response measurement with fluorescent probes, 478, 481–482
  injection into plant cells, 477–478
  photolysis conditions, 478–480
  structure, 475
  synthesis, 475–477
Sarcoplasmic reticulum/endoplasmic reticulum calcium ATPase
  calcium oscillation role, 439
  2,5-di(*tert*-butyl)hydroquinone caged compound studies
    calcium studies with fluorescent indicators, 435–437, 439
    cell loading, 435
    synthesis, 432
SERCA, *see* Sarcoplasmic reticulum/endoplasmic reticulum calcium ATPase
Single-photon photolysis, *see* Laser photolysis
Sodium, calcium-exchanger
  DM-nitrophen caged calcium studies
    electrophysiological measurements, 302–306
    photolysis kinetics, 301
    rate constant determinations, 305–306
    regulatory site studies, 301–302, 304–305
    stoichiometry, 303
Sodium, potassium-ATPase
  electrophysiological measurements with caged ATP
    bilayer experiments, 291–295
    binding of caged compounds, 293
    patch clamp studies, 295–298
    photolysis conditions, 291, 293
    time-resolved measurements, 293–294, 297–298
  ion specificity, 292–293
  physiological function, 290
  reaction mechanism, 290–291
Stimulated emission, theory, 179–182

T

Thiophosphoryl group, caged protein conjugates
  kemptide modification with 2-nitrobenzyl group, 126
  modification in presence of cysteinyl residues, 119
  selective reactivity in 2-nitrobenzyl modification, 126–127
  thiophosphorylation of peptides, 125
Thymidine triphosphate, caged compound synthesis, 489
Tubulin, *see also* Microtubule
  caged fluorescein-labeled tubulin
    applications in transport and microtubule dynamics, 348, 353, 356
    electron microscopy following photoactivation, 352–353, 355–356
    microinjection into neurons, 348, 350
    neuron culture for microinjection, 349–350
    photoactivation, 350–351
    preparation, 349
    video microscopy, 351
  microtubule preparation, 335
  preparation from bovine brain, 335
Two-photon photolysis
  caged calcium
    apparatus, 361–365
    applications, 356–357, 380
    calcium yield determination, 358, 369–374
    temporal behavior of calcium concentration distribution, 374–379
    uncaging action cross section determination, 365–369
  development, 176, 178
  laser systems, 178–179, 201–202, 357–358, 361–365
  pulse duration, 178
  spatial resolution, 359–361
Tyrosine, caged peptides
  applications, 78–79, 94
  design of inactive caged peptides, 88–90, 94
  extinction coefficient determination, 85
  kinetics of photolysis, 87–88
  quantum yield determination, 85–87
  synthesis
    2-bromo-α-carboxyl-2-nitrophenyl methyl ester, 80–82
    coupling reaction, 82

deprotection, 83
peptide synthesis, 83–85
protecting of tyrosine, 82
thiophosphoryltyrosine modification, 133

W

Waist position, Gaussian beam optics, 188, 190
Waist radius magnification, Gaussian beam optics, 189
Wave polarization, Gaussian beam optics, 190–191

X

X-ray crystallography, photolysis studies
caged enzymes, 255, 266
concentration requirements for caged substrates, 252–254
crystal size, 262
glycogen phosphorylase, 252–253
heat damage of crystals by flash, 216–218, 258
irreversibility of photolysis, 262–264
measuring compounds released by photolysis, 219, 253, 256–257
muscle fibers, 253–254
photolysis yield considerations, 254–255, 258–259
Ras GTPase studies in crystals
caged GTP affinity, 257, 259–260
data collection, 259
flash lamp photolysis, 215–216, 257, 261
rapid freezing followed by data collection, 261
Ras mutation for thermostability, 260
resolution of data, 259–260, 262
time-resolved measurements with Laue crystallography, 251, 255–256, 259, 264, 277
Xenopus oocyte
caged inositol 1,4,5-triphosphate studies of calcium release
caffeine studies, 393, 395
dose–response relationship, 391–392
InsP_4 agonist studies, 393
microinjection, 383–384
neurotransmitter receptor studies with caged compounds, 470–472

ISBN 0-12-182192-7